Notes and Exercises for Introductory Statistics and Probability

K. M. Brown   *(with contributions by C. P. Gregory and Y. Feinman)*

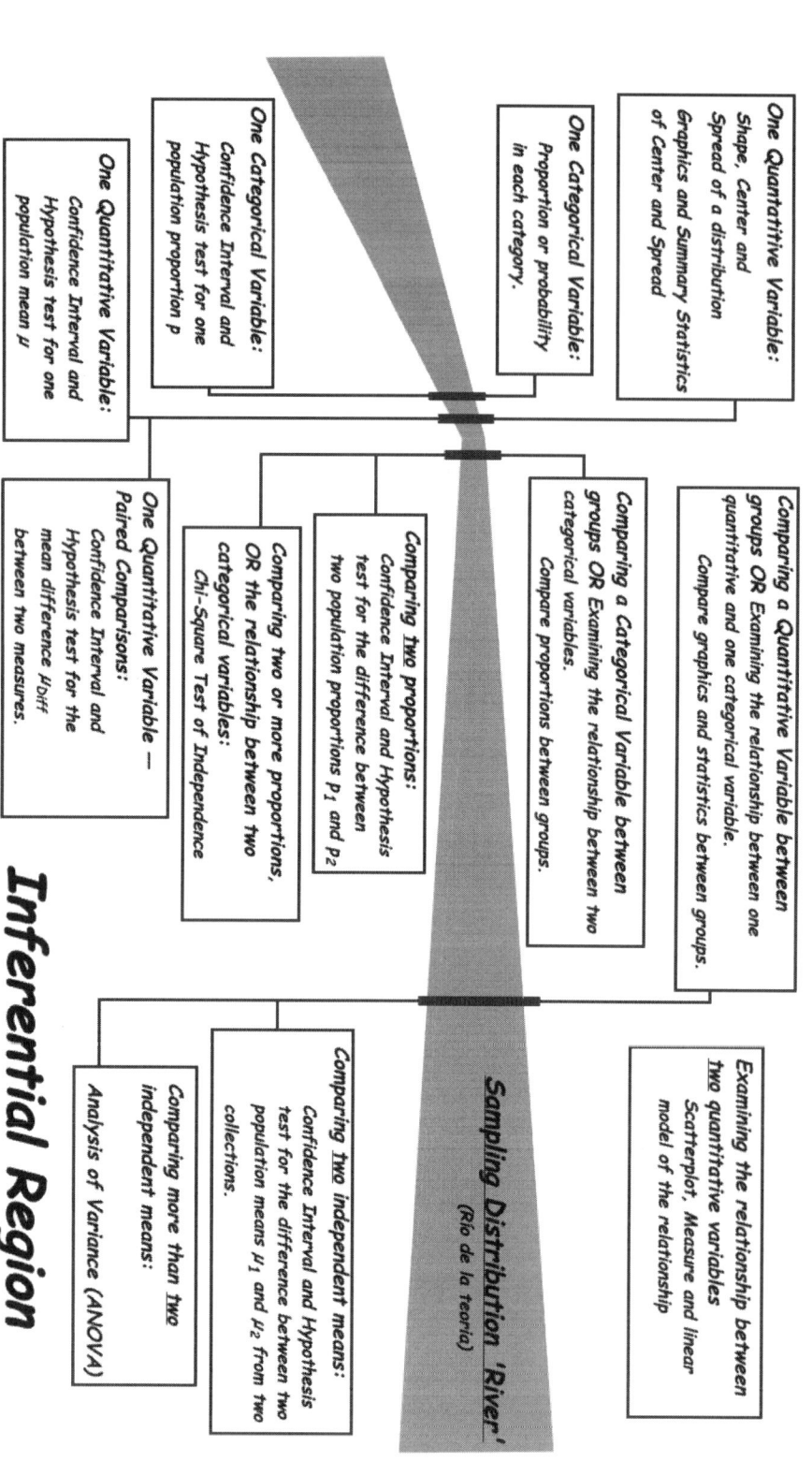

Table of Contents

Unit 1: Describing distributions

§1.1 Starting Out
- Notes — 1
- Exercises — 8
- Data Sheet: CombinedClassDataAut08 — 13

§1.2 Probability
- Notes — 14
- Exercises — 26

§1.3 Distributions
- Notes — 35
- Exercises — 42

§1.4 Shape, Center and Spread
- Notes — 51
- Exercises — 58

§1.5 Measures of Center/Location
- Notes — 62
- Exercises — 70
- Special Exercise A: Means, Medians and Mammals — 75
- Special Exercise S: Summation Notation — 76

§1.6 Measures of Spread
- Notes — 79
- Exercises — 87

§1.7 Models for Distributions
- Notes — 94
- Exercises — 103
- Special Exercise A — 110
- Special Exercise B — 112
- Revision: §§1.1 – 1.6: Hobbits and Men — 114

Unit 2: Describing relationships between variables

§2.1 Comparisons, Variables and Relationships
- Notes — 117
- Exercises — 122
- Analysis and Writing Assignment — 125

§2.2 A Graphic, a Model and a Measure: Scatterplots and Correlation
- Notes — 135
- Exercises — 143

§2.3 Making the Most of the Model: Best Fitting Lines
- Notes — 149
- Exercises — 155

§2.4 Is the Model Good and Useful?
- Notes — 164
- Exercises — 173

Unit 3: Beyond description: trusting data

§3.1 Can we trust data? Getting bad and good data
- Notes — 178
- Exercises — 190

§3.2 Trusting data, part 2: Sampling Distributions
- Notes — 209
- Exercises — 220

§3.3   Trusting data, part 3: Binomial Distributions
    Notes   *225*
    Exercises   *236*

§3.4   Trusting data, part 4: Sampling Distributions for Proportions
    Notes   *249*
    Exercises   *259*
    Special Exercises A, B   *266*

Unit 4: Inference for Categorical Variables

§4.1   Politics and Confidence: Estimating a Proportion
    Notes   *268*
    Exercises   *276*

§4.2   Hypothesis Testing: Are we less or more right-handed
    Notes   *284*
    Exercises   *296*

§4.3   Comparing Proportions, or what is the difference?
    Notes   *306*
    Exercises   *315*

§4.4   Do we have independence? Chi-square
    Notes   *321*
    Exercises   *332*

§4.5   Hypothesis test, confidence interval, both or neither?
    Notes   *339*
    Formula sheet for inferences for proportions   *347*
    Exercises   *350*

Unit 5: Inference for Quantitative Variables

§5.1   Inference for Quantitative variables: the t distribution
    Notes   *357*
    Exercises   *369*

§5.2   Hypothesis Testing for One Mean
    Notes   *375*
    Exercises   *385*

§5.3   Comparing Means: Two Measures, One Collection
    Notes   *393*
    Exercises   *400*

§5.4   Comparing Means: One Measure, Two Collections
    Notes   *405*
    Exercises   *410*

§5.5   Analysis of Variance
    Notes   *416*
    Exercises   *423*

§5.6   Reading a Mental Map
    Notes   *433*
    Formulas for t procedures   *439*
    Exercises   *441*

Tables

    Normal Distribution   *447*
    Chi-Square Distribution   *449*
    t-distributions   *450*

Preface: Some Questions about the Course in Statistics

Why is there a sea star on the cover?

Well, it's a nice picture. But it is also because statistics is about everything – sea stars, muscles and anemones included, which is probably why your university system has statistics as a requirement for your program of study. Even without the sea star, think of all of the information (we call it data) that is collected from people when they use social media, shop on-line, even talk. Everywhere you turn, there are data --- masses of data. *The purpose of statistics is to make sense of these data, to see if the data answer questions, to see if the data tell a story.* Tons of data just sitting there do not do it, do not show what patterns or relationships there may be lurking there. Making pictures (a better term is graphics) is a good start. A good graphic may show patterns otherwise hidden, and we now have good software to make the graphics easily. But that is jumping ahead just a bit.

Is statistics hard?

A much better and more useful question is to ask what kinds of challenges you are likely to encounter (how the course is "hard"). Here is a list of them:

- There is a body of technical terminology that must be mastered – a new language, in a sense. Statisticians use what look like common words, but give those words a technical sense connected that deviates somewhat from the "normal" usage. An example is the word "random."
- Statistics uses mathematical symbolism, including some symbols that may be new to you. Having said that, being able to get the answers using formulas is *definitely not* the most important thing, as it may have been in previous mathematics courses. We have software to do calculation; we only need to know what the software is doing, which is harder. Still, the advice is: do not fear symbols; indeed, embrace symbolism!
- What *is* important, and harder is to be able to say what the numbers that are calculated and the graphics that are produced *mean* in the light of questions that we are asking. We call this *interpretation*. This text is focused on interpretation; it gives examples in the **Notes** section, and the **Exercises** give practice.
- Finally, expect some abstract ideas, and some logic that may take days – perhaps even weeks – to understand. These difficult ideas and logic come after some weeks into the course, and have to do with the thorny question of whether we can actually trust the data that are collected to represent reality faithfully.

What will we be doing?

The core of the course is in the **Exercises**. They are designed to lead you through how statisticians analyze data (make sense of data) by actually analyzing data. The idea is that one learns by actually using the techniques that statistician use with data. The **Notes** provide backup with definitions of technical terms, and explanations. Probably a good strategy is to read through the **Notes** for a particular section before doing the exercises, without worrying too much if you do not understand it all at first reading; the questions in the **Exercises** will probably address some of the possible confusions.

Practical Requirements for the Course

- **Online Resources** It is likely that your college or university, as a part of its course management system, has a site devoted to the course that you are taking; that site should be your first recourse for information. Specifically, there may be online exercises and quizzes. Otherwise, the website for these materials is found at: http://www.cleonestats.blogspot.com/.

- **Fathom Required** To use the Notes and Exercises requires the statistical analysis package **Fathom**, which was developed by KeyCurriculum Press and sold by McGraw Hill Education. See the websites:

    www.mheonline.com/program/view/2/16/2645/0000FATHOM#program   or
    www.keycurriculum.com/products/fathom/fathom-pricing-and-purchasing

- **Data Sets**: You will be analyzing data – lots of data. The data sets are *not* included with the Fathom application. Rather, they will be found either at the course site at your college or university's course management system, or at http://www.cleonestats.blogspot.com/

For Instructors

This work is copyrighted by the primary author under a Creative Commons Attribution-Share Alike 3.0 Unported License. You are free to *Share* (to copy, distribute and transmit the work) and to *Remix* (to adapt the work) if you attribute the work faithfully. More detailed information can be found at the website mentioned above.

§1.1 Starting Out

First Steps

Statistics is for asking and answering questions about **collections**. More specifically, statistics asks and answers questions about collections of **data**. There can be collections of people, collections of places, such as towns or villages, collections of times, collections of animals, collections of good things to eat (apples, pears, apricots, etc.). Collections can be just about anything you can name.

Here is an example: here are data on a collection of students in four statistics classes. The data are probably similar to some data you may collect on the first day of class, and they are shown in a spreadsheet format in the statistical analysis package called Fathom.

Combined ClassData Aut 09	Gender	AgeYears	Height	Mother...	Poltical...	Langua...	Number...	Number...	Number...	Instruct...	Tattoo	Domina...
1	F	26	165.0	35	Liberal	2	6	7	2	60	Y	Right
2	M	19	166.0	27	Moderate	1	5	5	12	55	N	Right
3	F	19	154.0	32	Moderate	1	4	3	6	52	N	Right
4	F	19	165.0	24	Moderate	2	5	6	6	60	N	Right
5	M	22	175.3	32	Moderate	2	4	4	6	62	N	Right
6	M	19	187.0	35	Moderate	1	3	3	9	54	N	Right

Terminology: Cases, Variables (Quantitative and Categorical), Values

We need some terminology to navigate our way through what we see here. We refer to the objects in a collection as **cases**. For this collection, the cases are students taking statistics. In the spreadsheet format, each row is a different case. Here, each row (a row runs horizontally) represents a different student. It is very important to be clear in your mind about the cases for any collection; otherwise, you will not know what you are talking about.

The numbers or the words that you see in the spreadsheet shown above are *data*; other words that could be used for this general term *data* are *measurements* or *observations*. The *data* for the very first case—or student—in the collection above shows that the student is female, twenty-six years old, 165 centimeters tall, whose mother was thirty-five when she was born, etc. More importantly, for the cases in a collection, **data** are collected on **variables**. A **variable** measures some aspect of each case, and that aspect potentially varies from case to case. (Fathom calls variables **attributes**, but in these **Notes** we will use the term "variable.") The height of a student measured in centimeters (*Height* in the spreadsheet) is one example of a variable; gender is another example of a variable, and the number of languages a student speaks (*Languages* in the spreadsheet) is another. The *age* of the student is yet another variable. Notice that in spreadsheet format, the variables are the columns (a column runs vertically).

The **values** of a variable are the possibilities for that variable. It is obvious that the values for the variable *Gender* are male and female. For the variable *Age* the values can theoretically be any number, but for college students we do not expect to see any ages less than about fourteen, and we expect to see very few above fifty or so. That is, we can say that values of the variable *age* that are below fourteen and above fifty are not very *likely*. On the other hand, what *is likely* is that we will find students who have ages between nineteen and twenty-five. So, if we were asked to list the possible values for student ages, to be completely safe, we could write something like: "10, 11, 12…89, 90" with the understanding that the three dots (the ellipses) show that any number between twelve and eighty-nine can occur. Or, we could write "any age greater than 10 and less than 90." If we want to list the *likely values,* we may write something like "16, 17, 18, 19, 20…48, 49, 50."

What about the variable *height*? If the cases are students, and we are measuring in metric units, we do not expect to see many cases with values less than about 120 centimeters (about three feet, eleven inches) or greater than 220 centimeters (about seven feet, two inches). These values for short and tall people are not *likely*, although these values are possible. For possible values, we could write "any number between 120 cm and 220 cm." If our collection were elementary school students rather than college students then we would expect to see a different range of heights.

Variables can be either **quantitative** so that the values are measured with numbers, or they can be **categorical** where the values are essentially not numerical. Obviously *height* and *age* are both quantitative variables, whereas *gender* and whether or not a student has a tattoo (the variable *Tattoo* in the spreadsheet) are categorical variables. Categorical variables can have more than two categories; the variable *PoliticalView* has the categories "liberal," "moderate," and "conservative." The distinction between quantitative and categorical variables is very important because what we do with quantitative variables is often different from what we do with categorical variables.

> **Types of Variables**
> A **quantitative variable** has values that are *numerical* whereas a **categorical variable** has values that are essentially *non-numerical*.

Statistical Questions

We said that statistics is for answering questions about collections of data. But what kinds of questions do we typically ask? We will call the kinds of questions we ask **statistical questions**. Here are some examples of statistical questions, using the data above about statistics students.

1. Are male students or female students more likely to have a tattoo?
2. On average, are students in an evening section of a statistics course older than students in a day section of statistics? (We think of comparing two collections—the day students with the evening students.)
3. What percentage of students speaks just one language? What percentage speaks two languages? What percentage speaks three languages? Four? Five?
4. It makes sense that the more people there are in a household, the more cars there will be in that household. But is it true that for each additional person in a household there is an additional car? Probably not. So, can we say how many additional cars there will be in a household if another person is added?

We could go on with additional examples of statistical questions just for these data. One of the most important goals for this course is that you learn to use statistics to answer statistical questions and be able to express the answer to the questions in language that can be understood. At this point, however, notice two things about our four example statistical questions.

Some of the questions involve a comparison between parts of the collection—sub-collections—or a comparison between two collections. The first question compares the male and the female students (comparing sub-collections), and the second question compares the day students and the evening students (two collections.) We will see that comparing two collections or two sub-collections involves the same thinking and many of the same procedures.

Secondly, all of these questions are about the *collection* and not about the *individual* students, who are the cases in this collection. We typically do *not* ask, "What is the age of the oldest stats student?" Or "Who got the highest test score, and what was that score?" Or "What is the biggest household that any student lives in?"

That we do not often ask questions about the oldest or the biggest—questions about extremes—may come as a surprise because questions involving the biggest or smallest or greatest or least are the first kinds of questions that occur to some people. The reason that the questions about extremes come to mind is probably because we are interested in who is best or who is greatest, especially in the sports or entertainment world ("Which player scored the most goals?" "Which movie is the most successful this summer?"). Likewise, in the consumer economy, we are concerned with getting the best deal. What you will encounter in this course largely ignores these questions about extremes or questions about single cases; rather, we will be much more interested in describing the collection as a whole or in comparing parts of a collection. Or we will be interested in looking at the relationship between variables in the collection. (The fourth statistical question in our list above is about the relationship between variables. If you cannot see how you would approach this question, do not worry; we will get to it.)

So, in using statistics, we are typically not very interested in the individual cases. In particular, we are typically not interested in the extreme cases. We are interested rather in describing the collection as a whole. You may in fact come to have philosophical objections to statistics' concern with collections at the expense of the individual. Fair enough. However, describing and generalizing about collections is what statistics is about.

How do we answer statistical questions? Here are some examples.

Answers to Statistical Questions: Example 1

Our first statistical question was:

Are male students or female students more likely to have a tattoo?

How can we answer this question? We have to have some data. Each student was asked whether or not he or she had a tattoo. So, as a first step, we do some *calculation.* We count the number of male students who have tattoos and the number of female who have tattoos, and, in fact, in one of the exercises you will do that first for some data collected in 2008. Calculation is the first step, but we will see it is not the only step. We will illustrate this and the following steps using similar data from Pennsylvania State University.

At Pennsylvania State University, statistics students were also asked whether they had a tattoo or not. Here are the counts. The counts of students who have and did not have tattoos are shown in a Fathom Summary Table here. So eighteen female students had tattoos and thirteen male students had tattoos. Does that mean that that the females are more *likely* to have a tattoo? No!

PennState2		Tattoo		Row Summary
		No	Yes	
Sex	Female	119	18	137
	Male	55	13	68
Column Summary		174	31	205
S1 = count ( )				

Notice that in these statistics classes there were more females than males; there were 137 females and sixty-eight males. There are different numbers (or counts) of male and female students, but that does *not* stop us from answering our question. We can still answer our question even though the total counts of males and females are different. We use the idea of a **proportion** or **percentage**.

A proportion looks at a count in a specific category *relative to* a total number. A proportion is a fraction. To get a proportion we divide our number of "successes" (here, females with tattoos) by the total number of females. If we want percentages, we will multiply these proportions by one hundred. Hence, for the females we calculate $\frac{18}{137} \cdot 100 \approx 13.14\%$. After **calculation**, we **interpret** our calculations. We can say that 13.14% of the females in the Penn State collection have tattoos. Another way to talk about this is that out of one hundred female students, we expect about thirteen of them to have tattoos. Or we can say that the proportion of female students who have a tattoo is 0.1314. Or (as we shall learn later) we can say that the probability that a female has a tattoo is .1314 or 13.14%.

What about the males? For the males we calculate $\frac{13}{68} \cdot 100 \approx 19.12\%$. What can we say? We can say that 19.12% of the male students have tattoos. Or we can say that out of one hundred males, we expect about nineteen of them will have tattoos. Or we can say that the proportion of males that have tattoos is 0.1912. Or we can say that the probability that a male will have a tattoo at Penn State is 0.1912 or 19.12%

Proportion and Percentage

A **proportion** relates a count in a specific category to a relevant total count by dividing a specific count by the total count. Proportions must be between zero and one.

A **percentage** expresses a proportion as a quantity out of one hundred, so a proportion multiplied by one hundred gives the same fraction as a percentage. Percentages must be between zero and one hundred.

Our answer? From our calculation and our interpretation of the numbers, we can say that at Penn State, males are more likely than females to have a tattoo.

Answers to Statistical Questions: Calculation, Interpretation, and Truth

Notice that the answer to our statistical question involved two things: **calculation** and **interpretation**. The *calculation* was to divide the number of females who had tattoos by the total number of females (and multiply by one hundred) and to divide the number of males with tattoos by the total number of males (and multiply by one hundred). But we do not stop with calculation.

The next step is **interpretation**. *Interpretation* requires saying what the results of the calculation *mean* in the context of our original question. All of the sentences above that have "we can say" can be part of an interpretation. In the end, notice that it takes a number of sentences saying something about the numbers so that we can get to the answer to our question. Our conclusion is that at Penn State, males were more likely to have a tattoo than females, at least when these data were collected. To say this, we also have to say how the calculations back this up.

Answers to **statistical questions** always involve both some *calculation* (sometimes much, sometimes little) and some translating the calculations into language to answer the question. An answer to a statistical question is not complete without *interpretation*. You have been warned!

Two more issues. First, when we interpret, how far can we generalize? Will what we have found for the Penn State students necessarily apply to students in California, for example? The answer is that we cannot generalize unless certain conditions are met. What those conditions are, and how we use them, is the subject of the second part of the course. That part of the course is entitled **inferential statistics,** and in that part we learn when and how we can generalize or, in statistical language, *infer*.

The second issue has to do with the truthfulness of the Penn State students. Suppose some of them are not telling the truth and are saying they have a tattoo when they do not have one or *vice versa*. If some of them are not telling the truth, we are in trouble. Our calculations and our interpretations are only as good as our data, and lying makes bad data. Lying is just one way that data may be made bad.

Answers to Statistical Questions: Example 2

Our third statistical question was:

> *What percentage of students speaks just one language? What percentage speaks two languages? What percentage speaks three languages? Four? Five?*

Once again, as an illustration, we will show a similar question with a similar collection and allow you to answer the third statistical question from some data that has been collected. As a part of the Census At School project, students in secondary schools in Australia were asked how many people lived "at home." These data show the number of people in the student's household, including the student. So, our question is:

> *What percentage of students lives in households of two people? What percentage lives in households with three people? What percentage lives in households of four?*

What should we do with the data? What calculations should we make? Households with zero people do not make sense (vacant houses make sense but not vacant households), but we can have households with values of 1, 2, 3, 4, 5, etc. people. Hence, a good way to proceed will be to count the number of cases where the variable *NumberHousehold* takes on the values 1, 2, 3, 4, 5, etc. It will be convenient to make a table to hold the results of this calculation. How far should we go? Looking over the data reveals that there is one student with thirty people in her household and two students who reported nine people in their home. Those are the biggest household sizes. Let us make a table with cells for what we think we see. Then we laboriously make a tally, going through the data and recording every single instance of 1, 2, 3, etc.

Number Household	1	2	3	4	5	6	7	8	9	30	Total
Number of cases	|	||||||	|||||||	Etc.	Etc.				||	|	

Below are the numbers that we will see after quite a bit of work. (Later, but soon, we will get software to do this tedious work.)

Number Household	1	2	3	4	5	6	7	8	9	30	Total
Counts	1	10	63	142	112	49	12	6	2	1	398

Once again, what we really want for comparison are proportions or percentages. If we add all the numbers of cases, we see the collection contains 398 cases. To get the percentage of students who live in households of one, two, three, etc. people, we divide each of the counts by the total count of 398 and multiply by one hundred. Here is what we get.

Number Household	1	2	3	4	5	6	7	8	9	30	Total
Counts:	1	10	63	142	112	49	12	6	2	1	398
Percent of cases:	0.25%	2.51%	15.83%	35.68%	28.14%	12.31%	3.02%	1.51%	0.50%	0.25%	100.00%

If you add the percentages, we should get 100% since our tally includes *all* the students in the collection. (Sometimes the sum will be very, very close to 100% but not exactly 100% because we round to, say, two decimal places; for this particular calculation, we do in fact get 100%.)

That is the *calculation;* but what about the *interpretation?* What do the calculations mean?

Perhaps the first thing that draws your attention is the student whose household has thirty people. We call such data **outliers** because the data point lies outside the main body of the data. The figure of 30 could be a mistake (it could be "3") or perhaps this student lives in some kind of commune or perhaps she was not taking the questionnaire seriously. Statisticians are typically not so much interested in the extremes as in the main body of data. Statisticians do pay attention to outliers, but they are not usually the main focus of attention.

What can we say? One thing is that the great majority of secondary students live in households that have between three and six people (about 92%), and a second thing is that over 60% live in households that have either four or five people. A relatively smaller proportion of Australian students live in households of more than six people; the number is about 18%.

Notice that these (60%, 18%, etc.) figures do not really have meaning until we compare them with something else. Would we get the same kinds of numbers for secondary level students in North America rather than in Australia? (My guess is yes, but we would have to see.) For students in Western Europe? For students in West Africa? Southeast Asia? The Indian subcontinent? Suppose we could compare these numbers with the sizes of households two centuries ago? Lesson: *interpretation* involves comparison, either between collections or within a collection (are there differences for different parts of Australia?), or to a standard. Interpretation is based upon comparison.

Embracing Symbols—and the Meaning of "n"

Why do we have symbols? For fast recognition—at least that is what the commercial world thinks. Students often see symbols as a burden, something extra to learn. No! Learn symbols—indeed, embrace symbols–as a way to make a mental map of statistics. Embrace symbols as another connector relating what are otherwise entirely disconnected things. Here is your first symbol, n, and its definition.

> **Notation for the number of cases in a collection**
>
> **n** stands for the number of cases in a collection.

For the Penn State collection of students, $n = 205$, and for the Australian students, $n = 398$.

Summary to Starting Out: Terminology and Perspective

The main idea of this section is to introduce essential and basic *terminology* and essential *perspective* in order to be able to continue. Here is a summary of the terminology introduced above and some advice: good students seek to understand how these terms relate to each other.

Terminology:

- **Cases**   The objects (people, places, things, or times) about which we have *data* in a *collection*
- **Data**   The measurements or observations on *variables* for the *cases* in a *collection*
- **Collection**   The sum total of the data for all the cases together. Collections may be divided into sub-collections, where the data for the sub-collections are compared (e.g., comparing males and females).
- **Variable**   An aspect or characteristic of *cases* in a *collection* whose *value* potentially varies from case to case and which can be measured numerically (a *quantitative variable*) or by categories (a *categorical variable*)
    - **Attribute**   Another name for variable
- **Value of a variable**   The numbers or categories that are logically possible for a *variable*
- **Quantitative variable**   A variable whose *values* are expressed as numbers or numerically
- **Categorical variable**   A variable whose *values* are expressed non-numerically
- **Statistical Question**   A question about a *collection* of *cases* that can be answered by doing *calculations* on *data* and then *interpreting* the results of the calculation to answer the question
- **Notation**   Symbols that stand for quantities or operations that are part of statistical calculations

Perspective:

- The goal of statistics is to answer questions about data—that is, about measurements on variables for cases in a collection or in several collections.
- The focus of statistical analysis is to say something about collections and not necessarily about specific members or elements of the collection.
- The language of *probability* pervades statistical thinking and analysis, and this language is discussed in the next section. Statistical analysis deals in *likelihoods* ("How likely is it that a student from Australia lives in a household of five or more people?") rather than in certainties.
- Statistics always involves both **calculation** and **interpretation**.

§1.1 Exercises on Starting Out

1. California Statistics Students Here is part of a Fathom case table (a spreadsheet) and the description of variables for data collected in four statistics classes at a college in California in 2009.

	Gender	AgeYe...	Height	MothersAge	PoliticalView	Langua...	NumberHou...	NumberCars	NumberStates	Instructor...	Tattoo
80	M	20	169.0	26	Moderate	2	4	5	2	48	N
81	F	18	160.5	30	Liberal	1	5	3	4	50	N
82	M	19	183.0	27	Moderate	2	5	4	7	52	N
83	F	20	175.0	24	Moderate	1	3	1	6	55	N
84	F	17	173.0	34	Liberal	1	5	4	4	21	N
85	F	25	155.0	34	Liberal	2	3	3	4	62	Y
86	M	21	173.0	23	Liberal	3	2	2	11	55	Y

Variables:	Description
Gender:	Gender of the student: male or female
Age Years:	Age of the student (in years)
Height:	Height of the student, measured in centimeters
Mother's Age:	Age of the student's mother when the student was born
Political View:	Self-reported "political leaning" of the student: liberal, moderate, conservative
Languages:	Languages spoken by the student
Number Household:	Number of people in the household in which the student lives
Number Cars:	Number of cars in the household in which the student lives
Number States:	Number of states in the USA the student has visited or lived in
Instructor's Age:	"Guesstimate" by the student of the age of the instructor
Tattoo:	Whether or not the student has a tattoo: yes or no

Carefully read the description of the variables given above and answer the questions below.

a. List which variables listed in the description are *quantitative* and which variables are *categorical*.

b. What are *possible* values for the variable *Number Household*? (See the **Notes** for a definition of *values*. Also, notice that this question asks about *possible* values; that means what is possible for the variable.)

c. A student answers part b by listing: 2, 3, 4, 5 because that is what the student sees in the case table above. Why is ("2, 3, 4, 5") *not* an answer to the question about *possible* values?

d. What values for the variable *Number Household* do you think are likely, and what values do you think are not at all likely? Explain in complete sentences, giving brief reasons for your answer.

e. What are *possible* values for the variable *Number States*?

f. What values for the variable *Number States* do you think are likely and what values not likely?

g. According to the variable description, what are the possible values for the variable *Political View*?

		Gender		Row Summary
		M	F	
PolticalView	Liberal	35	51	86
	Moderate	38	31	69
	Conservative	5	7	12
	Column Summary	78	89	167
S1 = count ( )				

h. Calculate the proportion of male students who have "moderate" political views (answer: 48.7%).

i. Calculate the proportion of female students who have "moderate" political views (answer: 34.8%).

j. Use the results of parts h and i to give an answer to the *statistical question*: "Are male or female students in this collection more likely to have 'moderate' political views?"

2. Far West Colleges and Universities. Here is a spreadsheet showing another collection. This time, we have a collection of colleges and universities from the far West of the United States. (The collection has colleges and universities from the states of Alaska, California, Hawaii, Nevada, Oregon, and Washington.)

	Name	St...	FullTim...	IntlStu...	FTE	PctOverAg...	PctFullTime	Freshman...	Graduation_rate	TuitionF...	Sector
361	Wenatchee Valley College	WA	1952	1	2342	40	63	60	36	2541	Public 2-..
362	West Hills Community College	CA	2158	51	3059	37	45	64	40	624	Public 2-..
363	West Los Angeles College	CA	2121	113	4354	54	24	58	20	810	Public 2-..
364	West Valley College	CA	3314	62	5406	41	35	66	41	708	Public 2-..
365	Western Nevada Community...	NV	954	0	2292	57	19	56	20	1618	Public 2-..
366	Western Oregon University	OR	3800	57	4017	14	88	67	43	4488	Public 4-..

Variables	Description
Name	Name of the college or university
State	State in which the college or university is located
Full-Time Students	Number of full-time students enrolled
Int'l Students	Number of international students enrolled
FTE	Full-time equivalent number of students (If a college or university has both full-time and part-time students, a formula is used to express the participation of the part-time students as "full-time equivalent" students. For example, two students each taking six units might equal one full-time twelve-unit student.)
Pct Over Age 24	Percentage of the students who are over twenty-four years of age
Freshman Retention Rate	Percentage of the first-year students who continue to the second year
Graduation Rate	Percentage of all students who graduate
Tuition Fees	Total yearly fees for tuition
Sector	Whether the college or university is "four-year public," "four-year private," or "two-year public"
Av Debt	The average debt that a student at the college or university has incurred at the end of study
Pct Grad Debt	Percentage of graduates who have incurred debt by the end of their study
Selectivity	A measure of how selective the institution is for those colleges that area not "open admission." The categories are "minimally," "moderately," or "very selective."
Pct Hispanic	Percentage of students in the college or university who are of Hispanic origin
Pct White	Percentage of students in the college or university who are "white"
Pct Black	Percentage of students in the college or university who are "Black"
Pct Asian	Percentage of students in the college or university who are of Asian or Pacific Island origin

   a. What are the cases for this collection? (See the **Notes** for a definition of **cases.**)
   b. State which variables listed above are quantitative and which are categorical.
   c. List the possible values for the variable "**State**" for *this* collection.
   d. List the possible values for the variable "*Freshman Retention Rate.*" (Notice we are asking about "possible" values and not "plausible" or "likely" values.)
   e. In the **Notes,** there is a section about **Statistical Questions.** One of the points made there is that we usually ask *statistical questions* about *collections* and not so much about the individual cases in the collection. Which of the following statistical questions (on the next page) are questions about the collection (Far West Colleges) and which questions focus more on individual cases within the collection? Give a reason for your answer. (You will soon learn to find answers to such statistical questions.)

(i) On average, are tuition fees higher among the "four-year public" colleges and universities or among the "four-year private" colleges and universities?

(ii) Is there more variability in the tuition fees among the "four-year public" colleges and universities or among the "four-year private" colleges and universities?

(iii) Which university has the highest tuition fees?

(iv) Is it true that the higher the tuition fees in a college or university, the higher the average debt will be?

3. **Potatoes** You have a collection of potatoes; the cases are potatoes, and you want to study them.
   a. List three (or more) quantitative and two (or more) categorical variables that come to mind for the collection. (See the **Notes** for a definition of **quantitative** and **categorical** variables.)
   b. For each of the variables, list possible values for the variable. (See the **Notes** for a definition of **values**.)
   c. Write three good statistical questions about the collection of potatoes. (See the section in the **Notes** about **statistical questions**.)

4. **Questionnaire Questions to Variables.** Often data are collected from people (the cases) using a questionnaire (such as you may have done on the first or second day of class). Here is such a questionnaire item: "With which political party would you say you most closely identify?"
   a. State some possible values for the variable that this question is measuring.
   b. Is the variable categorical or quantitative? Give a reason for your answer.
   c. Give a name for this variable that describes what is being measured. (The name should not be the same as the values; the name should describe what varies.)

5. **Questionnaire Questions to Variables.** Here is another questionnaire question: "Estimate the amount of money you spent last month going to movies at theaters."
   a. State *all* the possible values for this variable. (Is $547.17 possible? Is $547.17 likely?)
   b. Is the variable categorical or quantitative? Give a reason for your answer.
   c. Give an appropriate name for the variable (The name may be more than one word.)

6. **Answering Statistical Questions.** You have been given the data for the students from several California statistics classes (**CombinedClassDataAut08**, which you will find at the end of this section of exercises). The variables measured in that collection are described in Exercise 1 and on the data sheet you have been given. Here is the statistical question we want to answer from these data.

   Are male students or female students more likely to have a tattoo?

   You will want to look at Example 1 as part of this question depends on that question.

   a. Make a table similar to the one on the right and from the data sheet tally the responses by gender to the question of whether a student has a tattoo. Or, get Fathom to make a Summary Table.

		Tattoo		Row Summary
		No	Yes	
Gender	Female	47	16	63
	Male	55	8	63
Column Summary		102	24	124

   b. Calculate the percentage of females that have a tattoo. Calculate the percentage of males that have a tattoo. Calculate the overall percentage of students in this collection that have a tattoo.

c. Using the results of part b, compare the California students with the Penn State students. Use complete sentences in writing your interpretation about any differences you find between the Penn State students and the California students in having a tattoo.

   d. In making an interpretation, a student writes, "The differences in the percentages of students having tattoos in Pennsylvania and California are because Easterners are so much more conservative than West Coast students." The student's instructor writes on the paper: "Be careful about unwarranted inferences." What do you think the instructor had in mind when writing this comment?

7. **Answering Statistical Questions.** Using the same collection of data on California statistics students (**Combined Class Data Aut 08**, at the end of this section of exercises), we want to answer the following question, which happens to be the third statistical question on our list.

   What percentage of students speaks just one language? What percentage speaks two languages? What percentage speaks three languages? Four? Five?

   You will find it convenient to consult Example 2.

   a. From the data sheet, make a table (like the one in Example 2) to tally the number of students who speak one, two, three, etc. languages.

   b. Calculate percentages of those who speak one, two, three, etc. languages.

   c. *Interpretation* Look at the percentage of the students that are monolingual (speak just one language). Reflect on whether you think that this number is a relatively big percentage or a relatively small one. Then think of what other information you would like to have to help you decide whether the percentage is relatively big or small.

   d. Write another good statistical question about the variable *Languages* (i.e., the number of languages a student speaks).

8. **Answering Statistical Questions.** In our **Combined Class Data Aut 08** data set for statistics students (at the end of this section of exercises), we have data on the size of the household, so we can answer the same kinds of questions we did with the Australian student data set in Example 2. *What percentage of students lives in households of two people? What percentage lives in households with three people? What percentage lives in households of four?*

   (NOTE: The number of people in the household includes the student.)

   a. From the data sheet, make a table (like the one in Example 2) to tally the number of students who live in households that have one person, two people, three people, etc. The data are in the variable *Number Household*.

   b. Calculate percentages of those who live in households of one, two, three, etc. people.

   c. *Interpretation* Compare your percentages with the ones we found for the Australian secondary students. Are they similar or much different?

   d. You should have found that about 2.4% of the California students live alone and another 9% live in households of size two, whereas only about 2.75% of the Australian secondary school students live in households of size one or two. Does this make sense in light of the fact that one collection is made up of secondary school students and the collection is made up of college students? Explain briefly but well.

9. **Les Écossais.** Here is a spreadsheet in Fathom of a collection of data from the 1851 Canadian Census for a small part of the province of Québec where (at that date) a large proportion of the population were Highland Scots. Each row in the spreadsheet is a household, and there are just three variables.

   Num HH — The number of people in the household
   Age — The age of the head of the household
   Place — Whether the household was in Newton or in the municipality of Saint-Télésphore

   a. What are the **cases** for this collection of data? Be specific.
   b. Which variables are **quantitative** and which are **categorical?**
   c. What are possible **values** for the variable *Num HH?*
   d. Make up a **statistical question** that could be answered with these data.

10. **Les Ozz** Here are more data from the Australian Secondary Student collection. Here we have two new variables: one variable (*Whre Live*) records the state or territory within Australia where the student lived. The states for this collection are:

Qld	Queensland
SA	South Australia
Vic	Victoria
WA	Western Australia

   The students were asked about their favorite type of music and were given a list of various types of music. If a student said that reggae was a favorite type of music then "yes" was recorded for the variable *Mus Regg*. If the student did not choose reggae then for that student "no" was recorded for *Mus Regg*.

   Our statistical question is:

   Is there a difference in the popularity of reggae among students from different parts of Australia?

   a. Explain why this statement is not true: "Reggae is more popular for students from the state of Victoria since there are nineteen students who like reggae from that state in our collection and that is more than the numbers liking reggae who are from the other states."
   b. The percentage of students from the state of Victoria who list reggae as one of their favorite types of music is 16.10%. Show how this was calculated.
   c. Do three other calculations like the one in part b to *calculate* the percentages of students liking reggae for the students from each of the four Australia states.
   d. Write an *interpretation* of the calculations you made in part c to answer the statistical question.

Combined Class Data Autumn 08

Case Number	Gender	BirthMonth	BirthYear	Age	Height (cm.)	Mother'sAge	PoliticalView	Languages	NumberHousehold	NumberCars	NumberStates	Instructor'sAge	Tattoo	DominantHand	Class
1	F	Nov	1982	26	181.0	28	Liberal	1	3	3	7	52	Y	Right	AA
2	F	Jun	1985	24	172.0	32	Liberal	1	3	3	12	50	Y	Right	AA
3	F	Nov	1990	18	156.0	26	Liberal	2	9	2	2	53	N	Right	AA
4	F	Jun	1985	24	179.0	28	Liberal	1	4	4	12	55	N	Right	AA
5	F	Nov	1987	21	156.0	24	Liberal	2	4	5	14	55	N	Right	AA
6	F	Jul	1986	23	168.0	23	Liberal	1	5	5	6	47	Y	Right	AA
7	F	Jul	1989	20	166.0	35	Liberal	2	4	4	4	57	N	Right	AA
8	F	Jun	1989	20	160.0	31	Liberal	2	4	4	12	54	N	Ambide	AA
9	F	Feb	1990	18	151.0	35	Moderate	2	5	8	2	53	N	Right	AA
10	F	Aug	1988	21	168.0	30	Moderate	2	4	4	6	52	Y	Right	AA
11	F	Apr	1989	19	174.0	31	Liberal	2	4	2	6	45	Y	Left	AA
12	F	Jan	1989	19	168.0	32	Liberal	3	4	4	3	46	N	Right	AA
13	F	Feb	1989	19	165.0	20	Moderate	4	4	2	2	52	N	Right	AA
14	F	Nov	1978	30	163.0	25	Liberal	3	1	1	4	55	N	Right	AA
15	F	Apr	1991	17	179.0	33	Liberal	3	3	1	12	50	N	Ambide	AA
16	F	Nov	1965	43	173.0	22	Liberal	3	4	4	13	52	N	Right	AA
17	F	Jan	1987	21	168.0	23	Liberal	2	4	3	6	50	Y	Right	AA
18	F	Dec	1987	21	161.0	26	Liberal	3	5	3	2	50	N	Right	AA
19	F	Jun	1989	20	155.0	25	Moderate	2	4	3	8	50	N	Right	AA
20	F	Oct	1986	22	170.0	24	Liberal	1	6	7	3	45	N	Right	AC
21	F	Feb	1976	32	160.0	30	Liberal	2	3	3	4	65	N	Right	AC
22	F	Jan	1990	18	165.0	33	Moderate	2	3	4	8	44	N	Right	AC
23	F	Oct	1957	51	165.0	34	Liberal	2	6	4	5	50	N	Right	AC
24	F	May	1987	21	163.0	28	Conservat	3	2	4	2	55	N	Right	AC
25	F	Nov	1990	18	160.0	31	Moderate	2	2	1	2	55	N	Right	AC
26	F	May	1989	19	161.0	22	Liberal	2	8	4	5	55	N	Right	AC
27	F	Mar	1990	18	167.0	24	Conservat	2	5	4	1	60	N	Right	AC
28	F	Sep	1988	20	170.0	21	Liberal	1	6	3	8	53	N	Right	AC
29	F	Jan	1988	20	156.0	18	Moderate	2	3	3	3	50	N	Right	AC
30	F	Mar	1990	18	155.0	35	Liberal	3	3	3	9	53	N	Right	AC
31	F	Jun	1990	19	166.0	28	Moderate	1	4	4	8	55	N	Right	AC
32	F	May	1990	18	157.0	33	Liberal	1	3	2	5	62	N	Right	AC
33	F	Jun	1989	20	168.0	24	Liberal	1	2	2	7	53	Y	Right	AC
34	F	Mar	1988	20	168.0	23	Liberal	2	4	7	6	62	N	Right	AC
35	F	Feb	1989	19	166.0	30	Moderate	1	5	3	3	55	N	Right	AC
36	F	Feb	1986	22	163.0	22	Liberal	2	6	3	8	54	Y	Right	AC
37	F	Apr	1990	18	170.0	25	Moderate	1	4	3	2	60	N	Right	AC
38	F	Oct	1990	18	177.0	21	Liberal	2	8	7	0	56	N	Right	AC
39	F	Jul	1987	22	167.0	38	Moderate	2	3	2	5	65	N	Right	AC
40	F	Apr	1989	19	157.0	32	Liberal	2	4	9	2	58	N	Right	AC
41	F	Dec	1982	26	161.0	30	Liberal	1	1	1	8	60	Y	Right	AD
42	F	Aug	1988	21	162.0	32	Liberal	1	5	7	8	65	N	Right	AD
43	F	Dec	1988	20	163.0	41	Moderate	2	4	3	4	64	N	Right	AD
44	F	Apr	1987	21	160.0	20	Liberal	2	4	3	2	63	N	Right	AD
45	F	Aug	1985	24	178.0	24	Moderate	2	5	4	10	65	Y	Right	AD
46	F	Jun	1991	18	167.0	30	Moderate	1	5	5	10	70	N	Right	AD
47	F	Jan	1983	25	170.0	23	Liberal	3	2	2	9	55	N	Right	AD
48	F	Sep	1989	19	168.0	28	Liberal	2	4	7	4	68	N	Right	AD
49	F	Mar	1988	20	161.0	35	Liberal	2	2	2	10	60	Y	Right	AD
50	F	Nov	1989	19	164.0	27	Liberal	2	7	6	9	60	N	Right	AD
51	F	Apr	1983	25	169.0	24	Moderate	1	5	3	9	42	Y	Right	AD
52	F	Nov	1989	19	170.0	28	Moderate	2	3	2	6	50	N	Right	AD
53	F	Jul	1987	22	176.0	39	Moderate	2	3	2	5	60	N	Right	AD
54	F	Mar	1988	20	164.0	31	Liberal	2	3	6	5	60	N	Right	AD
55	F	Jan	1990	18	162.0	29	Moderate	3	2	1	10	55	N	Right	AD
56	F	Aug	1990	19	152.0	32	Liberal	1	5	2	18	66	N	Right	AD
57	F	Dec	1981	27	163.5	30	Liberal	1	1	1	37	60	Y	Right	AD
58	F	Jan	1989	19	167.0	27	Moderate	1	4	4	2	58	Y	Right	AD
59	F	Jul	1962	47	165.0	24	Liberal	1	5	2	6	56	N	Right	AD
60	F	Jul	1988	21	155.0	28	Moderate	1	7	4	6	50	Y	Left	AD
61	F	Oct	1990	18	165.0	27	Conservat	1	6	6	8	64	N	Right	AD
62	F	Jan	1984	24	177.0	37	Moderate	1	3	3	19	56	Y	Right	AD
63	F	Jul	1990	19	157.2	27	Moderate	1	5	7	2	55	N	Right	AD
64	M	May	1989	19	167.0	29	Liberal	2	4	2	1	52	N	Right	AA
65	M	Nov	1988	20	167.0	31	Moderate	2	3	2	1	58	N	Right	AA
66	M	Apr	1991	17	169.0	33	Liberal	1	4	2	7	53	N	Right	AA
67	M	Oct	1988	20	154.0	23	Liberal	1	6	5	7	55	N	Right	AA
68	M	Dec	1987	21	178.2	22	Liberal	1	6	4	6	50	Y	Right	AA
69	M	Jun	1987	22	182.0	19	Liberal	2	4	5	3	35	N	Left	AA
70	M	Sep	1986	22	195.0	36	Liberal	1	2	2	8	56	N	Right	AA
71	M	Jun	1988	20	167.0	25	Moderate	1	6	5	1	48	N	Right	AA
72	M	Mar	1988	20	173.0	29	Moderate	2	3	2	8	58	N	Right	AA
73	M	Sep	1989	19	183.0	25	Moderate	1	3	2	6	53	N	Ambide	AA
74	M	Jun	1989	20	173.0	25	Liberal	1	4	4	11	52	Y	Right	AA
75	M	Nov	1987	21	181.0	28	Conservati	2	4	3	4	52	N	Right	AA
76	M	Feb	1984	24	179.0	39	Moderate	1	3	5	22	48	Y	Right	AA
77	M	Aug	1988	21	185.0	35	Moderate	1	4	3	7	50	N	Right	AA
78	M	Oct	1989	19	190.0	22	Liberal	1	5	4	4	60	N	Right	AA
79	M	Jan	1989	19	172.0	25	Conservati	1	3	5	5	49	N	Right	AA
80	M	Aug	1989	20	166.0	25	Moderate	2	4	4	5	44	N	Right	AA
81	M	Oct	1989	19	180.0	30	Moderate	1	4	7	5	54	N	Ambide	AA
82	M	May	1982	26	182.0	28	Liberal	2	4	2	4	50	N	Right	AA
83	M	Dec	1986	22	172.0	25	Conservati	1	4	4	2	52	N	Right	AA
84	M	Oct	1989	19	182.9	32	Conservati	1	4	2	19	58	N	Right	AC
85	M	Jul	1987	20	167.0	28	Liberal	2	10	6	3	55	N	Right	AC
86	M	Jun	1988	21	180.3	32	Liberal	1	3	3	14	58	N	Right	AC
87	M	Jul	1983	26	167.0	26	Moderate	2	3	5	13	55	Y	Right	AC
88	M	Jun	1989	20	170.0		Conservati	1	7	1	18	57	N	Right	AC
89	M	Jul	1982	27	182.0	32	Moderate	1	2	2	25	58	Y	Right	AC
90	M	Apr	1990	18	179.0	30	Moderate	2	4	1	4	60	N	Right	AC
91	M	Sep	1989	19	172.0	31	Liberal	1	4	4	3	55	N	Right	AC
92	M	May	1989	19	177.0	24	Moderate	3	4	3	5	55	N	Right	AC
93	M	Jul	1989	20	183.0	34	Liberal	1	4	3	17	52	N	Right	AC
94	M	Oct	1989	19	184.0	31	Liberal	1	5	5	12	53	N	Right	AC
95	M	Feb	1989	19	180.0		Liberal	1	5	3	17	48	N	Right	AC
96	M	May	1990	18	170.0	36	Moderate	2	2	2	2	54	N	Right	AC
97	M	Jun	1985	24	177.0	19	Liberal	2	3	3	3	52	N	Right	AC
98	M	Nov	1988	20	188.0	36	Moderate	2	4	4	17	56	N	Right	AC
99	M	May	1989	19	180.0	28	Moderate	1	4	3	6	61	N	Right	AC
100	M	Dec	1988	20	190.0	35	Conservati	2	5	3	8	63	N	Right	AC
101	M	Jan	1988	20	173.0	37	Liberal	2	4	6+	10	56	N	Ambide	AC
102	M	Sep	1989	19	124.5		Moderate	2	4	1	7	50	N	Right	AC
103	M	Mar	1988	20	170.0	23	Liberal	1	4	5	5	60	N	Right	AC
104	M	Jan	1986	23	175.0	28	Moderate	2	4	2	2	58	N	Right	AC
105	M	Mar	1983	25	175.0	19	Conservati	1	3	4	20	56	Y	Ambide	AC
106	M	Jun	1987	22	195.0	35	Liberal	1	2	2	10	60	N	Left	AD
107	M	Apr	1985	23	184.0		Conservati	2	5	5	1	100	Y	Right	AD
108	M	Aug	1985	24	188.0	27	Moderate	1	4	5	3	60	Y	Right	AD
109	M	Nov	1987	21	167.0	30	Moderate	1	4	3	10	58	N	Right	AD
110	M	Nov	1984	24	169.0	21	Conservati	2	5	4	4	47	N	Right	AD
111	M	Sep	1989	19	169.0	27	Liberal	1	4	4	7	62	N	Right	AD
112	M	May	1985	23	165.0	28	Conservati	2	5	4	7	64	N	Right	AD
113	M	Feb	1986	22	181.0	28	Moderate	2	5	1	6	60	N	Right	AD
114	M	Feb	1990	18	180.0	30	Liberal	3	5	4	3	48	N	Right	AD
115	M	May	1987	21	177.0	51	Moderate	2	3	4	2	60	N	Right	AD
116	M	Jun	1988	21	178.0	38	Moderate	1	2	4	12	60	N	Right	AD
117	M	Jan	1990	18	177.0	26	Moderate	2	4	2	2	58	N	Right	AD
118	M	Aug	1987	22	173.0	35	Moderate	1	4	7	16	58	N	Right	AD
119	M	Aug	1990	19	167.0	30	Moderate	2	3	5	4	55	N	Right	AD
120	M	Sep	1989	19	175.0	37	Liberal	1	3	3	8	65	N	Ambide	AD
121	M	Jun	1990	19	180.0	40	Moderate	1	4	4	16	67	N	Right	AD
122	M	Feb	1983	25	183.0	33	Moderate	1	4	4	13	57	N	Right	AD
123	M	Dec	1988	21	176.0	34	Moderate	2	4	6	11	60	N	Right	AD
124	M	Jul	1990	19	186.0	40	Liberal	1	4	3	22	65	N	Right	AD
125	M	Sep	1999	9	189.0	38	Liberal	2	3	1	4	50	N	Right	AD
126	M	Jan	1990	18	173.0	30	Moderate	1	3	3	9	56	N	Right	AD

§1.2 The Language of Statistics: Probability

Smuggled In

Our first statistical question (introduced in the last section) was:

Are male students or female students more likely to have a tattoo?

We answered this question in Example 1 for the Penn State students by calculating and interpreting the proportions of male and female students who had tattoos. For the females we calculated that $\frac{18}{137} \cdot 100 \approx 13.14\%$ had tattoos, whereas for the males we calculated that $\frac{13}{68} \cdot 100 \approx 0.1912 = 19.12\%$ had tattoos. We concluded that *for the Penn State students,* males are more likely to have a tattoo than females.

PennState2		Tattoo		Row Summary
		No	Yes	
Sex	Female	119	18	137
	Male	55	13	68
Column Summary		174	31	205

S1 = count ( )

In our statistical question we actually smuggled in the idea of probability by using the word "likely." We probably have some notion of how the word "likely" is used. One meaning that we will give to the word "probability" (or to the word "likely") is this: "If we chose a student completely at random from all the females on the Penn State campus, the probability that she would have a tattoo is 0.1314 or 13.14%." Or if we chose a student completely at random from all the male students at Penn State, the probability that he would have a tattoo is 0.1912 = 19.12%. Now what does "completely at random" mean? If students walked around the Penn State campus completely randomly then the chance that you—in your own random wandering—would meet a student (either male or female) with a tattoo would be $\frac{31}{205} \approx 0.1512$, or about 15.12%, since altogether there were thirty-one students out of 205 that had a tattoo. But students do not walk the campus completely at random; perhaps the ones with tattoos tend to be together and those without tattoos together. Our picture is flawed because no one actually walks around a campus completely randomly. But our definition of randomness demands just that.

Here is a better picture of randomness. If we were able to put all the Penn State students in a giant bin (like the lottery bins) and draw one out randomly then the probability that the student chosen would have a tattoo is about 15.12%. The ideas that we have from everyday life about probability, chance, and likelihood can carry us a certain distance, but we need specific terminology and notation to use probability without getting into trouble.

Language and Notation

We will begin with the example of students having (or not having) a tattoo. The first bit of terminology that we use is the idea of an **event.** An event can be nearly anything (not just something for which you buy tickets), as long as we can define it well. In our first example, we can define two events: "having a tattoo" and "not having a tattoo." In accord with what we see in the table, we will assign:

Y to be defined as the event: "Student has a tattoo"

N to be defined as the event: "Student does not have a tattoo"

Then, to express the idea of probability, we will use the notation $P(Y) = \frac{31}{205} = 0.1512$. This is read: "the probability of Y is 0.1512." Or, in more expanded form, we say: "The probability that a student has a tattoo is 0.1512." Notice that this *P(Y)* is similar to the function notation you learned in algebra: *f(x)* is read "*f* of *x*". Notice also that the calculation does not belong inside the parentheses. What goes inside the parentheses is the name of the event, often designated by a letter. Remember: embrace notation!

Events Can Also Refer to Numbers

In the last section we asked these statistical questions about Australian students:
What percentage of students lives in households of two people? What percentage lives in households with three people? What percentage lives in households of four? (The number of people in the household includes the student.)

We can now express these questions in probability terms as using the language of events:

What is the probability that an Australian student lives in a household of two people?
What is the probability that an Australian student lives in a household of three people?
What is the probability that an Australian student lives in a household of four people?

How do we use the probability notation to express the answers? Since the questions refer to different numbers, what we will usually do is to let a capital letter stand for the number, in this case the number of people in a household. So, in this example, we would have:

X = 2 to be defined as the event: "The Australian student lives in a household of two people."
X = 3 to be defined as the event: "The Australian student lives in a household of three people."

Then, if we have the data that were shown in Section 1.1, we would write: $P(X = 3) = \frac{63}{398} = 0.1583$ and read: "The probability that an Australian student (in this collection) lives in a household of three people is 0.1583." (Recall that the total number of students was $n = 398$.)

Number Household	1	2	3	4	5	6	7	8	9	30	Total
Counts:	1	10	63	142	112	49	12	6	2	1	398
Percent of cases:	0.25%	2.51%	15.83%	35.68%	28.14%	12.31%	3.02%	1.51%	0.50%	0.25%	100.00%

Notice that in the notation $P(X = 3) = \frac{63}{398} = 0.1583$ there is an "=" sign *inside* the parentheses and also *outside* the parentheses in this notation. The equals sign inside the parentheses refers to the event "household of three people," and the second equals sign is actually the verb "is" in our interpretation: "The probability that an Australian student (in this collection) lives in a household of three people *is* 63/398 which *is* 0.1583."

Example of Interpretation: The answer to our statistical question "*What is the probability that an Australian student in our collection lives in a household of four people?*" would be calculated (as before) as the proportion 142/398 and expressed as $P(X = 4) = \frac{142}{398} = 0.3568$, which is read, "The probability that an Australian student (in this collection) lives in a household of four people *is* 0.3568, or 35.68%"

Probabilities are proportions. You may well be more comfortable thinking in terms of percentages rather than in proportions. Get used to thinking in proportions; in essence, they are the same as percentages since percentages are just proportions multiplied by one hundred. We can see that probabilities are proportions in the way we calculate probabilities; look at our examples:

Tattoo example: $P(Y) = \dfrac{\text{Number of students who have a tattoo}}{\text{Total number of students}} = \dfrac{31}{205} = 0.1512$

Students in households of size four:

$P(X = 4) = \dfrac{\text{Number of students who live in households of size 4}}{\text{Total number of students}} = \dfrac{142}{398} = 0.3568$

That probabilities are proportions means that the smallest a probability can be is zero, and the largest a probability can be is 1, and these two ends of the interval of possible probabilities can be given meanings. If $P(E) = 0$ then the event E did not happen, and if $P(E) = 1$ then the event E was certain to happen. If we choose a student completely at random and we knew that $P(Y) = 0$ then we would know that no student had a tattoo, and if $P(Y) = 1$ we would know that *every* student had a tattoo. The box just below summarizes the terminology and notation about probability.

Definition and Notation for the Probability of an Event E:

We express the *probability* that event E happens using the notation and formula:

$$P(E) = \dfrac{\text{Number of Cases where Event } E \text{ is true}}{\text{Total number of Cases}}$$

$$0 \leq P(E) \leq 1$$

Sample Space: It is useful (but sometimes hard) to think of *all* the possibilities in a specific situation where we are applying probability. When we do this, we speak of the **sample space S** of all possible events. We often use braces $\{\cdots\}$ to indicate the elements in a sample space. In our tattoo example the student can either have a tattoo or not have a tattoo, so there were just two possibilities. This would be written in symbols as $S = \{N, Y\}$. For the example of the households for the Australian students, the sample space can be written $S = \{1, 2, 3, \ldots\}$ where the dots indicate that the numbers continue. Before we look at the data, we consider all the events that logically, and from our experience with similar data, could happen, even if we think the likelihood is very small. It is extremely unlikely that an Australian student lives in a household of 265 people, but it is possible, so we include this number in our sample space. It is not possible that the household size is a negative number, so we do not include negative numbers or numbers such as 3.658. In the context, only positive integers make sense.

The elements that we include in the sample space must also be what we call **mutually exclusive.** By mutually exclusive we mean that no two events in the list of events can happen at the same time. A student either has a tattoo or a student does not have a tattoo, so having and not having a tattoo *are* mutually exclusive. A student (at any particular point in time) lives in a particular size household; the numbers and hence events $X = 4$ and $X = 5$ should not happen at the same time for the same student.

The various sizes of households are mutually exclusive (we may have to be very careful how we ask the question about size of household).

One of our statistical questions was about whether males or female are more likely to have a tattoo. It would be quite *wrong* to list the sample space for this question as S = {N, Y, M, F} because obviously Y (has a tattoo) can happen at the same time as F (female), and so the events Y and F are *not* mutually exclusive. We shall see how we handle the situation of *non*-mutually exclusive events in the next section.

The elements of a sample space must also be **exhaustive,** which simply means that the sample space includes *all* the possibilities for our application of probability. It is for that reason that we listed the sample space for the number of people in a student's household as $S = \{1,2,3,...\}$. We have listed everything that is logically possible.

> **Sample space**
>
> A **sample space** is a complete listing (therefore, an **exhaustive** listing) of all the **mutually exclusive** events (therefore, events that cannot occur together) that are possible when applying probability language. The notation commonly used for sample space is $S = \{ \cdots \}$ where the events are listed in the braces.

Applying Probability to Our Questions. The application of our probability notation and calculation to the questions about the sizes of households for students is straightforward. Our statistical questions were:

What is the probability that an Australian student lives in a household of two people?
What is the probability that an Australian student lives in a household of three people?
What is the probability that an Australian student lives in a household of four people?

The calculated answers, in probability notation, and the written interpretations are:

$P(X = 2) = \dfrac{10}{398} \approx 0.0251$ The probability that we find an Australian high school student in a two-person household *is* (or equals) 2.51%.

$P(X = 3) = \dfrac{63}{398} \approx 0.1583$ The probability that we find an Australian high school student in a three-person household *is* (or equals) 15.83%.

$P(X = 4) = \dfrac{142}{398} \approx 0.3568$ The probability that we find an Australian high school student in a four-person household *is* (or equals) 35.68%.

However, our first statistical question about gender and tattoos is a bit more complicated because it involves two variables for each case (each student): the student's gender and whether or not the student has a tattoo—and we have already seen that the events associated with gender and the events associated with whether a student has a tattoo are not mutually exclusive. The questions about the size of households really involved just one variable measured for each student—namely, the size of the student's household, even though there were many different possible *values* for the variable. So, we need ways of using probabilities with events that are not mutually exclusive.

Let us return to our first statistical question and the table of data that goes with it, shown below.

Are male students or female students more likely to have a tattoo?

We have answered this question before by calculating the proportion for the females, $\frac{18}{137} \approx 0.1314$ and comparing this proportion with the proportion we calculated for the males, $\frac{13}{68} \approx 0.1912$. How do we apply the probability notation to this question?

Since there are two variables for each student, we must notice that there are four possible events in this case, the same number as the number of interior cells in the table. These four possible events are:

"F and N" 119 students Female and does *not* have a tattoo.
"F and Y" 18 students   Female and *does* have a tattoo.
"M and N" 55 students   Male and does *not* have a tattoo.
"M and Y" 13 students   Male and *does* have a tattoo.

PennState2		Tattoo		Row Summary
		No	Yes	
Sex	Female	119	18	137
	Male	55	13	68
Column Summary		174	31	205
S1 = count( )				

These four events are represented by the four cells of the table, and if we made a sample space for this application it would be $S = \{F \text{ and } N, F \text{ and } Y, M \text{ and } N, M \text{ and } Y\}$. Notice that every student can only be in one of these four interior cells (so these four combinations are *mutually exclusive*), and these four cells exhaust the possibilities for this application (so the four combinations are *exhaustive*). So how are we to handle the situation where we have two variables and the events are combinations of what were simple events?

"And," "Or," "If," and "Not" Probability Calculations and Interpretations

There are four different probability calculations that we can use, and for the impatient reader, we will reveal now that the one we want for our statistical question ("Are male students or female students more likely to have a tattoo?") is the "***if***" calculation. However, the other calculations are also very important.

And (Intersection) In some ways, the simplest calculation is the one that calculates the probability that a student "occupies" one of the four interior cells of the table. For example, the probability that a student is *both* female *and does* have a tattoo is $P(F \text{ and } Y) = \frac{18}{205} \approx 0.088 = 8.8\%$, and the probability that a student is both a male and *does* have a tattoo is $P(M \text{ and } Y) = \frac{13}{205} \approx 0.063 = 6.3\%$. If we compare these two probabilities, we see that the one for the females is larger, and this tells us that, for this collection, the likelihood of finding (or choosing at random) a tattooed female is greater than the likelihood of finding a tattooed male. However, this "and" calculation does *not* answer our question of whether the probability of having a tattoo is greater for males or females. In fact, this calculation confuses the issue because there are actually two probabilities at play. One probability is the likelihood that a student is a female rather than a male in this collection (remember we did not have equal numbers of males and females). The second probability is the likelihood of having a tattoo, which may be different for males and females. Useful as the calculation of "cell" probabilities is (the "and" calculation), the calculation does not answer our statistical question about the relative likelihood of a tattoo among male and female students.

The "and" is commonly referred to as the ***intersection*** of the two events, and you can see from the table relating gender and tattoos that this word makes some sense because the cell containing the eighteen tattooed females is the intersection of the "female" row and the "tattoo yes" column. The calculation of the probability of an *intersection* ("and") in general is given in the box below.

Intersection is also a good word to remember to guard against a common error: the first thing that comes to mind for some people with the word "and" is the word "addition." Banish the thought! "And" or intersection refers to the number where *both* event A *and* event B have happened.

There is a situation where $P(A \text{ and } B)$ *must* be zero, and that is if the events A and B are mutually exclusive events. It is impossible for a single student to both have and not have a tattoo at a single point in time, so $P(Y \text{ and } N) = 0$. Similarly, we know that $P(X = 4 \text{ and } X = 5) = 0$ because the events of household sizes of four and five are mutually exclusive. Our term *mutually exclusive* is reserved for events where it is impossible for the events to occur at the same time. However, if we found that $P(F \text{ and } Y) = 0$ for some collection of data, we would not conclude that the events F and Y were mutually exclusive; we would just know that there were no tattooed females in that collection.

> **Definition and Notation for the Probability of the Intersection of Two Events A and B:**
>
> $$P(A \text{ and } B) = \frac{\text{Number of cases where event } A \text{ and event } B \text{ are both true}}{\text{Total number of cases}}$$
>
> **Interpretation:** $P(A \text{ and } B)$ is read: "The probability that both A and B are true is…"

Or (Union) A second kind of probability calculation that we could make with more than one event is designated by the word "or" and is commonly known as **union**. The notation that we will use is $P(A \text{ or } B)$; as an example, we will calculate the probability $P(F \text{ or } Y)$. Your first inclination may be to think of "either F or Y". Not so! In probability

PennState2		Tattoo		Row Summary
		No	Yes	
Sex	Female	119	18	137
	Male	55	13	68
Column Summary		174	31	205
S1 = count ( )				

language and calculations, the word "or" does not have an exclusive meaning; rather, it will always mean "or including and." Hence the event "F or Y" in our example will include all of the females who also have tattoos (there are eighteen of them) as well as all the females who do not have tattoos (because they are females: there are 119 of them) and all of the males who have tattoos (because they have tattoos: there are thirteen of them). If you are not reluctant to write in your **Notes**, it may be a good idea to circle or shade in these numbers in the table. With these numbers, we can do the calculation: $P(F \text{ or } Y) = \frac{18 + 119 + 13}{205} = \frac{150}{205} \approx 0.732 = 73.2\%$. If we think of our "randomly choosing" interpretation for probability, we can read this as saying that "the probability of randomly choosing a Penn State student who is either female or has a tattoo or is both female and has a tattoo."

There is another way to calculate the probability of the union of two events that is sometimes useful. Notice that if we add 18 + 119 we get the total number of females, which is 137. And if we add 18 + 13 we get the 31 tattooed students. Now we can calculate the probability of choosing a female as $P(F) = \frac{137}{205}$, and we can also calculate the probability of having a tattoo, $P(Y) = \frac{31}{205} \approx 0.1512$. Now these two added together is $P(F) + P(Y) = \frac{18 + 119}{205} + \frac{18 + 13}{205} = \frac{18 + 119 + 18 + 13}{205}$, almost what we have in $P(F \text{ or } Y) = \frac{18 + 119 + 13}{205} = \frac{150}{205}$ *except* that we in our calculations have added the eighteen tattooed

19

females twice. But these eighteen tattooed females are the number in the intersection *F and Y*. So, we can do the calculation:

$$P(F \text{ or } Y) = P(F) + P(Y) - P(F \text{ and } Y)$$
$$= \frac{137}{205} + \frac{31}{205} - \frac{18}{205}$$
$$= \frac{168}{205} - \frac{18}{205}$$
$$= \frac{150}{205}$$

Recall the warning at the end of the last section about thinking that "and" must mean addition, when in fact it does not mean addition. The rule that we have shown above for calculating the probability of a union ("or") of two events is commonly called the **addition rule** since it involves a sum. The box below gives the rule in general. Whether you use this rule or simply determine what should be included in the union of two events A and *B* depends on the data that you have at hand.

The probability of a union of events is *not* what we want to answer with our question about the relative likelihood of tattoos between males and females. However, this "or" (union) is useful for other calculations. Think of our household collection of data again.

Number Household	1	2	3	4	5	6	7	8	9	30	Total
Counts:	1	10	63	142	112	49	12	6	2	1	398
Percent of cases:	0.25%	2.51%	15.83%	35.68%	28.14%	12.31%	3.02%	1.51%	0.50%	0.25%	100.00%

We may want to calculate the probability that an Australian student lives in a household of either size four *or* size five (household sizes that we think are fairly common), and we could write $P(X = 4$ or $X = 5)$. We know that $P(X=4) = \frac{142}{398} \approx 0.3568$, and $P(X=5) = \frac{112}{398} \approx 0.2814$, and we know that $P(X = 4 \text{ and } X = 5) = 0$ because these two events (household size of four and household size of five) are mutually exclusive. We can use the formula we used above and get:

$$P(X = 4 \text{ or } X = 5) = P(X = 4) + P(X = 5) - P(X = 4 \text{ and } X = 5)$$
$$= \frac{142}{398} + \frac{112}{398} - \frac{0}{398}$$
$$= 0.3568 + 0.2814 - 0$$
$$= 0.6382$$

Notice that if we know that two events are *mutually exclusive* then we know that the probability of their intersection must be zero, and we can ignore the intersection probability of the formula. We can extend this idea. Suppose we want to calculate the probability that an Australian student lives in a household that is smaller than four people. We can write this as $P(X < 4)$, but with our data, this means that we want to calculate $P(X < 4) = P(X = 1 \text{ or } X = 2 \text{ or } X = 3)$; household sizes of one or two or three are all that are possible that are less than 4. These events are all *mutually exclusive,* and so we can calculate:

$$P(X < 4) = P(X = 1 \text{ or } X = 2 \text{ or } X = 3)$$
$$= P(X = 1) + P(X = 2) + P(X = 3)$$
$$\approx 0.0025 + 0.0251 + 0.1583$$
$$= 0.1859$$

(We have used the calculations shown in the table above.)

Definition and notation for the probability of the union of two events—addition rule:

$$P(A \text{ or } B) = \frac{\text{Number of cases where either event } A \text{ or event } B \text{ or both are true}}{\text{Total Number of cases}} = P(A) + P(B) - P(A \text{ and } B)$$

Addition Rule for the Probability of the Union of Two Mutually Exclusive Events:

$$P(A \text{ or } B) = P(A) + P(B)$$

Interpretation: $P(A \text{ or } B)$ is read: "The probability that either A or b or both A and B are true is…"

If (Conditional) To answer our question about whether males or females are more likely to have a tattoo, we have to use a third (and extremely useful!) type of probability calculation called *conditional* probability. Once again, it will be helpful to refer to the table. What we really want to do is to *compare* the probabilities for females and for males, so it makes sense to consider *just* the females and then *just* the males. We have already shown this, but the notation showing that we are doing this is

PennState2

		Tattoo		Row Summary
		No	Yes	
Sex	Female	119	18	137
	Male	55	13	68
Column Summary		174	31	205

S1 = count( )

$P(Y \mid F) = \dfrac{18}{137} \approx 0.1314$. The vertical line in the notation indicates that we are just considering the females, and the probability can be interpreted in a number of ways. One interpretation is: "The probability that a student has a tattoo *given that* the student is female is 13.14%." Alternately, we could say, "The probability that a student has a tattoo *if* the student is female is 13.14%." There are other valid expressions in English, but languages are generally not as precise as the mathematical notation (which is one reason we have the notation!). We want to compare $P(Y \mid F)$ with $P(Y \mid M)$.

To calculate $P(Y \mid M)$, we *restrict* our denominator to the males and calculate

$P(Y \mid M) = \dfrac{13}{68} \approx 0.1912$, to get the probability that a student has a tattoo *if* (or *given that*) the student is a male. We can conclude (as we have done before) that for the Penn State data, males are more *likely* to have a tattoo than females.

Notice that for the calculation $P(Y \mid F) = \dfrac{18}{137} \approx 0.1314$, the numerator has just the eighteen tattooed females; that is, it has those students who are in the intersection of events F and Y. And in the denominator, we have just those students where event F is true. We can use these observations to make a rule for calculating conditional probability (see the box below), and we can use these to make a rule using other probabilities. Notice that $P(F \text{ and } Y) = \dfrac{18}{205}$ and also that $P(F) = \dfrac{137}{205}$, and so:

$$P(Y \mid F) = \frac{P(F \text{ and } Y)}{P(F)} = \frac{\frac{18}{205}}{\frac{137}{205}} = \frac{18}{205} \times \frac{205}{137} = \frac{18}{137} \approx 0.1314$$. Notice how the fractions work out so that the end of the calculation gives the number of "tattooed females" divided by "the total number of females." This kind of calculation can always be done; it is a rule, although if you have the data, it is often easier just to calculate the conditional probability directly as we did earlier.

Definition and notation for conditional probability:

$$P(A \mid B) = \frac{\text{Number of Cases where event } A \text{ and event } B \text{ are both true}}{\text{Number of cases where event } B \text{ is true}} = \frac{P(A \text{ and } B)}{P(B)}$$

Interpretation: $P(A \mid B)$ is read: "The probability that A is true *given* that B is true is…"

Not   Our probability toolkit has one more rule—one that is very useful. As an example, notice that from our table of male and female students who have tattoos, we could calculate the probability that a student does not have a tattoo in two ways. We could calculate $P(N) = \frac{174}{205} \approx 0.8488$, or we could get the same thing by calculating: $P(N) = P(\text{not } Y) = 1 - P(Y) = 1 - \frac{31}{205} \approx 1 - 0.1512 = 0.8488$. In this example, since there are just two events, "*not Y*" is just "*N*" but the same principle works if we have more than two events. In the last section we worked out $P(x < 4) = 0.1859$, the probability that an Australian lives in a household with fewer than four people. Now it is easy to get the probability that an Australian student lives in a household with four or more persons. The calculation of this probability is just $P(X \geq 4) = 1 - P(X < 4) \approx 1 - 0.1859 = 0.8141$. (So the probability that an Australian student has the experience of living with three or more other people is over .80.)

Definition and notation for the probability of "Not Event A"

$$P(\text{not } A) = 1 - P(A)$$

Interpretation: $P(\text{not } A)$ is read: "The probability that the event '*not A*' is true is…"

This rule works with conditional probabilities, with intersections, and with unions, although of course you have to work out the meaning of the *not* in these situations. Here is an example. We could calculate the probability that a student does not have a tattoo *if* that student is a female by calculating:

$$P(N \mid F) = P(\text{not } Y \mid F) = 1 - P(Y \mid F) = 1 - \frac{18}{137} \approx 1 - 0.1314 = 0.8686$$

Independence and Conditional Probability

We answered the statistical question "Are male students or female students more likely to have a tattoo?" by calculating and comparing the conditional probabilities $P(Y \mid F) = \frac{18}{137} \approx 0.1314$ and

PennState2		Tattoo		Row Summary
		No	Yes	
Sex	Female	119	18	137
	Male	55	13	68
Column Summary		174	31	205

S1 = count( )

$P(Y \mid M) = \frac{13}{68} \approx 0.1912$, and concluded that for the students in the Penn State collection, men are more likely to have a tattoo. There is another similar and very important use of conditional probability, in which we compare a conditional probability with the probability *without* the condition. In our example, we would compare the conditional probability of having a tattoo *given* that the student is female

$P(Y \mid F) = \dfrac{18}{137} \approx 0.1314$ with the probability that a student (any student, not just the females) has a tattoo, $P(Y) = \dfrac{31}{205} \approx 0.1512$. Since $0.1314 \neq 0.1512$, we say that the events Y (having a tattoo) and F (being a female student) are **not independent**. If these two probabilities had come out to be equal, we would say that the two events having a tattoo and being a female student were **independent.** The formal definition is:

Independent Events

Two events A and B are **independent** if $P(A \mid B) = P(A)$.

Two events A and B are **not independent** if $P(A \mid B) \neq P(A)$.

We will see that the idea of independence of events, and, by extension, independence of variables, is important in the second part of the course. Here are a number of comments about the idea.

First, if it is true that $P(A \mid B) = P(A)$ then it will also be true that $P(B \mid A) = P(B)$; the two events can be shown to be independent (or not independent) by using either event as the condition. Often, however, only one of the conditional probabilities actually makes sense.

Secondly, notice that our example used actual data and that we found the events Y and F to be *not independent*. With actual data it is very rare to find strict independence between events; that is, it is very rare to find that $P(A \mid B) = P(A)$. One of the most important uses of the idea of *independence* is in building a model of what the data in a collection would be if two variables (notice the extension beyond events) were independent. Then having built the model, we can compare the data we have to the "ideal" model of independent variables. However, we will wait until later sections to develop this.

The third comment is a kind of warning. It is common to confuse the notions of *mutually exclusive* and *independent*. If two events are *mutually exclusive* then $P(A \text{ and } B) = 0$. If two events are *independent* then $P(A \mid B) = P(A)$. These notions are quite different, and you can show mathematically that if two events are mutually exclusive, the events cannot be independent.

The fourth comment is also a warning. It is tempting to turn the idea of independence around and argue that *since* we cannot see any connection between events or variables *then* the events or variables *must* be independent. At this point, the independence of events for any given collection of data can only be shown by using the definition $P(A \mid B) = P(A)$ and not by any preconceived notions about the events involved. Here is an example.

Our statistical question is: "Are ambidextrous Australian students more likely to have a pet dog?" This sounds like a silly question; that is exactly why it was chosen. It *is* silly; we cannot think of any good reason why ambidextrous Australians would be more likely to have a pet dog.

CAS Australia 08 B

		PtsDog		Row Summary
		No	Yes	
Hand	Ambidextrous	13	40	53
	Left handed	16	35	51
	Right handed	171	295	466
	Column Summary	200	370	570

S1 = count( )

If the two events are "being ambidextrous" ($= A$) and "having a pet dog" ($= D$), then we would expect

$P(D \mid A) = P(D)$. Here are some data, and for these data $P(D) = \dfrac{370}{570} \approx 0.6491$ but $P(D \mid A) = \dfrac{40}{53} \approx 0.7547$, so the two events are *not* independent. When we get to §4.4, we shall develop a technique to determine whether this calculation really indicates that ambidextrous students are more likely to have a pet than, say, right handed students. Until that point, all we can say is that for our data the events A (being ambidextrous) and D (having a pet dog) are apparently not independent. We have deliberately chosen a silly example; there are other examples where the connection between the events (or variables) may or may not exist. A summary of the section is given on the next page.

Summary: Introduction to the Language of Probability

Probability is the language of statistics. You may regard this section as a kind of first introduction—even a kind of phrase book—to that language.

- ***Probability*** expresses the idea of *likelihood* of an *event E* using a number between 0 (completely unlikely) to 1 (certain), so that $0 \le P(E) \le 1$.

- $P(E)$ is read "the probability of event E…"

- With actual data, probabilities are calculated as fractions and expressed using the notation:

$$P(E) = \frac{\text{Number of Cases where Event } E \text{ is true}}{\text{Total number of Cases}}$$

- ***Sample space***   A complete listing (therefore, an ***exhaustive*** listing) of all the ***mutually exclusive*** events (therefore, events that cannot occur together) that are possible when applying probability language.

- ***Probability of the Intersection of Two Events A and B ("And")*** uses the notation and calculation:

$$P(A \text{ and } B) = \frac{\text{Number of cases where event } A \text{ and event } B \text{ are both true}}{\text{Total number of cases}}$$

- ***Interpretation of an "And" probability***

    $P(A \text{ and } B)$ is read: "The probability that both A and B are true is…"

- ***Probability of the Union of Two Events A and B ("Or")—Addition Rule***

$$P(A \text{ or } B) = \frac{\text{Number of cases where either event } A \text{ or event } B \text{ or both are true}}{\text{Total Number of cases}} = P(A) + P(B) - P(A \text{ and } B)$$

- ***Interpretation of an "Or" probability***

    $P(A \text{ or } B)$ is read: "The probability that either A or b or both A and B are true is…"

- ***Conditional Probability of Two Events A and B ("If")***

$$P(A \mid B) = \frac{\text{Number of Cases where event } A \text{ and event } B \text{ are both true}}{\text{Number of cases where event } B \text{ is true}} = \frac{P(A \text{ and } B)}{P(B)}$$

- ***Interpretation of a conditional probability***

    $P(A \mid B)$ is read: "The probability that A is true *given* that B is true is…"

- ***Probability of "Not Event A"*** $P(\text{not } A) = 1 - P(A)$

- ***Interpretation of a "Not" probability***

    $P(\text{not } A)$ is read: "The probability that the event '*not A*' is true is…"

- ***Independence of two events A and B***

    Two events A and B are ***independent*** if $P(A \mid B) = P(A)$.

    Two events A and B are ***not independent*** if $P(A \mid B) \ne P(A)$.

§1.2 Exercises on Probability

1. **California College Student Data on Gender and Tattoos:** In the **Notes,** we looked at whether men or women were more likely to have a tattoo; our collection of student data was from Penn State, and the data were collected in the 1990s. This exercise is about students in California in 2008 (**CombinedClassDataAut08**, found in paper form at the end of the exercises to §1.1). Our statistical question is still: *"Are male or female students more likely to have a tattoo?"* You answered a question similar to this one in the previous exercises, but here the answers focus on the calculation and language of probability. Here is the summary table that you should have.

    - Let **Y** be the event that the student has a tattoo;
    - Let **N** be the event that the student does *not* have a tattoo;
    - Let **M** be the event that the student is male;
    - Let **F** be the event that the student is female.

    Combined Class Data Aut 08

		Tattoo		Row Summary
		no	yes	
Gender	M	55	8	63
	F	47	16	63
Column Summary		102	24	126

    S1 = count ( )

    a. The probability that the person is a female is $P(F)$. Find this probability (by making a fraction) and express the answer as a number between 0 and 1. Round the answer to two decimal places.

    b. Find $P(Y)$ using a fraction and then express your answer rounded to two decimal places.

    c. Express in English the meaning of $P(Y)$.

    d. From the table directly, find the value of the probability $P(F \text{ and } Y)$ showing your calculation and give an interpretation for the number you calculated.

    e. Using the notation introduced for conditional probability, the probability that a person has a tattoo *given* that the person was female is $P(Y \mid F)$. Calculate $P(Y \mid F)$ using the numbers in the Summary Table for the California students shown above.

    f. Distinguish clearly between the meaning of $P(F \text{ and } Y)$ and the meaning of $P(Y \mid F)$.

    g. Using correct notation, find the probability that a person chosen has a tattoo *if* the person is male.

    h. From the analysis you have done for the California students, are females or males more likely to have tattoos? Give a reason for your answer.

    i. At the right is the Summary Table for the Penn State students in 1999. Compare the probabilities you calculated in parts e and g for the California students to the comparable Penn State probabilities. What differences (if any) do you notice?

    PennState2

		Tattoo		Row Summary
		No	Yes	
Sex	Female	119	18	137
	Male	55	13	68
Column Summary		174	31	205

    S1 = count ( )

    j. You have calculated $P(Y \mid F)$. Using the California data, calculate $P(F \mid Y)$. Look at the **Notes** at the definition of conditional probability and work out (and write down) a way to remember what number should be in the denominator of the fraction.

    k. Write down in English the difference in meaning between $P(Y \mid F)$ and $P(F \mid Y)$. **PTO**

1. A student calculates $P(N \mid F) = \dfrac{47}{102} \approx 0.4608$. Is this student correct? If you think so then give an interpretation of the number. If you think the student is wrong, explain the mistake that was made.

2. **Household Sizes** This exercise uses the variable *NumberHousehold*, the number of people in a student's household. This question uses the data in **CombinedClassDataAut08,** found in paper form at the end of the exercises to §1.1. Or you may do this exercise by getting the Summary Tables needed using Fathom. The instructions are shown as the bullets to the right of the vertical line.

    a. Remember that a *sample space* consists of all the possible simplest events for a variable. Even without looking at the actual data, think about the possible responses for *NumberHousehold*. Is "0" possible? Is "5" possible? Is "100" possible for a household? (Notice: the word is "possible," not "plausible.") Then write the sample space for these data using the brace symbols shown in the **Notes.**

- Either by opening **Fathom** and then using **File>Open** or by double clicking on the **Fathom file,** open the file **CombinedClassDataAut08.ftm.**
- Select the **Collection icon** (so that it has a blue border, like this on the right), and then drag down a **Table** from the **Shelf** to get the **Case Table** for these data, if it not there.
- From the **Shelf** drag down a blank **Summary** (that is, a **Summary Table**).
- Go back to the **Case Table** and select (so that the column becomes blue) the variable *Number Household* and drag it, *holding the shift key at the same time*—because this variable is quantitative—to the right-pointing arrow of the **Summary Table.**

    b. For all of the students (male and female together), calculate $P(X = 4)$ and give an interpretation of the number that you found.

    c. For all of the students, calculate the probability $P(x = 4 \text{ or } X = 5)$. [See the section on the definition of the probability of the union (or "or" probability) of two events in the **Notes.**]

    d. Give an interpretation of $P(x = 4 \text{ or } X = 5)$ in the context of "randomly choosing a student."

    e. Calculate the probability $P(X < 4)$. In the **Notes §1.2** (in the section *'Events Can Also Refer to Numbers'*) there is a similar calculation but for Australian high school students. Compare what you have found for this collection of California college students to the calculation of $P(X < 4)$ for the Australian high school students. Does the difference in the probabilities make sense to you, knowing that the Australian students are high school students?

    f. Again, for all of the students, calculate $P(X \geq 6)$. Compare what you get to $P(X \geq 6)$ for the *Australian* high school students shown and interpret in the context of the comparison.

- If you are using Fathom, drag the variable *Gender* (but do *not* hold down the shift key because this variable is categorical) to the down-facing arrow in the Summary Table.

    g. Calculate and give the meanings in English of the conditional probabilities $P(X < 4 \mid M)$ and $P(X < 4 \mid F)$. Your answer should begin: "$P(X < 4 \mid M) = \ldots$ shows the probability…".

h. Compare the probabilities $P(X < 4 \mid M)$ and $P(X < 4 \mid F)$. Is there a difference in these probabilities? If so, do you think that the difference is big? Or small? (One way to think of this is to consider whether you think you would get the same kind of difference if you had another collection of California students.)

i. For the data as they were collected, the events $P(X = 4)$ and $P(X = 5)$ are *mutually exclusive* since it is understood that the number in the household is the biggest number of people who normally live there. What should be the value of $P(X = 4 \text{ and } X = 5)$?

j. Compare $P(X < 4 \mid F)$ with $P(X < 4)$ and determine whether the events $X < 4$ and F are independent or not independent. (Refer in the **Notes** to the section on "Independence.")

k. Calculate the probability $1 - P(X \geq 6)$ and write the probability using just one $P(\quad)$ statement, without the "$1 - \ldots$"

3. **Political View and Gender** This exercise is to be done using Fathom using the file **CombinedClassDataAut10.ftm**. The Fathom instructions are shown below to the right of the vertical line. One of the variables measured is the *Political View* of students. Using the categories of that variable and the categories of *gender*, we can define these events:

L = Liberal,   **Mod** = Moderate,   **C** = Conservative,   **M** = Male,   **F** = Female

- Double click on the **Fathom file;** open the file CombinedClassDataAut10.ftm.
- Select the **Collection icon** (so that it has a blue border, like the icon shown on the right), and then drag down a **Table** from the **Shelf** to get the **Case Table** for these data.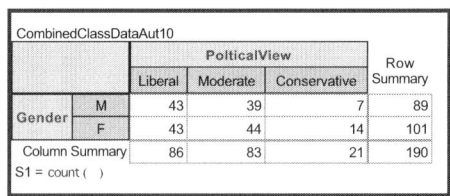
- From the **Shelf**, drag down a blank **Summary** (that is, a **Summary Table**).
- Go back to the **Case Table;** select (so that the column becomes blue) the variable *Political View* and drag it to the right-pointing arrow of the **Summary Table.**
- From the **Case Table,** select (so that the column becomes blue) the variable *Gender* and drag it to the down-pointing arrow of the **Summary Table.**

  The **Summary Table** should look like this one.

  Our statistical question for this exercise is: *Are male students or female students more likely to have "liberal" political views?* We will use this Summary Table to answer these questions.

CombinedClassDataAut10	PoliticalView			Row Summary
	Liberal	Moderate	Conservative	
Gender M	43	39	7	89
Gender F	43	44	14	101
Column Summary	86	83	21	190

  S1 = count ( )

  a. Confused Conrad (whom you will meet often in these exercises) says: "The answer is easy: since there are equal numbers of male and female students who have liberal political views, males and females have the same probability of having liberal political views." What is wrong with CC's thinking? Be very specific.

  b. Calculate the probability $P(L \mid F)$ and give an interpretation in the context of the data.

  c. Calculate the probability $P(L \mid M)$ and give an interpretation in the context of the data.

Check your numerical answers using Fathom by following the instructions below.
- Select the **Summary Table** so that it has a blue border.
- In the menu on the top of the screen, get **Summary> Add Formula.**
- In the dialogue box, find **Functions> SpecialMeasures>rowProportions** and double click on it, so that it comes up in the dialogue box. Select **OK.**

d. Check your answers. If a student's answer is 0.5 for both part b and part c, what conditional probabilities has this student actually calculated: $P(L \mid F)$ and $P(L \mid M)$ or $P(F \mid L)$ and $P(M \mid L)$? (This is a very common error.)

e. In the notation $P(L \mid F)$, what is the "given"? In the calculation, what total is used for the denominator?

f. Suppose a student's answer for both parts b and c is 0.2263; this is not correct. Your job is, by doing some calculations, to decide what mistake the student made. Then use the correct notation to show the probabilities that the student actually (but wrongly) calculated.

g. Use the correct probabilities and write an answer (in English) to the statistical question in italics next to the Summary Table above. (What we have found for this sample is contrary to what we usually find.)

h. Are the events L and F *mutually exclusive events* or *not mutually exclusive*? Give a reason for your answer.

i. Are the events L and F *independent events* or *not independent*? Give a reason for your answer. [Compare $P(L \mid F)$ and $P(L)$ to determine whether the events L and F are independent or *not* independent.]

j. Check the **Notes** for the addition rule and use that rule to calculate $P(M \text{ or } C)$. There should be three terms in your calculation. (Notice, we have asked for M "male" and not Mod "Moderate.")

k. Calculate $P(\text{not } L)$ using the formula for calculating "*not*" probabilities given in the **Notes**. Show your work completely. In terms of political views, what probability has the calculation of $P(\text{not } L)$ given you? Think in the context of the variable.

l. To the question "Find the probability that a student has conservative political views *given* that the student is a male," Confused Conrad writes: $P\left(\dfrac{7}{21}\right) = 0.33\overline{3}$. Find *all* the mistakes CC has made (and there is more than one), explain what they are, and give the correct answer to the question, using the correct notation (which CC has not done).

m. To the question: "Find the probability that a student both has *moderate* political views *and* also is a male student," a student writes: $P(Mod) + P(M) = \dfrac{83}{190} + \dfrac{89}{190} = \dfrac{172}{190} \approx 0.9053$. Find and explain *all* the mistakes and do the correct calculation.

4. **Number of States Visited** This exercise will also use the collection **CombinedClassDataAut10.ftm**. The variable we will analyze (*Number States*) is about the number of states California students have visited. Some students have never left California whereas others are very well-travelled.

   a. Without looking at the data, write the sample space for the variable *Number States* if the categories are to be regarded as events. Your answer should have $S = \{ \cdots \}$ and then the appropriate numbers inside the braces. Notice that you will want to use ellipses inside braces $\{ \ldots \}$ but notice also that you can be a bit more specific than you could be for the *Number Household* problem.

   b. Are the events listed in your sample space *mutually exclusive*? Give a reason for your answer.

   c. Is the variable *Number States* quantitative or categorical?

- Either by opening **Fathom** and then using **File>Open** or by double clicking on the **Fathom file**, open the file **CombinedClassDataAut10.ftm**.

- Select the **Collection icon** (so that it has a blue border), and then drag down a **Table** from the **Shelf** to get the **Case Table** for these data.

- From the **Shelf**, drag down a blank **Graph** (it will appear as an **Empty Plot**).

- Go back to the **Case Table;** select (so that the column becomes blue) the variable *Number States* and drag the variable to horizontal axis (the *x* axis) of the **Empty Plot**. As you do this, a welcoming space will open. What you have made is a **dot plot**, where each of the dots is the number of states visited by a specific student.

- From the **Case Table,** select (so that the column becomes blue) the variable *Gender* and drag this variable to the vertical axis (the *y* axis) of the **dot plot**. Spread out the dot plot so that you can see each dot clearly. The plot should look like the one on the right, but yours should be bigger.

   For these data, the number of males is $n_M = 86$ and the number of females is $n_F = 101$.

   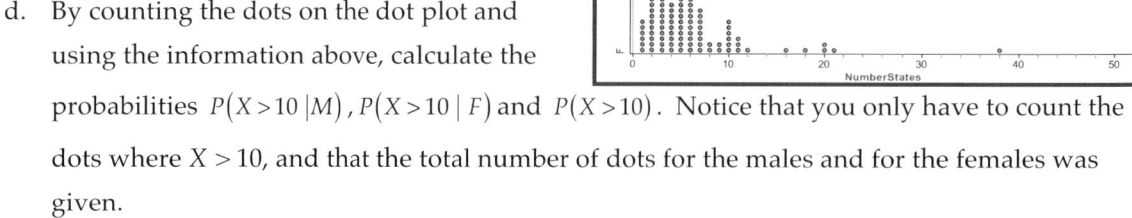

   d. By counting the dots on the dot plot and using the information above, calculate the probabilities $P(X > 10 \mid M)$, $P(X > 10 \mid F)$ and $P(X > 10)$. Notice that you only have to count the dots where $X > 10$, and that the total number of dots for the males and for the females was given.

   e. Interpret the probabilities you have just calculated to answer the question: "Are males or females more likely to have visited more than ten states?"

   f. On your dot plot, either by hand or using Fathom, color in the dots for $X > 10$ for the males and females. The bulleted instructions just below show how to do this using Fathom. (There is a reason for this exercise. Later in the course, probabilities will commonly be represented in this way.)

- From the **Case Table**, select and drag the variable **MoreThan10States** to the *body* of the **dot plot.** This will color in $P(X > 10 \mid M)$ and $P(X > 10 \mid F)$.

To answer the questions below and to check your answers, get the **Summary Table** shown here.

- From the **Shelf**, drag down **Summary** to get a **Summary Table**.
- From the **Case Table**, select and drag the variable **MoreThan10States** to the *right*-pointing arrow.

CombinedClassDataAut10

		MoreThan10States		Row Summary
		No	Yes	
Gender	M	62	24	86
	F	91	10	101
Column Summary		153	34	187

S1 = count ( )

- From the **Case Table**, select and drag the variable **Gender** to the down-pointing arrow of the **Summary Table**. Your Summary Table should look like the one here.

g. Calculate and give interpretations of $P(X>10 \text{ and } M)$, $P(X>10 \text{ and } F)$, $P(M)$ and $P(F)$.

h. Apply the formula $P(A \mid B) = \dfrac{P(A \text{ and } B)}{P(B)}$ to get $P(X>10 \mid M)$ and $P(X>10 \mid F)$. In the **Notes** see the subsection on **Conditional Probability ("If")**.

i. Find the probability you will get if you use the rule for **"Not"** and calculate $1 - P(X>10)$. Express this probability in symbols without the "1" and give the meaning in words.

j. Find the probability that a student has visited ten or fewer states if that student is male and express this probability using the correct notation.

k. Find the probability that a student is male and that student has visited ten or fewer states. Use the correct notation.

l. Find the probability that a student is male if the student has visited ten or fewer states. Use the correct notation.

5. **Bit of Review and Political Views and Tattoos.** This exercise reviews the material on probability, the analyses you made between gender and having a tattoo, and between gender and political views, and applies the ideas to the relationship between a student's political views and whether or not the student has a tattoo. Follow the Fathom directions indicated by bullets and get the **Summary Table** shown here.

- Either by opening **Fathom** and then using **File>Open** or by *double clicking* on the **Fathom file** (files that have a gold ball as their icon), open the file **CombinedClassDataSpr09.ftm**.
- Select the **Collection icon** (so that it has a blue border) and then drag down a **Table** from the **Shelf** to get the **Case Table** for these data.
- From the **Shelf**, drag down a blank **Summary** (that is, a **Summary Table**).
- Go back to the **Case Table**; select (so that the column becomes blue) the variable *Tattoo* and drag it to the right-pointing arrow.
- Then select the variable *Gender* from the **Case Table** and drag it to the down-pointing arrow. You should see this **Summary Table. PTO**

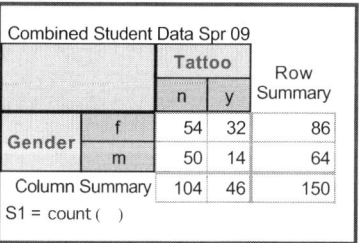

Combined Student Data Spr 09

		Tattoo		Row Summary
		n	y	
Gender	f	54	32	86
	m	50	14	64
Column Summary		104	46	150

S1 = count ( )

Combined Class Data Aut 08

		Tattoo		Row Summary
		no	yes	
Gender	M	55	8	63
	F	47	16	63
Column Summary		102	24	126

S1 = count ( )

a. Calculate $P(Y)$ and compare your results to what you found for **CombinedClassDataAut08** (the Summary Table is shown in Exercise 1 of this section) and for Penn State, discussed in the **Notes**. The two Summary Tables are shown here. Comment on the comparison.

b. In answer to our statistical question "Are males or females more likely to have a tattoo?" we got different results for the Penn State collection and the California collection. Use the correct notation and analyze these Spring '09 data to answer the same question. Answering the question means both doing the calculation and making the interpretation in the context of the question.

PennState2		Tattoo		Row Summary
		No	Yes	
Sex	Female	119	18	137
	Male	55	13	68
Column Summary		174	31	205

S1 = count ( )

c. In Exercise 3, you also looked at the relationship between political view and gender for the Autumn '10 sample, asking the question, "Are females or males more likely to have liberal political views?" We will look at this same relationship for Spring '09. Follow the directions in the bullet below then use the correct notation and analyze these Spring '09 data to answer the question about gender and political view. Compare your results to the results for Autumn '10.

- You can get a **Summary Table** with the variables *PoliticalView* and *Gender* by following the directions above, using *PoliticalView* in place of *Tattoo* or you may use the directions in the next bullet.
- In Fathom, select your **Summary Table** and then go to **Object>Duplicate Summary Table.** Then drag the variable *PoliticalView* and place it where the variable *Tattoo* resides.

d. Do you think that having liberal political views makes it more (or less) likely that a student has a tattoo? Or do you think it works the other way: having a tattoo means liberal political views are more (or less) likely? If you think "political views affect having a tattoo" then you will use $P(Y \mid L)$ compared with $P(Y \mid Mod)$ etc. If you think "having a tattoo affects political views," you will use $P(L \mid Y)$ compared with $P(L \mid N)$. By following the directions in the bullets above you should have the **Summary Table** for *PoliticalView* and *Tattoo*. Choose one of the options ("political views –> tattoo" or "tattoo –> political views"—you cannot opt out), do the calculations and give an interpretation.

6. **Real Estate Data: Style of House and Region in a County** The data for this exercise concerns houses that were sold in 2007–2008 in San Mateo County, California. There were a total of 3,947 houses sold in the period from June 2007 to June 2008, but the data for this exercise uses just $n = 400$ selected randomly from the 3,947. Part of the **Case Table** is shown below. Each row is a different house sold.

For this exercise, we will be only interested in the variables *Style3* and *Region*. *Style3* records for each house the style of the house in one of three categories: Traditional, Ranch or Other. All of the houses are in San Mateo County, but that county can be divided into four regions, namely: Central, Coast, North, and South. For this exercise, we give these letters to the categories:

Region: **C** = Central,   **CST** = Coast,   **N** = North,   **S** = South.
Style3:   **Con** = Contemporary,   **T** = Traditional,   **R** = Ranch,   **O** = Other.

a. What are the cases for these real estate data?

b. Are the variables *Style3* and *Region* quantitative or categorical?

- Double click on the **Fathom file;** open the file **San Mateo RE Sample Y0708.ftm**.
- Select the **Collection icon** (so that it has a blue border) and then drag down a **Table** from the **Shelf** to get the **Case Table** for these data.
- From the **Shelf,** drag down a blank **Summary** (that is, a **Summary Table**).
- Go back to the **Case Table;** select (so that the column becomes blue) the variable *Region* and drag it to the right-pointing arrow of the **Summary Table.**
- From the **Case Table,** select (so that the column becomes blue) the variable *Style3* and drag it to the down-pointing arrow of the **Summary Table.**

You should have this Summary Table.

c. Calculate each of the following probabilities and for each give an interpretation.

    i. $P(Con)$ ii. $P(N)$ iii. $P(Con \text{ and } N)$

    iv. $P(Con \mid N)$ v. $P(N \mid Con)$

San Mateo RE Sample 0708

		Region				Row Summary
		Central	Coast	North	South	
Style3	Contemporary	16	15	22	36	89
	Other	56	17	30	57	160
	Ranch	36	6	15	36	93
	Traditional	20	4	11	23	58
	Column Summary	128	42	78	152	400

S1 = count ( )

d. Using the probabilities that you have just calculated, you can determine whether the events **Con** and **N** are independent events or not independent events. Show your work and come to a conclusion.

e. Is it true that $P(Con \text{ and } N)$ is equal to $P(Con) + P(N)$? Show evidence for your answer.

f. Is it true that $P(Con \text{ or } N)$ is equal to $P(Con) + P(N)$? Show evidence for your answer.

g. Use correct probability notation to show that the proportion of Contemporary houses sold on the Coast is greater than the proportion of Contemporary houses sold in the North region.

h. To answer part g, a student calculates and puts down as an answer $P(15/400) \approx 0.0375$. There are many things wrong with what is written, both in the calculation that is done and in the use of the probability notation. Find and correct all of the errors.

7. **Size of Household in Different Places**   This exercise repeats and extends the analysis done in Exercise 2 but uses different collections. We think of the values of the variable—the household sizes—as events whose number is designated by X, so that X = 5 means a household of size five.

a. Is the variable **Number Household** a categorical variable or a quantitative variable?

- Open **CombinedClassDataSpr09.ftm**.
- To get this **Summary Table**, drag **Summary** from the **Shelf** and then select the variable *Number Household* from the **Case Table** and, *holding down the shift key* when dragging, place the variable on the right arrow of the **Summary Table.** Holding down the shift key is necessary because the variable *Number Household* is quantitative.

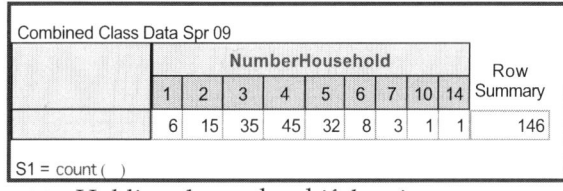

Combined Class Data Spr 09

	NumberHousehold								Row Summary
	1	2	3	4	5	6	7	10	14	
	6	15	35	45	32	8	3	1	1	146

S1 = count ( )

33

Below are three similar **Summary Tables:** One for high school students in South Africa, another for high school students in Australia, and a third for high school students in New Zealand.

Our statistical question is: *"Do the proportions of students who live in 'large households' differ among these places?"* Large households are defined as households of five or more people.

b. What is the appropriate type of probability calculation for this comparison: conditional ("if") probability, intersection ("and") probability, or union ("or") probability? Give a reason for your answer.

c. Calculate the probabilities and display them using the correct notation.

South Africa

CAS ZA 090427

NumberHousehold														Row Summary
2	3	4	5	6	7	8	9	10	11	12	14	15	18	
3	10	30	31	19	18	17	8	5	5	1	1	1	1	150

S1 = count ( )
gradeno > 8

Australia

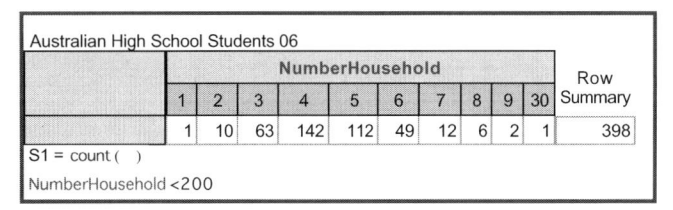

Australian High School Students 06

NumberHousehold										Row Summary
1	2	3	4	5	6	7	8	9	30	
1	10	63	142	112	49	12	6	2	1	398

S1 = count ( )
NumberHousehold <200

New Zealand

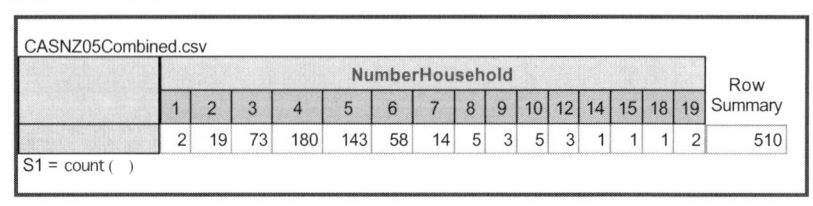

CASNZ05Combined.csv

NumberHousehold														Row Summary
1	2	3	4	5	6	7	8	9	10	12	14	15	18	19	
2	19	73	180	143	58	14	5	3	5	3	1	1	1	2	510

S1 = count ( )

d. Using the results of your calculations, write a clear, coherent answer to the statistical question stated just above part b (in italics). (You should be able to do this in one to three sentences.)

e. Suppose someone told you that the calculations you have done are not valid because there are different numbers of students surveyed in the different countries ($n_{Califronia} = 146$, $n_{South\ Africa} = 150$, $n_{Australia} = 398$ and $n_{New\ Zealand} = 510$). What would you say to this person? Would you agree? Make a case.

§1.3 Distributions and their pictures

The Important Idea of a Distribution

In the first section we looked at the basic goal of statistics, namely, to ask and answer questions about collections of data. To answer questions, we need some terminology, so we also looked at the ideas of **collections, cases, variables,** and **values.** To answer statistical questions, we saw that we must do both *calculation* and *interpretation,* where interpretation is saying what the calculations mean in terms of the question.

Now we come to what at first glance looks like more terminology; in fact, it is more than terminology. It is an idea that is "deep in the heart of statistics." This very important idea is the idea of a **distribution.** Here is a definition:

Definition of a distribution of a variable in a collection
A **distribution** consists of the number of cases at each value of a variable in a collection.

Grades Stats		
LetterGrade	A	6
	B	47
	C	85
	D	67
	F	28
Column Summary		233
S1 = count ( )		

If your first reaction to this definition is puzzlement, you should know that you are probably more familiar with distributions than you think. If you have been a student for some time, you have probably heard the term "grade distribution." Here is a grade distribution for some statistics classes taught by a professor in Costa Rica.

What we see fits the definition of a **distribution** because it shows the values of a variable ("A," "B," etc.) of the categorical variable *LetterGrade* and the number of cases (students) who earned each letter grade. Compare this sentence with the definition in the box above. We can (and will!) often show the proportions (or percentages) of cases at each value of a variable instead of the counts. When we use proportions we still have a distribution. We have a distribution because we have brought two things together: the values of the variable and the number or the proportion or the percentage at each value.

Combined Class Data Aut 08		
Languages	1	60
	2	54
	3	11
	4	1
Column Summary		126
S1 = count ( )		

You may see why we call this idea a distribution: we can think of the cases (the students) as being "distributed" into the different grade categories, much like mail is put into pigeonholes. (Think about this carefully, especially if you usually think that it is the grades that are distributed to the students!) Here the students are distributed to the grades.

We use the idea of a distribution for quantitative variables as well as categorical variables. In the exercises for §1.1 you calculated a distribution for a quantitative variable, and shown here is what you should have found. What this shows is that there were sixty students who spoke just one language, fifty-four students who speak two languages, etc. What you got fits the definition of a **distribution** because you have the counts of the cases (the number of students) at each value of the variable *Languages.*

Graphics for Distributions: Dot Plots

Here is still another example of a distribution of a quantitative variable. However, this time we have arranged the data so that we can compare two distributions for the variable *NumberHousehold* (= "Number of People in the Household"). One row shows the distribution for a collection of California Statistics students and the other row shows the distribution for the same variable for secondary students in Western Australia. The Summary Table also shows proportions for each household size for the students in each place.

CA and WA Comparison		NumberHousehold								Row Summary
		1	2	3	4	5	6	7	8	14	
Place	California	4	9	11	21	13	2	3	0	1	64
		0.0625	0.140625	0.171875	0.328125	0.203125	0.03125	0.046875	0	0.015625	1
	Western Australia	0	2	12	27	16	3	2	1	0	63
		0	0.031746	0.190476	0.428571	0.253968	0.047619	0.031746	0.015873	0	1
Column Summary		4	11	23	48	29	5	5	1	1	127
		0.0314961	0.0866142	0.181102	0.377953	0.228346	0.0393701	0.0393701	0.00787402	0.00787402	1

S1 = count( )
S2 = rowProportion

We are ready (that is, we have done the calculations) to answer this statistical question: *Are there differences in the distributions of household size when we compare students from California with students from Western Australia?*

It appears that a greater percentage of the California college students live in very small households (households of size one or two) compared with the Western Australian secondary students. Perhaps this greater percentage is because the California students are college students and are older so that some of them have set up their own households; they are not all living with their parents as the Australian secondary students are. Using the probability notation of §1.2, we can see that

$P(X < 3 \mid C) = \dfrac{13}{64} \approx 0.2031$ is bigger than $P(X < 3 \mid WA) = \dfrac{2}{63} \approx 0.0317$.

We can compare the numbers, and we should. However, oftentimes it is helpful to depict a distribution with a graphic. The simplest graphic we can make is called a ***dot plot***. Here is an example for the California students' data.

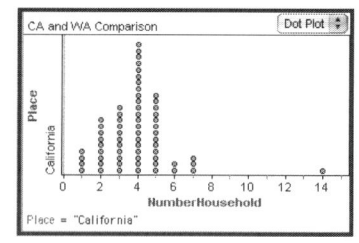

You can make a dot plot (by hand) by marking the values of the variable at equally spaced intervals on the horizontal axis (the *x* axis) and then stacking the dots (representing the cases) vertically for each value of the variable. Notice that a dot plot nicely shows our definition of a distribution in that a dot plot shows the number of cases at each value if you count the dots carefully.

Since we are comparing, it is often helpful to make a double (or triple, or more!) dot plot to show the comparison between two or more distributions. Here is a comparison for the household size data for the California and Western Australian students. It is fairly easy to see from the graphic that the Western Australia distribution for household sizes is just a bit to the right of the distribution for the California students.

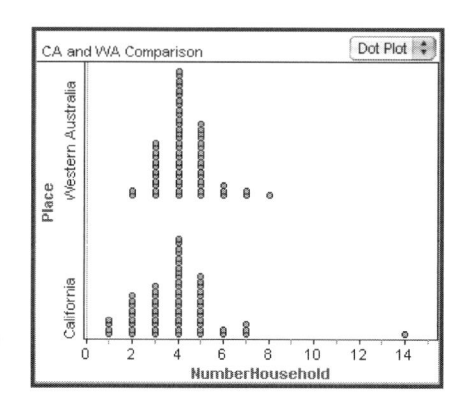

36

A dot plot is just one kind of graphic for a distribution. We will look at more graphics for distributions, and we will let software do the construction, usually.

Back of the Envelope Graphics: Stemplots

The idea of the *stemplot* is to organize—usually by hand—a collection of data so that it is easy to depict a distribution of a variable. Here is an example. We want to look at the distribution of the variable *Age* for just the twenty students who had tattoos in the **CombinedClassDataSpr08**. Here (on the right) are the data.

Age	17	18	19	20	21	22	Etc..	33	34	Etc	Total
Number of cases	I	I	IIII	IIII II	Etc.	Etc..	Etc..		I	Etc.	

CombinedClassdDataSpr08

	Gender	Age
1	F	19.0
2	M	18.0
3	M	19.0
4	F	20.0
5	F	20.0
6	M	20.0
7	M	19.0
8	F	21.0
9	F	20.0
10	M	45.0
11	F	30.0
12	F	20.0
13	M	19.0
14	F	20.0
15	F	21.0
16	F	20.0
17	M	19.0
18	M	21.0
19	F	34.0
20	F	17.0

Tattoo = "y"

We could do this, but there is a better way if we are working by hand. For each take each age, say 34, we consider the number in two parts: 3 | 4 . Mathematically, the "3" is the tens digit, and the "4" is the units digit. The tens digit we will call the *stem* and the units digit we will call the *leaf*. Then we make a vertical list of what we think are all the possible *stems*: in this example, we have students in their teens, in their twenties, in their thirties, etc., perhaps even one or two who are as old as fifty-something. We would make a list of stems (shown in the first step) with a bar on the right of the numbers.

First Step:
Make the stems
1 |
2 |
3 |
4 |
5 |

The second step (shown below) is to enter the *leaves* to the right of the bar but in the correct place. For the first six of the ages of the students with tattoos, the second step is shown below. Then the third step is to put the leaves in order. This can be done by re-arranging the leaves as they have been entered. Some people prefer to order the data first or to rank order the data as they are making the stemplot.

In any case, we will always want stemplots to be *rank ordered* —that is, with the numbers ranked from lowest to highest. The completed ordered stemplot for these data is shown here also.

Second Step:
Enter the leaves
1 | 9 8 9
2 | 0 0 0
3 |
4 |
5 |

Third Step:
Order the leaves
1 | 7 8 9 9 9 9 9 9
2 | 0 0 0 0 0 0 0 1 1 1
3 | 0 4
4 | 5
5 |

Notice that the plot we end up with shows the number of cases at each *group* of values. The stemplot is a graphic that shows a distribution with some of the values grouped together. Nonetheless, it is still a distribution, even with the values of the variable grouped together.

Split Stem Stemplots

Sometimes it is helpful to split the stems of a stemplot. We have not shown it here, but for the data on *Age*, we could make one line for all those who are twenty to twenty-four (i.e., "20, 21, 22, 23, or 24") and another line for those twenty-five to twenty-nine. We would do the same kind of thing for the ages ten to nineteen, thirty to thirty-nine, forty to forty-nine, etc. Notice that each

```
1 |
1 | 9
2 | 00113334444
2 | 555666667788
3 | 00112222334
3 | 555667777
4 |
4 | 5
5 |
5 | 6
```
Male Students' Mothers' Ages

of the stem-lines has five possible values assigned to it: we will call these **stems of five.** Here is an example of the use of split stems.

One of the variables measured for the **CombinedClassDataSpr08** was the age of the student's mother when the student was born. Here is a split stem stemplot for the variable *MothersAge* for just the male students from that collection.

Notice also that the stem-lines for 40–44 and for 50–54 are left blank, but the stem-lines are still shown in the stemplot. This shows clearly that there were no mothers aged 40–44 or 50–54, although there was one mother aged forty-five and another fifty-six. The stemplot, like the dot plot, shows gaps in the data.

We could also split the stems in another way; we could have five stem-lines with just two possible values for the variable for each stem-line. We will call these **stems of two.** The same distribution is shown here with the *stems of two*. There are no strict rules to decide between the stems with fives values and the stems with two values. It is up to the maker of the stemplot, and both too few and too many stems can conceal

```
1 | 9
2 | 0011
2 | 333
2 | 4444555
2 | 6666677
2 | 88
3 | 0011
3 | 222233
3 | 4555
3 | 667777
3 |
4 |
4 |
4 | 5
4 |
4 |
5 |
5 |
5 |
5 | 6
5 |
```
Male Students' Mothers' Ages

important features. What can we do with an ordered stemplot? We can answer questions about distributions. For example:

What percentage of the male students had mothers less than thirty years of age when the student was born?

Since the data are ordered, we can simply count the number of students whose mothers were under thirty, divide by the total number of students (here there are forty-six students), and multiply by one hundred. This comes out to $\frac{24}{46} \times 100 \approx 52.17\%$. Or, putting this calculation into the probability notation introduced in §1.2, we would have $P(X < 30 \mid M) = \frac{24}{46} \approx 0.5217$ where, in this case, X stands for values of the variable *MothersAge*.

More Bars in More Places: Histograms

A bar is a very common graphic device to show a quantity. Often the bars are vertical; sometimes they are horizontal. But bars are plentiful, graphically and otherwise. When the lengths of bars are used to show the frequencies of a value or a group of values in a distribution, then we have a **histogram.** To show the idea of a histogram, we have

```
1 |
1 | 9
2 | 00113334444
2 | 555666667788
3 | 00112222334
3 | 555667777
4 |
4 | 5
5 |
5 | 6
```
Male Students' Mothers' Ages

copied the stemplot for the male students' mothers' ages. Now imagine a bar made around each stem-line and imagine the entire graphic rotated ninety degrees, so that the variable is shown on the *horizontal* axis instead of on the vertical axis. Here is what we have, shown just to the right and below:

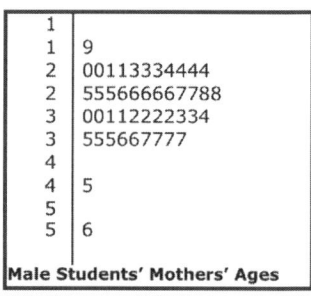

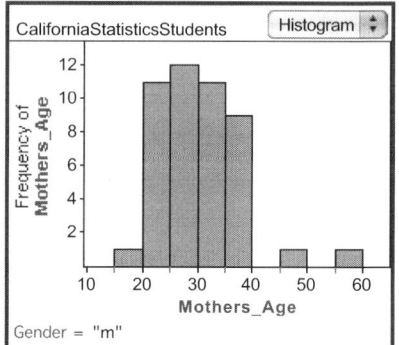

Then, on the *vertical* axis for this histogram, we have put a scale showing the **frequency** or **count** for each five-year grouping of the mothers' ages. If we look at the first bar we can see that there was just one student whose mother's age was in the interval 15 ≤ MothersAge < 20. The second bar shows that there were eleven students whose mothers' ages were in the interval 20 ≤ MothersAge < 25.

The interval "20 ≤ MothersAge < 25" means that the only mothers' ages in that bar are the ages that are greater than or equal to twenty but less than twenty-five, and hence not including age twenty-five. To determine that there was one student in the interval 15 ≤ MothersAge < 20 and that there were eleven students in the next interval of the variable 20 ≤ MothersAge < 25, it is helpful to draw a horizontal line along the top of the bar to the scale on the left hand side. With these data (where the n is small), we can be pretty accurate reading from the graphic and getting the numbers; with a histogram from a much larger collection, you may not be able to read the count exactly.

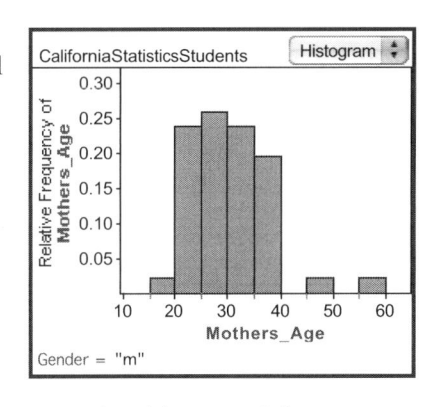

If we want the *percentage* or the *proportion* of students whose mothers were younger than twenty-five, we would first estimate the number of students (as we have done just above) and then divide by the total number of students. Here the total number of male students is $n = 46$, so we get

$\frac{12}{46} \cdot 100 \approx 26.09\%$, in proportions 0.2609, or in probability notation $P(X < 25 \mid M) = \frac{12}{46} \approx 0.2609$.

Most often, histograms are made using software, and it is easy to change the vertical scale from frequencies to **relative frequencies**, which are the frequencies divided by the total. In other words, the relative frequencies are the proportions. When this is done, you can read off the proportions for each bar.

Showing Proportions with Histograms The proportion of students whose mothers were younger than twenty-five (for just the male students) that we calculated,

$P(X < 25 \mid M) = \frac{12}{46} \approx 0.2609$ can be displayed on a

histogram by shading in the bars for mothers ages less than twenty-five. This is shown in the graphic here. Notice that the proportion 0.2609 is neither on the *x*-axis nor on the *y*-axis but rather refers to the *area* in the shaded bars.

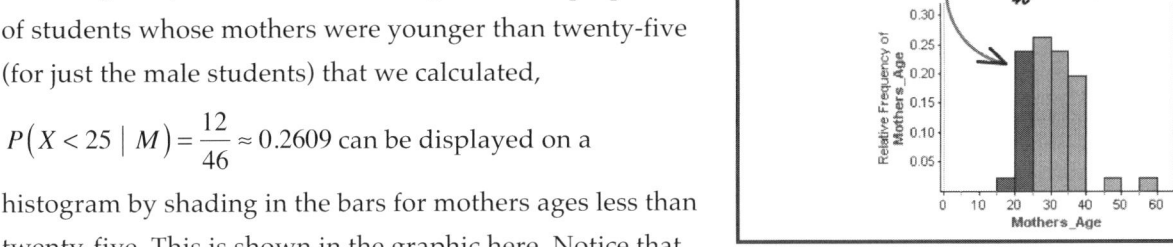

Histogram Bin Width The width of the interval 15 ≤ MothersAge < 20 as well as the interval 20 ≤ MothersAge < 25 is five years, as are the widths of all of the other bars in the histograms shown. This width—the width of the bars in a histogram—is known as **bin width**, and it can be changed.

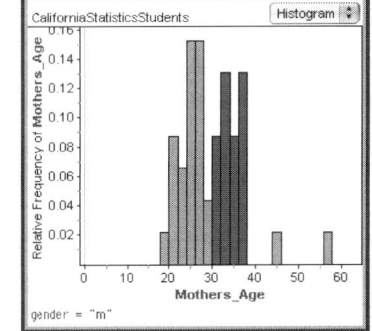

Here is the same distribution of *MothersAge* for the male students shown with bin width of width two. With a bin width of two, the age twenty-five will be in the middle of one of the bars if the bars start at age eighteen, and so we cannot show the proportion of mothers' ages less than twenty-five. What we can show (with this bin width) is the proportion of mothers of students in the interval $30 \leq \text{MothersAge} < 38$. Who decides on the *bin width*? If you are the researcher, you get to decide. Choosing too many bars or too few bars may not reveal the features of a distribution well. There is no one right answer as to the correct number of bars.

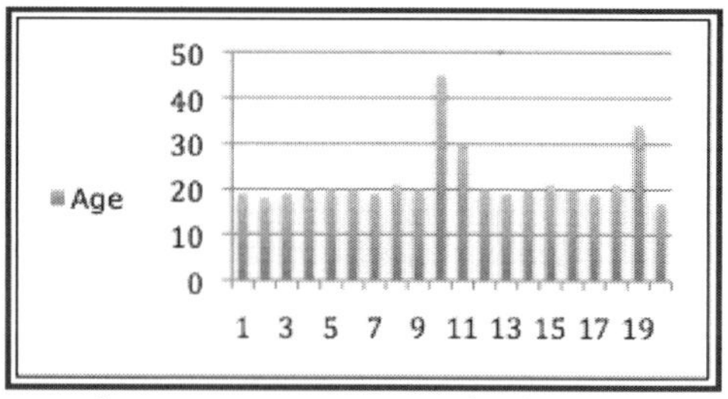

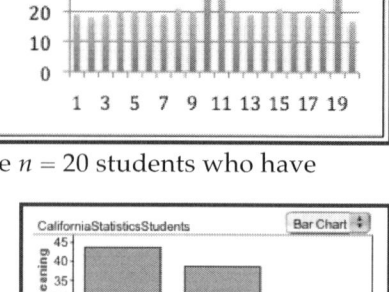

More Bars in More Places: Other Bar Graphics

Not every graphic that has bars is a *histogram*. A histogram is a graphic used to show a *distribution*; that is, it shows the values of the variable and the number (or proportion) of cases at each value or group of values. The graphic here (which was made with Excel®) shows the ages for just the $n = 20$ students who have tattoos. This is a bar graph (it has bars), but it is *not* a histogram. It does not show the number or proportion of cases at each value, but rather it shows the *value* for each case. What this graphic is telling us is that the tenth tattooed student has an age of forty-five. You may also see that the bars are separated rather than against one another, as in our examples of histogram. For *quantitative* variables, the bars are drawn (by convention) without spaces between them.

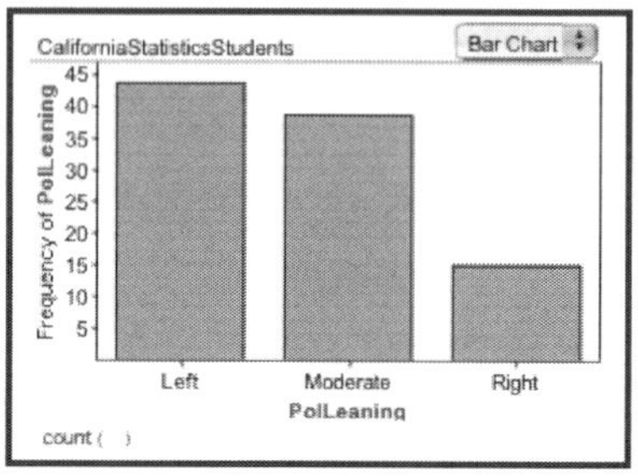

For categorical variables, a bar graph may also show a distribution, and when it does, the bars conventionally *do* have spaces between them. Here is an example that shows the number of students whose political leaning was "left," "moderate," or on the "right."

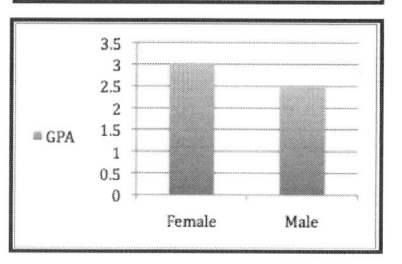

However, here is a bar graph used with categorical data where the graphic does *not* show a distribution; the variable on the vertical axis is GPA, and the chart shows the average GPA for males and female students. This kind of **bar chart** is common, is legitimate, and gives a good graphic presentation of the difference in average GPA between male and female students. However, what it shows is *not* a distribution either of the variable GPA or of the variable gender.

Example: Remember When You Were Thirteen—Interpreting Histograms

In their early teenage years, girls mature more rapidly than boys. You may recall a time in your school years when it seemed that many of the girls were as tall as the boys or taller. Perhaps this happened around age thirteen. Here is a statistical question that comes from this observation. *How do the proportions of students who are between 160 and 180 centimeters tall compare for male and female secondary students who are thirteen years old? Are the proportions similar?*

We can answer this question using the histograms for the male and female students from the Australian secondary students collection, but usually we need to determine from the graphic what the *bin width* is. By inspecting the bars, we see that the bin width is five centimeters, and that allows us to determine the boundaries of the interval 160 cm ≤ Height < 180. Here we have shaded in the interval. Notice that our histogram has relative frequencies so that we can read off the proportions from the vertical scale on the left. For the female students, we read off: 0.28 + 0.23 + 0.19 + 0.02 = 0.72 or in percentages, 72%, or $P(160 \leq \text{Height} < 180 \mid F) = 0.72$

40

You can confirm that, for the male students, reading off the heights of the bars sums up to about 0.69. We could also work with the actual numbers from the data and find that 75 out of 104 thirteen-year-old girls have heights in the interval 160 cm ≤ Height < 180, and that works out to the proportion $\frac{75}{104} \approx 0.721$. There are 45 out of the 65 thirteen-year-old boys with heights in the interval 160 cm ≤ Height < 180, or $\frac{45}{65} \approx 0.692$. The proportion of thirteen-year-old boys in the "mid-heights" (in the interval 160 cm ≤ Height < 180) is similar to the proportion of thirteen-year-old girls in the "mid-heights," and the histogram exhibits this feature. The sum of the heights of the bars for the heights of the thirteen-year-old boys and girls is similar. It is not the whole story, however, and you will get the chance to add to it in the exercises.

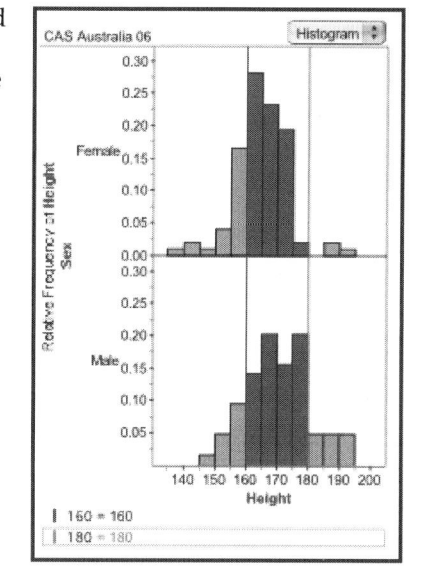

As you know from §1.2, we can interpret these calculations as probabilities: suppose we choose a thirteen-year-old Australian secondary female student completely at random from our collection. What is the *probability* that she is between 160 and 180 centimeters tall? Our answer is: $P(160 \leq Height < 180 \mid F) = 0.72$. For the boys, the *chance* or *likelihood* is about the same: we would calculate $P(160 \leq Height < 180 \mid M) = 0.69$ and say the probability that a thirteen-year-old Australian male student is at least 160 centimeters but not taller than 180 centimeters is about 69%.

Summary: Distributions and Their Pictures

- **Distribution of a variable**   The number of cases at each value of a variable in a collection. We may also speak of the proportion of cases at each value or group of values of a variable, and this is also a distribution.
- **Graphics for Distributions**   Distributions may be displayed by: **dot plots, stemplots, histograms.**
- **Dot plots and histograms** typically have the values of the variable on their horizontal axis, but the interval of values used for the bars of a *histogram* (the **bin width**) may vary.
- **Dot plots** typically do not have a scale on their vertical axis, but
- **Histograms** may have frequencies (counts), or relative frequencies, expressed as proportions or percentages on their vertical axis.
- Not all **bar graphs** are **histograms.**
- **Stemplots** should always display the data in rank order, but the choice of whether the stems are **stems of one, two, five,** or **ten** is left to the researcher.

§1.3 Exercises on Distributions

1. **California Statistics Students** Here is a table showing the counts for the number of states visited (*NStatesVisited*) by students in the collection of **CombinedClassDataSpr08.** The table shows that there were just six students who had only visited or lived in one state, thirteen students who had visited two states, fourteen who had visited three states, etc.

NStatesVisited	count
1	6
2	13
3	14
4	13
5	6
6	13
7	6
8	4
9	3
10	4
11	5
12	3
13	1
14	1
15	1
22	1
35	1
38	1
Column Summary	96

   S1 = count ( )

   a. What are the cases for this distribution? Are the cases students, or are they states, or are they visits, or something else? Give a reason for your answer.

   b. The definition of a distribution is given in the **Notes** (in a box). Write something that convinces you that what is shown here fits the definition of a distribution.

   c. *Review* Calculate the percentage of California students that have visited (or lived in) fewer than four states, using the probability notation of §1.2.

   d. *Review* Calculate the percentage of California students that have visited (or lived in) four or more states, using the probability notation of §1.2.

   e. *Review* If you add the probability in part c to the probability that you get in part d, you should get 1 (or something very near to it). Give a reason why you should get "1."

   f. In symbols, the answer to part e should be $P(X<4)+P(X\geq 4)=1$. Which probability rule of §1.2 ("and," "or," "if," or "not") does this illustrate? Explain your answer.

   g. **Bad Dot Plots and Good** The first dot plot shows an error sometimes made. The scale for the variable *NStatesVisited* must have equal intervals, unlike the bad example shown here. The scale shown here is terribly wrong, even though it is neat. The space between 15 and 22 should *not* be the same as between 14 and 15. By hand, make a dot plot of the distribution of *NStatesVisited*. The second dot plot shows what the scale should be, but the dot plot is not finished.

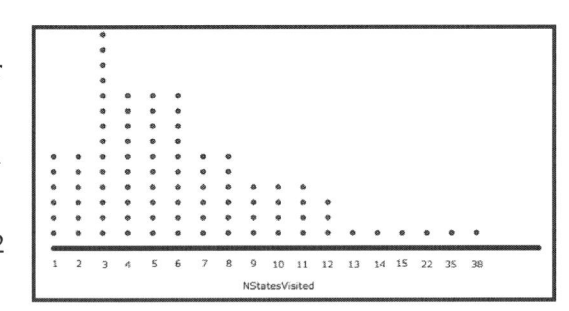

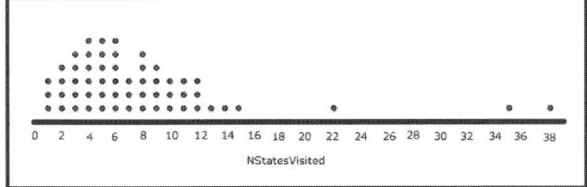

- Open the file **CombinedClassDataSpr08.ftm**.
- Get the **Case Table** for these data.
- From the **Shelf,** drag down a blank **Graph** (between **Table** and **Summary**).
- Go to the **Case Table,** select the variable *NStatesVisited*, and drag to the foot of the Empty Plot. You should be able to compare your handmade dot plot with the Fathom dot plot.
- Then from the right hand corner of the plot, change the output from a dot plot to a histogram.

h. Determine the *bin width* of the histogram that Fathom has produced. (Count the number of bars between the numbers "5" and "10" and then divide by 10 − 5 = 5.)

i. Copy the histogram (either by hand or print it) and *shade in* the bars that correspond to the answer to part c above. We will show probabilities as areas in this way very often.

2. **Students' Grades in Costa Rica** The distribution of letter grades for some students in Costa Rica was shown in the **Notes**. Here is a part of the spreadsheet for the entire collection.

Variables:	Description
Semester	Year and semester of the course
Sex	Gender of the student: male or female
Exam1	Score on Exam 1 (The scores are out of 100.)
Exam2	Score on Exam 2 (The scores are out of 100.)
Exam3	Score on Exam 3 (The scores are out of 100.)
Course_Grade:	Overall course grade, expressed as a percentage (The grade included homework and other assignments as well as the tests.)
LetterGrade:	Letter grade for the course: A, B, C, D, F

a. For this collection, what are the cases?

b. Which variables are quantitative, and which variables are categorical?

c. Calculate the percentage of students who earned each of the grades A, B, C, D, and F. You may wish to make a small table to organize the answer.

d. Use the probability notation of §1.2 correctly to show the probability of getting an A.

e. Make a statistical question that can be answered with just these data. You may consider all of the variables listed and not just what you see in the *LetterGrade* summary table.

3. **Household Sizes for California Students and Western Australian Secondary Students**

A natural statistical question is:

Are there differences in the distributions of household size comparing students in California and Western Australia?

a. Inspect the percentages and the counts in the table of the two distributions. Find one difference between the distributions and write what that difference is (using complete sentences, etc.).

b. Compare the proportions of students in the two places who live in households having five or more people. Use the probability notation of §1.2 correctly and interpret your results in the context of the data.

c. These two distributions have almost the same total count (64 for California and 63 for Western Australia). Would we be able to compare the two distributions by calculating proportions or percentages if the total counts were very different? (For example, suppose we had 180 students for Western Australia but still just 64 for California.) Give a reason for your answer.

4. **Household Sizes for California Students**

   Here are the distributions of the household sizes (*NPeopleHH*) for male and female California statistics students.

   a. By hand, make a double dot plot similar to the one shown in the **Notes** that shows the two distributions of household size for the male and the female students.

   b. In your judgment, are there differences in the two distributions or are they essentially very similar?

   California Statistics Students 1999

		Gender		Row Summary
		Female	Male	
NPeopleHH	1	1	3	4
	2	8	1	9
	3	6	5	11
	4	9	12	21
	5	6	7	13
	6	1	1	2
	7	2	1	3
	14	0	1	1
Column Summary		33	31	64

   S1 = count ( )

5. **Mothers' Ages of California Students: Stemplots** (Use the **CombinedClassDataAut08** sheet.)

   a. By hand, make an ordered stemplot of the distribution of the variable *MothersAge* for the *female* students. Use **stems of two** (see the **Notes** for an explanation of what stems of two are).

   b. By hand, make an ordered stemplot of the distribution of the variable *MothersAge* for the *male* students. Use stems of two as well.

   c. For each of the distributions shown in the stemplots (the *male* and the *female*) calculate the percentage of students whose mothers were age 30 or more when the student was born, using the conditional probability notation of §1.2 to write your results, using $X \geq 30$ in your notation. Not all the students gave information; use as denominators only the total for which you have information.

   d. *Interpretation:* Would you say that the percentages show that the mothers of *male* and *female* students are basically similar or would you say that there is an important difference?

   e. Use the stemplot for the female students you made to answer part a to create a dot plot (by hand) for the data on *MothersAge* for the *female* students.

   f. Use the stemplot or the dot plot you made (in part e) for the *female* students to make a histogram (by hand) with bin width of two years for the data on *MothersAge* for the *female* students.

   g. Is it possible from your *stemplot* to calculate the percentage of *female* students whose mothers were less than twenty-seven27 years old? If so, make the calculation. If it is not possible, say why it is not possible.

   h. Is it possible from your *histogram* to calculate the percentage of *female* students whose mothers were younger than 27 years old? If so, make the calculation. If it is not possible, say why it is not possible.

   i. Is it possible from your *dot plot* to calculate the percentage of *female* students whose mothers were younger than 27 years old? If so, make the calculation. If it is not possible, say why it is not possible.

- Open the Fathom file **CombinedClassDataAut08.ftm** and get a **Case Table**.
- Get a **Graph** from the shelf, drag the variable *MothersAge* to the horizontal axis, and drag the variable *Gender* to the vertical axis. You should see two dot plots on the same grid, and for the female dot plot you should be able to check your work in part f.
- Use the scroll on the **Graph** to change the dot plot to a histogram.
- j. Determine the *bin width* of these histograms. (Double click on the histogram to show the *bin width*.)
- k. Use the inspector (what you get if you double click on the histogram) to make the starting point age 18 (Fathom calls this bin alignment position) so that you can compare your handmade histogram with the one that Fathom has made.
  l. On your hand-drawn histogram for the female students, shade in the proportion of students whose mothers were 30 or older, which was $P(x \geq 30 \mid F) = 27/63 \approx 0.4286$.

7. **SF Espresso and Café Ratings**   There is an interesting website (*http://www.coffeeratings.com/*) that comes from the attempt by one man to "find, taste, and review every noteworthy espresso" he could find in the city of San Francisco. For each espresso and each café he uses a scale of 10 (10 being the best), and then for the overall score he averages the "espresso" score and the "café" score.

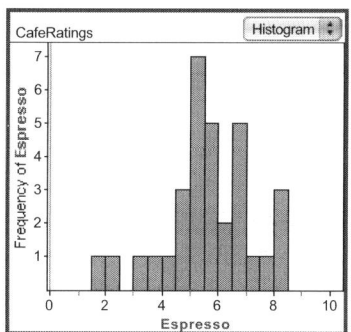

   a. Here is a histogram of the n = 32 **Espresso** ratings for just the *Embarcadero* neighborhood. What is the *bin width* for this histogram?

- Open a case table for the Fathom file **CoffeeRatingsSFEmbarcadero.ftm.**
- Drag down a graph from the shelf then drag the variable **Espresso** to the horizontal axis, and then change the dot plot to a histogram.
- To make your histogram like this one, select the graph, double click on the graph, and change binAlignmentPosition to 1.5. (Check your answer to 'a' as well.)

   b. Use the histogram you have made to find the proportion of cafés rated at 7 or more and express your answer in probability notation.

   c. If you add up the numbers for *all* of the bars in the histogram, what should the sum be? Why? Confirm your answer by tediously adding.

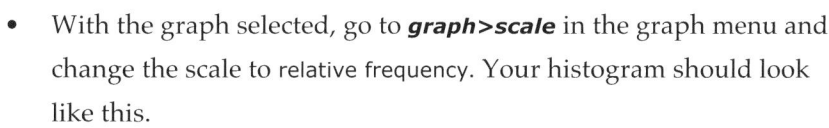

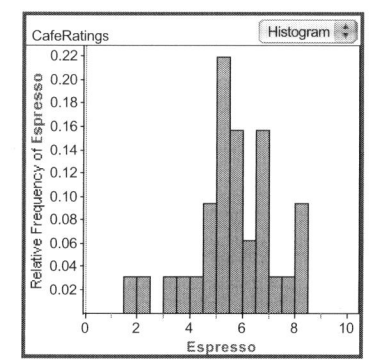

- With the graph selected, go to **graph>scale** in the graph menu and change the scale to relative frequency. Your histogram should look like this.

   d. Use this relative frequency histogram to estimate $P(X < 4)$. Show a sum of proportions from your histogram.

   e. If you add up the *proportions* for *all* of the bars in the histogram, what should the sum be? Why? (Note: With $n = 32$ cases, using the "frequency" rather than the "relative frequency" histogram is more accurate. However, the goal here is to get accustomed to relative frequency histograms, which have proportions on the vertical scale.)

45

8. **SF Espresso and Café Ratings (continued)** Shown below is a bar graph that is not a histogram. It is for the same data as the histogram in Exercise 7. Spreadsheet software will by default produce this kind of bar graph. To see the difference between a histogram and this kind of bar graph, answer these questions.
   a. What variable is on the *horizontal* axis in the histogram above? And what variable is on the *horizontal* axis in the bar graph shown here? 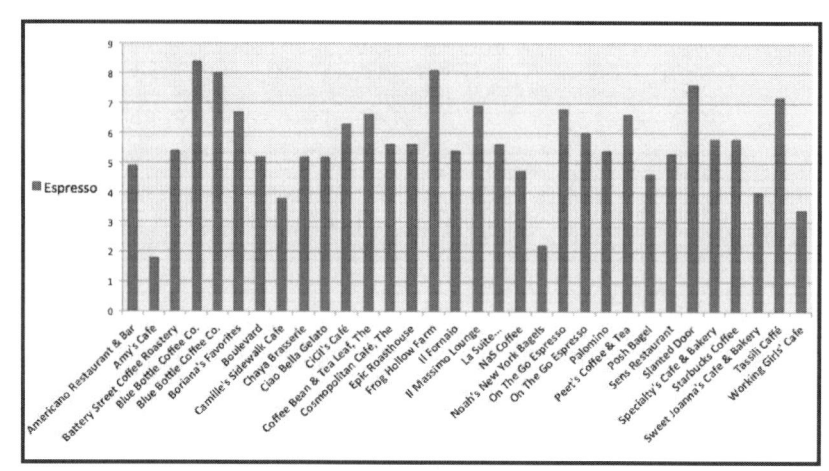
   b. What is on the *vertical* axis in the histogram? And what is on the *vertical* axis in the bar graph shown here?
   c. What does the *height* of a bar indicate for a histogram? And what does the *height* of a bar in the bar graph indicate. (To answer this question you may want to focus in on a single bar for both graphics: you could choose the right most one or the left most one in each of the two graphics.)
   d. Put into your own words the difference between a histogram and a bar graph such as the one here.

9. **Remember when you were fourteen** In Example 3 in the **Notes** for §1.3 we looked at the proportions of male and female Australian secondary students who were in the height interval 160 cm ≤ *Height* < 180 cm, and we found that the proportions in this "mid-height" range were quite similar. What happens if you look at fourteen-year-olds instead of thirteen-year-olds? Here are the histograms.
   a. From the histograms, calculate the proportion of male students and the proportions of female students who are in the height interval 160 cm ≤ *Height* < 180 cm.
   b. Write an interpretive sentence comparing what was found in Example 3 with what you have found here from part a.
   c. Express your answer to part a in probability notation.
   d. If you chose a fourteen-year-old *male* Australian secondary student from the collection, what is the probability that the student is 180 centimeters or taller? Use probability notation.

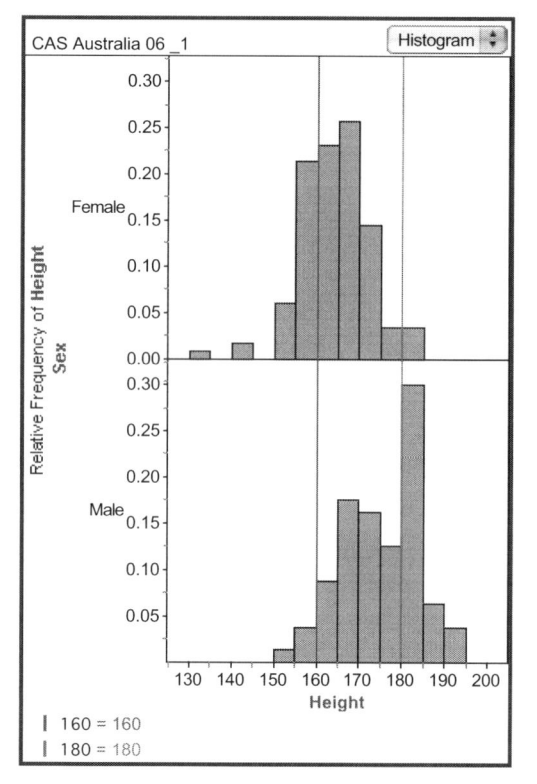

e. If you chose a fourteen-year-old *female* Australian secondary student from the collection, what is the probability that the student is 180 centimeters or taller? Use probability notation.

f. *Interpret* (i.e., put in a sensible English sentence or two) what you have found in parts d and e.

10. Time taken to travel to school in South and Western Australia The variable *TimeSchool* measures the time it takes a student (in Australia) to get from home to school.

a. If you add up the *relative frequencies* for all the bars shown, what answer should you get? Give a reason for your answer.

b. Add the relative frequencies to confirm your answer to the question in part a.

c. Find the probability that a student takes 40 or more minutes to get to school if that student is chosen at random from the South and Western Australian collection. Use probability notation.

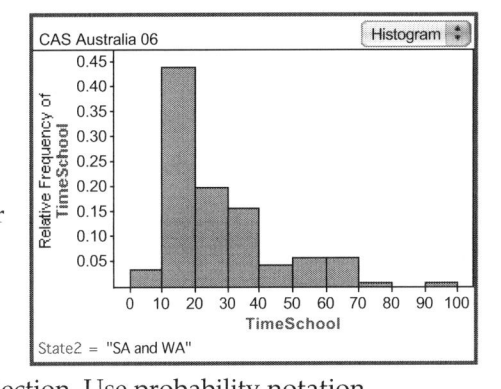

d. By doing as little work as possible, find the probability that a student takes less than 40 minutes to get to school. Show what you did.

11. Les Écossais Here are data showing the distributions of household size (*NumHH*) for households in two communities in Quebec in two census years in the nineteenth century. The cases are households; for the 1831 census there were $n_{1831} = 93$ households, and for the 1851 census, $n_{1851} = 97$. The variable is the number of people per household or *NumHH*. (Some background: Gaelic-speaking Highland Scots immigrants inhabited most of the households. The immigration had begun in the latter part of the eighteenth century but had continued during the nineteenth century.) Our statistical question is:

Is there evidence that the distribution of sizes of households for this community changed between 1831 and 1851?

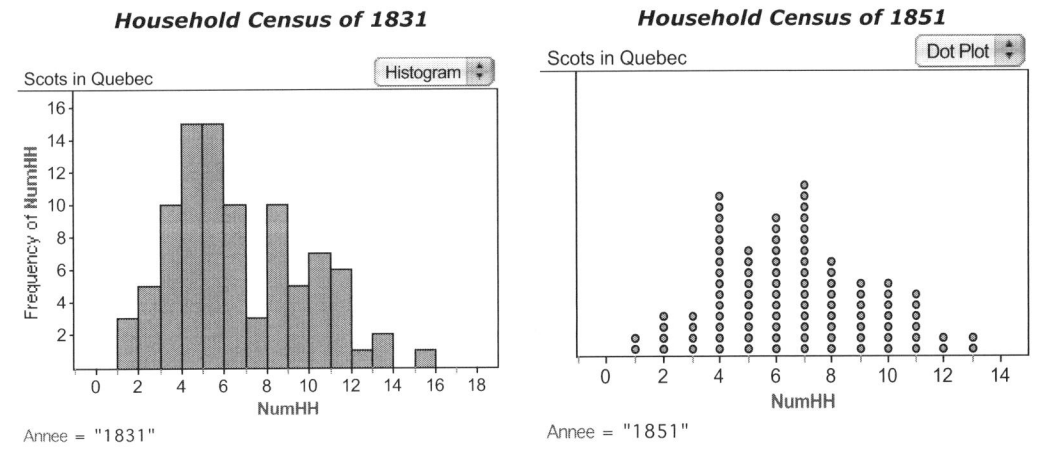

These distributions can be pictured using a histogram (1831) and a dot plot (1851). You now have a chance to compare the usefulness of each. (Actually you can easily make a dot plot, by hand, from the histogram or a histogram from the dot plot.) **PTO**

a. Make any kind of *calculation* (or calculations) you wish to make using the graphics you have been given in order to answer the statistical question that we have posed (in italics above) about whether there are any differences in the distributions of the sizes of households for 1851 and 1831.

b. Write a kind of mini-report giving your answer to the statistical question stated above about whether there are any differences in the distributions of the sizes of households for 1851 and 1831.

c. Which of the graphics did you find more useful: the histogram or the dot plot? Give a reason for your answer. <u>Note</u>: A good way to do this exercise is to use teamwork. Find a partner and together decide what kinds of calculations it would be good to have. Then one of you should do the calculations for the 1831 data and the other the calculations for the 1851 data. Then bring your calculations together and discuss what the two of you think they mean. One completely valid answer is that there is very little difference between the 1831 and the 1851 distributions of the number of people in households.

12. The Midge Question In 1981, two new varieties of a tiny biting insect called a midge were discovered by biologists W. L. Grogan and W. W. Wirth in the jungles of Brazil. They dubbed one kind of midge an Apf midge and the other an Af midge. The biologists found out that the Apf midge is a carrier of a debilitating disease that causes swelling of the brain when a human is bitten by an infected midge. Although the disease is rarely fatal, the disability caused by the swelling may be permanent. The other form of the midge, the Af, is quite harmless and a valuable pollinator. In an effort to distinguish the two varieties, the biologist took measurements on the midges they caught. The two measurements taken were of wing length (WL) and antennae length (AL), both measured in centimeters.

- Open the Fathom file on your computer entitled **Midges.ftm** and use the data to answer the question below in a short joint report. You will have to decide what to do and then do it with Fathom.

Question: Is it possible to distinguish an Af midge from an Apf midge based on wing and antenna length? Write a report that describes to a naturalist in the field how to classify a midge he or she has just captured. (This question came from Daniel J. Teague.)

[Reference: Grogan, William L., Jr. and Willis Wirth. 1981. "A new American genus of predaceous midges related to Palpomyia and Bessia (Diptera: Ceratopogonidae)." Proceedings of the Biological Society of Washington 94 (4): 1279-1305.]

13. **Real Estate Data** The data for this exercise concerns houses that were sold in 2007–2008 in San Mateo County, California. There were a total of 3,947 houses sold in the period from June 2007 to June 2008, but the data for this exercise uses just $n = 400$ selected randomly from the 3947. Here is a part of the Case Table for these data, where each row is a different house sold. For this exercise, we will be only interested in the following variables.

    Time_in_Market   records the time it took in days for the house to sell
    ListSale   is a categorical variable with three categories:
        "Over"   if the sale price of the house was more than the list price.
        "Same"   if the sale price of the house was the same as the list price
        "Under"   if the sale price of the house was under the list price

  a. What are the cases for these real estate data?
  b. Is the variable *Time_in_Market* a quantitative variable or a categorical variable?
  c. Is the variable *ListSale* a quantitative variable or a categorical variable?

- Open the Fathom file **San Mateo RE Sample Y0708.ftm** and get a **Case Table.**
- Get a **Graph** from the shelf, drag the variable *Time_in_Market* to the horizontal axis, and drag the variable *ListSale* to the vertical axis.
- Use the scroll on the **Graph** to change the dot plot to a histogram.
- With the graph selected, go to the **Graph** menu and to **Scale** and change to "Relative Frequency." You should have the graphic shown here.

  c. What is the *bin width* of these histograms? (Fathom will reveal the *bin width* if you double click on the histogram.)

  d. For each of the three categories of the variable *ListSale* ("Over" = O, "Same" = S, and "Under" = U) use the histogram to estimate the proportion of houses in the sample that spent less than 20 days on the market before selling. Express the results using good probability notation; i.e., for the "over" category, you should have a value for $P(X < 20 \mid O)$ and likewise for the other two categories.

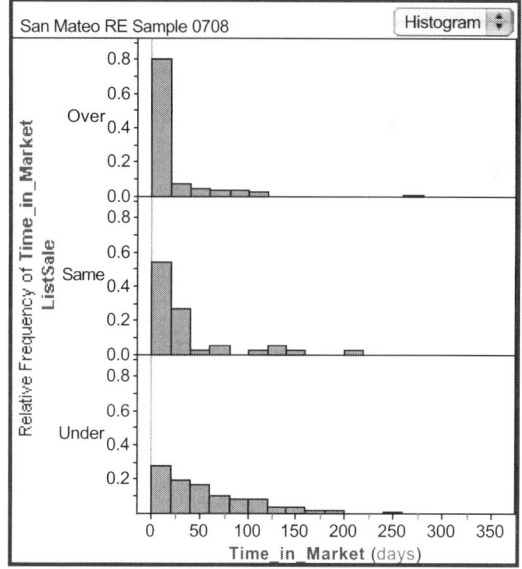

  e. Copy the Fathom output into a word-processing document and shade in the bar or bars on the plot that represent your answers to part d.

Follow the directions on the next page to check your answers.

- Get a **Summary Table** from the shelf; drag the variable *Time_in_Market* to the right-pointing arrow. The first number is the average (the mean) of the variable and the second number is the proportion of all houses that sold in less than 20 days. To get the second number, follow the next directions.

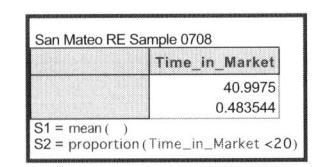

- Select the **Summary Table** and go to the menu **Summary: Add Formula.** When the dialogue box appears, type *proportion(Time_in_Market <20)* in the "open box" so that it looks like what you see on the right; the word "proportion" should be in blue, and the variable name should be in red. The **Summary Table** should look like the one here. The 40.9975 shows the average for the variable *Time_in_Market* for all the houses.

- Drag the variable *ListSale* to the down-pointing arrow to see the proportions that you estimated in part d.

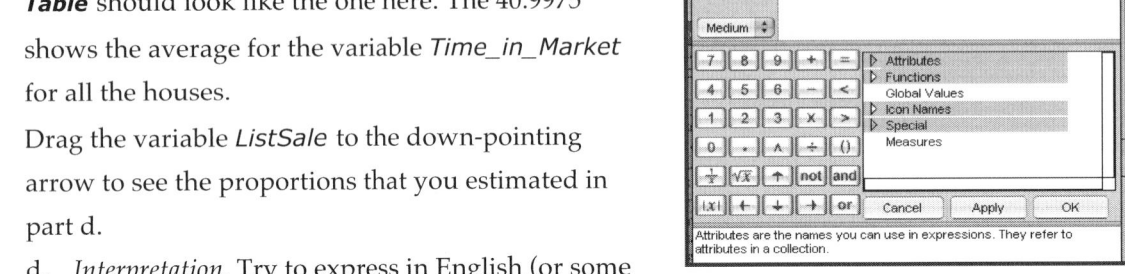

    d. *Interpretation.* Try to express in English (or some other language—but make it elegant) what the comparison of the three proportions means in the context of whether the houses sold for over, under, or the same as the list price.

    e. In either an easy or a tedious way, calculate $P(X \geq 20 \mid O)$, $P(X \geq 20 \mid S)$, and $P(X \geq 20 \mid U)$. (*Hint:* What probability rule are you using if you do this in an easy way?)

    f. Are the events $X < 20$ and $X \geq 20$ mutually exclusive? Give a reason for your answer.

    g. Are the events $X < 20$ and $X \geq 20$ independent? Give a reason for your answer.

- Select either the **Summary Table** or the **Graph** and go to the menu: **Object: Duplicate Summary Table** or **Object: Duplicate Graph.** You will see an exact duplicate of the graph you have made.

- Drag the variable *Region* to the duplicated **Summary Table** or the **Graph** you have made to *replace* the variable *ListSale*. Now you have the *Time_in_Market* related to the region in which the house was sold.

    h. Use the **Summary Table** or the **Graph** to explain in which regions the houses appear to sell fastest and in which regions the houses sell at a slower pace. Give evidence for your conclusions.

§1.4 Shape, Center, Spread of Quantitative Variables

A collection of data on colleges and universities is the subject of this section and our question is: how do colleges and universities differ? To answer this general question, we will look at three features of distributions of *quantitative* variables: **shape, center,** and **spread**.

In the last section, we looked at the important idea of a *distribution* and then we looked at various kinds of graphics to display distributions. You should be familiar with *dot plots, stemplots,* and *histograms,* and we will be using these, as well as the idea of probability covered in §1.2. But in this section, we have data on colleges and universities—that is, the cases are colleges and universities. Many of the variables that we look at, such as the tuition fees for the different colleges, are *quantitative* rather than categorical. For quantitative variables, we can analyze the **shape,** the **center,** and the **spread** of distributions. As we will see, these terms have specific, technical meanings within statistics, and the meanings are not necessarily what you would think they would be.

Shapes of Distributions: Right and Left Skewness

Every distribution has a form that you can see in a graphic; the technical term we use for this form is **shape**. We will use three terms to describe the shape of a distribution: *right-skewed, left-skewed,* and **symmetric**. Here is an example of a *right-skewed* distribution. It is called *right-skewed* because there is a "tail" on the right or high side of the values of the variable.

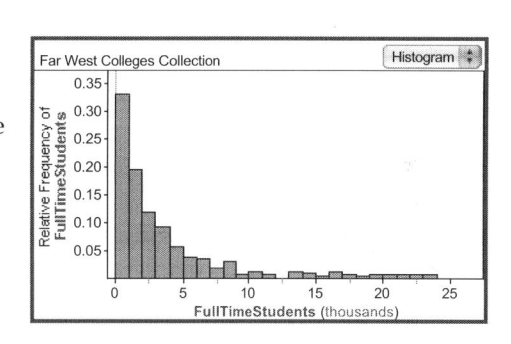

The cases in our collection are colleges and universities, and the histogram tells us that whereas most of the colleges have fewer than 5000 full-time students, there are some institutions—not many but some—where the number of full-time students is much greater, perhaps even over 20,000.

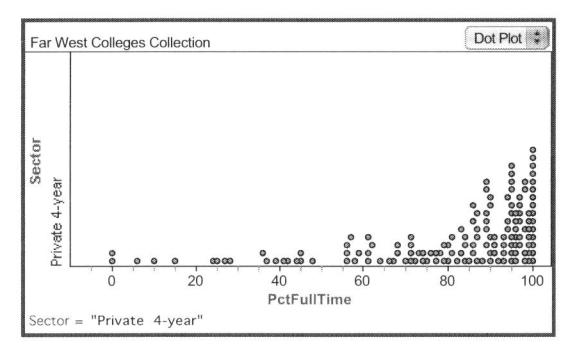

A **left-skewed** distribution has the "tail" on the left or low side of the variable. Here is an example of the variable "Percent Full-Time Students" for the private four-year colleges and universities. The graphic shows that most of the private colleges and universities have a relatively high percentage of their students as full-time students, but there are also some (but not many) private colleges or universities that have a much lower percentage of full-time students. (In one example we have used a *histogram* and in another a *dot plot*; either graphic can be used to show the shape of a distribution.)

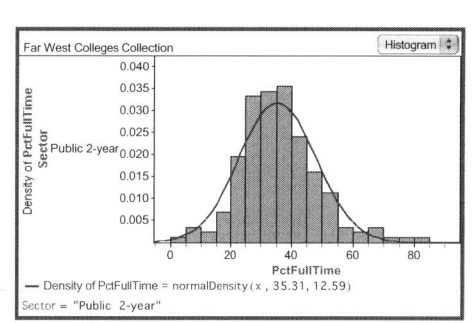

If a distribution is neither *right-* nor *left-* skewed then we may say that the distribution is **symmetric**. Distributions of real data seldom are exactly symmetrical, but they may be close to symmetrical. Here is an example of a distribution that is nearly symmetrical. About half of the two-year colleges have less than 35% of their students full-time, and about half of the two-year colleges have more than 35% of their students full-time. There is no appreciable skew either to the right or to the left for this distribution, although the tail on the right-hand side of the distribution is a bit longer.

The curved line that you see in the graphic is an example of a **density curve**. For now, think of a *density curve* as a smooth version of the *shape* of a distribution. A density curve can be thought of as a kind of idealization or as a **model** for the actual data. Like the histogram itself, the density curve's area includes 100% of the entire distribution; that is, if you calculate its area, the sum of the relative frequencies should be one, or 100%. The density curve that we have drawn for the distribution of *PctFullTime* is one that is perfectly symmetrical. This density curve is the one that we will see in §1.7 and is called the **Normal Distribution.**

Here (in the box) are three density curves that show left-skewed, right-skewed, and symmetric shapes of distributions and definitions for each of the shapes of the curves.

Shapes of Distributions

| LEFT-SKEWED | SYMMETRIC | RIGHT-SKEWED |

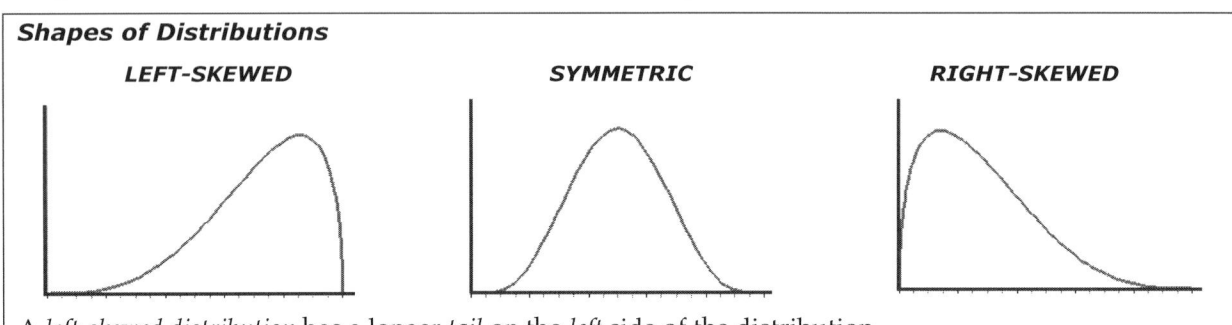

A *left-skewed distribution* has a longer *tail* on the *left* side of the distribution.
A *right-skewed distribution* has a longer *tail* on the *right* side of the distribution.
A *symmetric* distribution has roughly *equal* length *tails* on the right and left of the distribution.

All of these shapes are for distributions that have a single "peak." Sometimes the shapes of distributions do not fall into one of these neat categories. Sometimes we have too little data to say anything definitive about shape, and sometimes the shape is a mix. Quite often we encounter distributions that appear to have more than one peak. If a distribution appears to have just two peaks then that pattern is called **bimodal** (from two modes) and refers to a shape where there are two "humps."

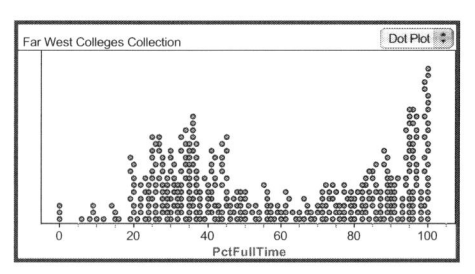

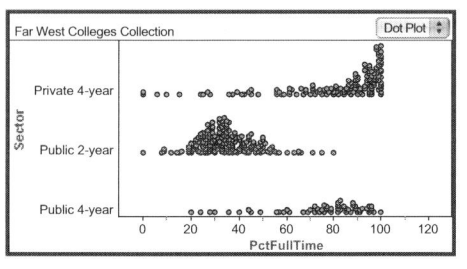

From the colleges and universities data, the variable *PctFullTime* provides an example of a *bimodal* distribution. A bimodal pattern may sometimes indicate that there are actually two groups in the data; by making plots of each of the groups, the bimodality can be explained. Of course, you have to have some idea what the groups are likely to be; for now, be assured that this knowledge is less of a problem than what you may think. Here, if we distinguish among public four-year, private four-year, and public two-year colleges then perhaps we can see what is happening with the *bimodality*. The right-hand "mode" comes from the left-skewed distribution of *PctFullTime* for the private four-year colleges and universities, while the left "mode" comes from the nearly symmetrical distribution for *PctFullTime* for the public two-year colleges.

Center or Location of a Distribution: "On Average"

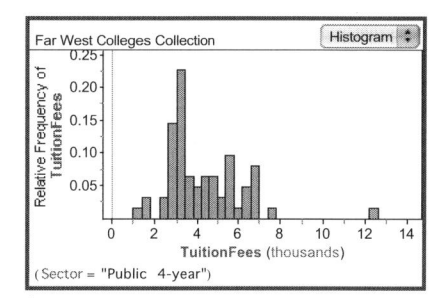

The idea of center When we say: "*On average*, tuition fees are about four thousand dollars per year for public four-year universities," we are speaking of what statisticians call the **center** of the distribution. We think of a value that characterizes the **location** of the distribution; we are answering the question: "Where–on the scale of the variable—*is* the distribution?" We look at center especially when we are comparing distributions and recall that comparison is important in statistics. Once again, the principle of comparison is important: "On average…four thousand dollars . . . $4000" really becomes important when "on average" the tuition fees for private schools are about $12,000.

Example: Let us look again at the distributions of the percent of full-time students (*PctFullTime*) for the three types of educational institutions; this time we will look at histograms, but we are looking at exactly the same three distributions as we did above just before the start of this section. The top histogram is for the private four-year colleges and universities; for this distribution, most of the colleges are toward the right end of the variable *PctFullTime*. That is, for most of these private four-year colleges, a high percentage of their students are full time. There are private four-year colleges that have a lower percentage of full-time students, but there are not many of these colleges. Therefore, we say that for the private four-year colleges, the distribution is mostly "centered" or "clustered" in the 90%+ area.

Now look at the histogram for the public two-year colleges (this is the second histogram). For them, most of the distribution is "centered" or "clustered" around 30% to 40%; the center is lower. Most of the two-year schools have only about 30% or 40% of their students full time rather than 90% or more full time.

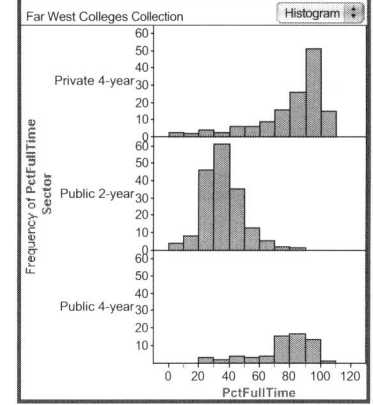

> Our question is: "*Where* on the scale of the variable is the distribution?"

> Our answer is: the distribution for the public two-year colleges is centered about 35%. We say that the center or location of the two-year college distribution is lower on the variable *PctFullTime* than the distribution for the private four-year colleges.

The distribution for the public four-year resembles the distribution for the private four-year colleges; the distribution is "centered" or "clustered" or "located" at around 80% full-time students. The comparison shows us that two-year colleges have a lower percentage of full-time students than do four-year educational institutions.

Higher, Lower, Bigger, Smaller, Greater, Lesser: Think Horizontally When we use these words to refer to the center of a distribution, and the values of the variable are on a horizontal axis, then a *higher*, *greater*, or *bigger* center means that the distribution is farther to the *right*—corresponding to larger numbers mathematically. A *lower*, *smaller*, and *lesser* center refers to a distribution primarily to the left. Think horizontally.

53

The height of the bars (or dots) does not mean the center is high. Look again at the histogram above for the *PctFullTime* for the public four-year colleges and compare it to the histogram for the two-year colleges. The histogram for the two-year colleges has the highest bars, but it is the type of college with the *lowest* percent full time, on average. The histogram for the public four-year schools is to the right on our variable, and so is higher. A high bar in a histogram does not necessarily tell us that the center is "high," and a high stack of dots in a dot plot does *not* tell us that the center of a distribution is high. The stack of dots just tells us that there happen to be many cases at that value.

The value with the greatest number of cases is known as the ***mode,*** but generally the *mode* is not a reliable indicator of the location of a distribution. A distribution may have several places with the same height, or mode. To know where the location or the center of a distribution is, you need to look at *where* the distribution as a whole resides on the variable.

The Spread of a Distribution and Its Meaning

The third thing we look at is the ***spread*** of the distribution. Distributions differ in their location—the center—but also in how *spread out* they are.

Here is an example: the dot plots show the number of full-time students (*FullTimeStudents*) for three types of colleges and universities. The plot shows that the distribution for the public four-year universities is considerably more spread out than the distributions for the other two types of schools. What does that mean? Since the cases are colleges, the meaning of a greater spread or variability is that the public four-year schools are "more different" (to use very inelegant English) from each other. Taken altogether, there are public four-year colleges and universities that range from extremely small to extremely large. The sizes of private four-year colleges and universities and of two-year colleges are not as variable.

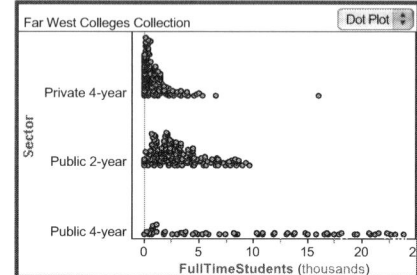

A start at understanding spread can be made by looking at the ***range*** of values for the variable. (The ***range*** is the highest number minus the lowest number.) We can see that the number of full-time equivalent students for the private schools is mostly from less than 100 full time students to about 8000 students (with one exceptional school), but for the public four-year colleges and universities, the numbers range from under 100 students to as many as 25,000 students.

In summary, there is a great *variety* or *diversity* in *FullTimeStudents* (the number of full-time students) for the public four-year schools and much more variety or "differentness" than with the private schools.

Outliers and Putting It Together

Notice that for the private four-year colleges there is one college that has a value for *FullTimeStudents* of about 16,000 and that the number of full-time students for this university is much more than most of the other private four-year institutions. (This happens to be the University of Southern California.) Cases that are far from the majority of the cases in a distribution are called ***outliers.*** There can be *outliers* on the high end of a distribution or the low end of a distribution, and, at this point, we will "detect" outliers just by looking; we will see a more exact way of detecting outliers later in §1.7.

Example and a warning What can we say about the *shape* and *center* of the three *FullTimeStudents* distributions? *Shape?* The distribution for the private four-year schools appears to be right-skewed, as does the distribution for the public two-year colleges. It is difficult from the graphic to say whether the distribution for the public four-year institutions is right-skewed, left-skewed, or symmetrical. *Center?* It looks as though the center or location of *FullTimeStudents* for the private four-year colleges is the least. We could say: "On average, the number of *FullTimeStudents* is the smallest for the private four-year colleges," even though we have not calculated the averages (we will do that in the next section). We could say that the center or location for *FullTimeStudents* for the two-year schools is somewhat higher because that distribution is generally somewhat to the *right* of the private schools distribution.

The ideas of *shape, center,* and *spread* are really aids to interpretation. They allow us to say something about the data. We will soon calculate numbers that indicate the centers and spreads of distributions, but, at this point, make certain to grasp the ideas and differences between *shape, center,* and *spread.* Here is the warning: work on making certain that you understand the differences between *shape, center, spread,* and the idea of a *distribution.*

Extended Example: Shape, Center, and Spread of Percent over Age Twenty-Four

Our overall statistical question is:

What differences are there in the distributions of the variable "percent of students over age twenty-four" (*PctOverAge24*) *between the private four-year colleges, public two-year colleges, and public four-year colleges, and what do the differences mean?* However, we will break this down into a number of smaller questions and show answers to these smaller questions (in "handwriting" font).

a. What are the cases for these data?

   The cases are colleges and universities.

b. What can you say about the shapes of the distributions?

   All three distributions appear to be somewhat right-skewed, since all three have a tail on the high side of the variable.

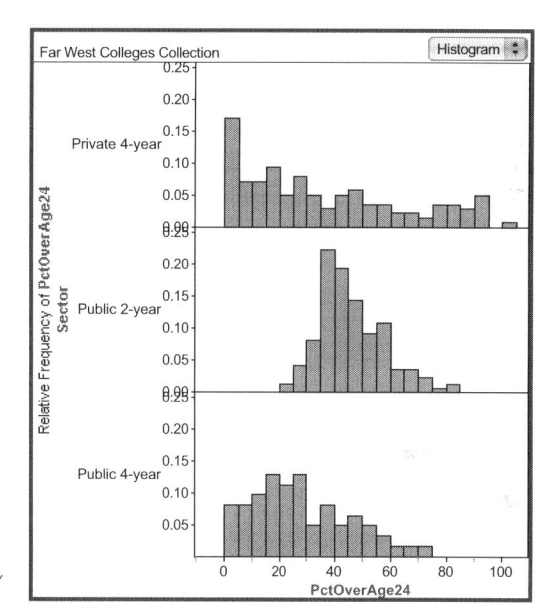

c. What can you say about the centers of the distributions?

   The distribution of the percentage over age twenty-four for the two-year schools is to the right of the distribution for the public four-year institutions. Since the spread of the distribution for the private colleges is so great, it is more difficult to say just from the graphic where the "center" is, but it is probably lower than for the two-year colleges. (We need some numbers!)

d. What can you say about the spreads of the distributions?

   We have already noted that the spread of the distribution for the private colleges and universities on the variable "Percent Over Age 24" is big. This means that there is great variety or diversity in the private educational institutions in the "percent of students over age twenty-four"; some private institutions have almost none of their students over twenty-four, whereas other schools have a high percentage over age twenty-four. The least spread appears to be for the two-year colleges, so that with respect to the percentage over twenty-four, these colleges are quite similar. The spread for the variable is greater for the public four-year schools but not as great as for the private institutions.

e. What does this all mean in the context of the data?

It makes sense that the percentage of students over age twenty-four in the two-year schools should be higher, on average, because some of the functions of two-year colleges are for retraining and also to give students who did not have the opportunity to go to college after high school a second chance when they are a bit older. That there is such a great variety (a large spread) in the private colleges suggests that some of these private institutions are vocational or professional in nature, whereas others in the private category are "traditional" liberal arts colleges that take students from high school.

Evenly Distributed and Warned

The exercises are designed to fend off common errors, but here are two or three warnings about common errors concerning shape, center, and spread.

First, make certain that you have the three features of distributions straight. Do not confuse spread with shape. Shape refers to the form of a distribution, whereas spread refers to how spread out over the variable the cases are.

Secondly, learn to use statistical language properly. Sometimes, people who are beginning to study statistics will say, "The data are widely distributed" when what they mean to say (and which they *will* say when they learn the language) is that "the distribution has a wide spread."

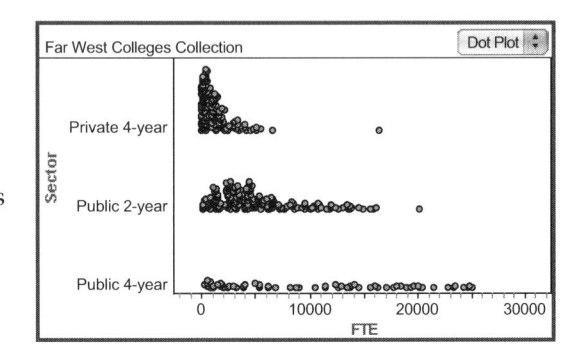

Third warning—about "evenly distributed": people who have learned statistics will almost never say, "The data are evenly distributed," but this is a very common statement among beginners. The statement may in fact be just completely wrong. The distribution of *FullTimeStudents* for the public two-year colleges is not at all "evenly distributed"; there are many colleges where the number of students is about three thousand to six thousand but *not* many colleges at all where the number of students is as high as fifteen thousand. If we chose at random, the probability of choosing a college where the number of students is somewhere in the range of three to six thousand students should be much higher than the probability that the college has over fifteen thousand. The probabilities are not the same.

The term "evenly distributed" may be more apt for something like the *FullTimeStudents* distribution for the public four-year colleges and universities. However, statisticians would say that those data appear to **uniformly** distributed and not "evenly" distributed. A distribution where the probability of each of the values (or ranges of values) is the same is called a **uniform** distribution and not an "even" distribution. A good example of a distribution that we expect to be at least fairly uniform is the month of the year in which people are born.

Summary: Shape, Center, and Spread of Distributions of Quantitative Variables

We use the ideas of *shape, center,* and *spread* to characterize the distributions of quantitative variables. Generally we use these ideas to compare distributions. We have concentrated on looking at pictures of distributions to get an idea of the concepts; in the next sections, we shall learn about numerical summaries of these characteristics.

- **Shape**   Whether a distribution is left-skewed, right-skewed, symmetric (if single-peaked), or bi-modal

    A ***left-skewed*** single-peaked distribution has a longer *tail* on the *left* side of the distribution.
    A ***right-skewed*** single-peaked distribution has a longer *tail* on the *right* side of the distribution.
    A ***symmetric*** single-peaked distribution has roughly *equal* length *tails* on the right and left.
    A ***bi-modal*** distribution is one with two distinct peaks.

- **Center or Location**   The location of a distribution of cases on the values of the cases
    - When the values of a distribution are graphed on a horizontal axis, greater means greater in the mathematical sense.
    - A distribution with a greater (or higher, or bigger) center is one to the *right* on the variable depicted horizontally.
    - The ***mode*** of a distribution is a value or groups of values with the highest count. The mode does not necessarily indicate that a distribution has a high center compared to other distributions.
    - A distribution may have several modes—that is, values or groups of values with the same count.

- **Spread**   The amount of variability, diversity, or "differentness" in a distribution compared with other distributions or to a standard

- **Outliers**   Cases that stand outside the general mass of the data

§1.4 Exercises: Shape, Center, and Spread

1. **Review and the Shape of a Distribution.** Here is a graphic of the distribution of the number of full-time students (the variable *FullTimeStudents*) for just the two-year colleges.

   a. *Review:* What are the cases for these data?

   b. *Review:* Is the variable *FullTimeStudents* a quantitative variable or a categorical variable?

   c. *Review:* What is the *bin width* for the histogram?

   d. *Review:* About what proportion of two-year colleges have fewer than 2000 students? You may express your answer as $P(X < 2000) =$

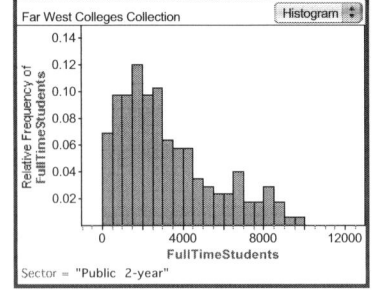

   e. Make a sketch of the histogram and shade in your answer to part d with together with an arrow pointing to the area the probability $P(X < 2000)$ represents.

   f. What name should we give to the shape of this distribution: *right-skewed, left-skewed*, or *symmetric?* Give a reason for your answer.

   g. At about how many *FullTimeStudents* does the center of this distribution appear to be? (The answer does not require any calculation at this point; just judge roughly where the center of the distribution is. You may well give your answer as a range of values.)

   h. What is the *range* of values for this distribution?

   i. In the **Notes**, there are several graphs showing the same variable *FullTimeStudents* but for the public four-year universities. Which distribution has the wider spread, this one or the one for the four-year universities? Give a reason for your answer.

2. **Selectivity and Tuition Fees**. (Note: this exercise continues on the next page.) The distributions shown are from the Far West colleges collection for just the private schools. We can compare the tuition fees for colleges and universities that are "very selective" in their admission policies with the tuition fees for institutions that are "minimally or moderately selective." Some numbers: there are $n_{vs} = 23$ for the "very selective" schools, and their average tuition fee is $27,609; there are $n_{mms} = 52$ for the "minimally or moderately selective" schools, and their average fee for tuition is $20,905.

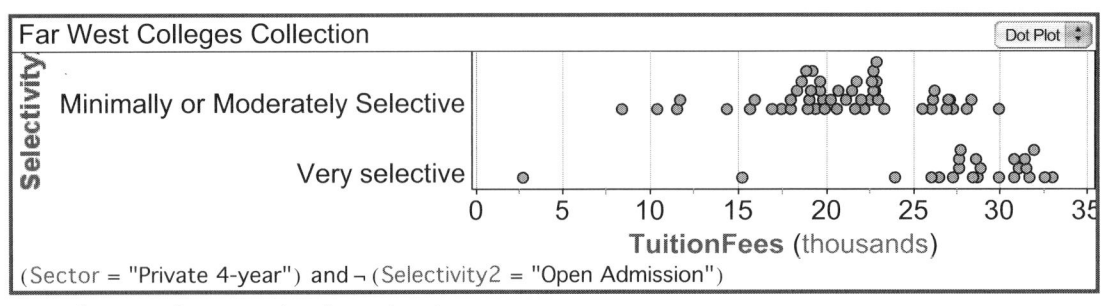

   a. What are the cases for these data?

   b. Judging from the dot plots and averages, which distribution has the higher center?

c. On average, judging by the plots and the numbers given, which type of college or university, the "very selective" or the "minimally or moderately selective," typically has the higher tuition fees? Give a reason (or reasons) for your answer.

d. Is the distribution of tuition fees for the "minimally or moderately selective" institutions *left-skewed, right-skewed, symmetric, bimodal,* none, or a mix of these? Give a reason for your answer, and if you answer "none" or a "mix," state which type of shape is in the mix.

e. For this and the next question (but not for the questions following that), disregard the two outliers for the "very selective" group. Disregarding these two outliers, which group has a distribution of tuition fees with the greater spread? Give a reason for your answer.

f. A common confusion is to think that if the center of a distribution is large (far to the right) then it must necessarily be that the spread of that distribution is big as well. Explain how part e shows that this confusion is not correct.

g. *Review:* Using the dot plots (counting dots), the notation "TF" for the variable *TuitionFees*, the $n_{vs} = 23$ ("very selective") and $n_{mms} = 52$ ("minimally or moderately selective"), calculate $P(TF \geq \$25000 \mid MMS)$, and $P(TF \geq \$25000 \mid VS)$. [*Hint:* There is a very easy way to get $P(TF \geq \$25000 \mid VS)$ using one of the probability rules.]

h. Express in English what $P(TF \geq \$25000 \mid MMS)$ and $P(TF \geq \$25000 \mid VS)$ tell you.

i. Explain how your answer to part h agrees with your answer to parts b and c (which are really the same question).

3. **Ages of People and Dogs.** The idea of this exercise is to be acquainted with the ideas of *shape, center,* and *spread* by thinking about where on a scale a distribution is, what shape the distribution has, and how spread out is a distribution that you can imagine. The variable is *age*, and in most of the parts of the exercise, the cases will be people. What you are to do is to sketch a rough **density curve** on a scale of ages in answer to each scenario. Here is an example: "Sketch the distribution of the ages of the students in your class." A good sketch would be like this one.

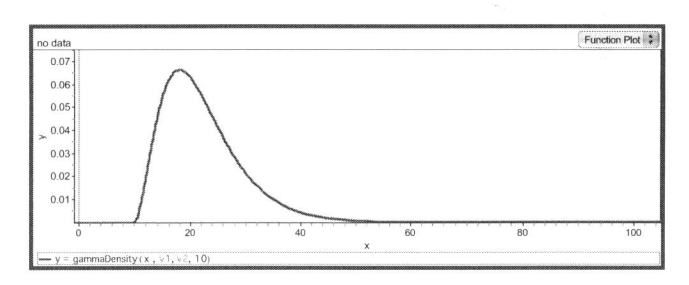

You may be given a sheet with a scale for the variable already drawn. Otherwise, you should draw a scale for each of the scenarios (a through f). Here are the scenarios:

a. The ages of the children in a fifth-grade class. (Think: what are the ages of the children likely to be?)
b. The ages of the people in a fifty-year reunion of a fifth-grade class.
c. The ages of people attending a performance of the latest hot rock band.
d. The ages of the residents of a retirement home.
e. The ages of the fans at a San Francisco 49ers game.
f. The ages of all the dogs in your county.
g. The age of death for classical music composers of the 1700s and 1800s.

4. **Ages of People and Dogs** Look at the sketches of the distributions that you made for the scenarios a through f in the exercise on "Ages of People and Dogs." In answering the questions below, also include the example distribution on the ages of the students in your class. For each of the questions below, give a reason for your answer in a complete sentence.
   a. Which of the distributions of *Age* is farthest to the right (and thus has the greatest center)?
   b. Which of the distributions of *Age* is farthest to the left (and thus has the lowest center)?
   c. Which of the distributions of *Age* has the greatest spread in the variable?
   d. Which of the distributions of *Age* has the least spread in the variable?
   e. Which of the distributions (if any) are likely to be right-skewed?
   f. Which of the distributions (if any) are likely to be left-skewed?

5. **Tuition Fees and Type of College** The data are from the Far West colleges collection and show the distributions of tuition fees for the three types of colleges and universities that we have been examining.
   a. What are the cases for these data?
   b. Put the three distributions (the three sectors) of tuition fees in order of smallest center (or location), middle center (or location) and biggest center (or location). Say how the ranking you have made in terms of center makes sense (or does *not* make sense) from what you know otherwise about the fees for different types of educational institutions. (If your ranking does not make sense then perhaps you should reconsider your ranking.)
   c. Put the three distributions (the three sectors) of tuition fees in order of smallest spread, middle spread, and greatest spread. Say how the ranking you have made in terms of *spread* makes sense (or does *not* make sense) from what you know otherwise about the fees for different types of educational institutions.
   d. Do any of the distributions appear *bimodal*? Hazard a guess as to why any *bimodality* you see should be there.
   e. Read the section entitled **Evenly Distributed and Warned** in the **Notes**. According to that section, state (in a sentence) how the distribution of tuition fees for the private four-year colleges and universities should be described.

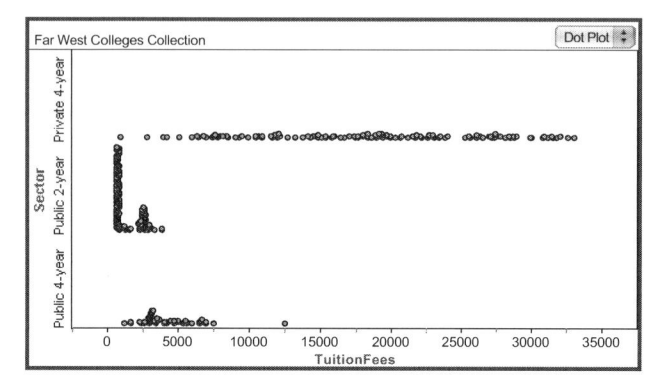

6. **Real Estate Data** (Note: This exercise continues on the next page.) The data for this exercise concerns houses that were sold in 2007–2008 in San Mateo County, California. There were a total of 3,947 houses sold in the period from June 2007 to June 208, but the data for this exercise uses just $n = 400$ selected randomly from the 3,947. To see the case table, see Exercise 13 of Section 1.3. For this exercise, we will be only interested in the variables *SqFt, List_Price,* and *Region*.
   - *SqFt* records the size of the living area of the house in square feet.
     List_Price is the price that the seller first asked for when the house was put on the market.
     Region San Mateo County in four regions: "Central," "Coast," "North," and "South."

a. Characterize the three variables we will analyze as either quantitative or categorical.
b. What are the cases for our data?

- Open the Fathom file **San Mateo RE Sample Y0708.ftm** and get a **Case Table.**
- Get a **Graph** from the shelf, drag the variable *SqFt* to the horizontal axis, and drag the variable *Region* to the vertical axis.
- Use the scroll on the **Graph** to change the dot plot to a histogram.
- With the graph selected, go to the **Graph** menu and to **Scale** and change to "Relative Frequency."

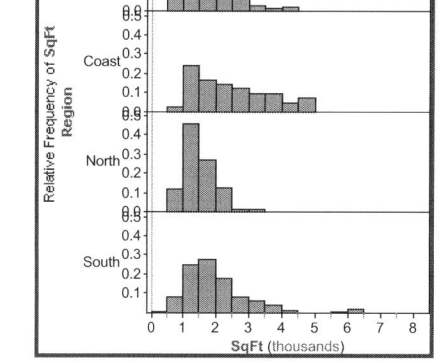

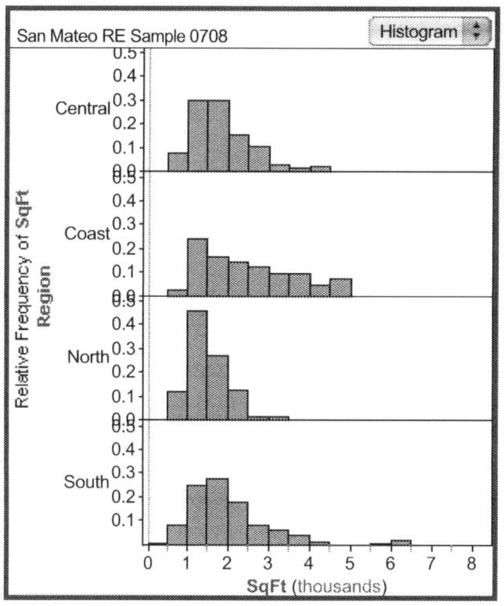

  You should have the graphic shown here. For the answers for parts c through f, each answer should have a reason given.

  c. What is the *bin width* of these histograms? (Fathom will reveal the *bin width* if you double click on the histogram. (The scale on the graph is expressed in thousands of square feet.)
  d. What are the most appropriate words to characterize the shape of these distributions? (See **Shape** in the **Notes.**)
  e. Which region appears to have the greatest variability in the sizes of houses (in of *SqFt*)?
  f. Can you say which region has the distribution of *SqFt* with the highest center or lowest center, or do the "centers" (locations) of the distributions appear similar?

- Select the **Graph** and go to the menu: **Object: Duplicate Grap.** to get a duplicate of the graph.
- From the **Case Table,** drag the variable *List_Price* to the **Graph** you have made to *replace* the variable *SqFt*. Now you have the *List_Price* related to the region in which the house was sold.
- Double click on the graph to get the Inspector: change the *bin width* to 250,000, and also change "ShowXGridlines" from "false" to "true."

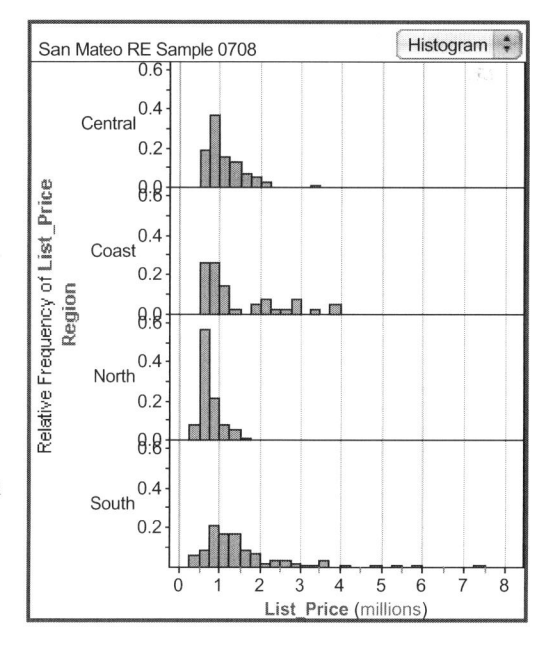

  You should have the graphic shown here.

  g. What are the most appropriate words to characterize the shape of these distributions? (See **Shape** in the **Notes.**) Give a reason for your answer.
  h. Which region appears to have the greatest spread or variability in the sizes of houses (in the distributions of *List_Price*)?
  i. Can you say which region has the distribution of *List_Price* with the highest center or lowest center, or do the locations of the distributions appear similar? Explain.
  j. Using the correct conditional probability notation and using the histograms, calculate the proportions of houses in each of the regions whose *List_Price* was less than $750,000. (You will have to look carefully at the histograms.)

61

§1.5 Numerical Summaries of Center/Location

The Idea of Center or Location: Where Is the Distribution?

We have already seen that the idea of the "center" of a distribution refers to its location on the scale of possible values. Suppose you were asked: "Comparing community colleges, four-year public universities, and four-year private universities, in which kind of institution are tuition fees the most and in which the least?" You would probably have a ready answer; you might even say, "On average, community colleges have the lowest tuition fees, private universities have the highest, and public four-year schools are between these." Here are three histograms on showing the distributions of tuition fees in colleges and universities in the Far West that backs up your idea.

The graphic backs up your idea because the distribution for the public four-year college tuition fees is generally to the right of the publically funded schools, and the distribution for the public two-year colleges tuition fees is to the left of the distribution for fees for public four-year institutions. (You may also notice that there are big differences in the *spreads* of the distributions of tuition fees; the privately funded schools have a huge amount of variability compared with the publically funded schools.)

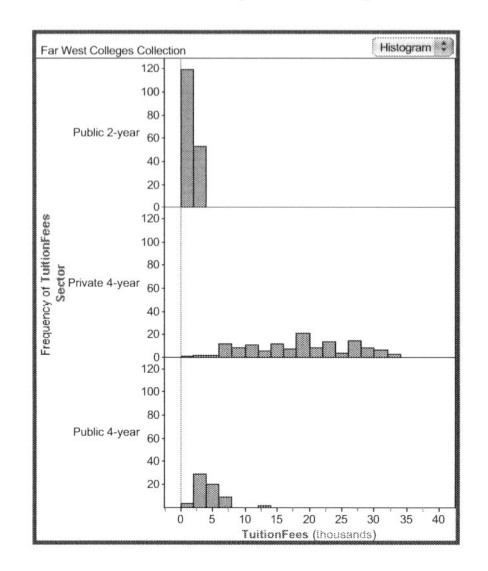

The fact that you may have answered "on average" may hint at the idea that we should be able to summarize the location (or center) of these distributions with a single number, and it may hint at the number you have in mind: the "average." This section is about averages, both the one you already know and another, about which you know less. The purpose of these averages is to summarize, with one number, the location of a distribution. These two averages, or measures of center, are called the **mean** and the **median.**

To illustrate the calculation of these averages, we will look at a second example with a smaller collection. The cases are again colleges and universities, but this time we are considering just the twenty-five publically funded undergraduate institutions in Oregon, and we are interested in two variables; one is the categorical variable that distinguishes the two-year (community colleges) from the four-year institutions, and the other is (for each institution) the percentage of students that are over 24 years old.

We expect that the *percentage over twenty-four years* will be higher on average for the two-year schools. The dot plot (below) appears to show this, since the distribution for the "Percent over 24" is generally to the right of the distribution for the four-year schools.

Mean, x-bar, and sigma

The average that you know about—the one that you learned in elementary school—is "the sum of all the things divided by the number of things." This measure is called the **arithmetic average** or the **mean** when we use it in statistics. The symbol that we use for the *mean* is $\bar{x}$, which is pronounced "*x-bar.*" The mean for the variable *PctOverAge24* for the eight four-year schools can be calculated as

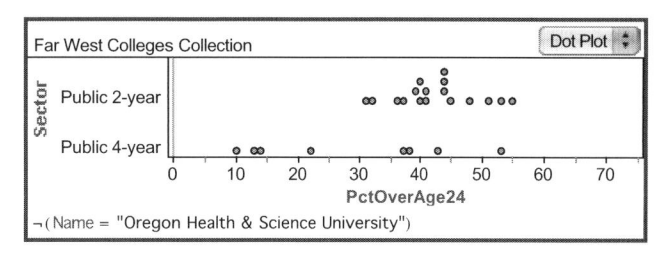

$$\bar{x} = \frac{43+37+13+53+38+22+10+14}{8} = \frac{230}{8} = 28.75.$$

For the four-year schools, on average, 28.75% of the students are over age twenty-four. In symbolic form (so that we can apply this formula to any data set), we write:

$$\bar{x} = \frac{x_1 + x_2 + \cdots + x_n}{n} = \frac{1}{n}\sum_{i=1}^{n} x_i$$

Using "x's" with the subscripts reflects our general way of writing the value for each of the cases in turn. In our example, $x_1 = 43\%$ and is the value for *percentage over age twenty-four* (*PctOverAge24*) for Eastern Oregon University, $x_2 = 37\%$, the value for Oregon Institute of Technology, etc. For this example, the sample size is $n = 8$; that is, there are eight cases in the collection.

The Greek letter $\sum$, pronounced "sigma," is used to show a long sum where we do not want to write out all of the x_i—especially where we are summing many of them. The definition is quite simple:

Num	Name	Pct Over Age24	Two-year or Four-Year
1	Eastern Oregon University	43	Four
2	Oregon Institute of Technology	37	Four
3	Oregon State University	13	Four
4	Oregon State University, Cascades Campu	53	Four
5	Portland State University	38	Four
6	Southern Oregon University	22	Four
7	University of Oregon	10	Four
8	Western Oregon University	14	Four
9	Blue Mountain Community College	40	Two
10	Central Oregon Community College	36	Two
11	Chemeketa Community College	45	Two
12	Clackamas Community College	44	Two
13	Clatsop Community College	51	Two
14	Columbia Gorge Community College	55	Two
15	Klamath Community College	31	Two
16	Lane Community College	44	Two
17	Linn-Benton Community College	37	Two
18	Mt Hood Community College	32	Two
19	Oregon Coast Community College	44	Two
20	Portland Community College	53	Two
21	Rogue Community College	48	Two
22	Southwestern Oregon Community College	41	Two
23	Tillamook Bay Community College	39	Two
24	Treasure Valley Community College	40	Two
25	Umpqua Community College	41	Two

$\sum_{i=1}^{n} x_i = x_1 + x_2 + \cdots + x_n$, where the little numbers at the foot and head of the $\sum$ tell the reader where the sum begins and ends. You may or may not have been introduced to this symbol in your previous mathematics studies, but you need to become familiar with it because we will use this symbol often in statistics. There is an exercise in this section that takes you through how this symbol is used.

Mean value for a distribution of a quantitative variable

$$\bar{x} = \frac{x_1 + x_2 + \cdots + x_n}{n} = \frac{\sum_{i=1}^{n} x_i}{n}$$

You should be able to confirm that the mean for the variable *Percentage over age 24* is $\bar{x} \approx 42.42$;the mean percentage of students who are over age twenty-four for the two-year colleges is higher than the mean percentage of students over age twenty-four for the four-year institutions

Median: the value of the middle case

The definition of the median value for the distribution of a quantitative variable is given in the box.

> **Median value for a distribution of a quantitative variable**
>
> The **median** is the value that divides a distribution so that half the values are less and half the values are more.
>
> To find the median M:
> 1. Rank order the data from lowest to highest.
> 2. If the number of cases n is odd then the median M is the value of the middle number, whose location from the smallest value will be found by the formula $\frac{n+1}{2}$.
> 3. If the number of cases n is even then the median M is the mean of the values of the two middle numbers, and the location from the smallest value will (still) be found by the formula $\frac{n+1}{2}$.

Here is the calculation of the median value for the percentage over age twenty-four for the $n = 17$ Oregon two-year institutions. The first step is to rank order the data. If you make an ordered stem plot of the data, you will be able to put the data in order easily. Then calculate the *location* of the median using $\frac{n+1}{2} = \frac{17+1}{2} = \frac{18}{2} = 9$, so the median M is to be found at the ninth case from the start, since the total number of cases $n = 17$ was an odd number. That case is shown on the stem plot in bold italics. The median value is $M = 41$ percent over age twenty-four. Confirm from the stem plot that this median does indeed divide the distribution into two equal halves; there are eight cases that are less and eight cases that are more.

```
1 |
1 |
2 |
2 |
3 | 12
3 | 679
4 | 0011444
4 | 58
5 | 13
5 | 5
```

There was an even number of four-year Oregon institutions. To find the median, again rank order the data from lowest to highest. A stem plot does this easily. Now calculate the location: $\frac{n+1}{2} = \frac{8+1}{2} = \frac{9}{2} = 4.5$. That the location is 4.5 shows that the median is between the fourth case and the fifth case since 4.5 is halfway between 4 and 5. The median is the mean of the values of these two. These two cases are shown on the stem plot in bold italics. The calculation of the median is thus $M = \frac{22 + 37}{2} = \frac{59}{2} = 29.5$ percent over age twenty-four. Usually, we will let software calculate means and medians, especially where the number of cases is large; it is good to know how the calculations are done.

```
1 | 034
1 |
2 | 2
2 |
3 |
3 | 78
4 | 3
4 |
5 | 3
5 |
```

Here is the Fathom **Summary Table** showing the mean (S1, the first number listed for each type of college), the median (S2, the second number listed for each type of college), and the count (S3, the third number listed for each type of college) of the *Percentage over age 24* distributions. For the public two-year schools, the mean is 42.41%, the median is 41%, and the number of colleges is $n = 17$, whereas for the public four-year schools, the mean is 28.75%, the median is 29.5%, and the number of colleges is $n = 8$. Both the mean and the median tell us

Far West Colleges Collection

		PctOverAge24
Sector	Public 2-year	42.4118 41 17
	Public 4-year	28.75 29.5 8
	Column Summary	38.04 40 25

S1 = mean ( )
S2 = median ( )
S3 = count ( )

¬(Name = "Oregon Health & Science Univer

that the average percentage of students over age twenty-four is higher in the two-year institutions than it is in the four-year institutions.

The Median Is Resistant but the Mean Is Not Resistant

Resistant to what? Answer: the median is a measure that is *resistant* to (or not affected by) to the influence of outliers and skewness, whereas the mean *is* sensitive to (or affected by) skewness and outliers. Here is why: the calculation of the mean involves a sum, whereas the calculation of the median does not. The sum in the calculation of the mean will be "pulled" in the direction of skewness: if a distribution is right-skewed, the sum in the mean will contain some very large numbers; if the distribution is left-skewed, the sum will contain relatively small numbers.

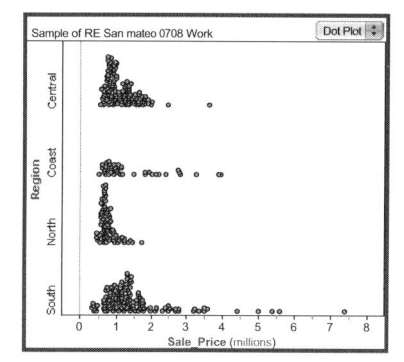

Here is an example showing the implications of the resistant nature of the median and the sensitive nature of the mean. The dot plots (above) show the sale prices of a sample of houses sold in San Mateo County in 2007–2008, categorized by regions within the county. All of the distributions are right-skewed but the Coast and the South especially so. Our question: on average, did houses sell for more in the Central region or the Coast region? It would be hard to make a judgment based on the dot plots; we need some numbers.

Here are the means and medians for the sale prices of houses in the four regions. Notice how if we compare the *mean* sale prices, we would get the impression that houses sold for more on the Coast compared with the Central region. However, that mean is being "pulled" by the skewness of the Coast distribution. A more appropriate measure for the location of the distributions is the *median* sale price, and that measure shows that the average prices are about the same.

The Shape of a Distribution, the Mean, the Median, and the Balance Point

Since the median is *resistant* to skewness and the mean is not *resistant*, we can work backwards and detect the shape of a distribution by looking at the mean and the median of a distribution.

Shape of a Distribution and Measures of Center

When a distribution is *right-skewed* then the mean > median.

When a distribution is *left-skewed* then the mean < median.

If a distribution is exactly *symmetrical* then the mean = median.

When working with actual data, we will very seldom encounter the situation where the mean and the median are *exactly* equal. However, if the mean and the median are close in value then we can say that the distribution is nearly symmetrical.

Compare the means and the medians of the sale prices of the houses in the San Mateo real estate sample shown above in the summary table and you will see that the distribution that appears to be least *right-skewed* (the one for the North region) is also the one that that has the smallest difference between the mean and the median.

Here is the plot of the sale prices for the South region showing the mean and the median.

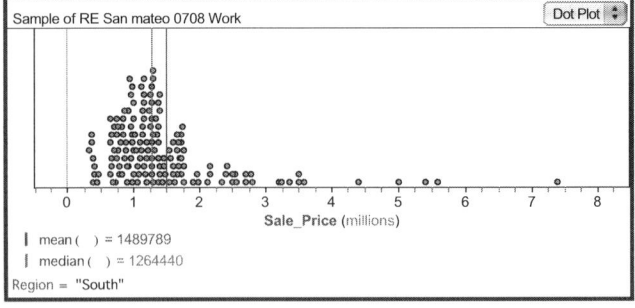

The median is the value that divides the distribution into two equal parts, so that there is the same number of cases above as below the median. In our probability notation $P(X < Median) = P(X > Median) = 0.50$, the probability that we find a case greater than the median is equal to the probability that we find a case less than the median. But what about the mean? What the mean shows for any distribution is the "center of mass" or (in other words) the balance point. If you had a huge heavy right-skewed distribution (like our friend Hiroyuki here), you would be able to carry it by balancing the distribution on your shoulder at the mean. The tail on the right side would just balance the clump on the left. Of course, if the distribution is symmetric then the balance point is at the center, since the tail on the left equals the tail on the right. For skewed distributions, the tails determine where the balance point is.

Extending the Summary: The Five-Number Summary

The median defines the location of a distribution by showing the value that divides the distribution into two equal halves. The median for the percent over age twenty-four for the two-year colleges is M = 41 and for the four-year schools it is M = 29.5. We can extend this idea by looking at the value that defines the point in the distribution so that 25% of the distribution is less than that point, and hence 75% is more than that point. We can call this **Quartile 1** or **Q1** because it is the point that defines the first quarter of the data. In a similar way, we define **Quartile 3** or **Q3** as the point where 75% of the distribution is less than that point and 25% of the distribution is greater than that point. What we are doing is breaking the data into four equal parts.

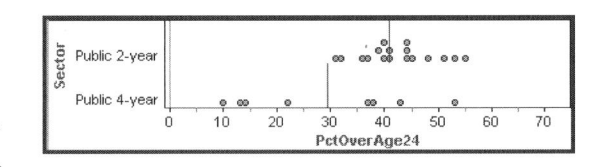

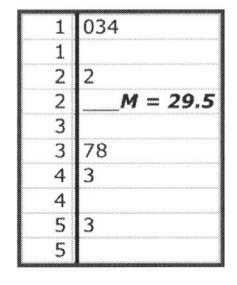

The way we will do this when we are working with small collections of data by hand is to recognize that one quarter is just one half of one half; we will work with the two halves of the data as divided by the median. We will find the median (one half) of the lower half of the data to get the **Q1**. To get the **Q3** we will get the median of the upper half of the data. Here is an example using the Oregon colleges and universities data.

To get the **Q1** for *Percentage over age 24* for the two-year colleges, we apply the rules for finding the median in the box above to the eight cases that are below the median, leaving out the median itself, M = 41. Since there are n = 8, we calculate the location $\frac{n+1}{2} = \frac{8+1}{2} = 4.5$ and this tells us that the

66

$Q_1 = \frac{37+39}{2} = \frac{76}{2} = 38$. For **Q3** we make a similar calculation for the eight cases *above* the median of $M = 41$. The location calculation again gives 4.5, which puts us between 45 and 48, so we have: $Q_3 = \frac{45+48}{2} = \frac{93}{2} = 46.5$. The calculation for the quartiles for the variable *Percentage over age 24* for the four-year schools follows the same pattern; we get the median of the lower half of the cases to get the **Q₁** and then get the median of the upper half of the data to get the **Q₃**.

> **Definition of Quartiles**
> **Q1** is the value that divides a distribution so that 25% of the values are less, and
> **Q3** is the value that divides a distribution so that 75% of the values are less.
> Hand calculation of **Quartiles**: find the median of the lower or upper half of the data.

Warning: Calculators and software (including Fathom) may use a more complicated formula than this hand formula, so your hand calculations may not match your software calculations; however, the differences will be extremely small. (Indeed, for our example, the Fathom results differ from what we have calculated.)

When we put together the minimum, the Q_1, the median, the Q_3, and the maximum for a distribution, we have what is called the **five-number summary.** The five-number summary for the distribution of the variable *Percentage over age 24* for the two-year colleges is: Min = 31, $Q_1 = 38$, $M = 41$, $Q_3 = 46.5$, Max = 55. For the four-year colleges and universities, the five-number summary is Min = 10, $Q_1 = 13.5$, $M = 29.5$, $Q_3 = 40.5$, Max = 53.

> **Five-Number Summary for the Distribution of a Quantitative Variable**
> Minimum, Q_1, Median, Q_3, Maximum

Had you reflected on the matter, you might have concluded that using just one number—the mean or the median—to characterize an entire distribution can be misleading. Using just one number may be "over-simplifying"; we may be able to have a better summary using five numbers.

A Graphic Based upon the Five-Number Summary: the Box Plot

One thing that we can do with these five numbers is to make a "schematic" graphic to show a distribution. That graphic is called a **box plot** or sometimes a **box and whisker plot.** Here are details.

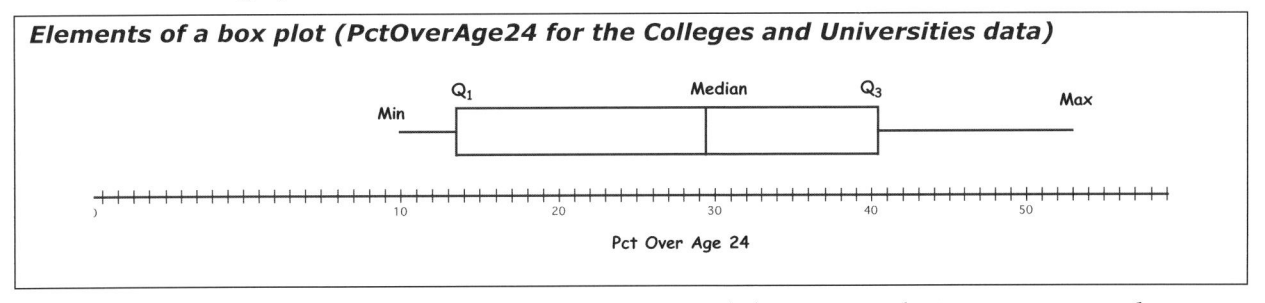

> **Elements of a box plot (PctOverAge24 for the Colleges and Universities data)**
>
> [Box plot showing Min, Q₁, Median, Q₃, Max along an axis labeled "Pct Over Age 24" ranging from 10 to 50]

The middle part of the box plot is called (quite naturally) the box, and the lines connecting the quartiles to the extremes are sometimes called "whiskers" (think cats). Box plots are especially useful for comparing two or more distributions, since you can see quickly the locations of the distributions on the variable and also get an idea of how much spread or variability there is in each of the two (or more) distributions. Here are box plots comparing the *Percentage over age 24* distributions for the two-year and four-year colleges.

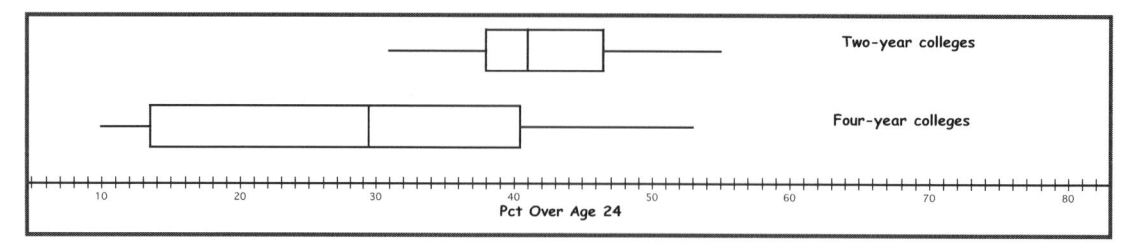

We can see what we have already concluded—that, on average, the percentage of students over age twenty-four is higher in two-year colleges than it is in four-year colleges. We can also see that there seems to be greater variability among the four-year institutions. Variability is very important topic and is where we are going next.

Summary: Numerical Measures of Center

The idea is to summarize the location of distributions by a just a few numbers. The measures that we use are the *mean (or arithmetic average)*, the *median*, and the *five-number summary*. The five-number summary also has the advantage of showing other features of distributions and can be turned into a graphic.

- **Mean** The "balance point" or center of mass of a distribution, calculated by:

$$\bar{x} = \frac{x_1 + x_2 + \cdots + x_n}{n} = \frac{\sum_{i=1}^{n} x_i}{n}$$

- **Median** The value that divides a distribution so that half the values are less and half the values are more. To find the median M:
    - Rank order the data from lowest to highest.
    - If the number of cases n is odd then the median M is the value of the middle number, whose location from the smallest value will be found by the formula $\frac{n+1}{2}$.
    - If the number of cases n is even then the median M is the mean of the values of the two middle numbers, and the location from the smallest value will (still) be found by the formula $\frac{n+1}{2}$.

- **Resistant measures** are measures *not* influenced by skewness or outliers in a distribution.
    - The *median* is a *resistant* measure of center and not influenced by skewness or outliers.
    - The *mean* is *not resistant* to skewness and outliers and it works most accurately for symmetric distributions.

- **Shape of a Distribution and Measures of Center**
    - When a distribution is *right-skewed* then the mean > median.
    - When a distribution is *left-skewed* then the mean < median.
    - If a distribution is exactly *symmetrical* then the mean = median.

- **Quartiles**
    - Q_1 is the value that divides the distribution so that 25% of the values are less.
    - Q_3 is the value that divides the distribution so that 75% of the values are less.

- **Five-Number Summary for the Distribution of a Quantitative Variable**

    Minimum, Q_1,   Median, Q_3,   Maximum,   where

- **Box Plot** is a graphic of a made from the five-number summary, as shown here:

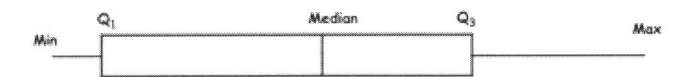

§1.5 Exercises on Measures of Center/Location

1. According to the **Notes**, there is a relationship between skewness and whether the mean or the median is larger for a distribution. Express in words what this relationship is.

2. **Mammals** These are data from the collection **Mammals2.ftm**. The data come from a study that related various characteristics of mammals to the number of hours the mammals typically sleep. The authors rated each mammal as to how likely the mammal was to be the prey of other animals. Here are the life spans in years of life (it appears to be maximum life span) of the thirteen mammals most likely to be preyed upon and the fourteen mammals least likely to preyed upon. So, $n = 13$ for the "most-preyed upon mammals" and $n = 14$ for the "least-preyed upon mammals." The numbers for life span are rounded to nearest whole year of life, although in the Fathom file the numbers are given to tenths of a year.

    Mammals Most Likely to be Prey

Number	Species	Life Span
1	Chinchilla	7
2	Cow	30
3	Donkey	40
4	Giraffe	28
5	Goat	20
6	Ground squirrel	9
7	Guinea pig	8
8	Horse	46
9	Lesser short-tailed shrew	3
10	Okapi	24
11	Rabbit	18
12	Roe deer	17
13	Sheep	20

    Mammals Least Likely to be Prey

Number	Species	Life Span
1	Arctic fox	14
2	Big brown bat	19
3	Cat	28
4	Chimpanzee	50
5	Eastern American mole	4
6	Genet	34
7	Giant armadillo	7
8	Gorilla	39
9	Gray seal	41
10	Gray wolf	16
11	Human	100
12	Jaguar	22
13	Little brown bat	24
14	Red fox	10

    a. Do you think that on average the least-preyed upon mammals should have a longer life span, shorter life span, or pretty much an equal life span to the most-preyed upon mammals?

    b. Find the mean life span of the most likely to be prey and the mean life span of the least likely to be prey. Show your calculations in an organized way so that someone reading your answer will know what you have done. Do the results fit your expectations as expressed in part a?

    c. Make ordered stemplots for the life spans of the most likely to be prey and the life spans of the least likely to be prey and from these calculate the medians.

    d. For the least likely preyed upon, the human appears to be an outlier. Now, suppose humans lived to be only fifty-two years old rather than one hundred. If that were so, would the *mean* life span change for the distribution? Would the *median* life span change? Explain your answer. You can base your explanation on calculations, but you can also give an explanation based upon what is said in the **Notes**.

    e. Find the five-number summary for the life spans for the most-preyed upon mammals.

    f. Find the five-number summary for the life spans for the least-preyed upon mammals.

    g. Use your results from parts e and f to make a pair of box plots to compare life spans of the most- and least-preyed. (Your graphic should resemble the example in the **Notes** that compares the PctOverAge24 in the two-year and four-year institutions.)

    h. In the collection of data, "Predation Level" runs from Level 1 (least-preyed upon) to Level 5 (most-preyed upon). On the next page are box plots for three levels of predation—the two you have been working on, in addition to all the levels between. The dots indicate *outliers*; your box plot for the least-preyed upon should have the whisker extending to the human, with a life span of one hundred years. In the next section, you will learn how such box plots are made.

Here is the question: do the box plots show evidence that life span in mammals is affected by their predation level, or not?

3. **Gestation** "Gestation" simply means the typical length of pregnancy for a mammal. Here the length of time is expressed in days. The histogram shown shows the distribution of *Gestation* for the mammals in our data set.

- Open the Fathom file **Mammals 2.ftm** and get a **Case Table.**
- Get a **Graph** from the shelf; drag the variable *Gestation* to the horizontal axis.
- Use the scroll on the **Graph** to change the dot plot to a histogram.
- With the graph selected, go to the menu to **Graph> Scale** and change to "Relative Frequency."

You should have the graph shown.

a. What are the cases for these data?

b. *[Review]* What is the width of the bars (the *bin width*) of this histogram?

c. *[Review]* Use the histogram to estimate $P(X \geq 250 \text{ days})$, where X stands for the length of pregnancy.

d. *[Review]* Use the histogram to estimate the proportion of mammals that have lengths of pregnancies (gestation) less than one hundred days. Use the correct probability notation.

e. *[Review]* Either make a sketch of the histogram from Fathom or make a copy of the histogram and shade in and label the answers to parts c and d with the probability notation pointing to the shading.

f. For this distribution of gestation, which measure of center will be bigger, the mean or the median? Give a reason for your answer. The Fathom instructions below will get the mean and the median.

- Get a **Summary Table** from the shelf; drag the variable *Gestation* to the right-pointing arrow. The number that appears is the mean.
- Select the **Summary Table** so that it has a blue border around it and go to the menu **Summary>Add Formula**.
- In the dialogue box next to *S2*, type *median( )*. Press **OK**. (The dialogue box should look like the one shown here.)

g. Confirm that your answer to f is correct. Write the mean and the median using the notation that is used in the **Notes.**

- Select the **Summary Table** so that it has a blue border around it and go to the menu **Summary>Five Number Summary.**

h. Use the five-number summary to make a box plot by hand. Make certain that your scale is correct.

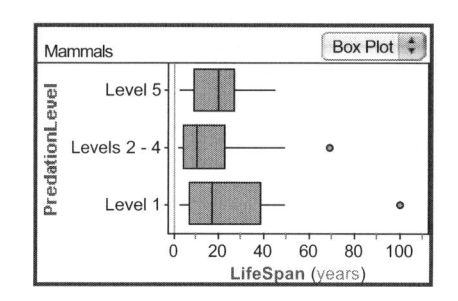

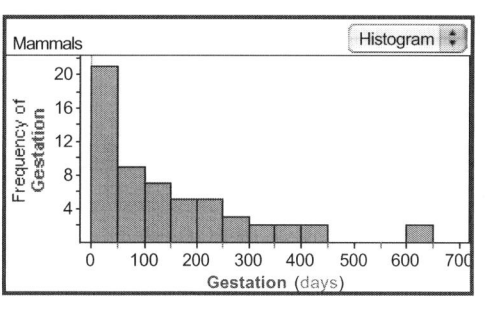

- Go back to histogram and scroll down the choices of graphics in the right-hand corner to make a box plot. The box plot that Fathom makes will show two *outliers*, and in the box plot you have made, the whisker should reach as far as the outliers. In a later section, you will see how to determine (mathematically) which cases should be classified as outliers and which not.

4. **A Wrong Idea** Is it true that the more cases there are in a collection, the bigger the mean or the median will be? *No!* It is not true. The mean and median do not get bigger because the number of cases n (the count) is bigger. Notice that in the formula for the mean

$$\bar{x} = \frac{\sum_{i=1}^{n} x_i}{n}$$ we divide by the number of cases (or count) n.

Here are more data on mammals. There are two collections of species of mammals. One collection includes mammals that are *more exposed* to danger from other animals (because they are bigger?) and there are $n = 9$ of these mammals. The second collection consists of mammals that are *less exposed* to danger from other animals. There are $n = 12$ of these mammals for which we have data. For both collections, we have the gestation period (length of pregnancy) in days recorded.

Species More Exposed to danger from Other Animals	Gestation (days)
Cow	281
Donkey	365
Giraffe	400
Goat	148
Horse	336
Okapi	440
Rabbit	31
Roe deer	150
Sheep	151

Species Less Exposed to Danger from Other Animals	Gestation (days)
Desert hedge hog	
Echidna	28
European hedge hog	42
Galago	120
Golden hamster	16
Owl monkey	140
Phanlanger	17
Raccoon	63
Rhesus monkey	164
Rock hyrax	225
Slow loris	90
Star nosed mole	
Tenrec	60
Tree shrew	46

a. Make (by hand or using Fathom) comparative dot plots for the variable *Gestation*.

b. For each of the two collections, calculate (by hand) the mean and median for the variable *Gestation*.

c. Is it true or false that the collection with the bigger count (the number of cases) has the bigger mean? Explain your answer using your calculations.

5. **Danger, Length of Pregnancy, and How to Read Fathom Summary Tables.** How to read first: Shown are means, the five-number summary, and counts for mammals at three levels of being exposed to (everyday?) danger.

Starting from the top number:

S1 = mean Gestation in days
S2 = min (minimum) Gestation
S3 = Q1, the first quartile of the Gestation data
S4 = Median
S5 = Q3, the third quartile
S6 = max (maximum) Gestation
S7 = count = number of cases in each of the categories

Mammals	Danger_Level			Row Summary
	High Danger	Intermediate Danger	Low Danger	
Gestation	222.5 d	104.136 d	102.235 d	142.353 d
	21.5 d	16 d	12 d	12 d
	112 d	28 d	38 d	35 d
	170 d	45.5 d	60 d	79 d
	365 d	140 d	120 d	210 d
	624 d	645 d	310 d	645 d
	19	22	17	58

S1 = mean( )
S2 = min( )
S3 = Q1( )
S4 = median( )
S5 = Q3( )
S6 = max( )
S7 = count( )

Reading down, the high danger mammals have a median gestation of 170 days, since S4 = median.

a. Looking at the means and the medians for the three danger levels, can you say that mammals in greater danger tend to have a longer or shorter length of pregnancy? Defend your answer with numbers.

b. Think about the definition of quartiles. For the low danger mammals, what number should replace the question marks in the probability statement $P(X \geq ???) = 0.25$ if X stands for gestation (days of length of pregnancy)? Say briefly why you chose the number that you chose.

6. **The Fellowship of the Ring** In J. R. R. Tolkien's classic, the heroes of the story are the hobbits. Soon after starting their journey from the Shire the hobbits came to a place named Bree. This town was in an area where both Men and Hobbits lived.

> The Big Folk and the Little Folk (as they called one another) were on friendly terms, minding their own affairs in their own ways, but rightly regarding themselves as necessary parts of the Bree-folk. Nowhere else in the world was this peculiar but excellent arrangement to be found.
>
> (J. R. R. Tolkien, *The Fellowship of the Ring*, Ballantine Books Edition, p. 206)

We have managed to get a sample of the heights (measured in cm) of Bree-folk who happened to be in *The Prancing Pony*, the inn in Bree. There are $n = 26$ Bree-folk represented, where $n_{Little} = 12$ and $n_{Big} = 14$. Here are the data.

Height (cm)
108
107
109
102
164
107
165
170
106
172
100
115
106
179
172
181
166
181
111
167

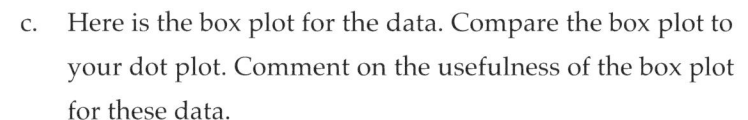

a. Make a dot plot of the data and comment on what you see.
b. Calculate the mean and the median for all of the data. Do the mean and the median make any sense in the context of these data? Explain. 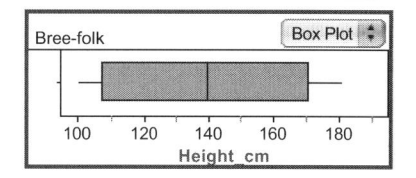
c. Here is the box plot for the data. Compare the box plot to your dot plot. Comment on the usefulness of the box plot for these data.
d. Calculate means and medians that *do* make sense to the Bree folk and explain their sense.

7. **Real Estate Data: Size of Houses** This exercise concerns a sample of data on houses that were sold in San Mateo County in the years 2007 to 2008. The collection you will use does not include *all* the houses that were sold; it is just a *sample* of the entire *population* of all houses that were sold. You will look at the variable *SqFt*. This variable records the amount of square feet in the living area of the house, so it is a measure of the size of the house. Here are some Fathom instructions and then some questions. You may have seen some of these already.

- Open the Fathom file **San Mateo RE Sample Y0708.ftm** and get a **Case Table**.
- From the shelf, drag down a **Graph** and drag the variable *SqFt* to the horizontal axis and the variable *Region* to the vertical axis. You should have four dot plots, one for each region.

a. What are the cases for this collection?
b. What is the most appropriate name for the *shape* of the distributions?
c. Which measure of center/location should give a more accurate notion of the location of the distributions, the mean or the median? Give a reason for your answer.
d. Judging from the plots alone, which region would you guess has the largest houses on average?

- From the shelf, drag down a **Summary Table;** drag the variable *SqFt* to the down-pointing arrow and the variable *Region* to the right-pointing arrow.

- With the **Summary Table** selected, go to the menu (at the top of the screen) and select **Summary>Add Five Number Summary**.

  e. Use the mean and the median to rank order the four regions in average size of house.

  f. *Looking carefully for a mystery:* Look carefully at the five-number summary; is there a figure there that either does not make sense or is highly unlikely if it is houses and not land that is being sold? Explain your findings. Think about what the numbers mean. (The "mystery" is that there is one house with zero square feet; they must have been selling the land without a house on it.)

- To fix the "mystery," select the summary table; go to the menu under **Object>Add Filter.** In the dialogue box, type in *SqFt > 0.* This will look at the data without the problem. Shown here, is the summary table that you should get.

  g. A student (your friend) persists in thinking that a higher number of cases will lead to a higher mean value. Write a short message to your friend using this summary table to show your friend that it is *not* so that a higher count leads to a higher mean or median.

Sample of RE San mateo 0708 Work					
	Region				Row Summary
	Central	Coast	North	South	
SqFt	1832.55	2427.64	1477.59	2003.03	1890.32
	128	42	78	151	399
	800	920	610	660	610
	1260	1430	1140	1350	1230
	1675	2210	1390	1760	1680
	2220	3270	1710	2320	2250
	4200	4890	3130	6400	6400

S1 = mean ( )
S2 = count ( )
S3 = min ( )
S4 = Q1 ( )
S5 = median ( )
S6 = Q3 ( )
S7 = max ( )
SqFt > 0

8. **Great Lakes Colleges and Universities I**   The same variables are measured as in the Far West collection. We will look at the variable *Freshman Retention Rate*, which is the percentage (it is given as a percentage) of first-year students who proceeded to the second year of study. Our statistical question will be how the *Sector* of institution (public two-year, public four-year, or private four-year) is related to *Freshman Retention Rate.*

- Open the Fathom file **GreatLakesCollegesUniversities.ftm** and get a case table.
- From the shelf, drag a **graph** and drag the variable *Freshman Retention Rate* to the horizontal axis and the variable *Sector* to the vertical axis. You should have a dot plot.
- Change the dot plot to a box plot by using the scroll in the upper right hand corner.
- From the shelf, drag a **Summary Table** and drag the variable *Freshman Retention Rate* to the downward-pointing arrow.
- With the **Summary Table** selected, go to the menu and get **Summary>Add Formula**. In the dialogue box that comes up, type in: median ( ) with nothing in the parentheses.
- Now drag the variable *Sector* to the right-pointing arrow in the **Summary Table.**

  a. What are the cases for these data? (That is, what is it that we are talking about?)

  b. Interpret what the means and the medians as well as the appearance of the box plots tell you about the differences in the average *Freshman Retention Rate* in the different types of institutions. Your interpretation should use the numbers but must say something about the type of institution and what the variable *Freshman Retention Rate* measures.

  c. What would be the most appropriate word for the shapes of the three *Freshman Retention Rate* distributions: right-skewed, left-skewed, or symmetrical?

  d. Explain how the relative sizes of the means and medians support your conclusion in part c.

Special Exercise A: Mammals, Means, Medians, and Resistance

The **Notes** have introduced two measures of the location of a distribution—the mean and the median. The median is *resistant* to skewness and outliers but the median is not. Here we investigate and show that the median is not affected by (that is *is resistant to*) outliers, but the mean *is* affected by outliers (is *not resistant to* outliers.)

- Use the Fathom file **Mammals 2.ftm**
- From the shelf, drag a **Graph** and then drag the variable *Bodyweight* to the horizontal axis. You should see this dot plot whose scale is in thousands of kilograms.
- Place the cursor on the most extreme outlier (it will turn red), and the name of the mammal and body weight (in kilograms) will appear at the foot of the screen.

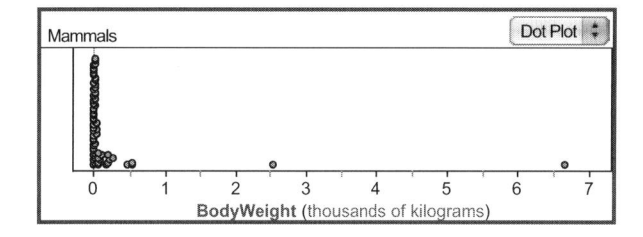

1. Record the names and the body weights of the two outliers in this collection. If you are not accustomed to metric measures, know that the conversion factor is 2.2 lbs/kg.

- From the shelf, drag a **Summary Table** (between **Graph** and **Estimate**) and then drag the variable *Bodyweight* to the right-pointing arrow.
- With the **Summary Table** selected, go to the menu at the top of the screen and get **Summary>Add Formula.** In the dialogue box that appears, type in median( ). Press **OK**.
- Repeat the instruction just above to get the count; you should type count( ) and press **OK.** You should see the summary table shown here.

Notice the huge difference between the values of the mean and the median. Fathom allows us to play with the data to see the effect of outliers and skewness in distributions on the mean.

- 2. We have a "body weight shrinking machine" (the ultimate slimming program) that we will apply to the body weight of the African elephant. In the dot plot, place the cursor on the African elephant and drag the dot so that the African elephant loses weight until he is just a bit heavier than the Asian elephant (the next outlier).

    a. Record the body weight of the mammals now that the African elephant has shrunk down. Compare it with the mean when he was heavy. (It is on this sheet.)

    b. Has the slimming program affected the median? Why is this?

    c. Continue to shrink the body weights of both elephants so that they are both under 1000 kilograms. Record the mean body weight and the median body weight. Has the median changed?

- To revert to the original data, either do **Edit>Undo Drag Point** many times or select the **Collection icon** and in **File>Revert Collection** and the two elephants will revert to their own weights.
- With the shift key down, select the two elephants and go to the menu **Edit>Cut Cases**. Now we have $n = 60$ mammals (instead of $n = 62$) in our collection.

3. a. Describe what has happened to the mean and median.

    b. Is the distribution of body weights without the elephants still right-skewed? Compare the mean and median.

Special Exercise S: Summation or Sigma Notation

What sigma: $\sum$ means:

- $\sum_{i=1}^{6} x_i$ means the same as $x_1 + x_2 + x_3 + x_4 + x_5 + x_6$. That is, $\sum_{i=1}^{6} x_i = x_1 + x_2 + x_3 + x_4 + x_5 + x_6$

- In English, $\sum_{i=1}^{6} x_i$ means: add the x's, starting with the first one x_1 and ending with the last one x_6.

- The x_i in the formula is just a "general term" for the values of the variable we are using, showing that the "i" can be 1, 2, 3, etc., up to the last one.

Example: Some very young children are just learning about geometry and measurement. Each of the children is given a small stick of a different length. Here are the measurements and the children's names. We are interested in the mean length of the sticks. The mean length of

Num	Name	Length of Stick (cm)
1	Adam	4
2	Bashir	7
3	Cindy	6
4	David	3
5	Eben	2
6	Fiona	8

the sticks can be written as $\bar{x} = \dfrac{\sum_{i=1}^{n} x_i}{n}$. In our example, since $n = 6$, this can be expanded as:

$$\bar{x} = \frac{\sum_{i=1}^{n} x_i}{n} = \frac{\sum_{i=1}^{6} x_i}{6} = \frac{x_1 + x_2 + \cdots + x_6}{6} = \frac{4+7+6+3+2+8}{6} = \frac{30}{6} = 5 \ cm$$

Facts about sigma for those seeing it for the first time:

- Sigma notation $\sum_{i=1}^{n} x_i$ does not mean $\sum$ multiplied by x.

- $\sum$ is a symbol that commands: "Add the x's."

Questions to answer:

1. Two more children are added to the group, and all the children are given new sticks. For this new set of data, find the mean length of the sticks and express your work showing both the sigma notation correctly and also the expansion of the notation as shown above.

Num	Name	Length of Stick (cm)
1	Adam	6
2	Bashir	8
3	Cindy	5
4	David	7
5	Eben	7
6	Fiona	8
7	Gina	9
8	Huw	6

2. A student writes the answer to question 1 as follows.

$$\bar{x} = \frac{\sum_{i=1}^{n} x_i}{n} = \frac{\sum_{i=1}^{8} x_i}{8} = \frac{\sum 6+8+5+7+7+8+9+6}{8} = \frac{56}{8} = 7 \; cm$$

However, there is one thing wrong with the answer, beside which the instructor writes, "Not needed." Determine what the error is and write for your own use (and to answer this question) what is wrong.

Order of Operations

In elementary algebra, you reviewed that the order of operations for arithmetic is: "Parentheses (or grouping symbols), Exponentiation, Multiplication/Division, Addition/Subtraction." Addition and subtraction come last. When sigma notation is used, the addition implied by the notation is often the last thing to be done. Here is an example; one of the formulas in the next section is called the *variance*,

$$s^2 = \frac{\sum_{i=1}^{n}(x_i - \bar{x})^2}{n-1}.$$ (We will study the meaning of this in detail in §1.6.)

Expanded, everything to the right of the sigma $\sum$ is done for each of the cases then the results are added. The subtraction inside the parentheses and squaring come first. So, for our example, since $\bar{x} = 5$ the expansion is:

$$s^2 = \frac{\sum_{i=1}^{n}(x_i - \bar{x})^2}{n-1}$$

$$= \frac{(x_1 - \bar{x})^2 + (x_2 - \bar{x})^2 + (x_3 - \bar{x})^2 + \cdots + (x_n - \bar{x})^2}{n-1}$$

$$= \frac{(4-5)^2 + (7-5)^2 + (6-5)^2 + (3-5)^2 + (2-5)^2 + (8-5)^2}{6-1}$$

$$= \frac{(-1)^2 + (2)^2 + (1)^2 + (-2)^2 + (-3)^2 + (3)^2}{5}$$

$$= \frac{1+4+1+4+9+9}{5} = \frac{28}{5} = 5.6$$

3. For the sticks for eight children where $\bar{x} = 7 \; cm$ calculate s^2 using

$$s^2 = \frac{\sum_{i=1}^{n}(x_i - \bar{x})^2}{n-1}$$ showing the expansion as above.

(The new data have been shown here once again.)

Num	Name	Length of Stick (cm)
1	Adam	6
2	Bashir	8
3	Cindy	5
4	David	7
5	Eben	7
6	Fiona	8
7	Gina	9
8	Huw	6

4. The **standard deviation** s is the square root of the variance, so $s = \sqrt{s^2} = \sqrt{\frac{\sum_{i=1}^{n}(x_i - \bar{x})^2}{n-1}}$. We will often want this rather than the variance. Notice here that since the square root applies to the entire sum, in the order of operations it comes last (everything inside the square root must be worked out first). For our example $s = \sqrt{5.6} \approx 2.37$, calculate s from s^2 for the eight children data.

5. Here are data for the number of bedrooms for n = 10 houses sold in 2007 to 2008 in one town in San Mateo County.

Address	Beds	Baths
384 ELM AV	2	2
2781 OAKMONT DR	4	2
2170 FLEETWOOD DR	3	2
35 TANFORAN AV	2	2
2990 SAINT CLOUD DR	3	2
2421 FLEETWOOD DR	4	2
546 5TH AV	3	2
2421 FLEETWOOD DR	4	2
1625 JUNIPER AV	3	2
2400 LEXINGTON WY	3	2

   a. Express the formula for the mean number of bedrooms for these houses using sigma notation but inserting the correct number for the size of the collection and then expand the formula (as you did above in the sticks problem) and work out $\bar{x}$. (The numerical answer is 3.1.)

   b. Express the formula for s using sigma notation; insert the correct number for the size of the collection and the mean, and then expand the formula to calculate s. (The numerical answer is 0.738.)

   c. Without really calculating, but by looking at the formulas and data, you should be able to state the values of $\bar{x}$ and s for the number of bathrooms for these data. Explain how you determined your answers.

6. **Common Student Errors.** Here are two errors that are made using the formula for the variance; the errors are illustrated using the number of bedrooms data. For each of the errors, put into words what the error is. (The answer has to be more specific than "the student did not use the formula correctly.")

   a. To calculate the variance using the formula $s^2 = \dfrac{\sum_{i=1}^{n}(x_i - \bar{x})^2}{n-1}$ a student writes what is written just below; what mistake is the student making about the formula?

   $$s^2 = \frac{(2-7)^2}{10-1} = \frac{(-5)^2}{9} = \frac{25}{9} \approx 2.78$$

   b. The mistake in the calculation below may be more difficult to spot, but it has to do with order of operations. What is the mistake?

   $$s^2 = \frac{\sum_{i=1}^{10}(x_i - 7)^2}{10-1}$$
   $$= \frac{(2+4+3+2+3+4+3+4+3+3)^2 - (7)^2}{10-1}$$
   $$= \frac{(31)^2 - (7)^2}{9}$$
   $$= \frac{961 - 49}{9}$$
   $$= \frac{912}{9}$$
   $$= 101.3\bar{3}$$

§1.6 Variation: Measurement and Interpretation

Variability (or Spread) and its challenges

Look again at the distributions of tuition fees for the three types of colleges and universities. It is clear that on average the tuition fees at the private colleges are higher than the tuition fees at the four-year public institutions, and the four-year public schools fees are on average somewhat higher than the tuition fees at community colleges. However, there is another thing to notice about the tuition fees at the private colleges. The tuition fees are much more spread or variable or diverse among the private colleges than they are for the public institutions. In other words (in extremely clumsy and poor English), the tuition fees are "more different" among the colleges.

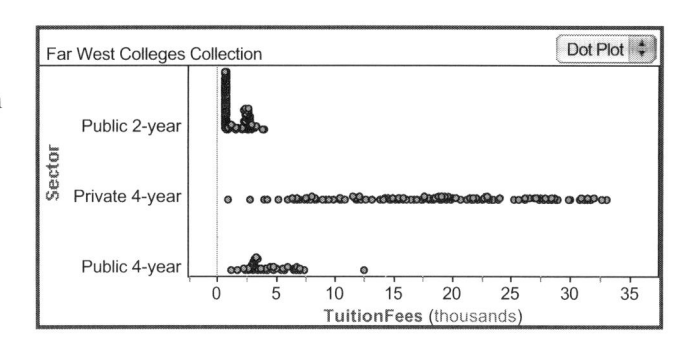

Try Exercise 1: Here is the question asked in Exercise 1:

"Imagine that you are explaining to a friend (perhaps by e-mail) what the graphic [the one above] tells you about tuition fees in public and private colleges. You can attach the graphic to your e-mail. You need to tell your friend what the dot plot shows about the center (or location) of the distributions but *also* about the differences in the *spreads* of the distributions between the private and the public institutions."

We will get to how you might have answered the question in Exercise 1 shortly. Let us say that there are good reasons for the greater variability in fees for the private schools. The state schools most often have their fees set or regulated by government in one way or another. The left "spike" of tuition fees you see in the two-year college distribution are the fees charged by California community colleges, and these fees are almost the same for every two-year college in the entire state. Private institutions, on the other hand, are more variable in many ways, including their sources of funding; private colleges range all the way from small church-related schools with just a few hundred students to major research universities such as Stanford.

So how did you explain to your friend what the graphic shows you? Was it like this?

"Fred [your friend's name], it is really hard to give you a figure for typical cost of a private college because the costs are all over the place, although on average they cost more than the state schools. The range of costs is enormous! There are some colleges that charge as little as $2500 per year for tuition, and some that cost over $30,000 a year—and everything in between. For state schools, the range is much smaller; you can expect to pay between about $1000 to $8000 per year for a state four-year school in the West."

The challenge of explaining variability. We think that you should have found the questions in Exercise 1 challenging and difficult. People are not as comfortable describing *variability* as they are describing the *location* of distributions. To say "on average" comes fairly naturally; talking about variability does not. So here is a warning. Of the things to learn in this section—terminology, calculation, and interpretation—it will be interpretation, or "explaining what the numbers mean" in the context of the data that you will find the hardest. However, we will try to give some guidance; as starters, know that you will find words such as "variability" and "diverse" and "different" to be useful.

A good *start* to interpretation is to use the idea of a range, as in: "the range of fees for the state schools is between about $1000 and $8000 for tuition." That is: the range is $7,000 = $8,000 – $1,000 for the state schools, but for the private colleges, the range is $27,500 = $30,000 – $2,500. The idea of range is a good start; we will go on to use more accurate measures of variability.

Measures of Spread: Range and Inter-Quartile Range

We will measure the spread or variation in a distribution in three different ways.

The simplest way to measure spread is to use the range, which has been introduced in the example.

Definition of Range

$$Range = Maximum\ value - Minimum\ value$$

Here is a **Summary Table** showing the means and the five-number summaries for the *TuitionFees* for the three types of colleges and universities. From the definition of range as *Maximum value – Minimum value* you should easily be able to get the ranges from the five number summaries for the three categories of colleges and confirm that the private colleges have the greatest range, at $32,127.

Far West Colleges Collection

	Sector			Row Summary
	Private 4-year	Public 2-year	Public 4-year	
TuitionFees	18501.6	1340.67	4285.26	8225.21
	885	624	1214	624
	11820	672	3030	799
	19083	788	3656	3035
	25244	2383	5496	14775
	33012	3882	12506	33012
	139	172	62	373

S1 = mean( )
S2 = min( )
S3 = Q1( )
S4 = median( )
S5 = Q3( )
S6 = max( )
S7 = count( )

Range, however, has a major disadvantage: range is not *resistant*; outliers affect its calculation. Our casual explanation above from the dot plot said that the range or the public four-year colleges was about seven thousand dollars, and yet for this example our definition of the range gives:

$$Range = \$12,506 - \$1,214 = \$11,292$$

There is one state school whose fees are considerably greater, and this makes the range as a measure of spread greater than what we would casually see from the dot plot. Range is a bad and misleading measure of variability in this instance.

Warning about range

Range is a bad measure of variability because it is not resistant to outliers.

A second (and better) measure of spread that avoids the outlier problem also uses the five-number summary. It is called the **Inter-Quartile Range** or **IQR.** It is nearly as easy to interpret as range.

Definition of Inter-Quartile Range (IQR)

$$IQR = Q_3 - Q_1$$

Interpretation of the IQR

The IQR is the Range of the Middle 50% of a Distribution.

As you can confirm from the five-number summary, the **IQR** for the four-year state schools is $IQR = 5496 - 3030 = 2466$, for the two-year public schools it is $IQR = 2383 - 672 = 1711$, and for the private colleges it is $IQR = 25244 - 11820 = 13424$. What does this give us?

The meaning of the IQR can be seen better by looking at box plots. The box in a box plot incorporates the middle 50% of the distribution, since 25% of the distribution is to the left of Q_1 and another 25% of the distribution is to the right of Q_3. In our example of the tuition fees, notice how clearly the bigger spread of the private colleges is seen in the graphic: the box for the private four-year colleges (showing its IQR) for tuition fees is much wider than the IQRs for the other two types of institutions.

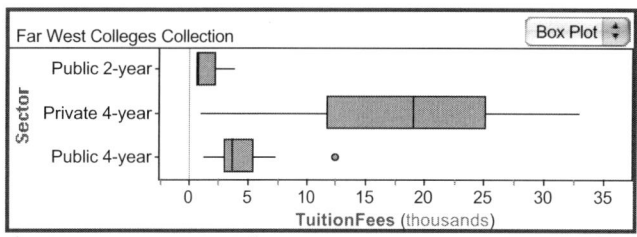

Measures Based on Distance from the Mean: Variance and Standard Deviation

The most common measures of variation are based on getting "something like" an average distance of the data in a collection from the mean of the data. We will use the data on the Oregon two-year and four-year schools and the percentage of students over age twenty-four. Here are dot plots of the data, showing the location of the means for the two groups of colleges. The names of the formulas for measuring spread from the mean are **variance s^2** and **standard deviation s**.

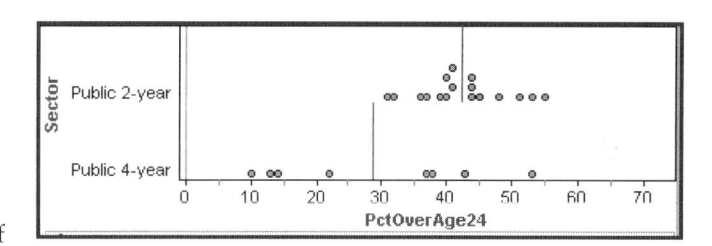

Definition of variance

$$s^2 = \frac{1}{n-1} \sum_{i=1}^{n} (x_i - \bar{x})^2$$

Definition of standard deviation

$$s = \sqrt{\frac{1}{n-1} \sum_{i=1}^{n} (x_i - \bar{x})^2}$$

Example. As you can see, there is basically just one formula since the standard deviation s is just the square root of the variance s^2. The way the formula works is shown (below) for the computation of the variance of *percentage over age 24*. The calculation starts by getting the difference between the value of each case x_i and the mean for the data: $x_i - \bar{x}$. For these data $\bar{x} = 28.75$. Since in this example $n = 8$, there will be eight of these differences. Then each of these differences

Num	Name	Pct Over Age24	Two-year or Four-Year
1	Eastern Oregon University	43	Four
2	Oregon Institute of Technology	37	Four
3	Oregon State University	13	Four
4	Oregon State University, Cascades Campus	53	Four
5	Portland State University	38	Four
6	Southern Oregon University	22	Four
7	University of Oregon	10	Four
8	Western Oregon University	14	Four

is squared and then these squares are summed up and the sum divided by $n - 1$, which in our example will be 7. What we get from this calculation is the variance s^2; the standard deviation s is found by getting the square root of this number. (In the second line of the calculation on the next page we have not shown two terms of the sum so that the sum will actually fit on the page.)

$$s^2 = \frac{1}{n-1}\sum_{i=1}^{n}(x_i - \bar{x})^2$$

$$= \frac{1}{7}\left[(43-28.75)^2 + (37-28.75)^2 + (13-28.75)^2 + \cdots + (22-28.75)^2 + (10-28.75)^2 + (14-28.75)^2\right]$$

$$= \frac{1}{7}\left[(14.25)^2 + (8.25)^2 + (-15.75)^2 + (24.25)^2 + (9.25)^2 + (-6.75)^2 + (-18.75)^2 + (-14.75)^2\right]$$

$$= \frac{1}{7}\left[203.0625 + 68.0625 + 248.0625 + 588.0625 + 85.5625 + 45.5625 + 351.5625 + 217.5625\right]$$

$$= \frac{1807}{7} \approx 258.2143$$

The variance for the *percentage over age 24* for the four-year schools is $s^2 = 258.2143$, and the standard deviation is $s = \sqrt{s^2} = \sqrt{258.2143} \approx 16.07$.

Looking at the variance formula closely and thinking graphically

The standard deviation is close to being the average distance of the data from the mean, and can be thought of as: "the average amount that a case value differs from the mean."

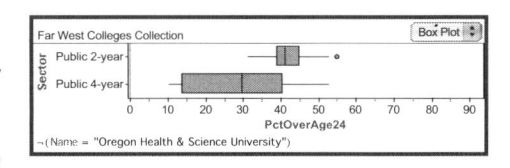

We will look at an example to develop this idea: the distances (either negative or positive) for our Oregon two- and four-year schools example are shown in the graphic below. (The boxplots are shown here.)

We have said that the spread or variation for the two-year colleges is smaller than the spread for the four-year schools for the variable *percentage over age 24.* Look at the distances (the "lines") in the plot below; the "lines" for the four-year schools look longer—on average—than the "lines" for the two-year colleges. The two-year colleges have a few "deviations from the mean" that are big, but even these are not as big as the ones for the four-year schools, and the two-year schools have many very short deviations. Each of these lines—the deviations or distances—is one of the $x_i - \bar{x}$ in the formula.

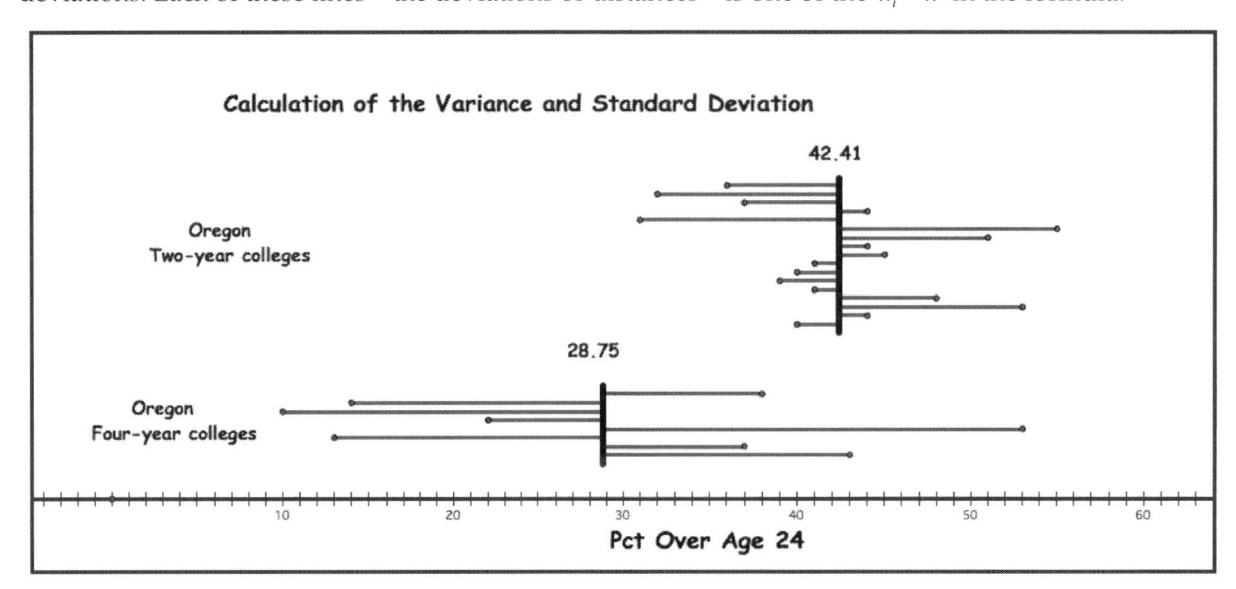

Why $\sum_{i=1}^{n}(x_i - \bar{x})$ is zero and why deviations are squared If we try to get the average of these distances, we will find that $\sum_{i=1}^{n}(x_i - \bar{x})$ will always be equal to zero, and so getting the average deviation $\dfrac{\sum_{i=1}^{n}(x_i - \bar{x})}{n}$ will always be zero. One way of seeing that the sum will always be zero is to see that the mean is the balance point of the data. That is, the sum of the lengths of the deviations on the left of the mean will just equal the sum of the lengths on the right-hand side of the mean. Another way is to work out: $\sum_{i=1}^{n}(x_i - \bar{x}) = \sum_{i=1}^{n} x_i - \sum_{i=1}^{n} \bar{x} = n\bar{x} - n\bar{x} = 0$

The solution to our dilemma is to square each of the deviations, so we are getting the average of the squared deviations. By taking the square root of the variance s^2 we get the standard deviation s so that we get a measure that has the same units as our original data.

Why the sum is divided by n − 1. In the formula $s = \sqrt{\dfrac{1}{n-1} \sum_{i=1}^{n}(x_i - \bar{x})^2}$ we divide by $n - 1$, and not n. Now, if the number in our collection is big, dividing by $n - 1$ will come out to be almost the same as dividing by n, so numerically there is not much difference. However, the question remains as to *why* we divide by one less than n. The complete answer can better be appreciated at the end of the course, or perhaps not even until into the next statistics course. For now, we can only give a hint of the reasoning. Later in the course, we will be concerned with not only what we actually see from our data but also with "what could possibly be." If our data have a mean $\bar{x}$ and this mean is calculated by the formula $\bar{x} = \dfrac{1}{n} \sum_{i=1}^{n} x_i$ then for that particular mean $\bar{x}$ and for n cases, the sum $\sum_{i=1}^{n} x_i$ is fixed. If the sum is fixed then only $n - 1$ cases in the data can vary freely.

Interpretation: the important and hard part

Should the variance and standard deviation for the *percentage over age 24* for the *two-year schools* be bigger or smaller? Looking at the box plots it seems that the amount of spread or variation in *percentage over age 24* is less, so the s^2 and the s should be smaller. A smaller number for the variance and for the standard deviation shows a smaller spread—a larger number shows a larger spread or variation. Notice that the *IQR* for the two-year schools is also smaller than the *IQR* for the four-year schools. Notice also that although the measures of center/location for the two-year schools are big (these colleges have on average a greater percentage of students over twenty-four), the measures of spread are smaller than for the four-year schools, where average percentage over age twenty-four is lower. The two-year schools have on average a higher

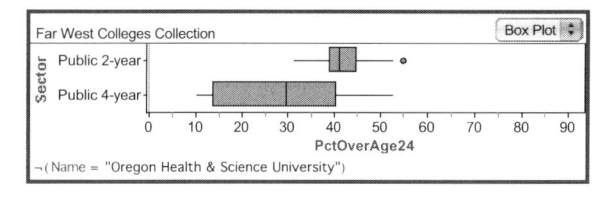

Far West Colleges Collection

	Sector		Row
	Public 2-year	Public 4-year	Summary
PctOverAge24	42.4118	28.75	38.04
	41	29.5	40
	6.73664	16.069	12.1603
	6	27	8
	17	8	25

S1 = mean( )
S2 = median( )
S3 = s( )
S4 = iqr( )
S5 = count( )

¬(Name = "Oregon Health & Science University")

percentage over age twenty-four, but there is much less variability among these schools. The four-year schools have a lower percentage on average, but there is much more variability amongst the schools in *PctOverAge24*.

Even without graphics to give us a visual picture, we should be able to use the numbers to compare the location and the spread of

Far West Colleges Collection				
	Sector			Row Summary
	Private 4-year	Public 2-year	Public 4-year	
PctOverAge24	34.539	45.4686	27.0323	38.3677
	26	43	24.5	39
	28.9486	11.343	17.3091	21.6583
	43	16	24	28
	141	175	62	378

S1 = mean ( )
S2 = median ( )
S3 = s ( )
S4 = iqr ( )
S5 = count ( )

several distributions. Here are the numbers not just for the Oregon colleges and universities but also for all the Far West institutions on the same variable *PctOverAge24*.

From these numbers we can say that the percentage of students over age twenty-four is higher on average for the two-year colleges, but the amount of variability among these colleges is smaller than it is for the four-year institutions. The mean and the median (S1 and S2 in the Fathom summary table) for the two-year colleges are both higher than for the four-year schools, but the standard deviation (S3) and the *IQR* (S4)—the measures of variation—are smaller than those measures for the four-year schools. The spread of the percentage over age twenty-four is especially great for the private colleges, with an inter-quartile range of 43%, whereas the *IQR*s for the other two types of institutions are considerably smaller. In general, the larger the value of our measures of spread, *IQR*, *s* (the standard deviation) and s^2 (the variance), the more variability there is in the data and the more spread out the data are.

Interpretation of Measures of Variation
The *bigger* the value of the standard deviation or inter-quartile range, the *more variability* the data have.

Resistance Again

Recall that the mean $\bar{x}$ is *not resistant*; the value of the mean can be affected by skewness and outliers. The variance and standard deviation are also sensitive to outliers and skewness; part of the reason that this is so is that the variance and the standard deviation use the mean in their calculation.

However, notice that in the formula for the standard deviation $s = \sqrt{\frac{1}{n-1}\sum_{i=1}^{n}(x_i - \bar{x})^2}$ deviations from the mean get squared. This means that outliers (which are far from the mean) or values in tails get exaggerated by squaring in the calculation of the variance and standard deviation. The consequence of the sensitivity of the variance and standard deviation to skewness is that the *IQR* is a more useful measure of spread since it can be used for highly skewed distributions. You may well ask at this point: why do statisticians use them at all (and why are they so important)? The answer is that the standard deviation makes complete sense for an important *model* distribution, the **Normal Distribution.** In the next section, we will look at *model distributions* and the *Normal Distribution* specifically.

Resistance and Measures of Spread
The *range, standard deviation,* and *variance* are *not resistant* to skewness and outliers.
The *Inter-Quartile Range* is *resistant* to skewness and outliers.

IQR and the identification of outliers by "Lower and Upper Fences"

Up to now, we have identified outliers by noting whether these data are "outside" the main body of data. There is a stricter definition that uses the *IQR*. We calculate what are sometimes called **fences;** any data point that is *outside* these fences is deemed an outlier and shown as a dot in a box plot; data that are within the fences are included in the whiskers or the box of a box plot. Here are the definitions of the upper and lower fences.

Definition of Lower and Upper Fences:

$$\text{Lower Fence} = Q_1 - 1.5 * IQR \qquad \text{Upper Fence} = Q_3 + 1.5 * IQR$$

Example. The **Summary Table** for the distribution of *TuitionFees* for public four-year colleges is shown below. From the numbers there, we can calculate that the *IQR* = 5496 − 3030 = 2466, and with the *IQR* and the numbers for the Q_1 and Q_3 we can calculate the lower and upper fences.

Lower Fence = $Q_1 - 1.5 * IQR = 3030 - 1.5 * 2466 = -669$, and

Upper Fence = $Q_3 + 1.5 * IQR = 5496 + 1.5 * 2466 = 9195$.

There are no outliers on the left since all the data are within the lower fence, which is negative here; on the right end, the maximum value is greater than the upper fence and is a dot. The whiskers are drawn to the biggest (or smallest) values that are within the fences. Here is graphic showing the fences and the data as well as the box plot. Usually, what we will see displayed is just the box plot and not the fences.

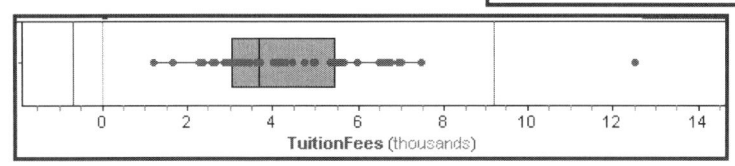

Far West Colleges Collection
TuitionFees
1214
3030
3656
5496
12506
62

S1 = min( )
S2 = Q1( )
S3 = median( )
S4 = Q3( )
S5 = max( )
S6 = count( )

sector = "Public 4-year"

Summary: Measures of Spread

Measures of spread are numerical measures of the variability of data in distributions. There are four different measures that we use: the *range, the Inter-Quartile Range (IQR), variance (s^2),* and *standard deviation (s)*.

- **Range** = Maximum value – Minimum value
- **Inter-Quartile Range** $IQR = Q_3 - Q_1$
- **Variance** $s^2 = \dfrac{1}{n-1} \sum_{i=1}^{n} (x_i - \bar{x})^2$
- **Standard Deviation** $s = \sqrt{\dfrac{1}{n-1} \sum_{i=1}^{n} (x_i - \bar{x})^2}$
- **Interpretation of Measures of Variation** The bigger the value of the standard deviation (s) or inter-quartile range (IQR), the more variability the data have. This interpretation is true for all of the measures. However, see the comments on *resistance* below.
- **Resistance of Measures of Spread**
  - The *range, standard deviation* and *variance* are *not resistant* to skewness and outliers.
  - The *Inter-Quartile Range* is *resistant* to skewness and outliers.

  Therefore, the *range* may be a bad measure of variability, even though the concept is familiar. The *standard distribution* and *variance* are more appropriate for symmetrical distributions.
- **Lower and Upper Fences**
  - Lower Fence = $Q_1 - 1.5 * IQR$
  - Upper Fence = $Q_3 + 1.5 * IQR$

  Whiskers are drawn to largest or smallest values within the fences; data outside the fences are shown by dots.

§1.6 Exercises on Variation: Measurement and Interpretation

1. **Explaining variability.** Read the first paragraph of the **Notes** for this section. It is clear that the tuition fees higher among the private colleges and universities, but it is also clear that there is much more spread or variability among what is charged by the private colleges.

   a. Imagine that you are explaining to a friend (perhaps by e-mail) what this graphic tells you about tuition fees in public and private colleges. You can attach the graphic to your e-mail. You need to tell your friend what the dot plot shows about the center (or location) of the distributions but *also* about the differences in the *spreads* of the distributions between the private and the public institutions.

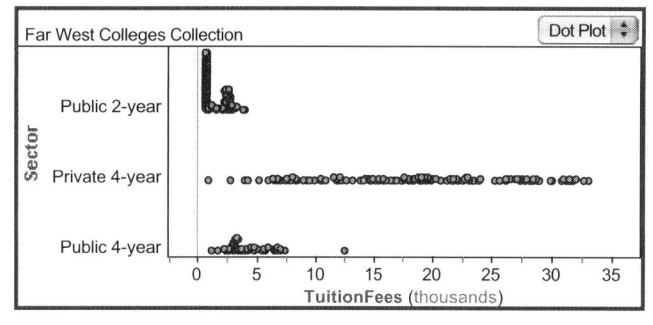

   b. Now assume that it is *not* possible to attach the graphic; explain the features of the graphic without your friend being able to see it. How will you explain the fact that the distribution of tuition fees at private universities is much more spread out or varied than are the fees at the public institutions?

2. **Calculating variance by hand.**

   a. There is a worked example on page 84 of the calculation of the variance. In that example, two of the terms (the things that get added) were left out so that the calculation would actually fit on the page. Your task (the "answer" to this question) is to work out and write down the two missing terms.

   b. Some people find it easier to calculate the variance by hand if the data are arranged in a chart. Here are the data for *percentage over age 24* for the two-year schools, showing in schematic form the calculation of the sums in the

i		x_i	$(x_i - \bar{x})$	$(x_i - \bar{x})^2$
1	Blue Mountain Community College	40	-2.41	5.8081
2	Central Oregon Community College	36	-6.41	41.0881
3	Chemeketa Community College	45	(i)	6.7081
4	Clackamas Community College	44	1.59	(ii)
5	Clatsop Community College	51	8.59	(iii)
6	Columbia Gorge Community College	55	12.59	158.5081
7	Klamath Community College	31	-11.41	130.1881
8	Lane Community College	44	(iv)	(v)
9	Linn-Benton Community College	37	(vi)	29.2681
10	Mt Hood Community College	32	-10.41	108.3681
11	Oregon Coast Community College	44	1.59	2.5281
12	Portland Community College	53	10.59	112.1481
13	Rogue Community College	48	5.59	31.2481
14	Southwestern Oregon Community College	41	(vii)	(viii)
15	Tillamook Bay Community College	39	(ix)	(x)
16	Treasure Valley Community College	40	-2.41	5.8081
17	Umpqua Community College	41	-1.41	1.9881
	SUMS	721	0.03	726.1177
	x-bar	42.41		

calculation of the variance. Fill in the missing entries. (Use the small Roman numerals to refer to the cells.)

   c. Use the sums given to calculate the variance s^2.

   d. Calculate the standard deviation s.

e. One of the columns shows $(x_i - \bar{x})$. The sum is shown at the foot of that column. What should the value of $\sum_{i=1}^{n}(x_i - \bar{x})$ be? The value shown here is not what the value should be because we have introduced some rounding error. (See the section in the **Notes: Looking at the variance formula closely and thinking graphically.**)

3. **Interpreting Measures of Variation**

   a. Confused Conrad (who has not been reading the **Notes**) says, "If the mean is bigger then the standard deviation must also be bigger." Use the numbers in the **Notes** for the example of *percentage over age 24* for either Oregon or for the Far West colleges to show CC his error.

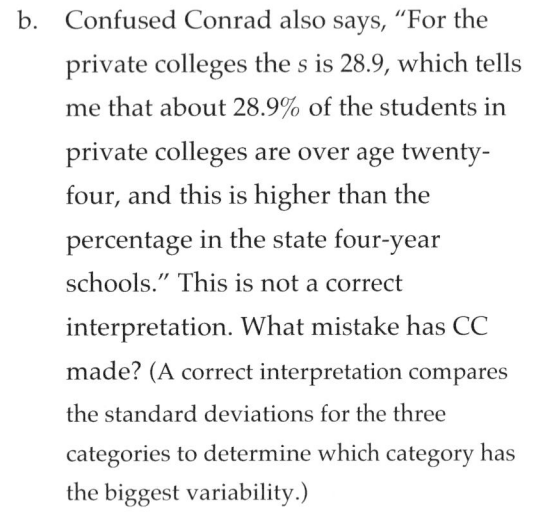

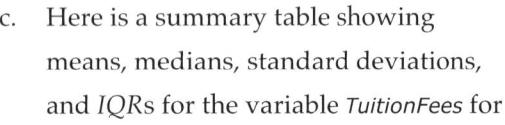

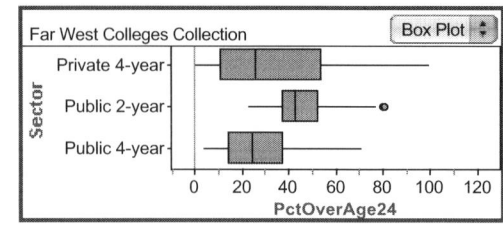

   b. Confused Conrad also says, "For the private colleges the *s* is 28.9, which tells me that about 28.9% of the students in private colleges are over age twenty-four, and this is higher than the percentage in the state four-year schools." This is not a correct interpretation. What mistake has CC made? (A correct interpretation compares the standard deviations for the three categories to determine which category has the biggest variability.)

   c. Here is a summary table showing means, medians, standard deviations, and *IQR*s for the variable *TuitionFees* for the public two- and four-year schools and the private four-year schools. Interpret the measures of center/location in the context of the data (in context means: give an interpretation that says something about how tuition fees are different in the three types of colleges).

   d. Give an interpretation of the measures of variation in the context of the data.

4. **Students' Grades.** A professor at the Instituto Tecnológico de Costa Rica kept careful records of students' scores for a calculus course that he taught semester after semester. There were three tests (all marked out of 100) and then the final course grade that depended upon the tests but also upon quizzes and homework. These records are for several semesters, but the tests given across the semesters are comparable.

   a. Explain what the measures of center tell you about the three tests and the final course grade. Use the numbers.

b. Explain what the measures of spread (or variation) tell you about the three tests and the course grade.

c. A student writes to answer part b: "Exam 3 ranged from about 17 to 100, so Exam 3 had the biggest spread." How can you tell this student has not been reading the **Notes**? (There is good evidence that Exam 3 did have the greatest variability; what is that evidence?)

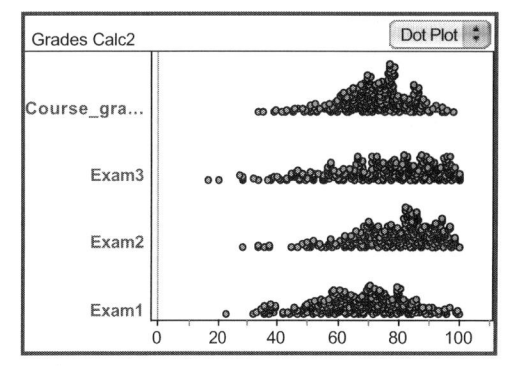

5. **Great Lakes Colleges and Universities.** We will look at the variable *Percentage Over Age 24* and its relationship with the type of college or university measured by the variable *Sector.* This relationship was examined in the **Notes** for the colleges in the Far West region. Our question is whether the relationships are similar for the institutions in the Great Lakes region.

- Open the Fathom file **GreatLakesCollegesUniversities.ftm** and get a case table.
- To print your graphics, open a Word document to transfer (and make smaller) the Fathom graphics.
- From the shelf, drag a **graph** and drag the variable *Percentage Over Age 24* to the horizontal axis and the variable *Sector* to the vertical axis. You should have a dot plot. If you wish, you may change your graphic so as to display box plots.
- From the shelf, drag a **Summary Table** and drag the variable *Percentage Over Age 24* to the downward-pointing arrow.
- With the **Summary Table** selected, go to the menu and get **Summary>Add Formula**. In the dialogue box that comes up, type in: median ( ) with nothing in the parentheses. It should appear as S2 in the list of measures. In the same manner add the measures standard deviation [type in s( )] and the inter-quartile range [type in iqr( )], which should be S3 and S4.
- Drag the variable **Sector** to the right-pointing arrow in the **Summary Table.**

    a. Which type of college has the highest center on the variable *Percentage Over Age 24?* Back up your answer with reference to numbers from your summary table.

    b. Which type of college has the smallest amount of variability in the variable *Percentage Over Age 24?* Back up your answer with reference to numbers from your summary table.

    c. Which type of college has the largest amount of diversity on the variable *Percentage Over Age 24?* Back up your answer with numbers from the summary table.

- Change the dot plot to a box plot and get the five-number summary for the **Summary Table.** (Get **Summary>Add Five Number Summary.**)

    d. Use the five-number summary for the public four-year colleges to calculate the **Lower** and **Upper Fences** and explain why no outliers are shown for this distribution on the box plot.

    e. Use the five-number summary for the public two-year colleges to calculate the **Lower** and **Upper Fences** for this type of college. Notice that there are data beyond the fences.

- Select the **Summary Table;** go to the menu and get **Object>Duplicate Summary Table.** This makes a copy of our summary table. Sometimes we need more than one; this is one of those times.

- Using **Add Filter** in Fathom. When we want to restrict our calculation to only certain cases in Fathom, we use **Add Filter.** If we want to "filter" the summary table to show the calculations for *just* the two-year colleges and for *just* the data that are *not* outliers, we select the summary table and get: **Object>Add Filter.**
- In the dialogue box (using "and" from the palate), type: *(Sector = "Public 2-year")* and *(PctOver24 > 20)* and *(PctOver24 < 68)*. Take care; computers are picky about quotation marks.
    f. This new summary table excludes the eight outliers. Compare the standard deviation *s* and *IQR* with the values that include the outliers. Which has changed the most, *s* or *IQR*? Cite what feature of the measures of variation mentioned in the **Notes** this comparison illustrates.
    g. True or false, and explain: "In 25% of the two-year colleges in the Great Lakes region (with the outliers included), half of the students are over age twenty-four."

6. **Real Estate Data: Time in the Market**   In this exercise you will look a sample of data on houses that were sold in San Mateo County in the years 2007 to 2008. The collection you will use does not include *all* the houses that were sold; it is a *sample* of the entire *population* of all houses that were sold. Our statistical question is whether there are differences among the regions of San Mateo in the length of time houses stay in the market before they are sold. The variable that measures the interval between the date the house was put on the market and the date the house was sold is called *Time_in_Market.*

- Open the Fathom file **San Mateo RE Sample 0708.ftm** and get a **Case Table.**
- From the shelf, drag down a **Graph** and drag the variable *Time_in_Market* to the horizontal axis and the variable *Region* to the vertical axis. You should have four dot plots, one for each region.
- Get a **Summary Table**, and drag the variable *Time_in_Market* to the downward-pointing arrow.
- With the **Summary Table** selected, go to the menu and get **Summary>Add Formula**. In the dialogue box that comes up, type in: median ( ) with nothing in the parentheses. It should appear as S2 in the list of measures. In the same manner add the measures standard deviation [type in s( )] and the inter-quartile range [type in iqr( )], which should be S3 and S4.
- Drag the variable *Sector* to the right-pointing arrow in the **Summary Table.**
    a. Judging from the shape of the distributions, which measures of center and spread would be best to use? Give a reason for your answer based upon the **Notes.**
    b. Use the measures that you chose in part a to explain in the context of the data any differences or similarities in *center* in the variable *Time_in_Market* among the four regions for the sample of houses that were sold in 2007–2008. In which regions or regions do houses typically stay on the market longer? Make certain that you refer to the numerical summaries and do not depend only on the graphics for your explanations.
    c. Use the measures that you chose in part a to explain in the context of the data any differences or similarities in *spread* in the variable *Time_in_Market* among the four regions for the sample of houses that were sold in 2007–2008. In which regions or regions do houses typically have the greatest variety of times in the market? Make certain that you refer to the numerical summaries and do not rely solely on the graphics for your explanations.

- Select your **Summary Table** and go to **Object>Duplicate Summary Table.** Then select your **Graph** and in the same way duplicate your graph. (This saves much work!)
- To your newly created **Summary Table** and **Graph,** drag the variable *SalePrice* in place of the variable *Time_in_Market.*
- With the **Summary Table** selected, you may want to go to **Summary>Format Numbers** and choose "Fixed Decimal" and opt for two decimal places. You will avoid scientific notation.

   d. Answer question b for the variable *SalePrice* instead of *Time_in_Market.*

   e. Answer question c for the variable *SalePrice* instead of *Time_in_Market.*

   f. *Putting it together.* Imagine that you have a cousin from Seattle who is thinking of setting up a real estate business in San Mateo County. Write a letter to her describing the differences in the markets in the four regions from what you have learned. Translate the technical language you have used to "Realtor's language."

Exercises 7–10: Measures of Spread and Colleges and Universities

Colleges and Universities. The next four exercises use the data on colleges and universities also used in the **Notes.** The data shown in the case table are for the Far West.

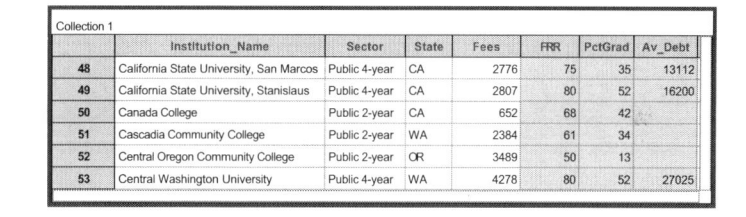

	Institution_Name	Sector	State	Fees	FRR	PctGrad	Av_Debt
48	California State University, San Marcos	Public 4-year	CA	2776	75	35	13112
49	California State University, Stanislaus	Public 4-year	CA	2807	80	52	16200
50	Canada College	Public 2-year	CA	652	68	42	
51	Cascadia Community College	Public 2-year	WA	2384	61	34	
52	Central Oregon Community College	Public 2-year	OR	3489	50	13	
53	Central Washington University	Public 4-year	WA	4278	80	52	27025

Some of the variables measured for each college or university are:

Sector	Whether the college or university is a "public two-year," "public four-year," or "private four-year" institution
State	The state in which the college or university is located: AK, CA, HI, NV, OR, WA
Fees	Total tuition fees for one year
FRR	Freshman retention rate: the percentage of freshmen in the college or university who continue
PctGrad	The percentage of students in the college or university who get a degree
AvDebt	The average debt of students at the college or university (measured for four-year institutions)

7. Interpreting Data on Fees for Four-Year Colleges

Our statistical question is:

How do fees differ between public and private institutions?

   a. What are the cases for these data?

   b. Before looking at the data, first write down some ideas about how you think the distributions of the variable *Fees* will differ between the private and public institutions.

   c. Compare the measures of center shown for the variable Fees and write what the measures tell you about the private institutions compared with the public institutions. A good start to what you write is: "On average the fees for private colleges and universities are…" In your sentence, you may wish to cite either the mean or the median or both.

   d. In general, what does a small value for a standard deviation tell you about a distribution? What does a big value tell you?

Collection 1

	Sector		Row Summary
	Private 4-year	Public 4-year	
Fees	17893.4	4221.37	13770.1
	18155	3580	13815
	7159.51	1840.59	8735.47
	11789	2475	15472
	132	57	189

S1 = mean( )
S2 = median( )
S3 = s( )
S4 = iqr( )
S5 = count( )
¬(Sector = "Public 2-year")

91

e. Now look at the measures of spread for the two types of institutions. For the private colleges, both the standard deviation and the iqr are much bigger than they are for public colleges and universities. What does that tell you? Good words to use in your interpretation are: "variable" or "diverse" or even "more different"—although that is slightly clumsy English.

f. Confused Conrad writes, in answer to part c, "The comparison of the measures of spread tells me that the fees are much higher at the private colleges and universities than for public colleges and universities." Explain how CC is confused.

g. True or false, and explain: "The reason that the measures of center (mean and median) and measures of spread are bigger for the private colleges and universities is because the count is bigger."

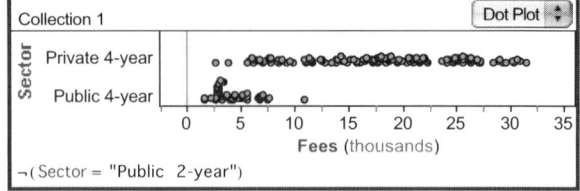

h. Explain in ordinary English that would be understandable to a fellow student not taking statistics what it would mean to a student looking for a college or university that the diversity or variability of fees is greater for the private colleges.

8. **Calculations on community college fees in Hawaii** There are just seven community colleges in Hawaii. Here are the data for the fees that they charge for a year.

a. Find the five-number summary and from that the Inter-Quartile Range (IQR) for these data.

b. Determine whether there are outliers for these data by calculating the Lower and Upper Fences.

c. Construct a box plot for the distribution of fees.

d. Find the mean annual fees charged for the community colleges in Hawaii. (Answer: $1175.86)

e. By hand, calculate, showing all of the steps, the standard deviation for these data using the formula for the standard deviation. The answer is $30.50.

Name	Fees ($/year)
Hawaii Community College	1240
Honolulu Community College	1158
Kapiolani Community College	1188
Kauai Community College	1158
Leeward Community College	1153
Maui Community College	1166
Windward Community College	1168

9. ***How do tuition fees in community colleges differ in different states?***
   a. A common mistake is to put too much importance on the number of cases in different categories. Generally, the *counts* do not tell you much about a distribution. However, here, the numbers corresponding to *count (.)* do tell you something. What do they say about community colleges in the three states?
   b. Look at the measures of center (mean and median). Write what these tell you about the distributions of fees in the three states.
   c. Look at the measures of spread. Write what these tell you about the distributions of fees in the three states. (Good words to use are "variability" or "diversity.")
   d. You have a friend who has been your fellow student at a community college in California. Now this friend is planning to move to the Pacific Northwest (either Oregon or Washington) and after gaining residency there, plans to continue studying in a community college. What can you tell your friend based upon these data? (Do not forget to speak to your friend about diversity in fees.)
   e. Based upon what you did in question 8, compare the distribution of fees for the community colleges in Hawaii with the distributions for fees on the Pacific Coast.

Far West Colleges

		Fees
State	CA	722.468
		732
		71.3914
		128
		109
	OR	2638.07
		2604
		359.045
		349
		14
	WA	2419.47
		2382
		264.874
		164
		34
Column Summary		1260.79
		780
		833.327
		1578
		157

S1 = mean ( )
S2 = median ( )
S3 = s ( )
S4 = iqr ( )
S5 = count ( )

(sector = "Public 2-year") and

10. ***How does the percent graduating differ for private and public four-year colleges?***
    a. What do the measures of center tell you about the percent graduating in private and public institutions?
    b. What do the measures of spread tell you about the percent graduating in private and public institutions?
    c. Comment on this assertion in light of the numbers and the box plots below: "If you attend a private school, you definitely have a better chance of graduating."

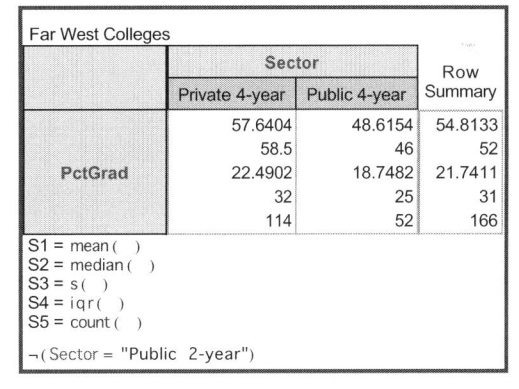

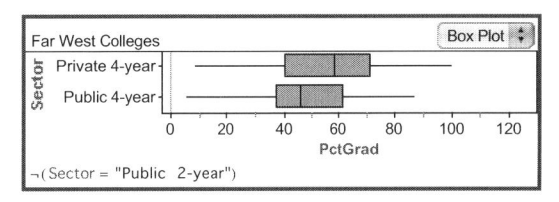

Far West Colleges

	Sector		Row Summary
	Private 4-year	Public 4-year	
PctGrad	57.6404	48.6154	54.8133
	58.5	46	52
	22.4902	18.7482	21.7411
	32	25	31
	114	52	166

S1 = mean ( )
S2 = median ( )
S3 = s ( )
S4 = iqr ( )
S5 = count ( )

¬(Sector = "Public 2-year")

Far West Colleges — Box Plot

Sector: Private 4-year, Public 4-year
PctGrad axis: 0, 20, 40, 60, 80, 100, 120

¬(Sector = "Public 2-year")

§1.7 Models for Distributions: The Normal Model

A model airplane or model train is a scaled-down version of the real airplane or train that hopefully preserves all of the features of reality. In mathematics and statistics, a model is often a function used to portray or replicate in succinct form some part of reality. We use models in statistics.

Distributions have *shape, center,* and *spread*, and we have found graphics (dot plots, histograms, box plots) to display and compare the shape, center, and spread of different distributions. We have also calculated measures so that we can compare the centers or locations of distributions (mean and median) and compare the spreads of distributions (standard deviation and inter-quartile range). But if we had a model of a distribution that we could use to compare with our actual real data distributions, we might be able to say: "Well, our distribution looks like a…" and then give a name that would be meaningful to others. By having a name (and perhaps some of the measures), these fortunate people would know pretty much what the distribution looked like. That is the idea of a **model distribution**; a model distribution has a known shape, a known center, and a known spread, which are determined by a mathematical formula. The model that is the most important for this course is called the **Normal Distribution**, although you may have thought of it as the **bell curve.**

Example: Heights of Elderly English Women (EEW)

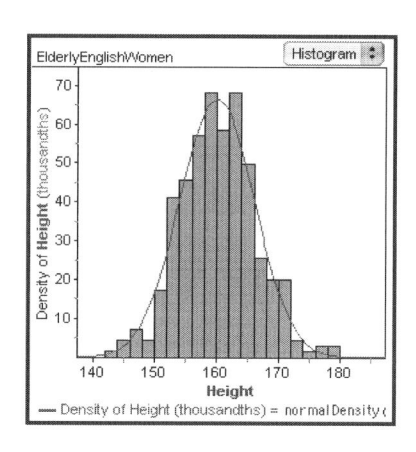

The histogram shows the distribution of the heights of $n = 351$ elderly English women who were part of a study of osteoporosis; they are short—their mean height is 160 centimeters. California and Penn State female students have a mean height of 165 centimeters, and Australian female high school students have a mean height of 164 centimeters. (To put all this in feet and inches, divide by 2.54 cm/in and get $160/2.54 \approx 63$ in. and $165/2.54 \approx 65$ in.—so 5' 3" and 5'5"). The smooth curve on top of the histogram is the **Normal Distribution** with mean 160 centimeters and standard deviation of 6 centimeters. From the graphic we can see that the Normal distribution model does not fit the histogram of the heights of the elderly English women *exactly*, but the model fits well.

Our Normal distribution smooth curve is one example of what is known as a **density curve;** it is a smooth curve and not a histogram because the normal distribution is a mathematical model—you can think of the density curve as the shape that you would get if you had a huge collection and a histogram of heights with an extremely large number of narrow bars, as in this example.

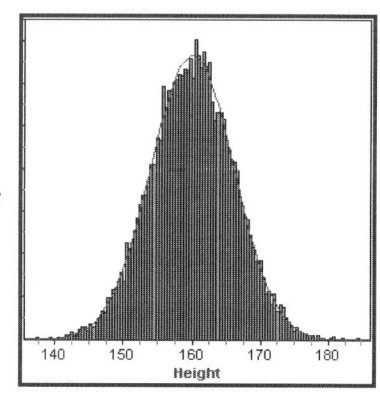

Histograms show an entire distribution; the bars include 100% of a collection, and typically we use them to determine the part or fraction or proportion of a distribution is above or below or between certain values of the variable, as you have done in the exercises. **Density curves** are designed for the same purposes, and so the area between the curve and the horizontal axis is 100%, or in terms of proportions, 1.

The Center and Spread of Normal Distributions

Our example collection of the heights of elderly English women (EEW for short) has mean 160 centimeters and standard deviation 6 centimeters. So, to this data we will fit a Normal model with mean $\mu = 160$ and standard deviation $\sigma = 6$. We use Greek letters here (μ is Greek "m" and σ is Greek small "s") for the mean and the standard deviation because we are referring to the model distribution that we are fitting.

Notation for Normal Model Distributions

μ is the mean σ is the standard deviation

Center The **Normal Distributions** are perfectly symmetrical, so the mean μ is in the center of the distribution, with 50% of all of the distribution (think of the histogram again) on the left of the mean and 50% on the right of the mean. Hence, for our model for the heights of EEW, the mean is in the center of the graph.

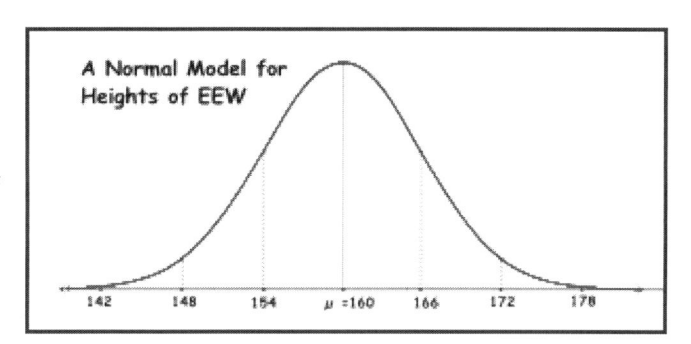

Spread What about spread? It is with this model that the standard deviation as we calculated it comes into its own. For these **Normal Distributions** the standard deviation measures the distance from the mean μ to the *inflection point* of the Normal density curve. The inflection point of the Normal curve is the point at which the slope changes from getting steeper and steeper to getting less and less steep. (To understand this, you can talk with little Udo, who has a Normal slide.) We can apply this to our model for the elderly English women. Since the standard deviation $\sigma = 6$ cm, we can draw on the horizontal scale the height that represents the mean plus one standard deviation, or $\mu + \sigma = 160 + 6 = 166$ cm. And, on the side of the mean that represents the EEW shorter than the mean, we can put $\mu + \sigma = 160 - 6 = 154$ cm. Then using this same distance (which is defined by the inflection point), we can specify the next standard deviation away from the mean on either side, and these will be $\mu + 2\sigma = 160 + 2 \cdot 6 = 160 + 12 = 172$, and $\mu - 2\sigma = 160 - 2 \cdot 6 = 160 - 12 = 148$.

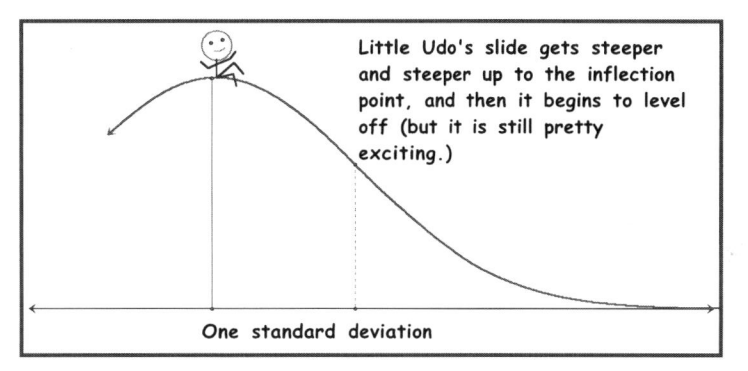

The Shape of the Normal Distributions and the Questions We Ask

For the *Normal Distributions* there is more to shape than symmetry. All Normal distributions have exactly the same shape, even though their pictures may make them look either tall and narrow or flattened out; the normal models have basically the same shape because their shape is determined by a mathematical formula.

To get an idea of what "the same shape" means, recall how we have been shading histograms to show the proportion of cases less than (or greater than) a particular value. We answered questions like this: "What proportion of EEW are shorter than 154 centimeters tall?" Or "What is the probability that an EEW is 172 centimeters or taller?" For the first question we looked at the heights of the bars of the histograms that included the women shorter than 154 centimeters, and we expressed that using our probability notation. If we did this for the EEW histogram, we would get $P(X < 154) \approx 0.15$, or about 15% of the EEW are shorter than 154 centimeters. For the question about the probability of a woman being 172 centimeters or taller we would do the same calculations, and finally get $P(X \geq 172) \approx 0.02$ —that the probability that an EEW is 172 centimeters or taller is only about 2%.

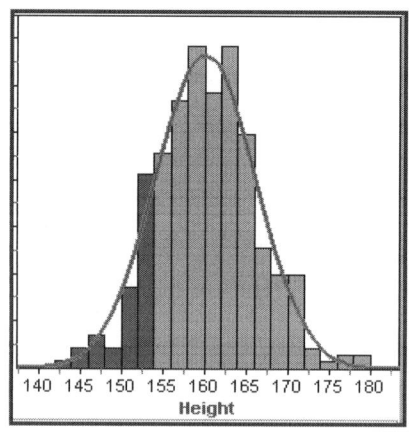

Think about what we are doing. We are using the space (actually the area of the histograms bars) to picture the answers to proportion or probability questions. *That is also what we will do with the Normal model.* Look at the graphic above; what can we say from the normal distribution model? We can say (because of the symmetry of the model) that 50% of the EEW have heights over the mean 160 centimeters and 50% less than 160 centimeters. We can actually say more because the model has a specific mathematical form. For any normal distribution we can specify that the proportion of data within one, two, and three standard deviations from the mean is approximately 68%, 95%, and 99.7%. This specification is called the **Empirical Rule.**

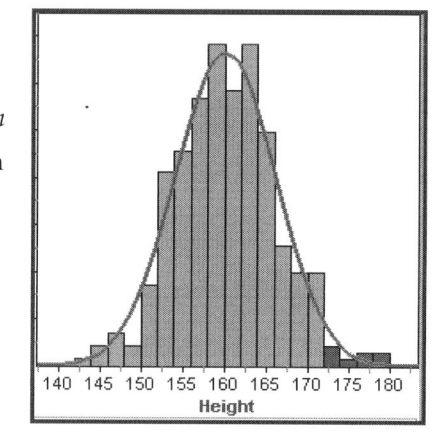

Empirical Rule for any Normal Distribution:

Approximately 68% of the distribution is between $\mu - \sigma$ and $\mu + \sigma$; i.e. within one sd of the mean.

Approximately 95% of the distribution is between $\mu - 2\sigma$ and $\mu + 2\sigma$; i.e. within two sd of the mean.

Approximately 99.7% of the distribution is between $\mu - 3\sigma$ and $\mu + 3\sigma$; i.e. within three sd of the mean.

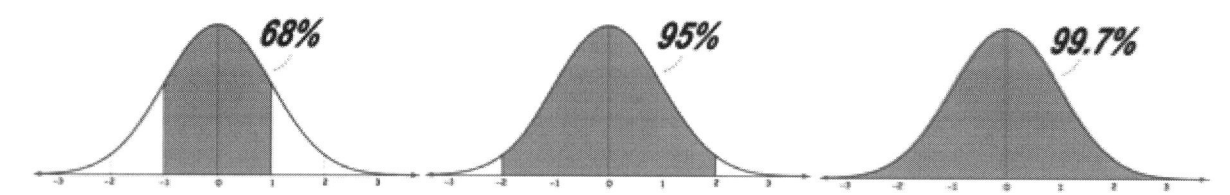

(We use the word "approximately" in the box above because the percentages are rounded; we will be able to get more exact numbers.) For our example of the elderly English women, this means that we can expect approximately 95% of the women to be between 148 centimeters tall ($\mu - 2\sigma$) and 172 centimeters tall ($\mu + 2\sigma$), according to the *Empirical Rule*. We can go farther than this. Since 95% of the distribution is between 148 and 172 centimeters, it follows that 5% is outside this interval. That 5% we can equally

96

divide between the women shorter than 148 centimeters and the women taller than 172 centimeters since the normal distribution is symmetrical. Therefore, we can say that the normal distribution model shows us that about 2.5% of the women should be 172 centimeters or taller. In symbols, our prediction from the model is that $P(X \geq 172) \approx 0.025$. This estimate from the normal distribution model is close to what we calculated from the data, and that shows that the model is a good fit.

Example 1: Using z Scores and the Normal Probability Chart

The proportion of EEW shorter than 151 cm? You should be able to see (using the same reasoning as just above) that about 2.5% of the EEW are shorter than 148 centimeters tall. However, if we want to know what proportion of the women are shorter than 151 centimeters, our *Empirical Rule* (68%-95%-99.7%) is not going to help us because 151 is somewhere between exactly one and two standard deviations shorter than the mean (see the graphic). The proportion of women shorter than 151 centimeters must be bigger than 2.5%, but what is it? We make a sketch (always!). In the sketch, the value of 151 centimeters is shown the thing we want. We want the proportion (or probability) of EEW who are shorter than 151 centimeters and that is shown the shaded region. We have a value (151 cm) and we want to know the area shaded in. Using our probability notation, we want to know $P(X < 151)$ and we are relying on the Normal model to describe the heights of EEW. To proceed, we need to calculate something called a *z score*.

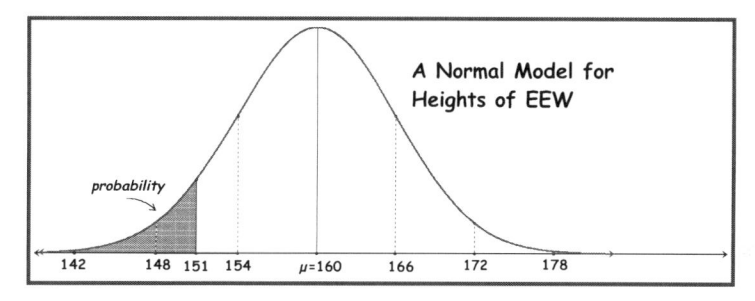

z scores. What we must do is to translate our question about the proportion of heights less than 151 centimeters into a question that can be answered with the **Standard Normal Distribution**; this is the Normal distribution that has mean $\mu = 0$ and standard deviation $\sigma = 1$. The 1, 2, and other numbers you see on the graphic are standard deviations (notice that the numbers -1 and 1 are at the inflection points). We do the translation using the formula for the **z score**; we calculate

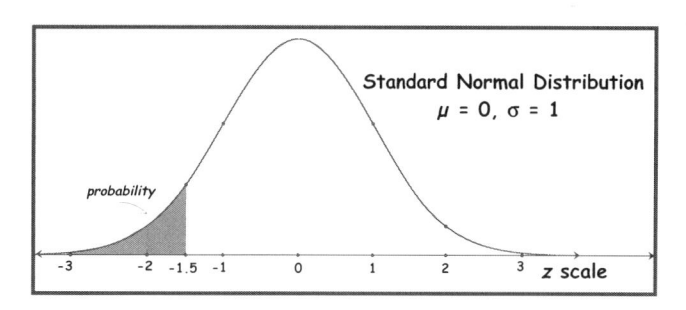

$$z = \frac{x - \text{mean}}{\text{standard deviation}} = \frac{151 - 160}{6} = \frac{-9}{6} = -1.51$$ since the Normal model that we are using for the heights of the women has $\mu = 160$ cm and $\sigma = 6$ cm. What this calculation tells us is that our height of $x = 151$ is $z = -1.50$ standard deviations from the mean. We can say that EEW who have a height of 151 centimeters are 1.5 standard deviations lower than the mean height of our model for the heights of the elderly English women.

Standard Normal Distribution

is the Normal distribution that has mean $\mu = 0$ and standard deviation $\sigma = 1$.

Definition and Formula for z score

The *z score*: $z = \dfrac{x - \text{mean}}{\text{standard deviation}}$ translates an *x* value on the variable being measured to a *z* value on the *Standard Normal Distribution*.

A *z score* shows the value of *x* in standard deviation units from the mean of a *Normal Distribution*.

Normal Probability Chart. The chart has the title **Standard Normal Probabilities** and can be found as the first chart in the **Tables** section after the end of the **Notes.** The title refers to the numbers in the *body* of the table that give the proportion or the probability to the left of a given *z score*; the units digit and the tenths (the first number to the right of the decimal point) of *z scores* are given in the left-hand column. The second decimal of *z scores* is given in the row at the top of the chart. So, for our z = -1.50, the chart shows that the proportion of the shaded region is .0668. We conclude that our Normal model says that about 6.68% of the EEW have heights less than x = 151 cm.

Shading and its meaning. We will often show a probability as a shaded area as in the graph above. This may be a new idea for you; you may well be more comfortable putting a numerical value on a scale (e.g., on the *x* axis). Our advice: get comfortable with shading; that is how we show probabilities in distributions.

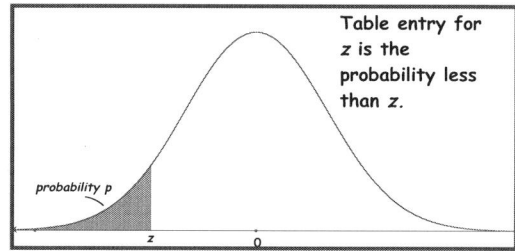

Example 2: Given a Value—Find a Proportion

The type of problem we worked in the example (at the same time introducing the *z* score and the *Normal Probability Chart*) asks for a proportion when we have (or are given) a value; hence we call it a **Given a Value—Find a Proportion** problem. Our question is very similar: what proportion of elderly English women (EEW) has heights less than 168 cm? In our probability notation, we want to find $P(X < 168)$.

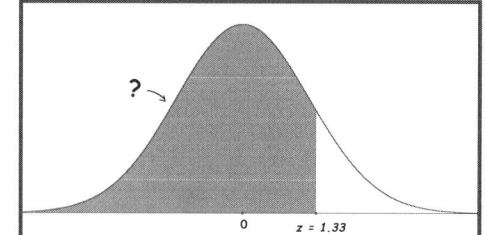

1. Sketch a Normal curve showing what you are trying to find; the "question mark" alerts us that we want a probability. (We do not have to place *z* on our graph exactly; since 168 is bigger than 160, our *z* will be positive.)

2. Calculate $z = \dfrac{x - \text{mean}}{\text{standard deviation}} = \dfrac{168 - 160}{6} = \dfrac{8}{6} \approx 1.33$ using $\mu = 160$ cm. and $\sigma = 6$ cm. (Now we can put z =1.33 on our sketch.)

3. Consult the **Normal Probability Chart** for z = 1.33. Now we have to use the second decimal place at the top of the chart. For the row 1.3, and the column .03, (shown here) we see $P(X < 168) = P(z < 1.33) = .9082$.

98

4. Interpret the result in the context of the question. "According to the Normal model, we can expect that nearly 91% of elderly English women will be shorter than 168 centimeters."

Actually, when we use the Normal model, the answer to the question "what proportion of these women have heights *168 centimeters or less* (including the $x = 168$ cm)?" will be the same as the answer to the question "what proportion of women have heights *less than* 168 centimeters (excluding the $x = 168$ cm)?" This is because when we use the Normal model, we are using a continuous density curve. When we get our answer from a histogram or a dot plot, or from a collection of data, there *will* be a difference in the answers to the questions "what is $P(X \leq 168)$?" and "what is $P(X < 168)$?"

Example 3: Given a Value—Find a Proportion

Our question is: "According to the Normal model, what is the probability that an elderly English woman in our collection is taller than 170 centimeters?" We are given a value and want to find a proportion—which is here interpreted as a probability. Read the question carefully; here we are asked about heights *greater than* ("taller") 170 centimeters rather than "less than." In symbols we want $P(X > 170)$. That makes this question slightly different. Here is the solution in steps.

1. Make a sketch. The sketch incorporates what we want: "what is the proportion greater than 170 centimeters?" (We have shown the Normal sketch with the heights for 1, 2, etc. standard deviations; you may make the sketch with or without these shown. The sketch is meant to help.)

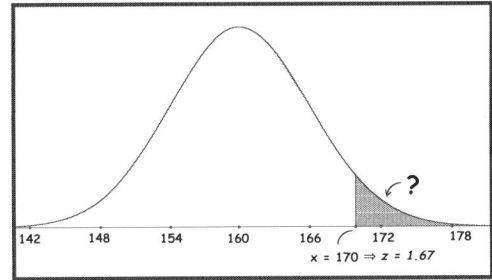

2. Calculate $z = \dfrac{x - \text{mean}}{\text{standard deviation}} = \dfrac{170 - 160}{6} = \dfrac{10}{6} \approx 1.67$ which we can now put on the sketch.

3. Consult the **Normal Probability Chart**. The number that corresponds to $z = 1.67$ is .9525. This cannot be the answer to our question! That would be saying that 95% of elderly English women are *taller* than 170 centimeters. Remember that the chart gives the proportion *less than* a z; in other words, the chart is telling us that $P(z < 1.67) = .9525$. We want the other side—the "greater than" side. This looks like a "not" probability problem; indeed it is. The solution is:

   $P(z > 1.67) = 1 - P(z < 1.67) = 1 - .9525 = .0475$ or, in words, "the probability that a woman is taller than 170 centimeters" *is* 1 - "the probability that a women is *not* taller than 170 centimeters." Comparing the number we got with the size of the shaded area, we see that 4.75% makes sense. [Remember that in using the Normal model, $P(z < 1.67) = P(z \leq 1.67)$.]

4. Interpret in context. "According to the Normal model, we can say that about 4.75% of the elderly English women are taller than 170 centimeters."

Checking Hand Calculated Answers with Software

There are many software applications that can do all these calculations. Here is a solution using **DistributionCalculator.ggb**, one of the easiest applications to use and included with the course materials. (The suffix *.ggb* means it comes from GeoGebra.) This application calculates many different distributions, and the

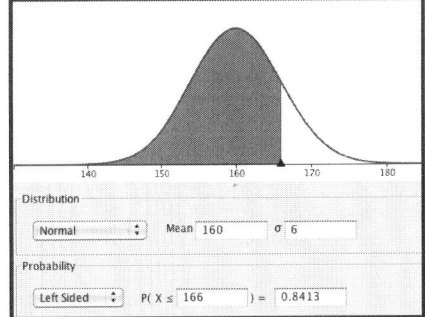

99

choice of distribution is controlled by the list just under the word "Distribution." Once "Normal" is chosen, there are spaces to enter the mean and standard deviation. Then, since we want $P(X < 168)$, the probability we want is "left-sided." That is, we want the probability that the height is less than, or to the left of, 168 centimeters. It is very important to see how the $P(X < 168) = .9082$ is displayed by the shading under the curve. That shading *is* the probability; if the probability were bigger, there would be more shading.

Example 4: Given a Proportion—Find a Value

We can use a Normal model to answer what is really the "reverse" question. What values of our variable correspond to a given proportion of the data? When we have this situation— we are given a proportion or probability, and what we want is a range or interval of values—we call this a **"Given a Proportion — Find a Value"** problem. Here is a simple example: "Using the Normal model with $\mu = 160$ cm. and $\sigma = 6$ cm, what are the values of the middle 95% of the distribution of the heights of elderly English women?"

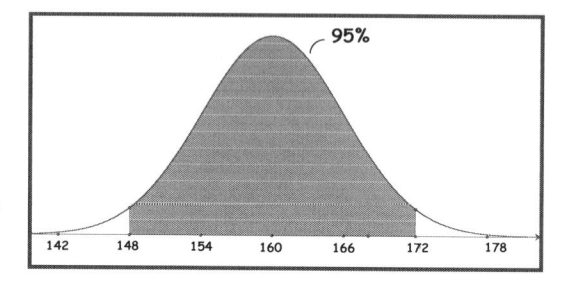

The answer to this question is not difficult; it can be derived from the *Empirical Rule*. Since about 95% of a Normal distribution is within two standard deviations on either side of the mean, we can say that about 95% of the women will have heights between 148 centimeters and 172 centimeters. Notice that the answer to this question is a range of values and not a probability; the probability (or proportion) was *given*. In symbols: $P(148 \leq X \leq 172) \approx 0.95$.

Example 5: Given a Proportion—Find a Value

Let us ask a more difficult question that cannot be answered with the *Empirical Rule*. We may want to know the *height of the tallest 5% of elderly English women*.

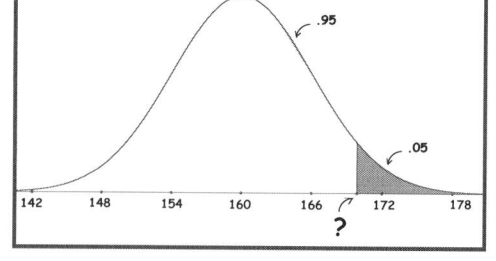

1. Make a sketch. Our sketch looks very much like the one for Example 3, but it is not the same; we see ".05" instead of a question mark pointing at the shaded region, showing that this region is what is *given* and *not* what is sought after. The question mark is pointing at a value for X, a height; this is what we want.

2. We work backwards; rather than calculate a *z score* using a value, we find the *z score* implied by our given .05. To find this we look at the body of the **Normal Probability Chart,** but remember that the chart gives cumulative proportions (proportions less than a *z*); the chart gives proportions to the left of a *z*. We need to look at what is to the left of our shaded area, and that is .95. So look at the *body* of the chart for .95. Alas, we do not see .9500, but we see two entries that are close to .95. One is .9505, and that entry corresponds to a *z score* of $z = 1.65$, and we see .9495, and that corresponds to $z = 1.64$. If our desired number were closer to either one of these, that

it the one we would choose for our z. Since each of these entries are exactly the same distance from .9500, we get the midpoint, and our z = 1.645,

3. Use the z score formula $z = \dfrac{x - \text{mean}}{\text{standard deviation}}$ and solve for what we want, x. We get:

$1.645 = \dfrac{x - 160}{6}$; multiplying, $6 \times 1.645 = x - 160$ and finally $9.87 + 160 = x$, we find $x = 169.87$ cm.

4. Interpret in context. "Our estimate according to the Normal model is that the tallest 5% of elderly English women have heights very near to 170 cm (169.87 cm)."

The Uses of Models and Specifically Normal Distributions

"Essentially, all models are wrong, but some are useful," wrote George E. P. Box.[1]

The *Normal distribution* for heights is a good example of this principle; as a mathematical model, the Normal curve never actually touches the horizontal axis. That means that according to the model, there is some extremely small probability that someone could be just one centimeter tall. Nonetheless, the Normal model is useful for our collection. Our example uses heights; however, other measures, such as foot lengths or arm spans, tend to have distributions that can be approximated by a Normal distribution. Normal distributions seem to fit biological measurements well (at least within the same species).

Here is another application; if you made repeated measurements of the same quantity (say, the weight of one single hefty Physics textbook) you would find that most of the measures of weight would cluster close to a peak (as the Normal distribution does), some of the measurements would be a bit off from this peak in one direction or another, and just a few of the measurements would be far off—either bigger or smaller.

The Normal model is also very useful in describing the distribution of averages of samples drawn randomly from the same population. It appears that Normal distributions describe situations where we have a great many small and independent "forces" operating, which is what we think is happening when we measure the same thing over and over again.

Of course, there are collections of data and there are situations where the distributions are not anywhere near to being Normal or even symmetrical; you have seen such skewed data. Many of the measures for the real estate data are extremely right-skewed; these data will not be well modeled by a Normal distribution. For skewed data, and for situations that do not lead to the Normal model, statisticians have other model distributions, and you will see some of these.

---

[1] Box, George E. P. and Norman R. Draper (1987), *Empirical Model-Building Response Surfaces*, New York: John Wiley and Sons, p. 424

Summary: Models of Distributions

- **Models of Distributions** are used to summarize the shape, center, and spread of actual distributions in a concise way. These model distributions are generally drawn from mathematical theory.
- **Density Curves** show the probability that a variable attains a range of values as the area between the curve and the horizontal axis if the variable has a (usually) model distribution. Since the sum of the probabilities for a sample space must sum to one, the area under a density curve must be one.
- **Normal Distribution** is a useful single-peaked model used extensively in statistics that:
  - Is perfectly symmetrical about its mean μ
  - Has its standard deviation σ located at the inflection points of the curve, such that
- **Standard Normal Distribution** is the normal distribution with $\mu = 0$ and $\sigma = 1$.
- **Empirical Rule** is a quick way of describing the shape of a Normal distribution; it says:
  - Approximately 68% of the distribution is between $\mu - \sigma$ and $\mu + \sigma$; i.e. within one sd of the mean.
  - Approximately 95% of the distribution is between $\mu - 2\sigma$ and $\mu + 2\sigma$; i.e. within two sd of the mean.
  - Approximately 99.7% is between $\mu - 3\sigma$ and $\mu + 3\sigma$; i.e. within three sd of the mean.
- **z score** $z = \dfrac{x - \text{mean}}{\text{standard deviation}}$ translates an x value to a z value on the *Standard Normal Distribution*.
  - A z score shows the value of x in standard deviation units from the mean of a *Normal Distribution*.
- **Normal Probability Chart** A chart showing the area (and therefore the probability) under the *Standard Normal Distribution* to the left of (or less than) a given value of z. Since the chart gives probabilities less than a value of z, probabilities greater than a value of z must be found by subtracting the probability found in the chart *from* 1, the total probability under the Normal density curve.
- The software application **DistributionCalculator.ggb** can be used to do the probability calculations and give a graphic showing the solution, where...
- **Shading indicates the size of a probability.** When working with density curves, a proportion (or probability) is shown by shading in an area of a curve.
- **Given a value, find a probability** If a particular value is known for a variable modeled by a Normal distribution with mean μ and standard deviation σ, find the z score and consult the Normal Probability Chart, keeping in mind that the chart gives probabilities to the left of the z score.
- **Given a probability, find a value** If a particular probability is given for a variable modeled by a Normal distribution with mean μ and standard deviation σ, work backwards by finding the z score for the relevant probability and then use the formula for the z score to find the value.
- **Normal Distributions** are used to describe some empirical distributions, errors in measurement, and random phenomena.

§1.7 Exercises: Models for Distributions

1. **Heights of Male California Students: Using Histograms and Drawing Pictures.** This exercise is based on the heights of male California students in statistics classes from spring 2007 through spring 2009. There were a total of $n = 210$ male students; their mean height (rounded to the nearest tenth of a centimeter) was 175.7 cm and the standard deviation of their heights was 8.1 cm, again rounded.

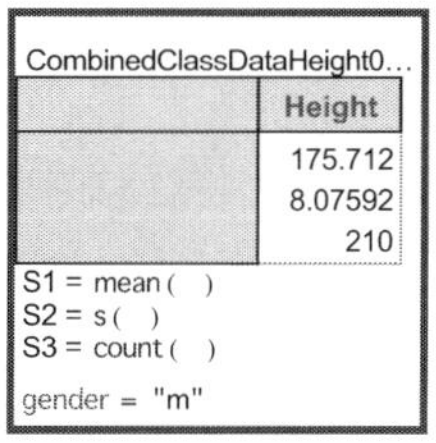

    a. *Using the histogram.* The histogram shown has the vertical axis in frequencies (counts). Use the histogram to estimate the proportion of male California students in this collection that were shorter than 170 centimeters. (To help: there were three students with heights less than 155 cm and there are $n = 210$ students. You can estimate the number of students between 160 cm and up to 165 cm and also the number of students between 165 cm and up to 170 cm by inspecting the graphic closely.)

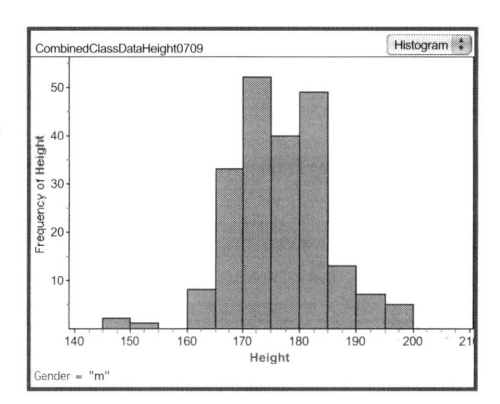

    b. *Using the histogram.* Express your answer in probability notation, using X to represent height.

    c. *Using the histogram.* Use the histogram to estimate $P(X \geq 185)$.

    d. *Sketching the Normal model* Make a horizontal axis like the axis shown here; work out the values for

    $\mu - 3\sigma, \mu - 2\sigma, \mu - \sigma, \mu, \mu + \sigma, \mu + 2\sigma,$ and $\mu + 3s$ where our Normal model has $\mu = 175.7$ cm. and $\sigma = 8.1$ cm; then using the fact that for a Normal distribution one standard deviation is at the inflection point of the curve, draw a nice sketch of the Normal model.

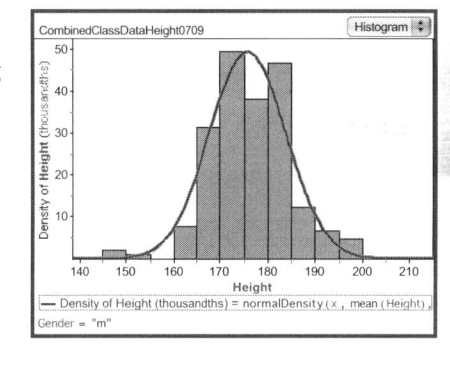

    e. On the sketch that you have drawn, shade in the proportion of students that are shorter than 170 cm.

    f. On the sketch that you have drawn, shade in $P(X \geq 185)$.

    g. Perhaps using a different kind of "shading," show on your picture $P(170 \leq X < 185)$.

    • Check your answers to parts e, f, and g using **DistributionCalculator.ggb**.

2. **Heights of Male California Students: Calculating the z score.** This exercise follows on from Exercise 1 and concentrates on the next steps in using the Normal model: calculating the *z score*. We are using the Normal model with $\mu = 175.7$ cm and $\sigma = 8.1$ cm.

    a. Find the *z score* for the height $x = 170$.

    b. Make a sketch of the **Standard Normal Distribution** (the Normal distribution that has $\mu = 0$ and $\sigma = 1$) and on that sketch locate the *z score* you calculated in part a. (Remember the inflection point.)

    c. Shade in on your sketch, $P(z < -0.704)$ [you should recognize $z = -0.704$ as the answer to part a].

d. Find the z score for the height x = 185.

e. On your sketch, show by shading the proportion of the Normal model distribution that is greater than the z score that you got as the answer for part d.

- Check your answers to parts b and c using **DistributionCalculator.ggb**. Use $\mu = 0$ and $\sigma = 1$.

f. Without really calculating, find the z score for the mean height of 175.7 cm. Explain briefly why your answer makes sense.

3. **Heights of Male California Students: Using the Normal Probability Chart.**

   a. Use the **Normal Probability Chart** to find $P(z < -0.70)$, which is the z score that you should have from Exercise 2a. You should have z = -0.704, but the chart can only handle two decimal places. This is the z score for the Normal distribution model with $\mu = 175.7$ cm and $\sigma = 8.1$ cm for the height of 170 cm.

   b. Your answer to Exercise 2b should resemble this Fathom-generated drawing of the *Standard Normal Distribution*. On your sketch, indicate *both* the z score, z = -0.704 and *also* the answer to part a for $P(z < -0.70)$. (*Hint:* One of these quantities should be on the horizontal axis, and the other as an "arrow" pointing to the shaded area. Get the two correct.)

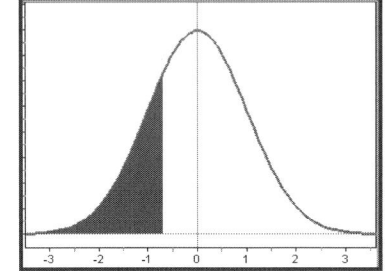

   c. Use the **Normal Probability Chart** to find $P(z \geq 1.15)$, which is the z score that you should have from Exercise 2d. Again you should have z = 1.148, but since the chart only goes to two decimal places, we use z = 1.15.

   d. A student writes as the answer to part c the number 0.8749, and that shows that the student is reading the chart correctly. But can this be the answer to the question, "What is the probability that $z \geq 1.15$?" or "find $P(z \geq 1.15)$?" Here is the picture of the **Standard Normal Distribution** that you should have from Exercise 2e. Explain (perhaps with the aid of the sketch) what is wrong with the wrong "answer" 0.8749.

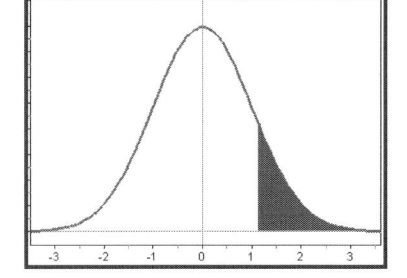

   e. On your sketch (that resembles the one shown here) show *both* the z score of 1.15 and *also* the answer to $P(z \geq 1.15)$.

   f. Use the **Normal Probability Chart** (or your work above) to find $P(z < 1.15)$ and explain what this number means in terms of the heights of male California students.

   g. Express the answer to part f in probability notation using X to indicate height.

   h. Make a new **Standard Normal Distribution** sketch and shade in $P(-0.70 < z < 1.15)$ [this sketch is not shown here, but you can check it with the **DistributionCalculator.ggb**].

   i. Use the **Normal Probability Chart** (or your work above) to find the numerical value of $P(-0.70 < z < 1.15)$. (*Hint:* You may find it helpful to perform a subtraction.)

   j. Express the meaning for the heights of male California students of the value you found in part I, and the shading you made in part h. You may find it helpful to use the probability notation using X to indicate height. (Again, you can check it with **DistributionCalculator.ggb**.)

k. **Jack the Giant**. Jack is 210 centimeters tall. Find Jack's *z score* if we use the model $\mu = 175.7$ cm and $\sigma = 8.1$ cm.

l. **Jack the Giant** Locate Jack on one of your sketches. (You should have found a *z score* bigger than any shown on the **Standard Normal Distribution** drawings above.

m. **Jack the Giant** Use the **Normal Probability Chart** (or by thinking) to find the probability that a male California student is as tall or taller than Jack.

n. **Jack the Giant** A calculator gives $P(X \geq 210) \approx 0.00001146$ using the Normal model $\mu = 175.7$ cm and $\sigma = 8.1$ cm. Does this agree with what your sketch and the **Normal Probability Chart** tell you? Explain.

4. **Heights of Australian High School Students: What about outliers?** This exercise uses data from the Census at School data collected from Australian high school students.

- Open the Fathom file **CAS Australia 08 A.** and get a box plot (shown here) with the variable *Height* on the horizontal axis and the variable *Sex* on the vertical axis.

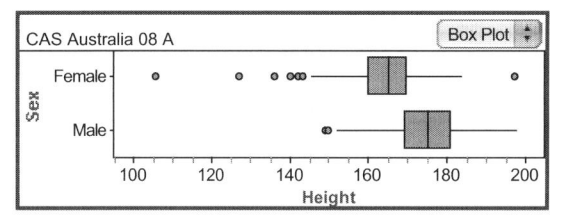

    a. [*Review*] Are the heights of the male students or the heights of the female students more variable (more spread out)?
    - First, use the **Range = Max − Min** to answer the question about the differences in variability in height between males and female students.
    - Then use the **IQR**s to answer the question about the difference in spread in height between male and females. You can estimate the *IQR*s from the graph or get it using Fathom
    - Then say why the two measures give you different answers.

- Get a **Summary Table** and drag the variable *Height* to the downward-pointing arrow.
- With the **Summary Table** selected, use **Summary>Add Formula** to get the standard deviation and **Summary>Add Five Number Summary**. (Also get the *IQR* by using **Summary>Add Formula**.)
- Drag the variable *Sex* to the rightward-pointing arrow of the Summary Table. You should have all the numbers to compare the male and female Australian high school students.

    b. [*Review*] Judging from the *standard deviations*, which height distribution has the greater *spread*, the heights of the male students or the heights of the female students? Cite the correct measures.

    c. [*Review*] Use the numbers in your Summary Table to calculate the **Lower Fence** and the **Upper Fence** for the *female* high school students. The numbers you get should agree with what you see in the box plot.

- Recall that standard deviation is most appropriate for symmetrical distributions, but the height distribution for the females has outliers. We will look at *female* height *in* the data without the outliers. Select the **Summary Table** and go to **Object>Add Filter**.

105

- You will see a *dialogue box*; in the "scrollable" list, find *Functions>Logical>inRange* and type: inRange(Height,145,185) and Sex = "Female". Fathom will automatically place parentheses around Sex = "Female". Press "OK". (The numbers 145 and 185 are the answers for part c.)

    d. Rounded to one tenth of a centimeter, the mean height of the female Australian high school students (without the outliers) is 165 cm and the standard deviation of 7.4. Make a nice sketch of the Normal model with $\mu = 165$ and $\sigma = 7.4$, showing the Height values at -3, -2, -1, 0, 1, 2, and 3 standard deviations from the mean.

    e. Use the Normal model to find the proportion of Australian female high school students whose height is less than 151 centimeters. Go through all the steps; show the sketch, do the calculation, consult the **Normal Probability Chart,** and interpret your result. (*Hint*: Which type of problem is this?)

    f. For the female Australian students, find and interpret $P(X \geq 170)$ using the Normal model. Draw a picture, do a calculation, and interpret. (*Hint*: Which type of problem is this?)

    g. Comment how your results from questions e and f compare with the proportion of elderly English women whose height is less than 151 centimeters and the proportion whose height is greater than 170 centimeters.

    h. Find the height for the *tallest 10%* of the Australian female high school students using the Normal model. Draw a picture, do a calculation, and interpret. (*Hint*: Which type of problem is this?)

    i. [*Review*] Use the Fathom Summary Table (not the Normal model) to calculate a good measure of the range of the heights of the "middle" 50% of Australian female high school students.

5. **Heights of Female Australian High School Students.** Fathom will allow you to bypass the use of the **Normal Probability Chart**. Fathom and the chart both give the **cumulative probability**; that means the probability is cumulated or added up from the *left* up to the *z* that you have specified. Here is how to use Fathom to answer 4e above, which was to find the proportion of female Australian high school students having heights less than 151 cm using the Normal model with $\mu = 165$ and $\sigma = 7.4$.

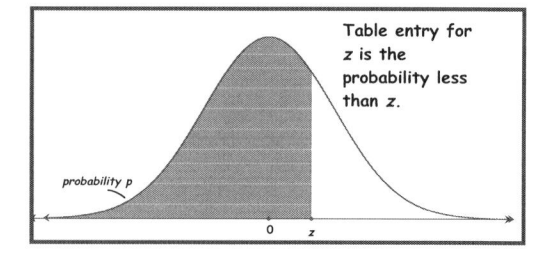

- Get a Fathom **Summary Table** and with it selected, go to **Summary>Add Formula.**
- In the *dialogue box* select **Functions>Distributions>Normal>normalCumulative**. Double click on this so that it comes up in the dialogue box. Read the instructions at the foot to know what to type in the parentheses that are provided. All of the numbers you need are given in the paragraph just above in this question; you just have to put the correct number in the correct place. If you do it correctly, the Summary Table should show the value for the probability to be 0.0292527.

    a. For the answer to question 4e above, you should have $z \approx -1.89189$, which you rounded to $z \approx -1.89$ so that you could consult the chart. The chart gave you the probability 0.0294, so that you could write $P(X < 151) \approx 0.0294$. Why does Fathom give the slightly different answer $P(X < 151) \approx 0.0293$?

b. Follow the instructions above to get Fathom to find $P(X<170)$ with the Normal model with the mean $\mu = 165$ and standard deviation $\sigma = 7.4$. Then use that answer to get $P(X \geq 170)$. Compare you answer to what you got for question 4f.

- For a "given a proportion—find a value" problem, such as question 4h above, go to **Summary>Add Formula** with the **Summary Table** selected.

- In the *dialogue box* select **Functions>Distributions>Normal>normalQuantile**. Double click on this so that it comes up in the dialogue box. Read the instructions at the foot to know what to type in the parentheses that are provided. Remember that we want the value for the *tallest* 10%. Your answer should be very close to the correct answer to question 4h. If it differs you may have typed normalQuantile(.10,165,7.4) instead of normalQuantile(.90,165,7.4). Fathom cumulates from the left, just as the Normal chart does.

c. Follow the directions above (using the Normal model with $\mu = 165$ and $\sigma = 7.4$) to find the height of the female Australian high school students so that 25% of the students are shorter and 75% are taller.

d. Follow the directions above (using the Normal model with $\mu = 165$ and $\sigma = 7.4$) to find the height of the female Australian high school students so that 75% of the students are shorter and 25% are taller.

e. Compare your answers to parts d and e with the Q_1 and Q_3 given in the Summary Table. How well to you think the Normal model fits these data?

f. The **Lower Fence** for the female students was 145 cm. Use Fathom and the Normal model with $\mu = 165$ and $\sigma = 7.4$ to find the proportion for the Normal model that are "outliers" in a tail.

6. **Heights of Male Australian High School Students.** Exercises 4 and 5 were about the heights of female Australian high school students; this exercise is similar, but it is about the male Australian high school students. If you have the Fathom file open for Exercise 4, you can skip the first bullets.

- Open the Fathom file **CAS Australia 08 A.** and get a box plot (shown in Exercise 4) with the variable Height on the horizontal axis and the variable Sex on the vertical axis.

- Get a **Summary Table** and drag the variable Height to the downward-pointing arrow.

- With the **Summary Table** selected, use **Summary>Add Formula** to get the standard deviation and **Summary>Add Five Number Summary**. (Also get the IQR by using **Summary>Add Formula**.)

- Drag the variable Sex to the rightward-pointing arrow of the Summary Table. You should have all the numbers to compare the male and female Australian high school students.

- If you have the **Summary Table** for the females, select it and go to **Object>Remove Filter** to get the complete **Summary Table**.

a. Look at the **Summary Table** for the male students and round the mean and standard deviation to the nearest tenth of a centimeter (the nearest millimeter). Make a nice sketch of the Normal model with μ and σ equal to these values showing the Height values at -3, -2, -1, 1, 2, and 3 standard deviations from the mean.

CAS Australia 08 A

Height	Sex Female	Sex Male	Row Summary
	164.391	174.703	169.353
	9.07594	9.47501	10.6007
	105.5	149	105.5
	160	169	162.5
	165	175	169
	170	181	176.5
	197	198	198
	289	268	557
	10	12	14

S1 = mean( )
S2 = s( )
S3 = min( )
S4 = Q1( )
S5 = median( )
S6 = Q3( )
S7 = max( )
S8 = count( )
S9 = iqr( )

b. **Tall Guys** Decide how tall a "tall" male student is. (If you think in feet and inches, express your idea first in inches and then use the multiplication factor 2.54 cm/inch to get the figure in centimeters.) Round to the nearest tenth of a centimeter. Using the Normal model, find the proportion of male Australian high school students that are shorter than the height you think is "tall." Follow the steps: make a sketch (or use the one you made), do a calculation, get the proportion from the chart, and interpret. (Or, instead of using the *z score* and chart, use Fathom or **DistributionCalculator.ggb;** but you still need to draw a picture and interpret.)

c. **Tall Guys, con't.** Use your answer to part b to calculate and interpret the probability that an Australian male student is taller than the height you think is "tall." Use probability notation in your answer.

- Either get a new graph of heights or use the one that you initially for Exercise 4; with the graph, use **Object>Add Filter** and type in: *Sex = "Male".* Remember that you can change the **bin width** by double clicking on the graph and that you can change from **Frequencies** to *relative frequencies* by going to **Graph>Scale.**

  d. **Tall Guys, con't.** Use the *histogram* or the Fathom instruction just above to estimate or find the proportion of male Australian high school students that are taller than the height you think is "tall."

  e. **Short Guys** Decide how tall a "short" male student is. Express this height in centimeters (rounded to the nearest tenth of a centimeter) and using the Normal model for the male Australian high school students, find the probability that a male student is shorter than this height.

  f. **Short Guys, con't.** Use your answer to part b to calculate and interpret the probability that an Australian male student is taller than the height you think is "short." Use probability notation.

  g. **Ordinary Guys** Find the proportion of male Australian high school students who are neither "tall" nor "short" according to the Normal model. Express your answer using probability notation. Remember that when using a continuous Normal model, there is no difference between "less than" and "less than or equal to."

7. **Grape Harvests in Europe from the fifteenth through the nineteenth centuries.** We introduced the Normal distributions as a *model* to fit an empirical distribution. Here is an example of an empirical distribution for which a Normal distribution fits very well, partly because the data are composed of means. The data show the mean start of harvest after September 1 for northern and central France, Switzerland, and Alsace, and cover the years 1484 to 1879.

The variable *StartHarvest* refers to *days after September 1*. So, if the grape harvest started on September 3, the value for *StartHarvest* would be 2, and if the harvest started on October 1, the value would be 30, since thirty days hath September.

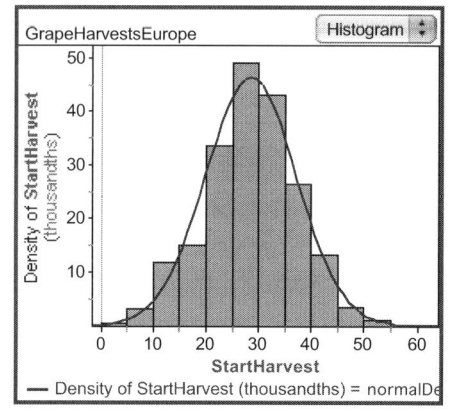

- Open the Fathom file **GrapeHarvestsEurope.ftm** and get a **Case Table** as well as **Graph>histogram** of the variable *StartHarvest*.
- Select the histogram; get **Graph>Scale>Relative frequency** to change the vertical scale to proportions.
- Make the **bin width** equal to five days if it is not already.
  a. What are the cases for these data? Look at the case table but mostly just think about the data.
  b. Use your histogram to estimate the proportion of years that the grape harvest took place before September 21 so that *StartHarvest* = 20.
  c. Use your histogram to estimate the proportion of years that the grape harvest started at least forty days from September 1.
- Get a **Summary Table** showing the mean and standard deviation for the variable *StartHarvest*.
  d. Round the numbers for mean and standard deviation to the nearest *tenth* of a day and sketch a nice Normal distribution model using the mean for μ and the standard deviation for σ. Show the values for -3, -2, -1, 1, 2, and 3 standard deviations on your graph.
  e. Using the Normal model, find the proportion of years that this model predicts that the grape harvest will begin sooner than September 21. (*Hint:* Which type of probability problem is this?)
  f. Using the Normal model, find the probability that the grape harvest happens at forty days or later after September 1. Show your work.
  g. Using the Normal model, find $P(20 \leq X < 40)$ where X stands for the days after September 1 that the grape harvest started (i.e., the number of days in *StartHarvest*). Sketch a picture of your answer.
  h. There is a number of days after September 1 where the grape harvest started in fewer than 5% of the years recorded. Find this number of days and round to two decimal places. (*Hint:* Is this a "given a value—find a proportion" problem or a "given a proportion—find a value" problem? Decide and draw a sketch, do a calculation, and interpret.)
  i. Express your answer (and therefore the question) in part h in probability notation using X to denote the days in the variable *StartHarvest*.
- Check your answers to parts e, f, g, and h using **DistributionCalculator.ggb**.
- Time travel back to the fifteenth and sixteenth centuries to check the accuracy of the data.

Special Exercise A: Normal Foot Lengths In a previous class, we collected data on the length of each student's right foot, measured in centimeters. Here is the histogram showing the distribution of right-foot lengths.

- You can get this histogram by opening ***CombinedClassDataSpr07.ftm*** and dragging the variable *Length_of_Foot* to a ***graph*** and (with the graph selected) going to the menu ***graph>scale>relative frequency.***

1. ***Some review using histograms*** On the histogram, the bars start at 16.5 cm. Hence the first bar shows the students whose right feet were in the interval: 16.5 cm ≤ *Length_of_Foot* < 17.5 cm
   a. What is the width of each bar in the histogram?
   b. Use the histogram and the relative frequency scale on the left side of the graph to find the percentage of students who have right foot length greater than or equal to 26.5 cm. For your answer, show a sum of proportions. Make a sketch of the histogram shown here or get your own histogram from Fathom and shade in the bars for *Length_of_Foot* ≥ 26.5.
   c. Use your answer to part b to find the percentage of students who have right foot length *less than* 26.5 cm. (There is a very easy way to do this.)
   d. Use the histogram and the relative frequency scale on the left side of the graph to find the percentage of students who have a right foot length less than 21.5 cm. (You should be adding the heights of four, not five, bars; why?)
   e. We want to know the percentage of students whose right foot is in the interval:
      21.5 cm ≤ *Length_of_Foot* < 26.5 cm.
   f. Use your answers to parts c and d (and some subtraction) to get the percentage of students whose right foot length is in the interval: 21.5 cm ≤ *Length_of_Foot* < 26.5 cm.

2. ***Using the Normal Model*** We can answer the same kinds of questions using our Normal model for the distribution.
   a. The ***Density Curve*** on top of the histogram is the Normal distribution that has $\mu = 23.9$, and standard deviation $\sigma = 2.4$. The choice of μ and σ is from the mean and standard deviation of the data. Work out the values for $\mu - 3\sigma, \mu - 2\sigma, \mu - \sigma, \mu + \sigma, \mu + 2\sigma,$ and $\mu + 3s$, and make a sketch of a Normal distribution (like the one here) with the values you have calculated, as well as μ. (See ***Special Exercise B*** 1a for guidelines about drawing.) 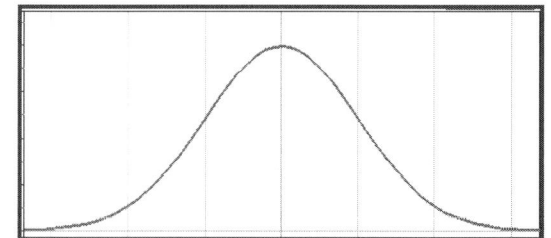
   b. Using the Normal model, find the percentage of students whose right foot length is less than 21.5 cm. Which type of probability question is this? As shown in the ***Notes,*** shade in and label with a question mark on the graph what it is that you are asked to find. See Example 2 in the ***Notes.*** 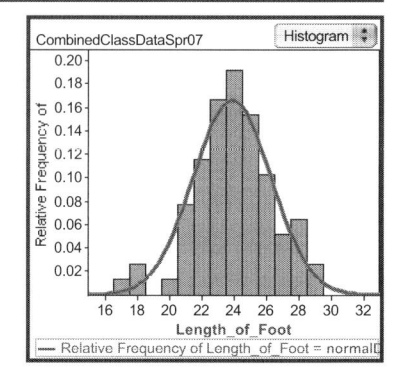
   c. For $x = 21.5$, calculate the z score: $z = \dfrac{x - mean}{SD}$

d. Use the z score you have calculated and the **Normal Distribution Chart** you have been given to find the proportion of students whose right foot is less than 21.5 cm long. Express the answer using the probability notation $P(X < 21.5) = $ _____. (The answer may be checked using **DistributionCalculator.ggb.**)

e. Multiply your answer to part d by 100 to get the percentage of students whose right foot is less than 21.5 cm long according to the Normal model and write a short sentence (in the context of the problem) interpreting what you have found.

3. **Using the Normal Model Again** Now our question is: "What is the percentage of students whose foot length is greater than or equal to 26.5 cm?"

   a. Is this question a "given a value—find a proportion" problem or a "given a proportion—find a value" problem? On the graph you made, label what it is that you are asked to find, using shading.

   b. Using the same mean and standard deviation (it is the same model!) $\mu = 23.9$ and standard deviation $\sigma = 2.4$ find the z score for this question.

   c. Your z score should be positive. By looking at the graphic, explain in the context of the problem why it makes sense that the z score is positive.

   d. Look at the Normal table and read off the number for your z score. You should get 0.8599, which translates to 85.99%. This is *not* the answer! *Why* is 0.8599 not the answer? That is, explain why such an answer makes no sense in the problem and also why the table gives you this answer.

   e. So, to get the answer to the question, what do you have to do with the 0.8599? Do what you have to do and get the answer.

   f. Put your answer in probability notation.

   g. Interpret the answer in the context of the problem.

4. In one "given a value—find a proportion" problem, you had to subtract what you get from the **Normal Distribution Chart** from 1, and in the other problem you did not. What will you look for in future problems to determine when and when not to subtract from 1? To answer this question, you are to make your own rule to help you remember when and when not to subtract from 1. But it better work! It is actually very logical.

5. **Using the Normal Model Again** We want to know the proportion of students whose right foot is in the interval: 21.5 cm ≤ Length_of_Foot < 26.5 cm and we want to use the Normal distribution model to find the answer.

   a. Is this question a "given a value—find a proportion" problem or a "given a proportion—find a value" problem?

   b. You should be able to use the results of questions 2 and 3 (with some thought about what you have found) to calculate the solution, and you should be able to do this without calculating a z score or looking at the **Normal Distribution Chart.**

   c. Express your answer using the probability notation, using X to represent Length_of_Foot.

6. Questions 2, 3, and 5 repeated what you did in question 1 with the histogram. Compare your results using the Normal model with what you got using the histogram.

Special Exercise B: Normal Foot Lengths

1. **Being an artist** In the small table are the answers to question 2a in the first Special Exercise on Foot Lengths. That question asked you to calculate $\mu - 3\sigma$, $\mu - 2\sigma$, $\mu - \sigma$, $\mu + \sigma$, $\mu + 2\sigma$, and also $\mu + 3s$ for the **Normal distribution** with $\mu = 23.9$ and standard deviation $\sigma = 2.4$—the one we have been using to model the foot lengths of the collection of students' feet. **PTO**

$\mu - \sigma$	21.5	$\mu + \sigma$	26.3
$\mu - 2\sigma$	19.1	$\mu + 2\sigma$	28.7
$\mu - 3\sigma$	16.7	$\mu + 3\sigma$	31.1

   a. Use these numbers to make your own beautiful sketch of a Normal distribution. Your sketch
      - should show the location of the mean μ of the model we are using for foot length and also the values for $\mu - 3\sigma$, $\mu - 2\sigma$, $\mu - \sigma$, $\mu + \sigma$, $\mu + 2\sigma$, and $\mu + 3\sigma$ (Remember about the inflection point.)
      - should look like a Normal curve and not like a tent or a mound. Here are two bad examples.
      - need *not* have a scale on the *vertical* axis (as the Normal curves produced by Fathom have.)

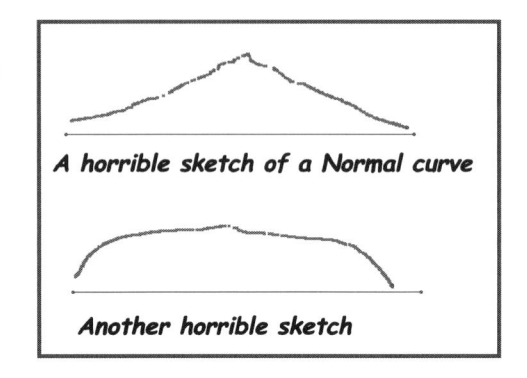

      A horrible sketch of a Normal curve

      Another horrible sketch

   b. Use the **Empirical Rule** to estimate approximately what proportion of students' feet should be within the interval 21.5 cm and 26.3 cm if the foot length distribution is modeled by a Normal distribution with $\mu = 23.9$ cm and standard deviation $\sigma = 2.4$ cm.

   c. Use the **Empirical Rule** to estimate approximately what proportion of students' feet should be 26.3 cm or longer if we use a Normal distribution with $\mu = 23.9$ cm and standard deviation $\sigma = 2.4$ cm.

   d. Is the question asked in part c a "given a value—find a proportion" problem or a "given a proportion—find a value" problem? Give a very short reason for your answer.

   e. Use the **Normal Distribution Chart** to determine the proportion of students' whose feet are 26.3 cm or longer, if the foot length distribution is modeled by a Normal distribution with $\mu = 23.9$ and standard deviation $\sigma = 2.4$. Recall the steps laid out in the **Notes:** make a sketch (you may use the one for part a), make a calculation, consult the chart and interpret your result.

2. **P'tite Feet: Using the Normal Model.** After taking this course, you plan to open a boutique in Healdsburg (Healdsburg is full of boutiques) for people with small feet. You will cater for the "smallest-footed" 10% of people. So your question is: "What is the foot length for the 10% of people with the smallest feet, according to the **Normal distribution** with $\mu = 23.9$ and $\sigma = 2.4$?"

   a. Is this question a "given a value—find a proportion" problem or a "given a proportion—find a value" problem? Give a reason for your answer.

   b. Make another sketch of a Normal distribution with $\mu = 23.9$ and standard deviation $\sigma = 2.4$ and indicate on this sketch what you are given and what it is you are being asked to find.

   c. Find the z score that you will use to solve this problem. (Remember that 10% is .1000.)

   d. Use the z score to find the numerical answer to the question. (The answer is 20.8 cm.)

   e. Give an interpretation in the context of the question asked.

3. **Sasquatch Ambition** Or, alternatively, instead of a boutique in Healdsburg for small-footed people, you could cater for the 5% of people with the biggest feet. Then your question is: "What is the foot length for the 5% of people with the biggest feet if we use the **Normal distribution** with $\mu = 23.9$ and standard deviation $\sigma = 2.4$ as our model?"
   a. Is this question "given a value—find a proportion" or "given a proportion—find a value"? Give a reason for your answer.
   b. Make another Normal distribution sketch using shading and labeling to show what you are given and what you are to find.
   c. Should the z score that you use be a positive or a negative number? Give a reason for your answer based on your sketch.
   d. Get the correct *z score* and do some algebra to find the numerical answer to the question.
   e. Forgetful Fiona gets 19.9 cm for her answer to part d. Convince FF that this cannot possibly be the answer (refer to your sketch) and determine what mistake she made. (Correct: 27.8 cm.)
   f. Put your answer into the probability notation that we have been using and interpret your answer in the context of the question.

4. **Just Normal** Probably the best business plan is to have shoes for 95% of foot lengths. We will again use the **Normal distribution** model with $\mu = 23.9$ cm and standard deviation $\sigma = 2.4$ cm.
   a. According to the **Empirical Rule,** what is the *approximate* interval of foot lengths that will accommodate 95% of the people if $\mu = 23.9$ cm and standard deviation $\sigma = 2.4$ cm?
   b. Is the question in part a "given a value—find a proportion" or "given a proportion—find a value"? Give a reason for your answer.
   c. Make another sketch of a Normal distribution and show the mean μ and the points for $\mu - \sigma$, $\mu + \sigma$ $\mu - 2\sigma$, and $\mu + 2\sigma$; show by shading the area of the graph that includes 95% of the data.
   d. The **Empirical Rule** gives an approximate answer of *two* standard deviations on each side of the mean, but there is a more exact answer, which you will now find using "given a proportion—find a value." (You will find that it is almost the same as the approximate answer.) Here are the steps to get this more accurate answer. From your sketch, you should have a shaded in portion in the center of 95%. What percentage does that leave on the left-hand side? Show this percentage and label it the "smallest ___ % of foot lengths" on your sketch.
   e. Your percentage should be 2.5%. Using the proportion .0250, find the *z score* from the body of the **Normal Distribution Chart** that corresponds to this and write it.
   f. Use your answer to part e and some algebra to get the value of the 2.5% smallest feet.
   g. For the right-hand side (the biggest feet), the fact that the **Normal Distribution Chart** gives cumulative proportions (proportions less than a z) means that we have to find the proportion to the left of the end of the 95% middle feet. What is that proportion? Use it to with the body of the chart to find the *z score*.
   h. Get the upper 2.5% of foot lengths, and thus the upper end of the 95% interval.
   i. You should have found that the answer to part g was positive but the same number as the answer to e. What feature of the shape of **Normal distributions** explains why this is so?
   j. Use the results to fill in the blanks: $P($ ___ cm. $\leq X \leq$ ___ cm.$) = 0.95$

Revision for §§1.1 – 1.6: Hobbits and Men

8. The Fellowship of the Ring In J. R. R. Tolkien's classic, the heroes of the story are the hobbits. Soon after starting their journey from the Shire the hobbits came to a place named Bree. This town was in an area where both Men and Hobbits lived.

> The Big Folk and the Little Folk (as they called one another) were on friendly terms, minding their own affairs in their own ways, but rightly regarding themselves as necessary parts of the Bree-folk. Nowhere else in the world was this peculiar but excellent arrangement to be found.
>
> (J. R. R. Tolkien, *The Fellowship of the Ring*, Ballantine Books Edition, p. 206)

We have managed to get a sample of the heights (measured in cm) of Bree-folk who happened to be in *The Prancing Pony*, the inn in Bree. There are $n = 26$ Bree-folk represented, where $n_{Little} = 12$ and $n_{Big} = 14$. Here are the data. (It was not easy to get the data!)

a. In the data shown on the right, the first entry is 165 cm. Is this number:

   i) a variable,   ii) a value of a variable   or   iii) a case

b. By hand, make a dotplot of the height data. Start by drawing a line; divide the line into equal intervals, and label the line appropriately. (The resulting dot plot should be neat; do not worry about getting something like 181 exactly correct.)

c. Calculate the mean height for *all* 26 of the Bree-folk together.

d. Show the symbol and the formula for the mean you have calculated in part c.

e. The correct answer to part a is ii) "value of a variable". Explain why *variable* is the wrong answer.

f. Explain why *case* is the wrong answer to part a.

g. Make an ordered stem plot using stems of five, and from that calculate the Five Number Summary for the heights of *all* the Bree-folk together. For the stems, divide the numbers between the tens and ones digits, so that the stems are 10, 10, 11, 11, . . . ,18, 18.

h. From your answer to c, make a boxplot. Comment on the usefulness of the boxplot as compared with the dot plot in the context of Bree-folk. (Does the boxplot obscure features of the data?)

	Height	Gender
1	165	Female
2	102	Female
3	111	Male
4	164	Female
5	170	Male
6	107	Male
7	109	Male
8	107	Male
9	181	Male
10	104	Male
11	108	Male
12	181	Male
13	174	Male
14	172	Male
15	106	Male
16	115	Male
17	102	Male
18	166	Female
19	104	Female
20	170	Female
21	175	Male
22	106	Female
23	172	Male
24	179	Male
25	172	Male
26	167	Male

- Open the Fathom file: **BreeFolkExpanded.ftm** and get a case table, by highlighting (not clicking on it) the collection icon, and dragging down a table.

- From the Shelf, drag down a **Graph**. To the horizontal axis drag the variable *Height*. You should see a dot plot, and you should be able to compare it with your hand drawn dot plot.

- On the top right hand corner of the dot plot, there is a pull down menu that will allow you to change the dot plot to a box plot. Do that, and compare the box plot that you have made to Fathom's.

i. Calculate means and medians for the height data that make sense to the Bree-folk. (*Hint:* From the quotation at the head of the exercise, it is clear that the Bree-folk make a clear distinction between Big Folk and Little Folk.)

j. By hand, calculate the Five Number Summary for the heights of the Little Folk.

- To check your answers to part i, drag down a **Summary.** Drag the variable *Height* to the down facing arrow, and then the variable *Folk* to the right facing arrow.

- To check your answer to part j, select the **Summary.** Go to the menu and choose: **Summary>Add Five Number Summary.**

- Select the Fathom box plot for *Height*, and drag the variable *Folk* to the vertical axis. You should see these boxplots. To the Bree-folk, this plot better shows what they know about those in their community.

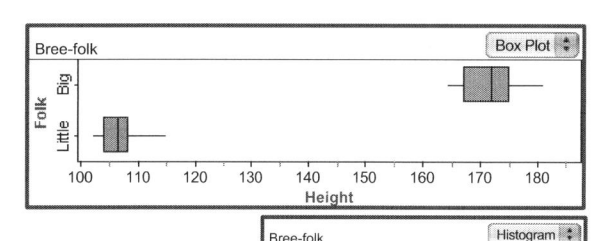

k. Here is a histogram showing the heights of the $n_{Little} = 12$ Little folk, the Hobbits. For this histogram, what is the bin width?

l. Find the probability that a Hobbit is 105 cm or taller, and express that probability using the correct notation. (Here, "given" notation is optional)

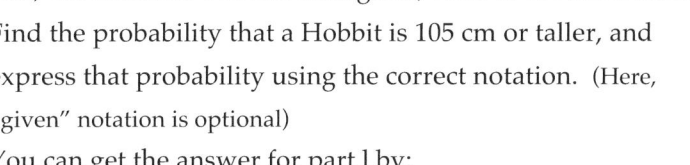

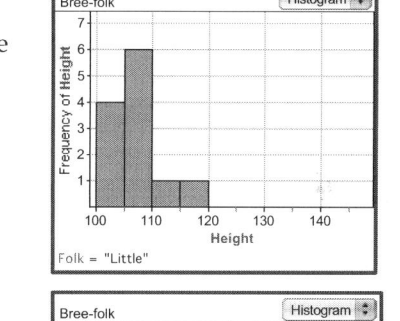

m. You can get the answer for part l by:

$1 - P(X < 105) = 1 - \frac{4}{12} = 1 - \frac{1}{3} = \frac{2}{3} \approx 0.67$. You are using one the probability rules; what is the rule called?

n. Here is a histogram for the heights of the "Big Folk." Notice that the vertical axis has relative frequency rather than frequency. Using this histogram, express the probability that a Big Folk is shorter than 175 cm. and use the correct notation. (You will have to approximate.)

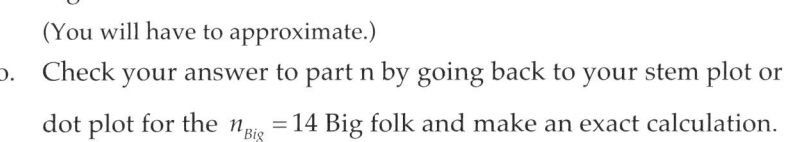

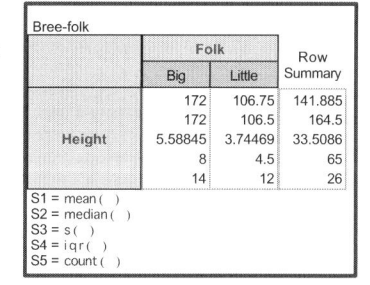

o. Check your answer to part n by going back to your stem plot or dot plot for the $n_{Big} = 14$ Big folk and make an exact calculation.

p. Here are measures of center and measures of spread for the heights of the Big Folk and the Little Folk. Which are the measures of center and which are measures of spread (or variability)?

q. Interpret the measures of *variability* in the context of the data. That is, you should cite which measure is being used, and what these mean in terms of the heights of the Big folk and Little folk.

r. If you calculate $\sum_{i=1}^{n}(x_i - \bar{x})$ will you get: i) the mean, ii) the standard deviation, or iii) zero? Why?

115

s. The first two heights for the Big folk are 165 and 164, and the last height is 167 of the $n_{Big} = 14$ heights. Using these numbers, set up (but do not complete) the calculation of the standard deviation using the correct formula. Use " ... " to show the missing parts, but put the numbers in the correct places.

t. Using the correct notation, find the probability that someone in the *Prancing Pony* is both a male and one of the Big folk, using the table here. [Use F, M, B, L for the events that a female, etc. are in the *Prancing Pony*.]

Bree-folk

		Gender		Row Summary
		Female	Male	
Folk	Big	4	10	14
	Little	3	9	12
Column Summary		7	19	26

S1 = count ( )

u. Calculate $P(F|B)$ and give a meaning to the calculation.

v. Find the probability that if a male is found in the *Prancing Pony*, that male is one of the Big folk. Express the probability using the correct notation.

w. Consider all of the Bree-folk found in the *Prancing Pony*. Calculate probabilities to compare whether females are more likely amongst the Big folk or amongst the Little folk.

x. Determine, with the correct calculation, whether the events F and B are independent or not.

y. Are the events F and B mutually exclusive? Give a reason for your answer.

§2.1 Comparisons, Relationships, and Interpretation

Review Example: Using the tools to calculate and interpret

This section begins by reviewing what we have done in the previous sections. The first examples are similar to the exercises for this section. At the same time, we will look at some statistical questions in some new ways.

Data on used BMWs, Audis, etc. In this data set, each case is a used car that had been placed on the website www.cars.com in Boston, Chicago, Dallas, or the San Francisco Bay Area. We chose to look at only five different models that compete in the "imported" sport sedan sector; the five are the Audi A4, BMW 3 Series, Mercedes C-class, Infiniti G35, and Lexus IS. The data were collected from the website in summer 2009. All of the cars were chosen of the makes listed just above that were advertised on the site for the four places listed (Boston, etc.). We can compare the cars or we can compare the places. The data look like this; the variables and their definitions are listed below.

	Make1	Place1	Year	Price	Miles	Distance	Age	Convert...	NoPrice	Body	Dist
605	BMW 3 S...	SF Bay A...	2008	31788	13194	33	1.33333	Not Conv...	Price Given	Sedan	33 mi.
606	BMW 3 S...	SF Bay A...	2008	30988	13656	33	1.33333	Not Conv...	Price Given	Sedan	33 mi.
607	BMW 3 S...	SF Bay A...	2008	30788	14573	33	1.33333	Not Conv...	Price Given	Sedan	33 mi.
608	BMW 3 S...	SF Bay A...	2008	30692	8040	11	1.33333	Not Conv...	Price Given	Wagon	11 mi.
609	BMW 3 S...	SF Bay A...	2006	30388	26787	33	3.33333	Not Conv...	Price Given	Sedan	33 mi.
610	BMW 3 S...	SF Bay A...	2006	29995	30862	27	3.33333	Convertible	Price Given	Convertible	27 mi.
611	BMW 3 S...	SF Bay A...	2006	29995	24907	27	3.33333	Not Conv...	Price Given	Sedan	27 mi.

Make1:   Make of the car: Audi A4, BMW 3-Series, Mercedes C-Class, Infiniti G35, Lexus I.S
Place1:   Place the car was being sold: Boston, Chicago, Dallas, San Francisco Bay Area
Year: Model year of the car being sold
Price: Price that the seller was asking for the car
Miles: Miles that the car being advertised had been driven
Distance: Distance that the seller is from the zip code entered in the www.cars.com search engine
Age: Age of the car; depends upon the model year and time of year the data were collected
Convertible: Whether the car being sold is a convertible or not
NoPrice: Whether or not the price of the car is given
Body: Body style of the car: sedan, coupe, convertible, wagon, hatchback

First Question: Examining two categorical variables *In which city (Boston, Chicago, Dallas, or San Francisco) are convertibles more likely to be found in the cars for sale?* How do we answer this question? This looks like the kind of question that we answered in §1.2 when we looked at probability; we want to *compare* the different places by the proportion of convertibles for sale. Using Fathom, we would make a *summary table* such as the one shown. If we let C stand for the *event* that the car being sold is a convertible, we would then calculate the conditional probabilities

Summer 09 Used Cars

		Convertible	Not Convertible	Row Summary
Place1	Boston	17	338	355
	Chicago	60	631	691
	Dallas	36	558	594
	SF Bay Area	39	536	575
	Column Summary	152	2063	2215

S1 = count( )

for convertibles $P(C \mid Boston)$, $P(C \mid Chicago)$ etc. and compare these. Here are two examples for review: $P(C \mid Boston) = \dfrac{17}{355} \approx 0.048$ and $P(C \mid SF\ Bay\ Area) = \dfrac{39}{575} \approx 0.068$. It will be a good exercise to calculate and compare the other two conditional probabilities; you may be surprised at the result. Notice that the "conditions" (the "givens") are the categories of the variable *Place1*.

The lesson here is that if we want to examine the relationship between two categorical variables then the tool that we have at our disposal is to compare conditional probabilities. We can think of this as comparing one of the categorical variables (likelihood of a convertible) by the categories of the other categorical variable (the place).

Second Question: A quantitative variable by categories of a categorical variable

If we look just at the model year 2006 cars being sold, are there differences in the distributions of miles driven between convertibles and cars that are not convertibles? (In other words, are convertibles driven more or less?) This looks like the kind of question we answered when we were discussing the center and spread of distributions. Here are the kinds of analyses we would do. We would make both a graph showing the distributions and get a summary table showing the calculations. Our results show that the distribution for *Miles* for the convertibles is to the left the distribution for the non-convertibles, indicating that the convertibles were on average driven fewer miles. The mean and medians (numerical measures of location) agree with this assessment, since the numbers are higher for the non-convertibles than for the convertibles. The measures of spread (standard deviation and iqr) indicate that there is also somewhat *less variability* in the miles driven for the convertibles being sold than for the other body styles.

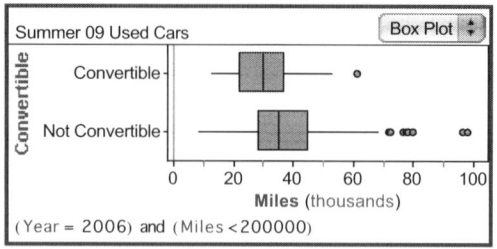

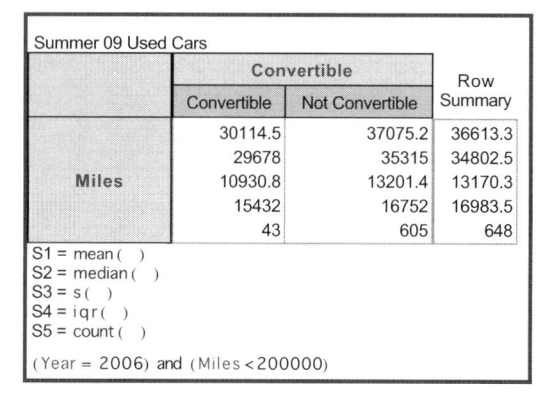

How do we decide what to calculate?

Think: Categorical and Quantitative Variables How did we decide that we should calculate means, medians, standard deviations, and IQRs in the *second question* and probabilities in the *first question*? The answer is that we looked at whether the variables were *categorical* or *quantitative*. In the *first question*, the two variables (*Place1* and *Convertible*) are both *categorical*; to handle the situation where we are comparing categories of one categorical variable (*Convertible*) according to the categories of a second categorical variable (*Place1*), we employ conditional probabilities. In the second question, one of the variables is *quantitative* (*Miles*), but the second variable (*Convertible*) is categorical. Since we have a quantitative variable, it makes sense to calculate means and medians and compare them according to the categories of the categorical variable.

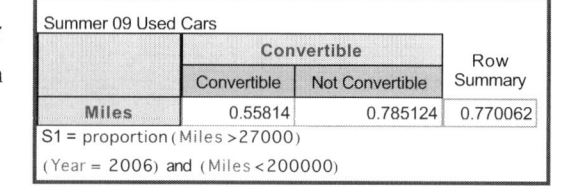

We can also use the tools of probability when we have a quantitative variable; for example we could calculate the proportion of 2006 cars for sale that have been driven more than (say) 27,000 miles for the convertibles and non-convertibles. The results can be written $P(Miles > 27000 \mid C) \approx 0.56$ and $P(Miles > 27000 \mid NC) = 0.79$. You can also think of making a categorical variable for *Miles* with two categories: miles driven greater than 27,000, and miles driven 27,000 or less. We are comparing just one of these categories for the convertibles and non-convertibles. The probability of finding a 2006 non-convertible for sale that had been driven more than 27,000 miles was higher than finding a convertible that had been driven more than 27,000 miles.

Think: Explanatory and Response Variables Notice that the phrase "according to the categories of the categorical variable" appeared twice just above. We call a variable **explanatory** when we have reason to think that a second variable (which we call a **response variable**) will vary according to the values of this **explanatory variable**. So, we suspect (though we do not know until we look at the data) that the likelihood that a car for sale is a convertible (the *response variable*) will vary according to the place (the *explanatory variable*) in which it is sold. That is, we think that the *miles driven* may vary among the various makes of car. Of course, there may be no substantial differences in the *miles driven* among the various makes of car, but we will not know until we look at the data. Here are some questions to consider:

- How do you decide which variable is the *explanatory* and which variable is the *response* variable? The answer depends on how we think one variable affects another or which variable is in some sense "prior." For example: we think that it may be possible that the convertibles are driven less over a number of years—so that the fact that a car is a convertible may affect its miles driven. However, the miles that a car is driven cannot affect whether it is a convertible or not.
- Can *quantitative* variables also be *explanatory*? Yes; a large part of this chapter is about quantitative variables that are explanatory. The age of a car we think affects the price of a car; in this example, age is a *quantitative* variable being used as an explanatory variable.
- Are there situations where the choice of one variable as *explanatory* and the other as *response* is either impossible or does not matter? Yes. If you had data on the exam scores on a collection of students who were all taking chemistry and history then we cannot say that chemistry affects history or history affects chemistry, even though there may be a relationship between the chemistry and history scores.
- Is the distinction between *explanatory* and *response* variables the same as the distinction between *independent* and *dependent* variables? Yes; in mathematical terms, the distinction is the same. The reason that different terms are used in statistics texts is that the terms *independent* and *dependent* have a different, and well-defined, meaning.

> **Explanatory and Response Variables**
>
> An **explanatory** variable allows us to predict in some way the values of a **response** variable.

Putting these bits of advice together, we offer these guidelines.

> **Guidelines for Calculation**
> - If the variables in the statistical question are both *categorical*, calculate *conditional probabilities* with the categories of the *explanatory variable* as the conditions.
> - If one variable is *quantitative*, make a graphic that compares the distributions of the *quantitative variable* according to the categories of the *categorical variable*.
> - If one variable is *quantitative*, calculate and compare measures of center and spread for the categories of the *categorical variable*.

Suppose the two variables are *both quantitative*? We have not considered that situation up to now.

Comparisons between Groups or...Relationships between Variables

Consider the example at the beginning of this section. The most natural way of thinking about the data in the first example—the one about the proportion of convertibles in the four places—is to think of four groups: the four different places the data were collected. We compare the proportions of convertibles being sold in these four places. However, there is another way to think to look at the same table.

We can *also* think of *Place1* and *Convertible* as two variables within one collection; after all, both of these are columns in the Fathom **caseTable,** and we think of the columns as variables and the rows as

Summer 09 Used Cars		Convertible		Row Summary
		Convertible	Not Convertible	
Place1	Boston	17	338	355
	Chicago	60	631	691
	Dallas	36	558	594
	SF Bay Area	39	536	575
Column Summary		152	2063	2215

S1 = count ( )

cases. If we think this way then the *summary table* shows the **relationship** between the variables. Whether we think of the categories as groups, or whether we think of them as values of a variable, the calculations we make are very often the same.

Although the calculations are the same, there may be differences in the way in which the data were collected that lead to differences in interpretation. Consider the cars example. We collected the data on the cars from four different places; the fact that we did leads us to think of comparing places. On the other hand, in those four places, *for each car* we recorded whether the car was a convertible and *for each car* we recorded the miles that the car was driven. The fact that both of these variables (*Convertible* and *Miles*) were recorded (or measured) *for each case* leads us to think of looking at the *relationship* between a car being a convertible and the miles that the car was driven. To analyze the proportion of convertibles by place, so far we have just one choice of the kind of calculation we will do: we will calculate conditional probabilities. However, we can *look* at these calculations from two different perspectives.

Calculation and then...Interpretation

You should already be aware that it is not sufficient just to quote the numbers; the numbers need to be related to the context of the data. However, it is also wise to keep in mind how the data were collected. Do the proportions of convertibles in Boston and San Francisco tell us that about 4.8% of the cars in Boston are convertibles and about 6.8% of cars in San Francisco are convertibles? It may be, but we cannot *infer* or generalize that from our data; recall that these data represent only cars being sold over the Internet and only five models of "imported sport sedans." It may even be that the proportion of convertibles *for sale* may be higher than the proportion of convertibles on the road. Keep in mind how the data were collected.

Interpreting calculations: how to complete the Writing Exercises successfully

The exercises for this section involve analyzing data and then writing about what you have found. You will find this a challenge because you probably have not done this in the past; you will make mistakes, and you will have to work at it and revise what you have done. Some advice:

- **Understand the context: the cases, the variables.** Make certain you understand clearly what your cases are, what the variables are, whether the variables are quantitative or categorical, and which variable is being considered as the explanatory and which variable is the response.
- **Correct calculations.** Make certain that the calculations that you are doing (or having Fathom do) are the calculations that are appropriate for the questions being asked.

- ***Consider your audience.*** Your audience is *not* your instructor. It is better to think of your audience as the readers of a student newspaper. A common mistake is to be too technical. You do not need to explain the standard deviation formula; you *may* need to explain to your readers, in the context of the data, what the standard deviation tells you.
- ***Beware writing "it" and "they."*** Writing about technical matters demands that we be very specific. A common writing mistake with statistics is to use "it" and "they" vaguely. Here is an example: "It is right-skewed." What is right-skewed? Australian students? Used cars? Mammals? (What does a right-skewed Australian student or used car or mammal look like?) You are of course referring to the *distribution* of a variable. Distributions of variables can be right-skewed: Australian students or mammals cannot be. Be specific: "The distribution of the variable [name the variable being analyzed] is right-skewed."
- ***Over-Inference***. Another common mistake is to read into the data more than the data can bear. "The students who play lots of computer games come from dysfunctional families, and so…" when in fact there are no data about family backgrounds and you have no way of knowing what you have written.
- ***Start early and check your work.*** This assignment is not one you can do well by starting the night before the assignment is due. If you try that, there is a high probability that you will fail (that is, $P(\text{Fail} \mid \text{Start Late}) > 0.8$ or $P(\text{Fail} \mid \text{Start Late}) > 0.9$).

You will find these exercises challenging; however, if you master the idea, you will ultimately find it very interesting to be able to take some data, do some calculations to analyze the data, and then make sense of what the data show you about a statistical question. Good luck!

Summary: Types of Variables, Calculation, and Interpretation

- ***Calculations depend upon the types of variables involved.***
    - Analyzing two *categorical* variables leads to using conditional probabilities.
    - Analyzing a *quantitative* variable by the categories of a *categorical* variable generally leads to comparing the graphics and the summary measures of the quantitative variable according to the categories of the categorical variable.
- ***Calculations depend upon which variable is explanatory and which is response.***
    - An **explanatory variable** allows us to predict in some way the values of a **response variable**.
    - The choice of which variable is *explanatory* and which variable is *response* depends upon how we think the variables are related.
    - It may be neither of the two variables can be assigned the role of *explanatory* or *response*.
- ***Interpretation of calculations***
    - Must take account of the statistical question being asked
    - Must take account of how the data were collected
    - Must take account of the audience to which the interpretation is directed

§2.1 Exercises: Calculation and Interpretation

1. Interpreting Shape, Center, and Spread: Steam Schooners, California History

At one time (about a hundred years ago), the primary means of transport to the communities on the coast one hundred miles or so north of San Francisco was by ship. Originally, the ships were schooners (sailing ships), but someone had the idea of installing a steam engine, so between 1875 and 1923 many wooden steam schooners were built. We have

Steam Schooners	Name	Tonnage	PlaceBuilt	YearBuilt	TimeBuilt	StateBuilt
1	Argo	210	Ballard, WA	1898	Early: 18...	WA
2	C. G. White	169	San Francisc	1884	Earliest: ...	CA
3	C. H. Wheele	371	Portland, OR	1900	Early: 18...	OR
4	Cleone	197	San Francisc	1887	Earliest: ...	CA
5	Daisy	621	Fair Harbor,	1907	Later: 19...	WA
6	Daisy Mitchell	612	Fairhaven, C	1905	Early: 18...	CA
7	Egaria	2360	Astoria, OR	1920	Latest: a...	OR

data on 216 of the wooden steam schooners that were built. (The data are from: Jack McNairn and Jerry MacMullen, Ships of the Redwood Coast, Stanford, CA: Stanford University Press, 1945.) Here is a description of the variables measured.

Tonnage   Measures the cargo capacity of a ship; the bigger the tonnage, the more the ship will carry.
PlaceBuilt   Place the steam schooner was built
YearBuilt   Year the steam schooner was built
TimeBuilt   Categorcial variable that groups the year the ships were built into four groups:
"Earliest: before 1896",   "Early: 1896–1905,   "Later: 1906–1915", "Latest: after 1915"

a. What are the cases for these data?

b. Which variables are quantitative and which are categorical?

In the **Summary Table** (below) there are measures of the center and the spread of the four distributions of tonnage.

c. Which variable in the *Summary Table* is the explanatory variable and which variable is the response variable? Give a reason for your answer. ("We think that…depends on…")

Steam Schooners	TimeBuilt				Row Summary
	Earliest: before 1896	Early: 1896-1905	Later: 1906-1915	Latest: after 1915	
Tonnage	284.977	536.949	758.074	1389.98	731.972
	258	469	749	1253	649
	106.582	209.087	154.057	543.118	470.837
	108	299	249	499.5	547
	43	59	68	44	214

S1 = mean( )
S2 = median( )
S3 = s( )
S4 = iqr( )
S5 = count( )

d. Shape of the tonnage distributions. What do the measures of mean and medians of *Tonnage* tell you about the shapes of the four tonnage distributions? Give reasons for your answer.

e. Comparing average tonnage of the ships among the time periods. Compare the measures of center for Tonnage across the four categories of *TimeBuilt*. What do these measures tell you about the average *Tonnage* for the ships built in the different periods?

f. Comparing variability in tonnage among the time periods. Compare the measures of spread for Tonnage and write what these measures tell you about the *Tonnage* of ships and the time they were built. Words that are good to use to describe difference amounts of spread are: "variability," "more or less different," "diverse." (How would the shipyards before 1896 be different from the shipyards after 1915 in the variety of ships being built?)

2. Household Size in Past Time and Now There are just two variables in this analysis.

Number_in_Household   Number of people in a household
Place   The places we are comparing are:
Belgrade, Serbia, in 1734 (See Laslett 1972.)
Bristol, Rhode Island, in 1689 (See Laslett 1972.)
Nishinomiya, Hama-issai-cho, in 1713 (See Laslett 1972.)
Québec Scottish communities, in 1851 (Census of Canada)
California Students' households in 2008–'09
Here are box plots and summary statistics.

a. You meet someone who has just arrived via a time machine from one of these places, and assuming that you can communicate (you have developed the ability to converse in Serbian, Japanese, or Scots Gaelic), and knowing what you know from the measures of center and spread from the summary tables and the box plots, explain to the time traveler how your experience is different from theirs. You can choose the place your time traveler comes from, but you need to identify it.

- How do the typical or average household sizes in the past differ from the present-day household sizes? (Measures of center)
- How do the household sizes vary in times past compared with modern-day household sizes? (Measures of spread or diversity)

b. The data from Québec are from the Canadian Census of 1851. The enumerators went from house to house and asked each household head about each person in the household. The data for the California students was collected from students. Are the data on household sizes comparable? Discuss why or why not.

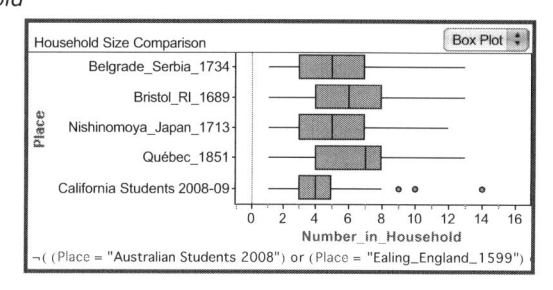

c. Which variable is the explanatory variable and which the response variable?

d. The *Notes* made the distinction between comparing groups and looking at relationships between variables. For these household data, which of the perspectives outlined in the *Notes* fits better? Give a reason for your answer.

Reference: Laslett, Peter (ed.), *Household and Family in Past Time,* Cambridge, Cambridge University Press, 1972

Household Size Comparison		Number_in_Household
Place	Belgrade_Serbia_1734	5.410115 / 2.622224 / 178
	Bristol_RI_1689	5.819446 / 2.749274 / 72
	Nishinomoya_Japan_1713	4.992425 / 2.509514 / 132
	Québec_1851	6.628877 / 2.754994 / 97
	Students California 2008-09	3.988974 / 1.54512 / 272
Column Summary		5.018644 / 2.467863 / 751

S1 = mean( )
S2 = median( )
S3 = s( )
S4 = iqr( )
S5 = count( )

¬((Place2 = "Australian Students 2008") or (Place2 = "Ealing_En

3. **Interpreting Measures of Center and Spread: Gearhead Cars** Road & Track is a magazine that is published for people who like cars. Hence, the cars they write about do not represent all cars that are being made but rather the cars that are interesting to "gearheads"—that is, people who like cars. Here are some of the variables measured:

Continent   Place manufactured: America, Asia, or Europe

Horsepower A measure of how powerful the engine is

MPG Miles per gallon: a measure of how much fuel is used. The larger the MPG, the less fuel is used.

Acc_060 The number of seconds taken to reach 60 mph from a standing start. The smaller the number, the quicker the car can accelerate.

a. What are the cases for these data?

b. Which of the variables are quantitative?

c. Which of the variables are categorical?

Road&Track Jan09 Cars

	Make	Model	Continent	Horse...	MPG	Acc_060...
11	Audi	S6	Europe	435	16	5.1
12	Bentley	Continen...	Europe	552	12.5	4.3
13	Bentley	Continen...	Europe	600	11	4
14	BMW	M3 Coup...	Europe	414	13.1	4.3
15	BMW	M3 Sedan	Europe	414	14.6	4.6
16	BMW	X6 xDri...	Europe	400	14	4.8

In the **Summary Table** you will see measures of the center and the spread of the three distributions of MPG for cars that come from America, from Asia, and from Europe for the cars featured in Road & Track. Answer the following questions based on this summary table.

d. What do the measures of mean and medians tell you about the shapes of the three MPG distributions? Give reasons for your answer.

e. Compare the measures of center across the three categories of Continent. What do these measures tell you about the average MPG for the made in the different continents and featured in Road & Track?

Road&Track Jan09 Cars

		Continent		Row
	America	Asia	Europe	Summary
MPG	17.26	22.0297	16.7413	18.4104
	17	20.4	16	17.8
	2.81603	6.44364	5.36998	5.79642
	2.9	3.6	6	5.9
	25	37	63	125

S1 = mean( )
S2 = median( )
S3 = s( )
S4 = iqr( )
S5 = count( )

f. Write what the measures of <u>spread</u> tell you about the MPG and where the cars were built (the three categories of the variable Continent). Words that are good to use to describe different amounts of spread are: "variability," "more or less different," "diverse."

Here is another summary table showing the differences in the variable Horsepower by Continent.

g. Use the medians to write an interpretation of the differences in average horsepower for cars from the different continents.

h. Use the information given to calculate the IQRs for Horsepower for the cars from the different continents and say what the numbers tell you.

i. For the analyses in this exercise, which variable is (or variables are) being considered as the response variable and which variable (or variables) as the explanatory variable? Give a reason for your answer.

Road&Track Jan09 Cars

		Continent		Row
	America	Asia	Europe	Summary
Horsepower	138	106	70	70
	263	200	265	250
	400.5	263	414	306
	540	291	507	480
	750	480	650	750
	26	37	63	126

S1 = min( )
S2 = Q1( )
S3 = median( )
S4 = Q3( )
S5 = max( )
S6 = count( )

> ### *Writing Assignment: Exercises 4–12*
>
> **Directions: The five instructions saying what you are to do:**
>
> 1. <u>Choose one</u> of the exercises in the group of Exercises 4–12. (However, if you have an idea that is comparable to these exercises but is different then consult your instructor.)
>
> 2. Follow the directions to <u>do the statistical analysis</u> using Fathom.
>
> 3. Create a Word (or Pages) document to keep the **graphics** and **Summary Tables** that you may need for your report and in which to write the report.
>
> 4. Based upon the statistical analysis that you have done, <u>write a two- to three-page report</u> explaining to a reader who has *not* had a course in statistics what the analysis tells you about the statistical questions. Some or all of the **graphics** and **Summary Tables** that you saved will be incorporated in the paper.
>
> 5. You will be given additional instructions about the following, which you should note:
>   - Whether the assignment is to be done individually or as a team, or in either mode
>   - The format of the report (double spacing, type of referencing, etc.)
>   - Mode of submission (whether electronically, or paper, or both)
>   - Due date or dates

Here are some guidelines to follow:

> - Think whether the statistical question involves two categorical variables or one categorical variable, and one quantitative variable and one categorical variable. Consult this section of the **Notes.**
>
> - Decide which of the variables is the *explanatory* variable and which variable is the *response* variable.
>
> - Even before looking at the Fathom output, sketch out roughly what you *think* about the statistical questions. Break down the questions. How do *you* think the *response* variable is related to the *explanatory* variable? You may wish to discuss with others.
>
> - To incorporate Fathom output in a Word or Pages document, follow these instructions:
>   - Highlight the Fathom **graphic** or **Summary.**

- **Go to Edit>Copy as picture** and paste what you have copied into your Word or Pages document.

> - Read **Interpreting calculations...** in this section of the **Notes**. Specifically, make certain that you are writing for the intended audience, who is not your instructor.
>
> - There is an example essay posted on-line that should show not only the process but will also give you some idea of what the essay should be at the end.

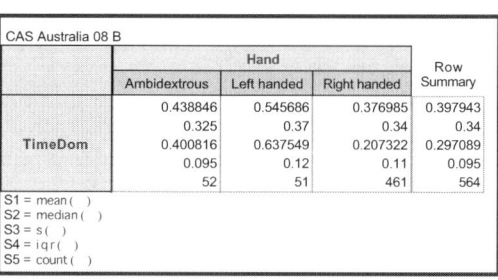

4. Australian Students' Handedness and Reaction Times

Introduction and Background The Australian Census @ School questionnaire was done on-line and included this question: [Image from www.abs.gov.au/websitedbs/CaSHome.nsf/Home/Students+Area, Review Questionnaire.]

> Use your **DOMINANT HAND** to test your reaction time.
> This question requires the use of a mouse and the ability to recognize a symbol. If you are unable to do this please skip to question 11.
>
> Press the Start button. As soon as you see a symbol appear in the box, click **Stop**.

The reaction time was recorded in seconds (or fractions of a second) and is recoded in our data in the variable *TimeDom*. The *dominant hand* for a right-handed person will be the right hand, for a left-handed person the left hand, and for an ambidextrous person (presumably) whichever hand the person chooses. Then the students were asked to do the same thing on the computer with their non-dominant hand. The variables that you will use for this exercise are:

TimeDom. Reaction time in seconds to a visual stimulus using one's dominant hand
TimeNon. Reaction time in seconds to a visual stimulus using one's non-dominant hand
Hand.   Whether the student was right-handed, left-handed, or ambidextrous
Sex.   Whether the student was male or female

Statistical Questions
1. Are there differences in the proportions of male and female students who are left-handed or ambidextrous?
2. Are there differences in the shape, center, and spread of the distributions of reaction times using the dominant hand among the three categories of the variable *Hand*?
3. Are there differences in the shape, center, and spread of the distributions of reaction times using the *non*-dominant hand among the three categories of the variable *Hand*?

Data Analysis
- Open the Fathom file and get a **Case Table** for **CASAustralia08B.ftm**
- Get a **Summary Table** that relates the variables *Hand* and *Sex*.
- Get a graph that shows the three distributions of *TimeDom* by the categories of the variable *Hand*. You may use dot plots or box plots; expect to see very right skewed distributions with many outliers.
- Get a **Summary Table** showing two measures of center and two measures of spread for the variable *TimeDom* broken down by the categories of *Hand*. You should get the table just above.
- Repeat the graphical and summary table analysis for the variable *TimeNon*. A fast and efficient way to do this is to select the graph or the summary table that you have made and in the menu go to **Object>Duplicate Graph or Object>Duplicate Summary Table** then replace the variable *TimeDom* with *TimeNon*.
- If you want to look at the analysis for just the males and then just for the females, you can select the graph or summary table and in the menu get **Object>Add Filter** and type: Sex = "Male" or Sex = "Female."

CAS Australia 08 B

	Hand			Row Summary
	Ambidextrous	Left handed	Right handed	
TimeDom	0.438846	0.545686	0.376985	0.397943
	0.325	0.37	0.34	0.34
	0.400816	0.637549	0.207322	0.297089
	0.095	0.12	0.11	0.095
	52	51	461	564

S1 = mean ( )
S2 = median ( )
S3 = s ( )
S4 = iqr ( )
S5 = count ( )

5. Australian Students' Concentration Game Response Times

Statistical Questions

1. Is there evidence that students who spend much time playing computer games will record shorter response times for a concentration game than students who play computer games less often?
2. Is there a difference between male and female students in the relationship between the response times and the reported hours per week playing computer games?

Background

The data come from Australian Census @ School questionnaire that is done on-line by thousands of primary and high school students. The survey included a "concentration question" as well as questions about the students themselves. You can experience what the students did by going to

http://www.abs.gov.au/websitedbs/CaSHome.nsf/4a256353001af3ed4b2562bb00121564/e80b600a036b28e5ca2574120011bcfd!OpenDocument

and scrolling down to Question 31. A tiny version of what you will see initially (before the game starts) is shown here. The reaction time was recorded in seconds (or fractions of a second) and is recoded in our data.

TimeConc.	Reaction time in seconds to a concentration game described above.
HrVidGme	Number of hours per week students report playing computer or video games
Games10	A categorical variable that divides HrVidGme into two categories: - "Less than 10 hours" per week - "Ten or More hours" per week
Sex.	Whether the student was male or female

Data Analysis

- Open the Fathom file **CAS Australia 08 B.ftm** and get a **Case Table**.

- Get a graph showing the distributions of the variable *TimeConc* broken down by categories of *Games10*. You may use dot plots or box plots, but the graph should resemble the one here.

- Get a **Summary Table** showing two measures of center and two measures of spread for the variable *TimeConc* broken down by the categories of *Games10*. It should resemble the one shown here.

- Get a **Summary Table** that relates the number of males and the number of females who play computer and video games either less than ten hours per week or ten or more hours per week.

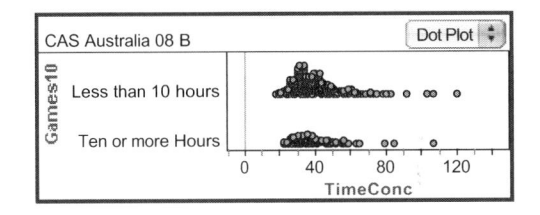

- When you want to look at the analysis for just the males and then just for the females, you can select the graph or summary table and in the menu get **Object>Add Filter** and type: *Sex = "Male"* or *Sex = "Female."* A fast and efficient way to do this is to select the graph or the summary table that you have made and in the menu go to **Object>Duplicate Graph** or **Object>Duplicate Summary Table**.

6. Internet Used Car Markets for BMW 3 Series in Boston and San Francisco

Statistical Questions These questions compare the markets for BMW 3 Series cars being sold on www.cars.com in Boston and the San Francisco Bay Area. The questions are whether there are differences in the ages, prices, and miles driven for the cars being sold through the Internet in the two places.

1. Do the distributions of the *ages* differ in center, spread, or shape between Boston and the San Francisco Bay Area or are they similar?
2. Do the distributions of the *prices* differ in center, spread, or shape for the two places or are they similar? Of course, if there *were* a marked difference in the ages the cars for sale, we would expect to see a corresponding difference in the prices. If the ages of typically lower in one place, we would expect the prices to be higher. Do the distributions of the miles driven differ in center, spread, or shape for the two places or are they similar?
3. Are there differences in the distributions of body styles for the cars being sold in Boston and SF?

Background and Variables

All of the information is from what is shown on the www.cars.com website. These data only include cars that had been listed in the two weeks prior to July 13, 2009, and within a hundred-mile radius of the zip code for CSM, 94402. Keep in mind that the data only refer to the cars that are being sold; the data do not necessarily represent *all* BMW 3 Series or Mercedes' C Class cars on the road. It may well be that the cars that are being sold differ from those that are not being sold.

Place1	Whether the car is being sold in Boston or being sold in the San Francisco Bay Area
Body	The body style of the car being sold: sedan, coupe, convertible, hatchback, and wagon
Age	Estimated age of the car being sold. The only information about age given is the model year. The age is calculated from this information. See the explanation under Question 7.
Price	Price of the car listed on the website
Miles	The number of miles the car for sale has been driven

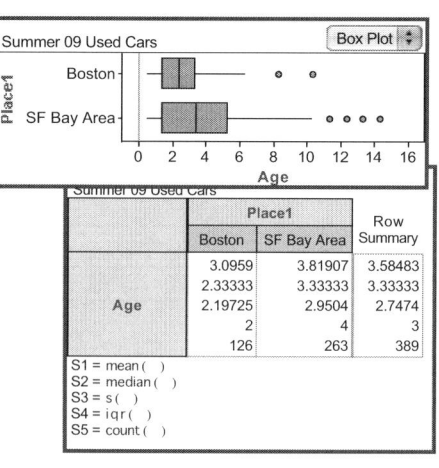

- Use the Fathom file **Summer 09 BMW Boston SF B.ftm**.
- Get a **Graph** showing the distributions of the variable *Age* broken down by categories of *Place1*. You may use dot plots or box plots, but the graph should resemble the one above.
- Get a **Summary Table** showing two measures of center and two measures of spread for the variable *Age* broken down by the categories of *Place1*. It should resemble the one shown here. Repeat this analysis for *Price* broken down by the categories of *Place1* and also for the variable *Miles* broken down by the categories of *Place1*. A fast and efficient way to do this is to select what you have made and in the menu go to **Object>Duplicate Graph** or **Object>Duplicate Summary Table**.
- It will be useful and informative to get the scatterplot and linear model for the relationship between *Age* and *Price*. Interpret the slopes and r^2 to say something about your statistical questions. 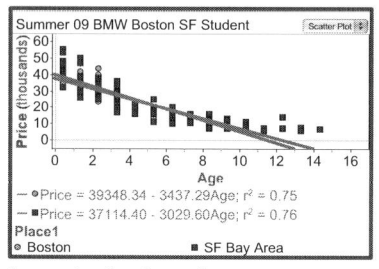
- Get a **Summary Table** showing the relationship between the body style and whether the car is in Boston or the SF Bay Area, and analyze the numbers using conditional probabilities.

7. Internet Used Car Markets for BMW 3 Series and Mercedes C Class

Statistical Questions These questions compare the markets for BMW 3 Series cars and Mercedes-Benz C Class being sold on www.cars.com in the San Francisco Bay Area.

1. Do the distributions of the ages and prices differ in center, spread, or shape between the BMW and the Mercedes?
2. Do the distributions of the prices by age differ for the two makes or are they similar? Of course, if there *were* a marked difference in the ages the cars for sale, we would expect to see a corresponding difference in the prices. If the ages of typically lower for one make, we would expect the prices to be higher. Get scatterplots to answer this question.
3. Do the distributions of the miles driven differ for the two makes or are they similar?
4. Are there differences in the distributions of body styles between the makes of car?

Background and Variables

See the comments for Exercise 6. Keep in mind that the data only refer to the cars that are being sold; the data do not necessarily represent *all* BMW 3 Series or Mercedes' C Class cars on the road. It may well be that the cars that are being sold differ from those that are not being sold. See the additional comments for Exercise 6.

Make1	Whether the car is a BMW 3 Series or a Mercedes-Benz C Class
Body	The body style of the car being sold: sedan, coupe, convertible, hatchback, and wagon
Age.	Estimated age of the car being sold. The only information about age given is the model year. Since a model year 2008 car (to take an example) may have been bought any time from approximately October 2007 to (approximately) October 2008, we can only get an estimated average age. We count the months from October 2008 (ten months) for the "youngest possible" 2008 car and the months from October 2007 to July 2009 (twenty-two months), get the average of these two (sixteen months), and turn this into years (1.3333 years).
Price	Price of the car listed on the website
Miles	The number of miles the car has been driven

Data Analysis

- Use the file **Summer 09 BMW Benz SF A.ftm**.
- Get a **Graph** showing the distributions of the variable *Age* broken down by categories of *Make1* and a **Summary Table** showing two measures of center and two measures of spread for the variable *Age* broken down by the categories of *Make1*.
- Get a **Graph** and **Summary Table** (showing two measures of center and two measures of spread) for the variable *Price* broken down by the categories of *Make1* and also for the variable *Miles* broken down by the categories of *Make1*. A way to do this is to select what you have made and go to **Object>Duplicate Graph** or **Object>Duplicate Summary Table**.
- It will be useful and informative to get the scatterplot and linear model for the relationship between *Age* and *Price*. Interpret the slopes and r^2 to say something about your statistical questions.
- Get a **Summary Table** showing the relationship between the body style and whether the car is a BMW 3 Series or a Mercedes Benz C Class; use probabilities to analyze this table.

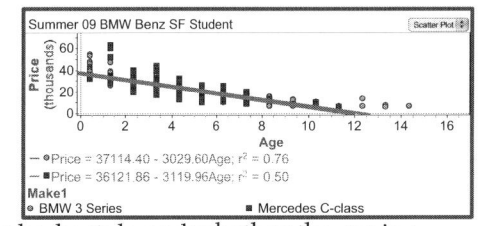

8. Real Estate Markets in San Mateo County between 2005 and 2008

Background and Variables You have already analyzed some of the data on the houses sold in San Mateo County in previous exercises. There are data for the houses sold between June 2005 and June 2006 (Y0506) and also for the houses sold between June 2007 and June 2008. The variables that we will analyze here are:

Year	Whether the house was sold in Y0506 or in Y0708
ListPrice	The price the seller originally asked
SalePrice	The price at which the house actually sold
SqFt	The living area of the house measured in square feet
SoldOverList	Two categories: if the SalePrice > ListPrice then "Over List Price" but if the SalePrice ≤ ListPrice then "Not Over List Price"

Statistical Questions

1. Are there differences in the centers and spreads of the distributions of the sizes of houses (measured by the variable *SqFt*) between the houses sold in Y0506 and the houses sold in Y0708?

2. Are there differences in the centers and spreads of the distributions of the *ListPrice* of the houses sold in Y0506 compared with the houses sold in Y0708? If there are differences, what are they?

3. Are there differences in the centers and spreads of the distributions of the *SalePrice* of the houses sold in Y0506 compared with the houses sold in Y0708? If there are differences, what are they?

4. Are there differences in the proportions of houses that sold "Over List Price" for the houses sold in Y0506 compared with the houses sold in Y0708?

Data Analysis

- Open the Fathom file **San Mateo RECompSampleBoth.ftm** and get a **Case Table**.

- Get a **Graph** showing the distributions of the variable *SqFt* by categories of *Year*. You may use dot or box plots; the graph should resemble the one here.

- Get a **Summary Table** showing two measures of center and two measures of spread for the variable *SqFt* by the categories of *Year*. It should resemble the one shown here.

- Get a **Graph** and **Summary Table** (showing two measures of center and two measures of spread) for the variable *ListPrice* by the categories of *Year*. A fast and efficient way to do this is to select what you have made and in the menu go to **Object>Duplicate Graph** or **Object>Duplicate Summary Table.**

- Get a **Graph** and **Summary Table** (showing two measures of center and two measures of spread) for the variable *SalePrice* by the categories of *Year*.

- Get a **Summary Table** showing numbers of houses that were sold over list price for the year Y0506 compared with the years Y0708 and from that table calculate proportions that give you some insight to question 4.

Here is the table that you should get.

Sample for San Mateo RE Comparison

		SoldOverList		Row Summary
		Not Over List Price	Over List Price	
Year	Y0506	210	284	494
	Y0708	235	121	356
Column Summary		445	405	850

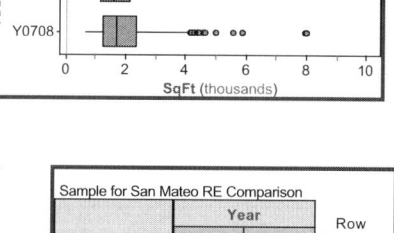

S1 = count ( )

9. Roller Coasters around the World

Background and Variables

There is a website (www.rcdb.com) that is a database for roller coasters. The website says that it "is a comprehensive, searchable database with information and statistics on over 2000 roller coasters throughout the world." The data in the Fathom file **Roller Coasters World.ftm** is from this website and includes data for roller coasters in North America, Latin America, and Europe. However, the data in the database is not complete; the database compilers are dependent on the information available from amusement park operators and manufacturers of roller coasters. Some of the variables are not measured for all roller coasters.

Here are some of the variables measured. Notice that the measurements are all in metric units, so length and height are measured in meters, and speed is measured in km/hr.

Construction	Whether the roller coaster is of steel construction or wood construction
Length	Length of the roller coaster in meters
Height	Height of the roller coaster in meters
Speed	Speed in kilometers per hour (km/hr)
Inversions	Whether or not the roller coaster has inversions
Duration	How long the ride is in seconds
Region	Three categories: "North America," "Latin America," and "Europe"

Statistical Question In this exercise, there is only one statistical question, but it needs to be broken down by the variables measured. The question is:

"What are the differences in the distributions by *Region* of length, height, duration, and whether there are inversions?"

To answer these questions, it will be necessary to break down the question into a number of sub-questions. (Make certain that you consider the spread of distributions as well as the center.)

Data Analysis

- Use the file: **Roller Coasters World. ftm** .
- Get a **Graph** showing the distributions of the variable *Length* broken down by categories of *Region*.
- Get a **Summary Table** showing two measures of center and two measures of spread for the variable *Length* broken down by the categories of *Region*.
- Get a **Graph** and **Summary Table** (showing two measures of center and two measures of spread) for the variable *Height* broken down by the categories of *Region*. Analyze the relationships between the variables *Speed* and *Region*, *Duration* and *Region* in a similar way. A way to do this is to select the graph or the summary table that you have made and go to **Object>Duplicate Graph** or **Object>Duplicate Summary Table.**
- A useful addition to these analyses is to use scatterplots and a linear model to show the relationships between quantitative variables.
- Use a **Summary Table** and probability calculations to answer the question about inversions.

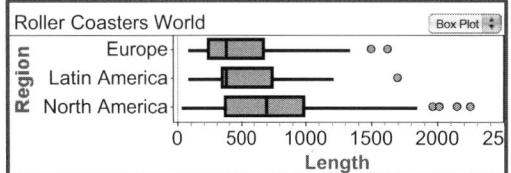

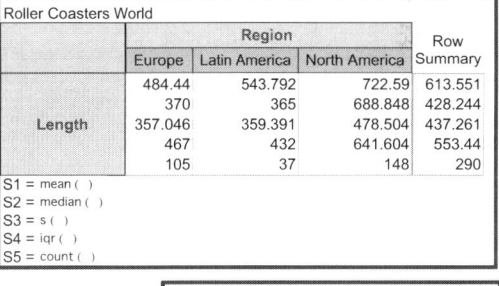

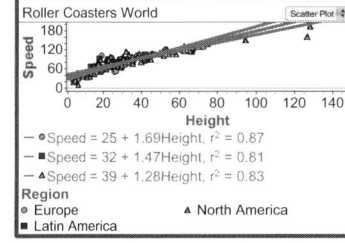

10. NHANES Health Data: Obese and Overweight Gender Differences

Background and Variables

The data are downloaded from www.eeps.com/zoo/index.html but are part of the National Health and Nutrition Examination Survey (NHANES). Both social and medical variables are recorded. For this exercise, we will concentrate on these variables, although there are data on many more variables.

Sex — Male or female
Waist — Waist circumference measured in cm
BMI — Body mass index $BMI = \dfrac{Weight}{(Height)^2}$, where weight is measured in kilograms and height is measured in meters. The idea is to measure "fatness" by taking account of height, since taller people will weigh more just because they are taller. It is widely used but not without controversy. It may not measure "fatness" as well as some other indices.
AgeCat — Age categories: "12 ≤ Age < 20"   "21 ≤ Age < 30"   "31 ≤ Age < 40"   "41 ≤ Age < 50"   "51 ≤ Age < 60"   "61 ≤ Age < 70"   "71 ≤ Age < 85"

Statistical Questions For these health data, it is best to analyze a range of ages. So you will first choose one of the age ranges listed above. The statistical questions are as follows.

1. What differences in center, spread, and shape of the distributions of *BMI* are there between males and females for the age category you have chosen?

2. The usual definition of obese is *BMI* > 30, and the usual definition of "overweight" is *BMI* > 25. Determine what differences there are between males and females in the proportion overweight.

3. What differences in center, spread, and shape of the distributions of *Waist* (waist circumference) are there between males and females?

Data Analysis

- Open the Fathom file **NHANES Data A.ftm**.
- Get a **Graph** showing the distributions of the variable *BMI* broken down by categories of *Sex*.
- To get the graph for just the age group you have chosen, select the graph and go to **Object>Add Filter** and type in: *Age>50* then use the *and* on the palette and then *Age<61* if you have chosen the "51 ≤ *Age* < 60" age category.
- Get a **Summary Table** showing two measures of center and two measures of spread for the variable *BMI* broken down by the categories of *Sex*. You will have to do the **Object>Add Filter** so that the information is just for your age category.
- Get a **Graph** and **Summary Table** (showing two measures of center and two measures of spread) for the variable *Waist* broken down by the categories of *Sex*. A fast and efficient way to do this is to select the graph or the summary table that you have made and in the menu go to **Object>Duplicate Graph** or **Object>Duplicate Summary Table**.
- To answer the question about the proportion of males and females (in your age category) who are overweight, etc. (question 2). Select your **Summary Table** for the variable *BMI*; get **Summary>Add Formula** and make a new formula by typing: *proportion(BMI>25)*. Do a similar new measure for the proportion obese.

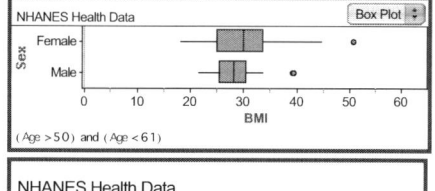

11. Birth Data Questions I: Influence of Smoking

Background and Variables

Almost every birth is recorded on a birth certificate, and the collection of data is available to the public, although it is not easy to download it. There are millions of records for one year. We have data on n = 100,000 births randomly selected for the year 2006. For this exercise, a random sample of n = 2048 has been drawn from the larger random sample; these data are available in the Fathom file **BirthSample06A.ftm** Here are the variables measured and included in this file.

Sex	Sex of the child
BirthWeight	Weight of the child at birth in grams
Plurality	Whether the birth was single birth (1), a twin (2), or triplet (3)
Gestation	Estimated length of pregnancy in weeks
Premature	Whether the birth was premature (gestation < 37 weeks) or full-term
TotalBirthOrder	For the mother, whether this birth was number 1, 2, 3, etc.
AgeMother	Age of the mother of the child in years
AgeFather2	Age of the father of the child in years
EducMother	Mother's education: (a) Less than HS, (b) high school, (c) college, (d) post-graduate
ParentsMarried	Whether the parents of the child are married
MotherSmokes	Whether the mother of the child is a smoker or not. The data on smoking is not collected on all birth certificates in the same manner. For this reason, the count for this variable is lower.

Statistical Questions on Mothers' Smoking

1. Is there a difference in average *BirthWeight* for *full-term* babies depending upon whether the mother smokes or not? Is there more variability in *BirthWeight* for full-term babies depending upon *MotherSmokes*?
   a. Is the relationship between *BirthWeight* and *MotherSmokes* the same for first births as for all births? (Filter for first births.)
   b. Is the relationship between *BirthWeight* and *MotherSmokes* the same for boy babies and girl babies? (Do the analysis separately for males and females.)
2. Are babies born to smoking mothers more likely to be premature? Is it equally true for boy babies and girl babies?

Data Analysis

- Use the Fathom file **BirthSample06A.ftm**.
- Get a **Graph** showing the distributions of the variable *BirthWeight* broken down by categories of *MotherSmokes*. Make certain that you filter for the full-term births only.
- Get a **Summary Table** showing two measures of center and two measures of spread for the variable *BirthWeight* broken down by the categories of *MotherSmokes*. It should resemble the one shown here.
- For the other analyses, it is helpful to use **Object>Duplicate Graph** or **Object>Duplicate Summary Table** and then filter as necessary.
- Get a **Summary Table** to show the relationship between *MotherSmokes* and *Premature*. Calculate probabilities correctly to answer the statistical question.

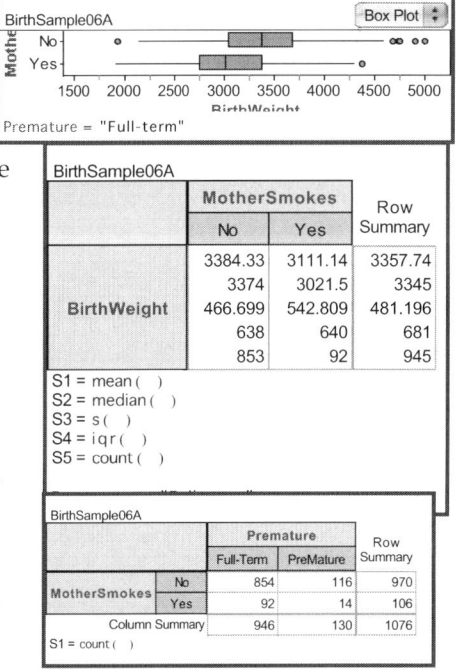

BirthSample06A — Premature = "Full-term"

BirthSample06A	MotherSmokes No	MotherSmokes Yes	Row Summary
BirthWeight	3384.33	3111.14	3357.74
	3374	3021.5	3345
	466.699	542.809	481.196
	638	640	681
	853	92	945

S1 = mean ( )
S2 = median ( )
S3 = s ( )
S4 = iqr ( )
S5 = count ( )

BirthSample06A		Premature Full-Term	Premature PreMature	Row Summary
MotherSmokes	No	854	116	970
	Yes	92	14	106
Column Summary		946	130	1076

S1 = count ( )

12. Birth Data Questions II: Influence of Education

Background and Variables

The data that you will use is the sample of information on births. The detailed definitions of the variables are shown in question 11 above.

Statistical Questions on Mothers' Education For these data are there many statistical questions that can be asked. Here are the questions to be answered for this exercise. (You may choose others; consult your instructor.)

1. Is there a difference in average *AgeMother* for these births depending upon her education? Are there differences in the variability of the age of the mother (that is, in *AgeMother*) by the education of the mother?

    a. Compare the average ages and the variability in ages of the mother among the different education groups for just the data on *first births*. (See the data analysis section below.) In other words, answer the questions about averages and variability for just first births.

    b. Are the answers about average age of mothers and the differences in variability in age of mother the same for the married mothers as for the unmarried mothers?

2. Is there a relationship between mothers' education and whether the mother is married or not?

Data Analysis

- Open the Fathom file **BirthSample06A.ftm**.
- Get a **Graph** showing the distributions of the variable *AgeMother* broken down by categories of *EducMother*. You may use dot plots, histograms, or box plots, but the graph should resemble the one here.
- Get a **Summary Table** showing two measures of center and two measures of spread for the variable *AgeMother* broken down by the categories of *EducMother*. It should resemble the one shown below.
- For other analyses, it is helpful to use **Object>Duplicate Graph or Object>Duplicate Summary Table** and then filter as necessary.
- Get a **Summary Table** to show the relationship between *EducMother* and *ParentsMarried*. Calculate probabilities correctly to answer the question about there being a relationship between mothers' education and whether the mother is married or not. (*Hint:* What statistical technique should be used to analyze two categorical variables?)

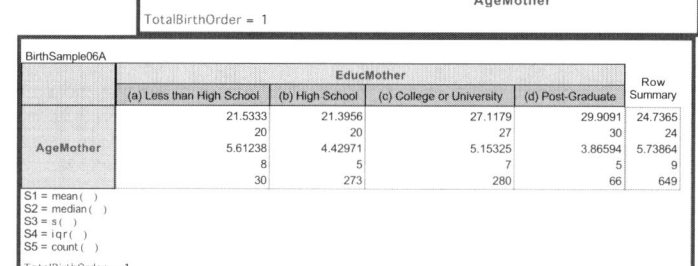

§2.2 A Graphic, a Model, and a Measure

Cars lose their value with age; a bigger house is more valuable

These are the kinds of relationships between variables we will explore in the remainder of the chapter. Notice that we have *two quantitative* variables, and we are looking at the relationship between them. How can we handle this new situation? Answer: in a way similar to what we did before—we will look at graphics, a model, and a measure. When we looked at a single distribution, we looked at dot plots (or box plots, or histograms)—in other words, *graphics*—then we calculated measures such as the median, IQR, or the mean and standard deviation, and, finally, for certain kinds of distributions, we considered the *Normal distribution* as a model of the data.

Graphic: a Scatterplot Here is an example of the $n = 23$ Lexus IS cars for sale in Boston in summer 2009. A **Scatterplot** is essentially an $X - Y$ coordinate system where x is the horizontal axis and y is the vertical axis. Each dot represents the **ordered pair** of the values on each of the two variables for each case, where here the cases are cars for sale on the Internet through www.cars.com. Here the ordered pairs represent the *ordered pair (Age, Price)* so that the most expensive Lexus for sale is plotted at (1.33, 46985), that is, was 1.33 years old (a 2008 car) and the price was $46,985. The oldest Lexus for sale was plotted at (7.33, 14995). However, it is the pattern of the data for all of the points plotted that interests us, and from the pattern we can see what we know: cars lose their value as they age. The higher the age, we see that the lower the price is, in general. Of course, there is some variability depending on other factors about the cars.

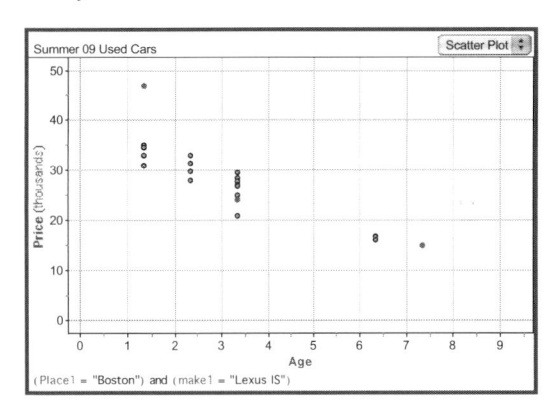

Scatterplots: Direction, Form, and Strength Here is another example of data plotted on a *scatterplot*. Here the cases (the dots) are houses that have been sold in San Mateo County, and the variables are a measure of the living area of the house for sale *SqFt* and the *ListPrice* of the house. The first thing we notice about the data is something we think we know: as the size of the house increases, so does the value (or listed price) of the house. There are many more houses than cars, but that fact is incidental.

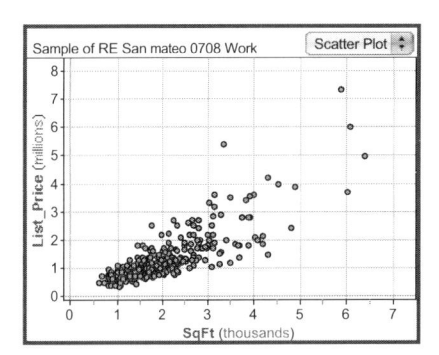

Direction With *scatterplots* we look first at the **direction** of the plot. A plot like the one relating *SqFt* and *ListPrice* shows a **positive** relationship; since as the x value increases, the y value also increases. A plot like the one relating *Age* and *Price* for used cars shows a **negative** relationship; as the x value increases, the y value *decreases*. It is possible for a plot to show a relationship that is neither positive nor negative; here is an example. Our question is: do older or younger houses sell more quickly so that their *Time in Market* is less? If there were a tendency for younger houses to spend less time on market, we would see a *positive* relationship (low age ~ short time on market/high age ~ long time on market), and if younger houses tended to take longer to sell, we would see a *negative* relationship. What we actually see for our sample of houses is that the *Age* of a house appears to have very little to do with the *Time in Market*. In such instances, we would say that there appears to be **no relationship** between the variables. The terms *positive* and *negative* follow the terminology you encountered in algebra when studying the slopes of straight lines.

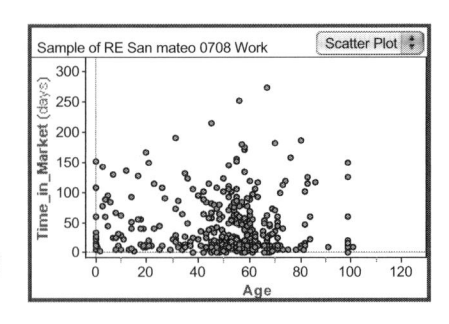

> **Direction**
> We call a *relationship* **positive** if as the values of *x* gets larger, the values of *y* also get larger.
> We call a *relationship* **negative** if as the values of *x* gets larger, the values of *y* get smaller.

Form The second thing we look at is the **form** of the relationship, and in studying the *form*, we are primarily interested in whether the plot shows a **linear** or straight-line pattern or not. Why? A straight-line relationship between two variables is extremely simple—much simpler than a quadratic or an exponential relationship; so if it is possible to use a straight-line model for the data, we will. The *scatterplot* of the negative relationship between *Age* and *Price* for the Lexuses in Boston (the plot on the previous page) appears linear, as does the positive relationship between *SqFt* and *ListPrice*, at least for houses with less than about 500 square feet of area. Not every relationship between variables is linear in form, however. Here is the scatterplot of the relationship between *Age* and *Price* for Porsche 911 sports cars for sale in the San Francisco Bay Area and in Los Angeles. The plot shows a curve rather than a straight line. (We sometimes call relationships that exhibit a curve **curvilinear;** in this example, you may recognize a pattern that may be *exponential decay*, at least for the cars younger than about twelve years old.)

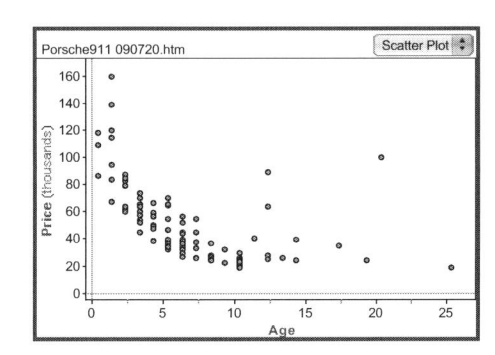

If the relationship in a scatterplot shows *no* relationship between the variables (such as for *Age* and *Time_in_Market* above) and resembles a cloud of points, we would *not* use either the term linear or the term curvilinear.

Strength The third aspect of scatterplots to look at is the most difficult to assess visually from a graphic; this third aspect looks at the **strength** of the relationship between the variables. By *strength* we mean (roughly, here) the amount of scatter around the basic pattern, where the basic pattern may or may not be linear. Here are two examples of a strong relationship, one very linear, the other not. The linear one is the relationship between the *ListPrice* and *SalePrice* of our sample of San Mateo County houses; the price at which a house sells should be not far from the original listed price.

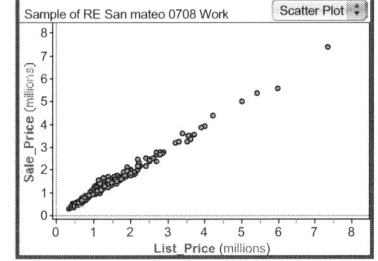

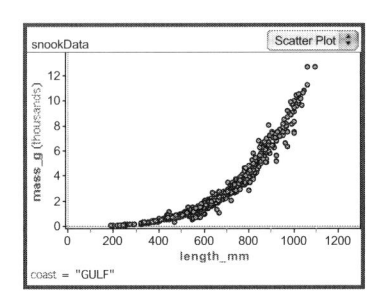

The non-linear example shows the *strong* relationship between the length (measured in millimeters) and the mass (measured in grams) of a fish popularly called snook (*Centropomus undecimalis*) found in warm waters in Florida, the Gulf Coast of the Southern states, and in Southern California. Longer snook have greater mass (hence a positive relationship), but the relationship is not linear; the slope relating the mass to the length of the longer snook is steeper than the slope for the shorter snook.

Our other examples are more linear, but the relationship is weaker. The relationship between *Age* and *Price* for the Porsche 911 is not as strong, and the relationship between *Age* and *Time_in_Market* is weaker still. Although it should be clear what a *strong* relationship is, and it is fairly clear that we should call relationships with a great amount of scatter "weak," the assessment of strength for scatterplots that are neither strong nor weak seems a bit subjective. Are they "moderately strong" or "moderately weak"? What we need is a measure of strength of relationship. After the next paragraph, we introduce such a measure.

Explanatory and Response Variables Again In the scatterplots shown above, did you notice that the variable that we think of as *explanatory* is regarded as the *x* variable and the variable we think of as *response* is regarded as the *y* variable? We think that the *Age* of a car influences the *Price* that the car can fetch in the market. Asking a high price for the car cannot change its age; it may delay the sale of the car while it grows still older, but it cannot make the car younger. Of course, there are situations when it is not clear which variable should be regarded as the *explanatory* and which the *response*. In those situations, the placement of the two variables on a scatterplot may not matter. However, for any specific pair of variables, we usually have a "natural way of thinking" about the relationship between the two variables, and the way we think of the relationship usually indicates which variable should be the *explanatory* and which variable the *response*.

The Correlation Coefficient: A Measure Based on a Model

As it happens, we do have a measure of the strength of association between two variables. This measure also shows the direction of the relationship, and it is based upon a linear model for the data. It is called the **correlation coefficient**. The symbol for this measure is *r*. The *correlation coefficient* measures the **strength of linear association** between two quantitative variables. It is calculated so that it varies between $r = -1$, which indicates a perfectly linear negative association, and $r = 1$, which indicates a perfectly linear positive association. Hence, $-1 \leq r \leq 1$ and values between these limits indicate a negative or positive linear association with some amount of scatter—or, it is important to note, a relationship that happens not to be linear. Here are some examples (we will actually look at the data we have been analyzing shortly).

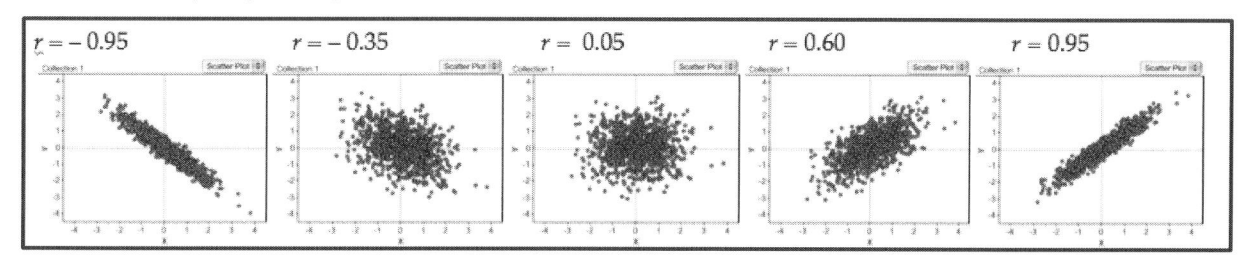

Correlation Coefficient, r

- Measures the *linear* association between two *quantitative* variables
- Has values in the interval: $-1 \leq r \leq 1$
- Is calculated by the formula: $r = \dfrac{1}{n-1} \sum_{i=1}^{n} \left(\dfrac{x_i - \bar{x}}{s_x} \right) \left(\dfrac{y_i - \bar{y}}{s_y} \right) = \dfrac{1}{n-1} \sum_{i=1}^{n} z_x \cdot z_y$

Using the Formula: Calculating the Correlation Coefficient

We will show an example that has to do with just nine houses for sale in Foster City in our real estate sample; we are looking at the relationship between the size of the house (measured by the variable *SqFt*) and the *ListPrice* of the house. Here is the plot for just these $n = 9$ houses. The vertical and horizontal lines that we have drawn on the plot show the means for the two variables *SqFt* and *ListPrice*; we will see that having these lines there gives us some insight to what the formula is actually doing. The formula uses the means and standard deviations for the two variables (shown in the *Summary Table*), so that here for *SqFt*, $\bar{x} = 1954.4$ and $s_x = 427.1$ and for *ListPrice*, $\bar{y} = 1048864$ and $s_y = 146217$. For these $n = 9$ cases, the *Correlation Coefficient* $r = 0.821$.

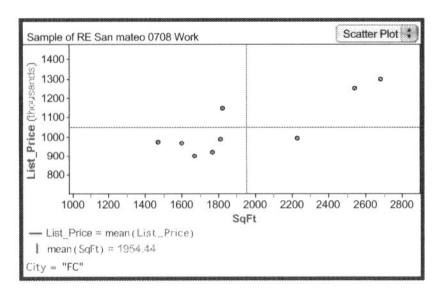

It may help to follow the calculations below by having the complete data in a table. The data for the nine houses is shown in the two columns on the left, and the results of the calculations of the formula are shown in the remaining columns, with the sums at foot of the table. Here is what the calculation looks like in "formula form."

SqFt x	ListPrice y	z(x)	z(y)	Product
1470	975000	-1.13420116	-0.50516698	0.57296097
1820	1150000	-0.31475932	0.69168428	-0.21771407
1600	968000	-0.82983705	-0.55304103	0.45893393
1810	985888	-0.33817194	-0.43070231	0.14565144
2540	1248888	1.37094962	1.36799415	1.87545105
1770	920000	-0.43182244	-0.88132023	0.38057385
2230	995000	0.64515827	-0.36838398	-0.23766597
1670	898000	-0.66594868	-1.03178153	0.68711355
2680	1299000	1.69872635	1.71071763	2.90604112
Sum		0.00009365	0.00000000	6.57134588

$$r = \dfrac{1}{n-1} \sum_{i=1}^{n} \left(\dfrac{x_i - \bar{x}}{s_x} \right) \left(\dfrac{y_i - \bar{y}}{s_y} \right)$$

$$\approx \dfrac{1}{9-1} \left[\left(\dfrac{1470 - 1954.4}{427.1} \right) \left(\dfrac{975000 - 1048864}{146217} \right) + \cdots + \left(\dfrac{2680 - 1954.4}{427.1} \right) \left(\dfrac{1290000 - 1048864}{146217} \right) \right]$$

$$\approx \dfrac{1}{8} \left[(-1.1342)(-0.5052) + (-0.3148)(0.6917) + (-0.8298)(-0.5530) + \cdots + (-0.6659)(-1.0318) + (1.6987)(1.7107) \right]$$

$$\approx \dfrac{1}{8} \left[(0.5729) + (-.02177) + (0.4589) + (0.1457) + (1.8755) + (0.3806) + (-0.2377) + (0.6871) + (2.9060) \right]$$

$$\approx \dfrac{6.5713}{8} \approx 0.8214$$

Zzzzzz: a picture of what the formula is doing The graphic below (if studied carefully) shows what is happening inside the formula. We have drawn the means of the two variables in the graph: the dashed vertical line is the *SqFt* mean and the dashed horizontal line is the *ListPrice* mean. Then we can show the distances of each data point from the mean of *x* (here *SqFt*) and from the mean of *y* (here *ListPrice*) in *standard deviation* units. These distances are actually z scores: that is $z_x = \frac{x_i - \bar{x}}{s_x}$ and $z_x = \frac{y_i - \bar{y}}{s_y}$. In the formula, it is these zs that get multiplied before the terms are added.

- **Point A** is the first point (the first house) and **Point B** the second point (house) in the list above; check that the z_x and z_y scores on the graph agree with the table.

- Points that are in **quadrant I** will contribute a positive product, that is $z_x z_y = \left(\frac{x_i - \bar{x}}{s_x}\right)\left(\frac{y_i - \bar{y}}{s_y}\right) > 0$) in the sum in the formula, and points that are in **quadrant III** will also have a positive product, since the two z scores will both be negative. This means that if our data are mainly in quadrants I and III, our sum will be positive, and our *r* will be positive. In our example, most of our points are in these two quadrants. Check out that the product of the z scores for point A is 0.573.

- Points that are in **quadrant II** (or **quadrant IV**) will contribute a *negative* product $z_x z_y < 0$ for the sum in the formula (as does our point B), and so if most of the data are in these two quadrants, the correlation coefficient *r* will be negative.

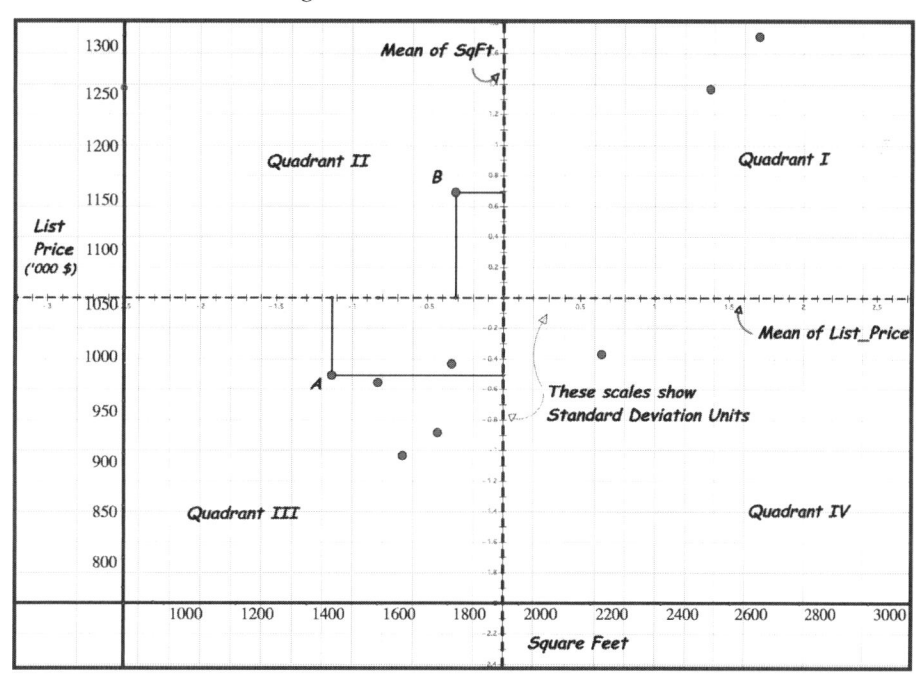

- If the data are scattered about in *all* the quadrants, the positive and negative products—when they are added—will tend to cancel each other out, and we will end up with a number near to zero. The number may be positive and may be negative, but it will be near to zero.

Correlations in Data: Used car data We return to our data sets to see how the measure of linear association actually works. Here are three examples from the used car study of Lexuses and Infinitis.

The first graph shows the relationship between *Age* and *Price* for all of the cars (not just the Lexuses from Boston). The association looks fairly linear and negative as we would expect (as the age of a car increases the value decreases) and with a bit of scatter; obviously, even with cars that are the same year model, there are differences in price asked. So about what value should the correlation coefficient be?

The second graph shows the relationship between *Miles* (miles that the car has been driven) and *Price*, whose association we would expect to be negative. In the center portion the relationship appears to be quite linear.

The third graph shows the association between *Age* and *Miles* driven, which we expect to be a positive association.

What we see in the graphs should be reflected in the numbers for the *correlation coefficients* between these variables. The values of the correlation coefficients r are shown in the summary table below.

Notice that the correlation of a variable (such as *Age*) with itself is $r = 1$. Secondly, notice that the correlation coefficient does not depend upon which variable we choose as x and which variable we choose as y. Look at the number for *Age* and *Miles* and then look at the number for *Miles* and *Age* in the summary table. The value of r does not depend on the choice of *explanatory* and *response* variables. We can see this in the form of the formula for the correlation coefficient, $r = \frac{1}{n-1} \sum_{i=1}^{n} \left(\frac{x_i - \bar{x}}{s_x} \right) \left(\frac{y_i - \bar{y}}{s_y} \right) = \frac{1}{n-1} \sum_{i=1}^{n} z_x \cdot z_y$ since before we sum, there is a product, and reversing the order of the product does not change its value.

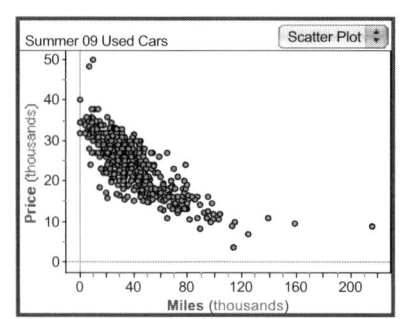

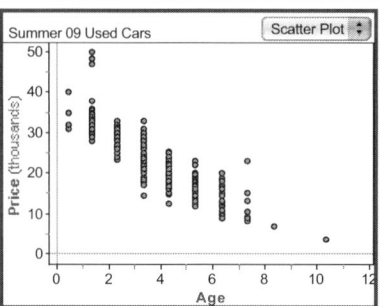

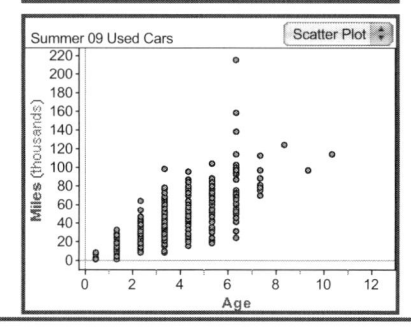

Summer 09 Used Cars

	Age	Miles	Price
Age	1	0.756543	-0.883391
Miles	0.756543	1	-0.799797
Price	-0.883391	-0.799797	1

S1 = correlation ( )

Real Estate Data Here is the summary table for the correlations for the real estate data. You should be able to check that $r = -0.061$ for the association between *Age* and

Sample of RE San mateo 0708 Work

	Age	List_Price	Sale_Price	SqFt	Time_in_Market
Age	1	-0.205108	-0.185829	-0.348588	-0.0608161
List_Price	-0.205108	1	0.994569	0.822379	-0.104747
Sale_Price	-0.185829	0.994569	1	0.807327	-0.138747
SqFt	-0.348588	0.822379	0.807327	1	-0.0396667
Time_in_Market	-0.0608161	-0.104747	-0.138747	-0.0396667	1

S1 = correlation ( )

Time_in_Market is sufficiently weak to fit what you see in the scatterplot for the two variables above. The $r = 0.995$ for *List_Price* and *Sale_Price* should accord with the graphic for those two variables.

Cautionary Comments on Correlation There are some facts about correlation that are sometimes misunderstood. Here are some of those facts, followed by the potential misunderstandings.

The correlation coefficient r is for quantitative variables only. The word "correlation" has entered into modern speech to such an extent that people easily say: "There is a correlation between gender and political party preference." In statistics, the word *correlation* is reserved for the association between quantitative variables, and neither gender nor political party preference is quantitative. It is quite correct to say, "There is an *association* between gender and political party preference" but not a *correlation*.

The correlation coefficient r measures a linear association between two variables. Recall the snook data where the relationship between the length and the mass appears to be without much scatter at all, but the relationship is curved and not linear. For these data, $r = 0.911$, whose proximity to 1.00 you may find impressive. However, if these data were "straightened out" but had the same amount of scatter about a straight line, the correlation coefficient would be about 0.99. The lesson is that a value of *r* measures either scatter about linearity or lack of linearity, or both! Studying the scatterplot should show what is happening.

Correlation does not imply causation. Many of the examples in this section have been chosen because the relationship between the variables "makes sense" in that we have some ideas about the connection between the variables. As cars grow older, people know that the cars are likely to have been driven much, that the car is more likely to cause trouble, that the car looks out of date, and so people are less willing to pay what they would for a younger car. So we say that *Age* affects *Price.*

It may be that there are some other variables affecting both of the variables we are analyzing and that there is not a direct connection between the variables. If we had a collection of students who were all taking (say) statistics and chemistry, and we looked at the correlations between these test scores in statistics and chemistry, we would probably find a fairly strong positive correlation; that positive correlation does not mean that taking chemistry helps students taking statistics or vice versa. There are other variables.

> **Cautions about Correlation**
> - The *correlation coefficient r* is only appropriate for quantitative variables.
> - The *correlation coefficient r* measures linear association between two variables.
> - *Correlation* between variables does not imply that one variable causes the other.

Summary: Analyzing the relationship between two quantitative variables

- A **scatterplot** is an X – Y coordinate system on which the ordered (x, y) pairs of the two quantitative variables are plotted to show the *direction, form,* and *strength* of the relationship between the two variables.
    - A **positive direction** indicates that as x is larger, y is also larger.
    - A **negative direction** indicates that as x is larger, y is smaller.
    - **Form** refers to whether the data in the plot appear to approximate **linear** form (a straight line) or not.
    - **Strength** refers to the extent to which the data in the plot appear to adhere to a particular form (hence, "stronger") or show scatter with respect to a particular form ("weaker").
- The **correlation coefficient,** whose symbol is **r,** is a numerical summary of the relationship between two quantitative variables that:
    - Is based upon a *linear* model for the data, and
    - May take on the values in the interval $-1 \leq r \leq 1$, where
    - The value $r = -1$ indicates a (perfect) negative linear relationship, the value $r = 1$ indicates a (perfect) negative linear relationship, and values near zero either no relationship or possibly one non-linear.
    - Is calculated using the formula $r = \dfrac{1}{n-1} \sum_{i=1}^{n} \left(\dfrac{x_i - \bar{x}}{s_x} \right) \left(\dfrac{y_i - \bar{y}}{s_y} \right) = \dfrac{1}{n-1} \sum_{i=1}^{n} z_x \cdot z_y$.
- **Correlation and causation** A *correlation coefficient r* (even one that is strong—near to $r = -1$, or $r = 1$) does not necessarily indicate a **causal relationship** between the two quantitative variables.

§2.2 Exercises on Scatterplots and Correlation

1. **Mexican Roller Coasters** The data for this exercise come from the Roller Coaster Data Base (www.rcdb.com). The data on roller coasters are very incomplete because the website depends upon the owners or the builders of the roller coasters for the information. There are certainly more than thirteen roller coasters in Mexico; there were just thirteen with complete information for our variables. The variables are:

   Height (measured in meters)
   Length (measured in meters)
   Speed (measured in km/hr)

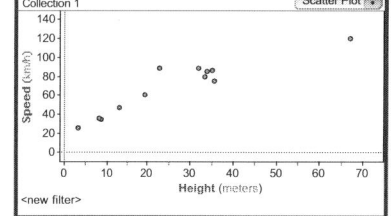

   We will examine the relationship between *Height* and *Speed* first.

   a. What are the cases for these data?

   b. Using the scatterplot, make a judgment about the *direction*, the *form*, and the *strength* of the association between the height of roller coasters and their speed.

   c. The table shows the calculation of the *correlation coefficient* $r = \dfrac{1}{n-1}\sum_{i=1}^{n}\left(\dfrac{x_i - \bar{x}}{s_x}\right)\left(\dfrac{y_i - \bar{y}}{s_y}\right)$ in the same fashion as in the subsection **Using the formula**. Label your answers "a" through "i."

	Name	Location	Height x	Speed y	z(x)	z(y)	z(x)*z(y)
1	Batman the Ride	Mexico City	33.3	80.0	0.50	0.47	0.23
2	Boomerang	Mexico City	35.5	75.6	0.62	0.32	a
3	Medusa	Mexico City	32.0	88.5	0.42	0.77	0.33
4	Roller Skater	Mexico City	8.5	34.9	-0.92	b	1.02
5	Superman el Último Escape	Mexico City	67.0	120.0	2.42	1.87	4.54
6	Tsunaumi	Mexico City	8.0	36.0	-0.95	-1.07	1.01
7	Montaña Infinitum	Mexico City	33.8	85.3	c	0.66	0.35
8	Ratón Loco	Mexico City	13.0	46.8	-0.66	-0.69	0.46
9	Catarina	Guadalajara	8.0	36.0	d	e	1.01
10	Titan Cascabel	Guadalajara	22.9	88.5	-0.10	0.77	-0.08
11	Catariños	Monterrey	3.3	26.0	-1.22	-1.42	f
12	Tornado	Guadalupe	19.5	60.0	g	h	i
13	Tsunaumi	Aguascalientes	35.1	86.9	0.60	0.71	0.43
	Sum		319.9	864.5	0.00	0.00	11.29597
	Mean		24.6	66.5			
	Standard Deviation		17.49	28.57			

   d. Write the formula for the correlation coefficient and indicate the part of the formula that corresponds to the number 11.29597—perhaps by enclosing that part of the formula with a kind of circle or enclosure.

   e. Use the number 11.29597 to get the value of the correlation coefficient, $r = 0.94133$. Comment on whether the number you got is in accord with your answer to part b.

- Use **RollerCoastersMexicanSample.ftm**, get a *graph*, and make the scatterplot above.
- With the graph selected, get (from the menu) **Graph>Plot Value** and, in the dialogue box, type in *mean(Height)*. A vertical line showing the mean of the explanatory variable *Height* should appear.

- With the graph selected, get (from the menu) **Graph>Plot Function** and, in the dialogue box, type in **mean(Speed)**. A horizontal line showing the mean of the response variable *Speed* should appear. The vertical and horizontal lines on your plot divide the plot into four quadrants, labeled I, II, III, and IV, as shown in the subsection **Zzzzzz**.

  f. Quadrant III has the roller coasters whose *Height* is _____ than the mean *Height* and whose *Speed* is _____ than the mean *Speed*. Fill in the blanks with the words "less" or "more."

  g. In quadrant III, the z_x scores will be _____ and the z_y scores will be _____ and therefore the contribution to the sum in the formula will be _____. Fill in the blanks with the words "positive" or "negative."

  h. One roller coaster makes a negative contribution to the sum in the formula for *r*. Find which one it is, either from the plot (you can get Fathom to highlight your choice by clicking on its dot) or from the chart on the previous page to determine the name of that roller coaster. Give a reason for your choice.

  i. Translate the value of the correlation coefficient *r* = 0.94133 into a statement about the height and speed of roller coasters without using the word "correlation" or "association" ("The higher a roller coaster is…").

  j. The relationship between *Height* and *Speed* for the Mexican roller coasters is strong and positive. Do you think that the relationship between *Length* and *Speed* will also be strong and positive? What about the relationship between the *Height* and *Length* of the roller coasters? Give reasons for your answers.

- Using the Fathom file **RollerCoastersMexicanSample.ftm**, get a **graph** and make a scatterplot with the variable *Length* on one axis and the variable *Speed* on the other axis. (On which axis should *Length* go? Why?)

- Using the Fathom file **RollerCoastersMexicanSample.ftm**, get a **graph** and make a scatterplot with the variable *Length* on one axis and the variable *Height* on the other axis. (On which axis should *Length* go? Why?)

- Get a **Summary Table** and drag the variable *Height* to the right-pointing arrow then drag the same variable *Height* to the down-pointing arrow. The number in the **Summary Table** is the correlation coefficient.

- Drag the variable *Speed* to the right-facing arrow and then again to the down-facing arrow. You should have the **Summary Table** shown here.

MexicanRollerCoasterSample	Height	Speed
Height	1	0.941331
Speed	0.941331	1
S1 = correlation ( )		

- Drag the variable *Length* to the right-facing arrow and then again to the down-facing arrow. You should be able to read off the correlations between any pair of the variables. The **Summary Table** you get should be an expanded version of the one shown here.

  k. The correlation coefficient *r* has the same value for two variables whichever variable is chosen as the explanatory variable *x* and whichever variable is chosen as the response variable *y*. State how the **Summary Table** shows this fact and also state why the formula for

  $$r = \frac{1}{n-1}\sum_{i=1}^{n}\left(\frac{x_i - \bar{x}}{s_x}\right)\left(\frac{y_i - \bar{y}}{s_y}\right)$$

  shows this fact. (*Hint:* If you switch the roles of the two variables would the calculations change?)

l. ***Interpretation.*** Now that you have the correlations between each pair of variables of the three variables, Height, Length, and Speed, you should be able to write what these correlations mean for visitors to amusement parks or for builders of roller coasters.

Questions in square brackets—[…], such as the next two—are usually optional questions; ask your instructor.

[m. *Review* There is a variable that distinguishes between the coasters in Mexico City and those in other parts of Mexico. Do an analysis to determine whether there is evidence that Mexico City has higher, longer, and faster roller coasters. Show your Fathom work and write up your conclusions.]

[n. *Review* Given what was said about the data in www.rcdb.com, explain why you may want to treat your conclusions stated in part m with some caution.]

2. **Mexican Roller Coasters <u>Not</u> in Mexico City** Here are the means and standard deviations for the variables Height and Speed for the $n = 5$ roller coasters for which we have data and that are outside Mexico City.

MexicanRollerCoasterSample	Height	Speed
	17.8 m	59.5 km/h
	12.6 m	28.6 km/h
	5.0	5.0

S1 = mean( )
S2 = s( )
S3 = count( )
MexicoCity = "Other Places"

Use the information on the previous page for these five roller coasters, this information on means and standard deviations, and the formula for the

$$r = \frac{1}{n-1} \sum_{i=1}^{n} \left(\frac{x_i - \bar{x}}{s_x}\right)\left(\frac{y_i - \bar{y}}{s_y}\right)$$ to calculate r. (Answer: $r = 0.930$)

3. **Road & Track *Gearhead Cars: Correlation between Horsepower and MPG***

Here are more data on cars tested by the car enthusiast magazine Road & Track. Remember that the cars in the data set do not represent all of the cars for sale but rather cars that would be of interest to "gearheads."

Horsepower	A measure of the power of a car; the higher the number, the more power
MPG	Miles per gallon. The number of miles that a car will travel on one gallon of gasoline; the bigger the number, the more economical the car
Cylinders :	Number of cylinders: (a) four (b) six (c) eight (d) more than eight cylinders

a. The cases are cars. Before looking at the data, but from what you know about cars, make a guess about the direction of the relationship between Horsepower and MPG—should the direction be positive or negative? Give a reason for your answer.

- Open the Fathom file **RoadandTrackMay09Europe.ftm.** This Fathom file has only the cars that come from Europe; it includes the six-cylinder cars that are a part of Exercise 2.
- Get a **Graph** and make a scatterplot with the variable *Horsepower* on the *x*-axis and the variable *MPG* on the *y*-axis.
- Make the **Summary Table** shown. Drag the variables *Horsepower* and *MPG* each to the *right arrow* in the summary table, and then with the summary table selected, use **Summary>Add Formula** to get the s( ) and the correlation(Horsepower,MPG).

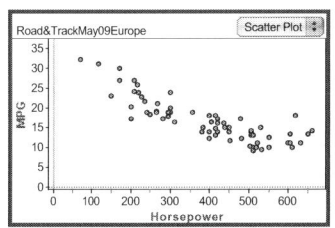

Road&TrackMay09Europe	Horsepower	MPG
	393.254	16.8224
	146.901	5.38153
	-0.848633	-0.848633
	67	67

S1 = mean( )
S2 = s( )
S3 = correlation(Horsepower, MPG)
S4 = count( )

- Drag the variable *Cylinders* to the *downward*-pointing arrow on the summary table. This **Summary Table** (not shown here) will have the means, standard deviations, and correlations for each of the categories of *Cylinders*.

    b. The correlations are not the same for the various numbers of cylinders. Which correlation is strongest and which is weakest? Give a reason for your answer.

- Drag the variable *Cylinders* to the **body** (not the *foot* or the *left edge*) of the scatterplot. You will see four different kinds of symbols corresponding to the four categories of the variable *Cylinders*.

    c. Describe what you notice about the distribution of the different types of cars (four, six, etc.) in the plot of the relationship between *Horsepower* and *MPG*.

    d. Use what you observed and described in your answer to part d to explain how it can be that the overall correlation $r = -.848$ can show a stronger negative relationship than the correlation for any of the categories that together comprise all the cars. (This question may appear to be difficult. You may want to check with your instructor about what you think is happening with the data.)

4. **Mammals: Amount of Sleep, Length of Pregnancy, and Other Tales**

    Here is a matrix of correlations and four scatterplots from the collection of mammals.

Mammals	Gestation	BodyWeight	NondreamingSleep	DreamingSleep	TotalSleep	LifeSpan
Gestation	1	0.651102	-0.594703	-0.450899	-0.631326	0.629089
BodyWeight	0.651102	1	-0.375946	-0.109383	-0.307186	0.305518
NondreamingSleep	-0.594703	-0.375946	1	0.514254	0.962715	-0.407499
DreamingSleep	-0.450899	-0.109383	0.514254	1	0.727087	-0.311601
TotalSleep	-0.631326	-0.307186	0.962715	0.727087	1	-0.437657
LifeSpan	0.629089	0.305518	-0.407499	-0.311601	-0.437657	1

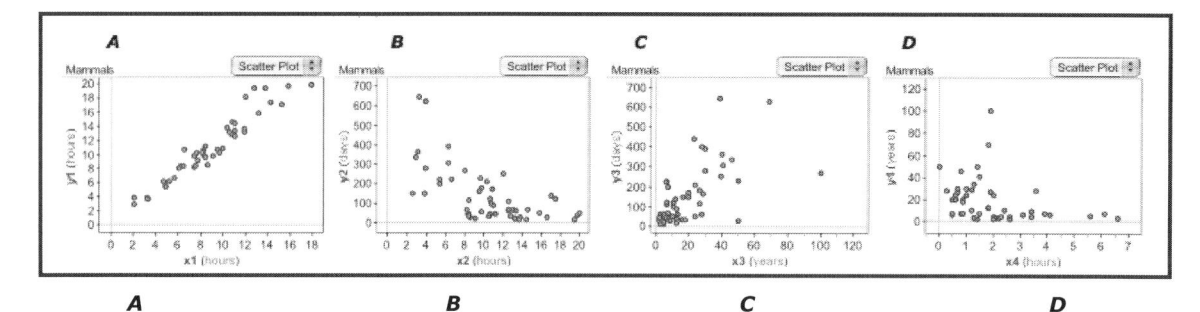

    S1 = correlation ( )

    A   **B**   **C**   **D**

    a. **Detective Work** The four plots show the relationship between four of the five pairs of variables listed just below. Your job is to determine which pair goes with which plot and which pair of variables has no plot. You should certainly use the matrix of correlations, but the units on the plots may also help you.

    I   *TotalSleep, Gestation*   **II**   *DreamingSleep, LifeSpan*   **III**   *LifeSpan, Gestation*

    IV   *NondreamingSleep, TotalSleep*   **V**   *DreamingSleep, NondreamingSleep*

    b. **Before the Judge.** Having come to a decision about the plots, show the court that you are right. In other words, give reasons for each of your choices.

    c. We are interested in the relationship between *BodyWeight* and *Gestation*. Which variable do you think should be the *explanatory* variable, and which variable should be the *response* variable? Does it matter for the calculation of the correlation coefficient? Explain.

d. Look at the correlation coefficient *r* given for *BodyWeight* and *Gestation* in the matrix above. Make a small sketch of what you think the scatterplot will look like for the *r* that is given.

- Open the Fathom file **Mammals 2.ftm.** You may make plots to confirm your answers to part a.
- Make the scatterplot of the relationship between *BodyWeight* and *Gestation.*

   e. Does the actual plot look like your idea? If not, try to explain the magnitude of the correlation coefficient from what you see in the plot and what was said about the correlation coefficient in the **Notes.**

   f. For the relationship between *TotalSleep* and *Gestation*, the correlation coefficient is $r = -0.63$. (You may wish to confirm the plot by getting the scatterplot with Fathom.) Comment on this interpretation of the correlation coefficient: "Sleeping longer (sometimes as much as 10 to 14 hours a day) *causes* the length of pregnancy to be shorter." (*Hint:* See in the **Notes** the section entitled **Correlation Cautionary Comments.**)

- Get a **Graph** and on the *x* axis put the variable *TotalSleep*, which records the (average) total amount of sleep a mammal gets. On the *y*-axis put the variable *Danger_Level.*

   g. Which of these interpretations is correct, according to the **Notes:** *I*, *II*, both, neither? Explain why.

   I "There appears to be a correlation between *Danger_Level* and *TotalSleep* for mammals."

   II "There appears to be an association between *TotalSleep* and *Danger_Level* for mammals."

5. **Roller Coasters: Height, Length, Speed, Duration, and G-Force.** This exercise uses the American Roller Coaster data set.

   a. Just below are two scatterplots and a number of alternate values for the correlation coefficient for these plots. For each of the alternates, state whether that option is:
   - *an impossible value* by the definition of the correlation coefficient
   - *a possible value* but *not likely* to be the value for either of the plots shown
   - *a possible value* and quite likely to be the value for one of the plots shown

   Give a reason for your answer for each alternate value and, of course, identify which option goes with which plot. Notice also that we do not have much data on *G-force*.

   I $r = -.94$
   II $r = .68$
   III $r = 1.27$
   IV $r = .14$
   V $r = -.15$

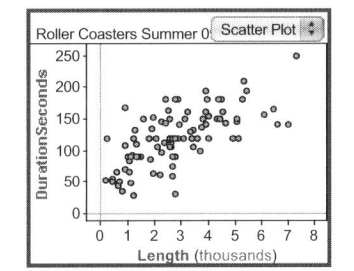

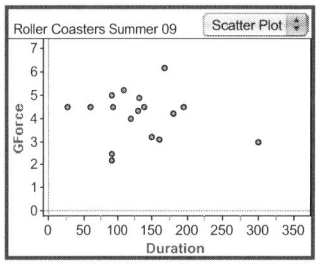

   b. Put into words what the scatterplot for the relationship between *Length* and *Duration* means for people comparing roller coaster rides. (*Hint:* A good way to start is: "There is a tendency…").

- Open the Fathom file **RollerCoasters Summer 09 A.ftm** and make a **Summary Table.**
- To make a *correlation matrix* (something like the one shown Exercise 4), drag each (in turn) of the variables **Length, Duration,** and **Gforce** to the *right-facing* arrow and then drag the same variables to the *down-facing* arrow. You should be able to check your answers to part a using your matrix.

c. The correlation matrix you have just made should also give you an *r* for the relationship between the length of a roller coaster (*Length*) and the variable *Gforce*. What can you say from the correlation coefficient about the relationship between *Length* and *Gforce* compared with the relationship between *Duration* and *Gforce*? Which relationship (according to the correlation coefficient) has the stronger linear association? Give a reason for your answer. (If you want to see the plot, you can make one in Fathom: get a **Graph** and drag one variable to the *x*-axis and the other variable to the *y*-axis.)

- Get a **Graph** and make a scatterplot of the relationship between the *Length* and *Height*.

  d. Describe the relationship that you see in the plot, using the ideas of *direction, form,* and *strength*. Make a guess about the value of the correlation coefficient. (You will get it shortly.)

- Add the variable *Height* to the correlation matrix to see the *r* for *Length* and *Height*.
- Select the scatterplot you have made and add a filter (**Object>Add Filter**) for just the wooden roller coasters (the name of the variable is *Construction*, so type the words: *Construction* ="W"). In the same way, add a filter for the wood roller coasters for the *correlation matrix.*
- Duplicate the scatterplot and matrix (**Object>Duplicate...**) and change the filters on both to see just the steel roller coasters (type: *Construction* ="S").

  e. Use the plots and the values of *r* you have just made to explain how the relationship between *Length* and *Height* for roller coasters differs between the wood and steel roller coasters.

§2.3 Best-Fitting Lines: Making the Most of the Model

More Statistical Questions about Used Cars and Used Houses

In the last section we looked at graphs showing how cars lose their value as they get older (not a big surprise) and the larger the house is, the more valuable it is (also not surprising).

In these scatterplots, the form of the data is fairly linear, although there is some scatter. For the relationship between the *Age* and the *Price* of used Lexuses and Infinitis, $r = -.88$, and for the relationship between *SqFt* and *List_Price* of houses in San Mateo County, we can calculate $r = .82$. However, we can go beyond the correlation coefficient r, a single measure of linear relationship. We get an estimate of *how much* a used Lexus or Infiniti loses each year it ages. We use our data to get an idea of *how much* a square foot of house in San Mateo County costs. We can get these things by fitting a straight-line model—a linear model—to the data.

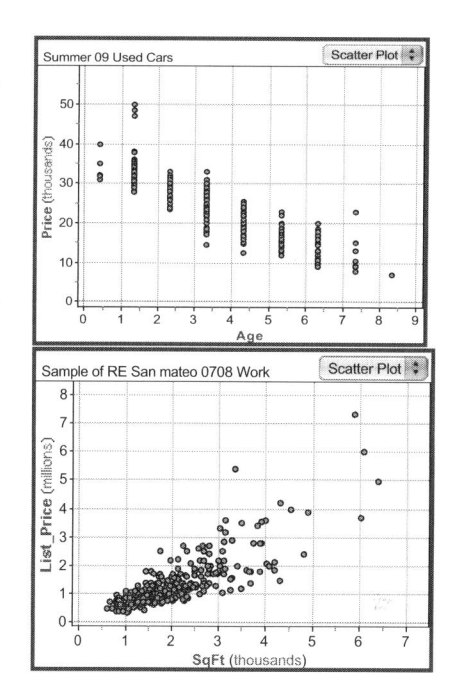

Remembering about straight lines Recall from algebra courses that a straight line can be expressed by the equation $y = mx + b$, where m is the slope and b is the y-intercept. Remember also that slope can be thought of as $m = \dfrac{\text{rise}}{\text{run}} = \dfrac{\text{change in } y}{\text{change in } x}$, so that $y = \dfrac{3}{4}x + 2$ means that for every 4 units running along the *x*-axis, the line rises by 3 units. Or, for every one unit running on the *x*-axis, the line rises ¾ of a unit. For our use in statistical modeling, the form of the equation and the idea of the slope are the most important ideas, but the notation is slightly different.

Notation for equations of lines in statistics

In statistics we express a straight line by the equation $y = a + bx$, where b is the slope and a is the y-intercept.

Example of a linear model for prediction and interpretation: used Infiniti cars The plot on the next page shows a line that we have drawn to model the relationship of the *Price* of a used Infiniti G-35 to its *Age*. Our line has the equation $\hat{y} = 36201 - 3516x$, where y is *Price* and x is the *Age* of the used car. The reason that we use the symbol $\hat{y}$ (pronounced "y-hat") is that we intend to use our equation to predict the value of y (in this instance, a price) from a value for x (here, age). Our prediction of the price $\hat{y}$ for the age $x = 3$ we get from the equation $\hat{y} = 36201 - 3516x$. When we put $x = 3$ into this equation, we get:

$$\hat{y} = 36201 - 3516(3) \approx 36201 - 10548 = 25653$$

We could say that we expect that in three years our new Infiniti G-35 has lost $10,550 and is now worth about $25,650. This predicted value $\hat{y}$ for $x = 3$ is *on* the line shown on the plot; indeed, *all* the predicted values are on the line we have drawn. We can do the same calculation for any other value of

Age and get a predicted value for the *Price*, as long as the value for *Age* is reasonable. Notice, however, that our linear model predicts that after about eleven years, the Infiniti G-35 will be worth nothing; what this shows is that our linear model is probably not good for values of *Age* beyond nine years or so. We have no data for the years beyond about eight years old, except for one car at age eleven years.

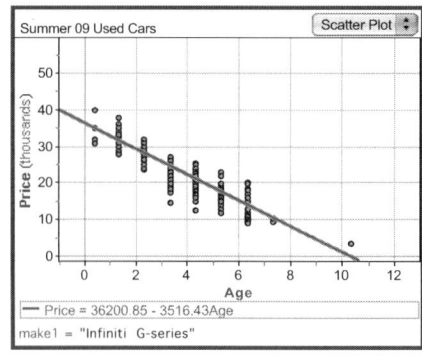

For a three year-old Infiniti G-35, we lose $10,550. Using our equation $\hat{y} = 36201 - 3516x$ we can also say that for *every year that Age increases for an Infiniti G-35 the Price will change by the value of the slope –3516*. To be specific, we can say that for every year that an Infiniti G-35 ages, it loses about $3,516. This sentence is an *interpretation* of the slope of our model.

We have used our linear model for both **prediction** and **interpretation.** We have used the equation to *predict* a value for price, and we have *interpreted* the **slope** to say how the price declines as a car ages. We may also sometimes be able to interpret the **y-intercept**, but not always. The y intercept of a line is the value of y where x = 0, so we could say that our prediction for the *Price* of a *new* Infiniti G-35 is about $36,200. However, notice that our data is about *used cars*; we do not have any data on *new* Infinitis—just used ones, so we should probably be a bit cautious about making this prediction.

Another Example of Prediction and Interpretation: Size and Price of House. Here is the plot showing the relationship between the size (*SqFt*) and the listed price (*List_Price*) for houses sold in San Mateo County in 2007–2008. We have fitted a linear model with the equation,

$\hat{y} = 25331.67 + 728x$, where x is *SqFt* and y is *List_Price*.

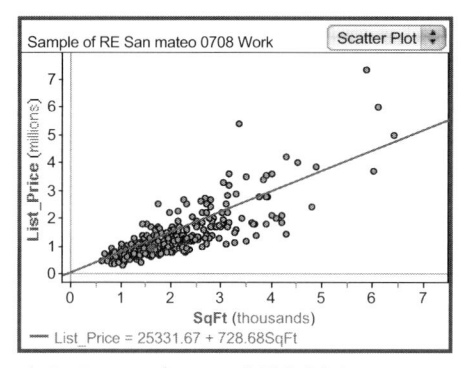

There was a house sold in Redwood City in the sample that had *SqFt* equal to x =1610. Using our equation $\hat{y} = 25331.67 + 728(1610) = 25331.67 + 1173174 = 1198506.47$, our predicted list price for this house is $1,198,506. How does this compare with the actual value of the list price (which we also have in the data)? The actual value of *List_Price* for this house is $949,000; what they got (in this case) was lower than what we predicted. When we compare the two, we calculate what we call the **residual,** which is the difference between the actual value and the predicted value. For this case we get the

$$residual = y_i - \hat{y} = 949000.00 - 1198506.47 = -249506.47 \text{ dollars}$$

The value is negative because the actual value is less than the predicted value and thus below our linear model, whose equation is $\hat{y} = 25331.67 + 728.68x$. Another house had x = 2131 square feet of living area; if you calculate the predicted value $\hat{y}$ for *List_Price* and compare it with the actual value of $2,250,000, you will get a *residual* of $671,851.25. This *residual* in this example is positive, showing that the actual value is above the predicted value on the line.

Definition of residual

Residual = **Actual value** – **Predicted value** = $y_i - \hat{y}$

Our *interpretation* of the slope tells as *that for every square foot in size that a house is bigger, we can expect that the list price will be about $729 more.* Notice that in our interpretation, it is important to mention the units for the *x* variable and for the *y* variable. We can generalize this interpretation.

> **Interpretation of Slope**
> The slope tells us the amount of change in the *y* variable for every *unit* change in the *x* variable. The interpretation is to be done in the context of the variables, using the units in which the variables are measured.

Explanatory and Response Variables Again In both our examples we have chosen one variable to be the *explanatory* variable—for the used cars example, the explanatory variable, denoted by *x*, was *Age*. That implies that the *response* variable was *Price*, denoted by *y*. It would be possible to get a linear model to predict the *Age* of a used car (making *Age* the *y* variable) from the *Price* (making *Price* the *x* variable), but that choice would make less sense to us. As a car grows older, we think the value tends to decrease; we do not think that making a car worth less (for example, by abusing the car in some way) will make the car older in age.

When we calculated the *correlation coefficient*, which also depends on a linear model, it did not matter which variable we chose to be *x* (the explanatory variable) and which variable we chose to be *y* (the response variable); the result was the same whatever choice of explanatory and response we made because *r* just looks for a linear relationship between the variables. In fitting a model for prediction and interpretation, it *does* make a difference because in our prediction and in our interpretation, we look at *how* one variable (the explanatory*)* affects the other (the response*.)* It should also be clear from looking at a scatterplot the variable chosen to be the response variable; the response variable is the one on the *y*-axis and the *explanatory* variable is the one on the *x*-axis.

Are there instances in analyzing data where it is difficult to choose which variable should be the *explanatory* and which variable the *response*? Or are there perhaps instances both choices make some sense? Yes, such situations exist. You may have run into that difficulty when you looked at the length and height of roller coasters; which variable has priority: length or height? Lesson: the application of statistics follows the way we think about things.

How do we get the straight line? Is there a "best-fitting" straight line?

There are many, many possible lines but one best-fitting line Here is a scatterplot from the used car data again; this plot shows the relationship between the *Price* (the *response* or *y* variable) of the car being sold and the *Miles* that car had been driven (the *explanatory* or *x* variable). Once again, our general idea is that there is a negative association—the greater the number of miles the car had been driven, the lower the price. In the plot we have drawn in a number of lines to represent the relationship; some of the lines obviously do not fit the data well at all. Each of the lines drawn here has a different slope and different *y*-intercept. We could draw an infinite number of different prediction lines, some of them good but most of them very bad as models. So is there a "best" one? And if so, how do we find it? There *is* a best-fitting line, and there is a specific formula

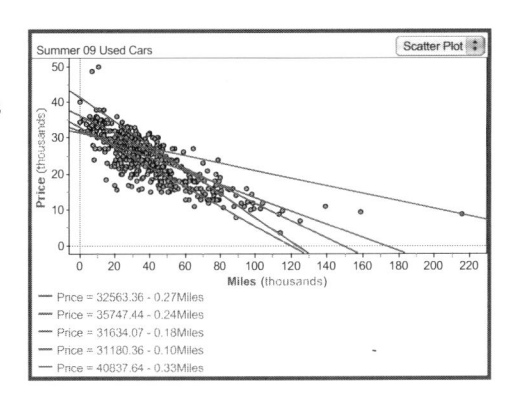

151

for the slope and the *y*-intercept to find it. But we need to clear away some wrong ideas first.

Some wrong but popular ideas followed by the correct idea. Here are two wrong ideas.

- "Of all the possible lines, the best-fitting line is the one that hits (or includes) the largest number of data points." No! In fact, the best fitting line may hit *none* of the data points and be better than a line that hits some of the data points.
- "You can get a good line by connecting the points at the ends of the graph and then finding the equation of that line." No! A look at our example above should convince you that this is not a good strategy; any line that includes the car with nearly 220,000 miles will result in a poorly fitting line.
- You can "eye-ball" a line that will fit the data. Perhaps so, but there are some simple formulas that will guarantee a line and an equation for that line that is better than any estimate that comes from just looking at the plot.

So, how do we decide which line, of all the infinite number of possible lines, is the best one? The theoretical answer is that we inspect the sizes of the *residuals* for *every* single data point for *every* possible line. Of an infinite number of possible lines, the one where the sum of squared residuals the smallest is the *best* line. Since there are an infinite number of lines to check, this looks impossible.

As it happens, we do not have to actually do this (to Caspar's relief!) because there is a formula for the slope and the *y*-intercept of the line that makes the *sum of squared residuals* the smallest. However, even though we are spared the toil of inspecting every possible line, the principle is a very important one and merits its own box showing the formulas for the slope *b* and *y*-intercept *a*.

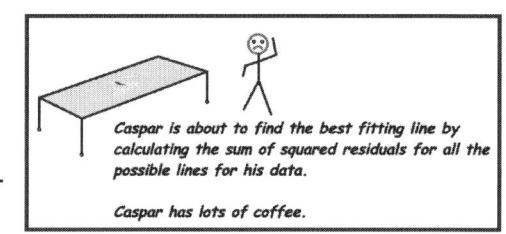

Caspar is about to find the best fitting line by calculating the sum of squared residuals for all the possible lines for his data.

Caspar has lots of coffee.

The Best-Fitting Line

- The *best-fitting* straight line $\hat{y} = a + bx$ to predict $\hat{y}$ from x is the line that makes the sum of squared residuals the smallest, or

- In symbols: the *best-fitting line* makes $\sum_{i=1}^{n}(y_i - \hat{y})^2$ as small as possible, and so

- The best-fitting line is also called the **Least Squares Regression Line** or **Least Squares Line**

The Least Squares Regression Line is $\hat{y} = a + bx$ where

- the slope is: $b = r\dfrac{s_y}{s_x}$

- the *y*-intercept is: $a = \bar{y} - b\bar{x}$

and where $\bar{x}$ and $\bar{y}$ are the means and s_x and s_y the standard deviations of the *explanatory* and *response* variables x and y, and r is the correlation coefficient for x and y.

Example: Using the formulas We will actually get the best-fitting (or *least squares*) lines for two cars (Lexus and Infiniti) so that we can compare the *two* cars and specifically their *rates of depreciation*. The scatterplot with the two *least squares regression lines* is shown on the next page; one of the lines is to predict the *Price* from *Age* for the Lexuses, and the other line is for the Infiniti G-35 cars. Here are the means, standard deviations for *Age* and *Price*, and the correlation coefficient for the relationship between them. To get the *slope* of the least squares line for the Infiniti, we calculate, using the number shown in the table for the Infiniti:

$$b = r\frac{s_y}{s_x} = -0.902579\frac{6515.4}{1.57391} \approx -3736.3402$$

Then, using this value for the slope, we calculate the *y*-intercept, using the formula:

$$a = \bar{y} - b\bar{x} = 22621.2 - (-3736.3402)(3.58847) \approx 36028.94.$$

Our equation for predicting the *Price* from *Age* for the Infiniti G-35 cars sold is $\hat{y} = a + bx = 36028.94 - 3736.34x$. For our interpretation of the slope, we can say that *for every year that an Infiniti G-35 ages, we expect the price to decrease by $3,736*. (The calculation of the regression equation for the Lexus IS cars is left as an exercise.)

Pondering the calculations. Our value for the *y*-intercept was not exactly what Fathom got with its calculation; we were two cents off. This is an instance of **rounding error**; in our calculations we used numbers with far fewer decimal places' accuracy than what Fathom (and other software) typically use.

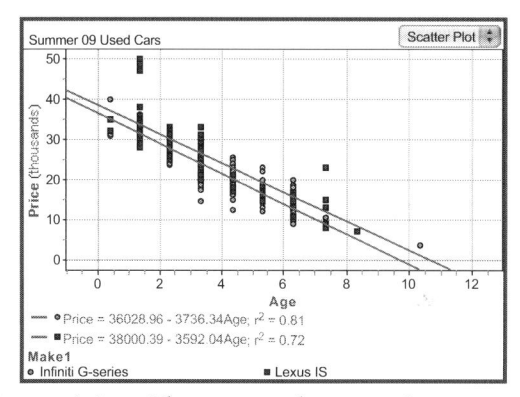

Notice that we use the *correlation coefficient r to get the slope and the y*. That means that in order to calculate our best-fitting line, we must first calculate the correlation coefficient as well as the means and standard deviations for the variables *x* and *y*-intercept of the *Least Squares Line*. Notice that we have to calculate the slope *b* before we can calculate the *y*-intercept $a = \bar{y} - b\bar{x}$ because the formula for the *y*-intercept uses *b*.

Pondering the meaning: Comparing the lines. Look at the lines in the plot and look at the equations. The lines are nearly parallel—not exactly parallel but almost, and you can see that the slopes are very similar. (If the lines were actually parallel, the slopes would be equal.) We can identify which car is which by noting that the *y*-intercept for the Lexus is about $38,000 whereas the *y*-intercept for the Infiniti is about $36000; therefore, the top line represents the best-fitting line for the Lexus and the lower one, the best-fitting line for the Infiniti.

What meaning can we draw from the fact that the slopes of the two lines are nearly equal? In the context of our used cars, our evidence is that the amount in value that these two competing cars lose per year is just about the same; we can say that their rates of depreciation (to use car jargon) are about

153

equal. That the slopes are nearly equal also means that one car is consistently more expensive; here it is the Lexus. However, the difference is not large—about two thousand dollars, on average. Our data come from just one collection of cars being sold at a particular time; our collection is merely one **sample** of many such samples that we might have downloaded from www.cars.com. Every sample that we choose will be different; there is variation from sample to sample. Hence, we cannot be completely certain that the difference of two thousand dollars between the Lexus cars and the Infiniti cars is just a feature of the particular sample we have chosen or whether it is something that we would see if we took sample after sample. It is possible that if we had access to *all* the Lexus IS cars and Infiniti cars sold over the Internet, we would find essentially no difference or that we would find the kind of difference that we have found in our sample. Can we generalize from our sample to say something about Lexus and Infiniti cars? These questions we will begin to take up in Unit 3 and especially in Units 4 and 5.

Summary: Fitting a Straight Line Model to Data

- The overall idea is to fit a **linear model**, $\hat{y} = a + bx$, which we use for **prediction** and **interpretation.**
  - We get a *predicted value* $\hat{y}$ using our linear model by inserting a value for x in $\hat{y} = a + bx$.
  - We *interpret* the slope b of the equation $\hat{y} = a + bx$ as the change we expect in y for a unit change in x.
  - We have chosen one of the two quantitative variables as the **explanatory** variable (so, in symbols, x) and the other quantitative variable as the **response** variable, y.
- The **residual** for any given data point is the difference between the actual and the predicted or, in symbols:

$$\text{residual} = \text{Actual value} - \text{Predicted value} = y_i - \hat{y}$$

- There are many possible linear models for any given set of data but only one model that we declare to be the **best-fitting line.**
  - The *best-fitting* straight line $\hat{y} = a + bx$ to predict $\hat{y}$ from x is the line that makes the sum of squared residuals the smallest, or
  - In symbols: the *best-fitting line* makes $\sum_{i=1}^{n}(y_i - \hat{y})^2$ as small as possible, and so
  - The best-fitting line is also called the **Least Squares Regression Line** or **Least Squares Line**
- The **Least Squares Line** has the equation $\hat{y} = a + bx$ where
  - the slope is: $b = r \dfrac{s_y}{s_x}$,
  - the y-intercept is: $a = \bar{y} - b\bar{x}$, where
  - $\bar{x}$ and $\bar{y}$ are the means and s_x and s_y the standard deviations of the *explanatory* and *response* variables x and y, and r is the correlation coefficient for the relationship between x and y.

§2.3 Exercises on Fitting Lines

1. **Used Cars, Including BMWs** This exercise uses the same data set that was used for the example in the **Notes** except that used BMWs have been added. Here is the same Summary Table that was presented in the **Notes** with the addition of the mean, standard deviation, and correlation coefficient for the BMW as well as for the Lexus and Infiniti.

   a. Using the information in the summary table and the formulas in the box in the **Notes** for the slope b and y-intercept of the Least Squares Regression Equation, do (and show) the calculations and get $\hat{y} = 38000.40 - 3592.04x$ for the equation predicting Price from Age for the Lexus IS.

   b. Careless Carrie has $b = -0.851356 \dfrac{1.52305}{6426.06} \approx -0.0002018$ for her calculation. How has CC been careless?

   c. Interpret the slope of the Least Squares Regression Equation in the context of the data collected. Make certain to include units.

- Open the Fathom file **Summer 09 BMWLexusInfinitiC.ftm** and get a scatterplot with Price as the response variable and Age as the explanatory variable.
- Drag the variable Make1 to the *body* of the scatterplot so that the three different makes of cars are shown.
- With the **Graph** selected, go in the menus to **Graph>Least Squares Lines.**

   d. The Lexus IS cars should be indicated by triangles. There is one Lexus IS that is 7.333 years old and is being sold for $22,900. (You may be able to see it on the scatterplot.) Use the Least Squares Regression Equation to find the predicted price for this Lexus IS and also find the *residual* for this car.

   e. The very last plot shown in the **Notes** shows the relationship between Price and Age for the Lexus and the Infiniti, and you should see that the Least Squares Regression Equations are the same as in the **Notes** for these two cars. How has the addition of the BMWs changed the appearance of the plot? (Think in terms of spread and also in terms of linearity.)

   f. From the Least Squares Regression Equations shown, what can you say in context about how the BMW loses value compared with the Lexus or Infiniti?

- Your answer to part e should have been to notice that the spread of the variable Age is greater for the BMW (all of the Lexus and Infinitis were less than eleven years old) and that the plot looks curved rather than linear, and, hence, the linear model may not be appropriate. The Least Squares Regression Equation for the BMW is being affected by the older, cheaper cars. Select the **Graph** and go to **Object>Add Filter** and type Age <11. What we have done is to compare the relationship for the BMWs younger than eleven years old with the Lexus and Infinitis (where all the cars for sale were less than eleven years old).

g. Compare the slopes of the *Least Squares Regression Equations* now and see if your answer to question e would change.

- Save your file (perhaps giving it your own name); you will use it in the exercises in the next section.

2. **Roller Coasters and the Least Squares Equation.** In this exercise you will use Fathom to try to "eye-ball" the best-fitting line. You will then compare your attempts to the line found by the equations for the least squares line. We will look at the relationship between the *Duration* of the ride, recorded in seconds, and the *Length* of a roller coaster, measured here in feet.

   a. Do you expect to see a positive or a negative relationship between the *Length* of a roller coaster and the *Duration* of the ride? As well as answering "positive" or "negative," put your idea in the form: "The greater the... the greater (or smaller) will be the..."

   b. Which variable should be the *explanatory* and which variable should be the *response*? Give a reason.

- Open the Fathom file **RollerCoastersSummer09A.ftm** and make a scatterplot relating the variables *Length* and *Duration* of the ride, using the correct assignment of explanatory and response variables to the axes.

- For simplicity, we will only look at the $n = 25$ wooden roller coasters; hence, with the **Graph** selected, go to **Object>Add Filter**. We want to filter for the wooden coasters; the variable that distinguishes wooden from steel is the variable *Construction*. Hence, type: Construction ="W".

Since the data appear roughly linear, our goal is to find a linear model that fits well to the data relating the *Length* of a roller coaster and the *Duration* of the ride. Follow the step-by-step directions to "eye-ball" your best line.

- With the **Graph** of the wooden roller coasters selected, go to **Graph>Add Movable Line.** Move the *Movable Line* until you get a line that *you* think fits the data well.

   c. Write the equation of the line that *you* have determined fits the data well in the form $\hat{y} = a + bx$.

- With the **Graph** of the wooden roller coasters selected, go to **Graph>Show Squares.**

   d. Copy the value of *"Sum of Squares"* (it may be a number something like 20,300) and explain what this *"Sum of Squares"* that Fathom has calculated actually *is*.

- The **Notes** said that the *Sum of Squares* that Fathom has calculated is a measure of how well the line fits the data and that the goal is to make the *Sum of Squares* as small as possible. Move the *Movable Line* so that your *Sum of Squares* is small—get it as small as possible. (You will do well to get it under 17,000, and you will not be able to go below 16,000—in fact, you will not be able to reach 16,000. Try anyway.)

   e. Write the equation of this "new" line and record its Sum of Squares.

   f. According to the **Notes,** the line that makes the *Sum of Squares* the *smallest* possible for a collection of data is the best-fitting line and is called the_____.

- With the **Graph** selected, go to **Graph>Least Squares Line.**

   g. Write the equation of the least squares line that Fathom gives you and the Sum of Squares. (The Sum of squares should be smaller than the Sum of Squares you managed to get.)

h. Calculate the equation of the least squares line by using the formulas for the slope and the *y*-intercept shown in the **Notes** with the numbers for the means, standard deviations, and correlation coefficient here.

i. Use the least squares equation to predict the *Duration* in seconds of the coaster called "Grizzly" (in Santa Clara, California) that has a *Length* of 3,250 feet and (actual) *Duration* of 160 seconds.

j. Find the residual for the prediction you found in part i.

k. Interpret the slope of the equation of the least squares line in the context of the data.

l. Interpret the *y*-intercept if you think that it makes sense, or say why it makes no sense.

Roller Coasters Summer 09

	Length	Duration
	3752.09	130.609
	1756.08	42.7516
	0.772724	0.772724
	23	23

S1 = mean( )
S2 = s( )
S3 = correlation (Length, Duration)
S4 = count( )
(Type = "W") and ¬(Length = "missing") and

3. **A Tale of Three Towns: San Bruno, Redwood Shores, and Hillsborough**

This exercise compares list prices and its relationship with the size of the house for three places in San Mateo County. The variable *List* is the variable *(List_Price/1000)*; the first house in the *Case Table* for the San Bruno data had a list price of $685,000 and that is recorded as 685. This exercise is a combination of calculation and interpretation, but the end product should be a comparison of the houses for sale in the three towns––a tale of three towns.

Three San Mateo County Towns

	SqFt	List
1	1160	685
2	1630	679
3	1410	710
4	720	459
5	1440	770
6	1390	700
7	950	399
8	1390	700
9	1370	695
10	1695	875

City = "SB"

a. Look at the plot showing the relationship between *SqFt* and *List* for the houses from the three towns and also compare the correlation coefficients that are given in the *Summary Table*. What do the three correlation coefficients and the plot tell you about the *linearity* and the *scatter* in the relationship between *SqFt* and *List* for the three towns?

[b. *Review; may be optional (ask your instructor).* The formula for the slope of the *Least Squares Regression Equation* given in the **Notes** is $b = r \frac{s_y}{s_x}$. If you are starting from scratch with data, the first thing that needs to be done to use this formula is to calculate the correlation coefficient *r*, and for that you need the means and the standard deviations for each variable. Calculate the correlation coefficient *r* for the San Bruno data using the formula for the correlation coefficient *r* and using the values of the means and standard deviations that are given here.]

Three San Mateo County Towns

City		SqFt	List
HIL		3516.91	2842.72
		890.439	691.1
		11	11
		0.653291	0.653291
RS		2079.09	1180.72
		414.812	187.718
		11	11
		0.936151	0.936151
SB		1315.5	667.2
		296.989	139.14
		10	10
		0.874031	0.874031
Column Summary		2334.72	1591.56
		1091.37	1030.55
		32	32
		0.91553	0.91553

S1 = mean( )
S2 = s( )
S3 = count( )
S4 = correlation(SqFt, List)

c. Use the formulas for the slope and the *y*-intercept of the *Least Squares Regression Equation* to get the best-fitting equations for the relationship between *SqFt* and *List* for the three towns. All the information that you need is given in the *Summary Table* above.

- Open the Fathom file **Three San Mateo County Towns.ftm** and produce either the plot shown here (or three plots, one for each town; in that case make certain that the scales are the same for each plot).
- Get Fathom to produce the *Least Squares Lines* on the plot and check your calculations in part c.
    d. Using the *Least Squares Regression Equation* for San Bruno, get the predicted value for the list price of the second house in the San Bruno sub-sample—the one that has 1,630 square feet and was listed at 679 (that is $679,000).
    e. Get the residual value for the San Bruno house whose predicted value you got in part d.
    f. Write a paragraph or two using the plots, the information on this page, and your calculations of the *Least Squares Regression Equations* to discuss the differences and similarities in the size and prices and the relationships between size and price for the houses being sold in the three places: San Bruno, Redwood Shores, and Hillsborough. Write your tale of three towns.

4. **Women's Heptathlon Results for 2008** [Fathom] The heptathlon is a women's event held in the Olympic games since 1984. It has seven events, and each competitor must participate in all of the events.

    - 100-meter hurdles; time measured in seconds   HundredMeterHurdles
    - High jump; distance measured in meters   HighJump
    - Shot put; distance measured in meters   ShotPut
    - 200-meter run; time measured in seconds   TwoHundredMeters
    - Long jump; distance measured in meters   LongJump
    - Javelin throw; distance measured in meters   JavelinThrow
    - 800-meter run; time measured in seconds   EightHundredMeters

    For each of these events, there is a score and a total that is the sum of these scores. The highest total wins.

    Above are the means, the standard deviations, and the correlation coefficient r for the relationship between performance in the *HighJump* and performance in the *LongJump*. [Notice that the *Summary Table* gets the means and standard deviations for just the cases where there are data for both; one of the competitors (Yana Maksimova of Belarus) did not complete the long jump, so the mean and sd for the *HighJump* also excludes her.]

    a. Does the direction of the correlation coefficient make sense in the context of these data? Explain briefly.
    b. Of the two variables, *HighJump* and *LongJump*, which do you think should be the *explanatory* variable and which should be the *response* variable? Or does the choice not matter, as neither is "prior" to the other? Choose one as the explanatory variable and the other as the response, and from the means, sd, and the r given, calculate the *Least Squares Regression Equation* for your choice.

c. Interpret the slope of the *Least Squares Regression Equation* you have just calculated in the context of the heptathlon competitors.

d. Calculate the *Least Squares Regression Equation* for the opposite choice of explanatory and response variables from the same information given above.

- Open the Fathom file **Heptathlon 2008 A.ftm** and get a scatterplot for *HighJump* and *LongJump*.
- On the scatterplot, get the *Least Squares Line* for your choice of explanatory and response variables on your scatterplot. You should be able to check your calculations for one of the equations you calculated.
- Drag the explanatory variable to the response variable axis; Fathom is programmed to switch the two. Now you should be able to check the other calculation. (You may have small rounding errors.)

e. Answer this "typical test question" based on what you have done in this exercise: "True or false, and explain your answer: 'For the relationship between any two variables, there is just one r but two possible least squares regression equations, although only one of the two may make sense.'"

f. Marie Collonvillé of France jumped 6.21 m in the *LongJump* and 1.86 m for the *HighJump*. Use the relevant *Least Squares Regression Equation* to predict Ms. Collonvillé's *HighJump* performance from her *LongJump* performance.

g. Find the residual for the prediction you made in part f. Does Ms Collonvillé jump higher than what you would expect from her *LongJump* performance?

h. Predict Ms. Collonvillé's *LongJump* from her *HighJump*, and find the residual.

[i. If Exercise 5 is done then write a short paragraph or two explaining any differences or similarities between the analysis you have done for the 2008 heptathlon and the analysis for the 1988 heptathlon.]

[Note: This exercise is paired with exercise 4.]

5. **Women's Heptathlon Results for 1988** [Fathom] The heptathlon is a women's event held in the Olympic games since 1984. It has seven events, and each competitor must participate in all of the events. See Exercise 4 for the variables measured. The analysis and questions for this exercise parallel the analysis and questions for Exercise 4. The goal is to compare the two heptathlons, twenty years apart.

On the right are the means, the standard deviations, and the correlation coefficient r for the relationship between performance in the *HighJump* and performance in the *LongJump*

a. Does the direction of the correlation coefficient make sense in the context of these data? Explain briefly.

b. Of the two variables, *HighJump* and *LongJump*, which do you think should be the *explanatory* variable and which should be the *response* variable? Or does the choice not matter, as neither is "prior" to the other? Choose one as the explanatory variable and the other as the response, and from the means, sd, and the r given, calculate the *Least Squares Regression Equation* for your choice.

Women's Heptathlon 1988	HighJump	LongJump
	1.782	6.1524
	0.0779423	0.474212
	25	25
	0.782442	0.782442

S1 = mean( )
S2 = s( )
S3 = count( )
S4 = correlation(HighJump, LongJump)

c. Interpret the slope of the *Least Squares Regression Equation* you have just calculated in the context of the heptathlon competitors.

d. Calculate the *Least Squares Regression Equation* for the opposite choice of explanatory and response variables from the same information given above.

- Open the Fathom file **Heptathlon 1988 A.ftm** and get a scatterplot for the relationship between *HighJump* and *LongJump*.
- On the scatterplot, get the *Least Squares Line* for your choice of explanatory and response variables on your scatterplot. You should be able to check your calculations for one of the equations you calculated.
- Drag the explanatory variable to the response variable axis; Fathom is programmed to switch the two. Now you should be able to check the other calculation. (You may have small rounding errors.)

e. Answer this "typical test question" based on what you have done in this exercise: "True or false, and explain your answer: 'For the relationship between any two variables, there is just one r but two possible least squares regression equations, although only one of the two may make sense.'"

f. Dong Yuping of China jumped 6.40 m in the *LongJump* and 1.86 m for the *HighJump*. Use the relevant *Least Squares Regression Equation* to predict Ms. Dong's *HighJump* performance from her *LongJump* performance. (She finished sixteenth overall.)

g. Find the residual for the prediction you made in part f. Does Ms Dong jump higher than what you would expect from her *LongJump* performance?

h. Predict Ms. Dong's *LongJump* from her *HighJump* and find the residual.

[i. If Exercise 4 is done then write a short paragraph or two explaining any differences or similarities between the analysis you have done for the 2008 heptathlon and the analysis for the 1988 heptathlon.]

6. **Women's Heptathlon Results for 2008** The heptathlon is a women's event held in the Olympic games since 1984. It has seven events, and each competitor must participate in all of the events. For the variables, see Exercise 4.

a. Before looking at the data, think about the data: do you expect a positive, negative, or no relationship between the variables *HundredMeterHurdles* and *HighJump*? Give a reason for your answer.

b. Which variable should be regarded as the explanatory variable and which variable should be regarded as the response variable here? Or does it not matter? Give a reason for your answer.

- Open the Fathom file **Heptathlon 2008 A.ftm** and get a scatterplot for the relationship between *HighJump* and *HundredMeterHurdles*. Put *HighJump* on the horizontal axis.
- On the scatterplot, get the *Least Squares Line* to predict time in the *HundredMeterHurdles* from the distance for the *HighJump*.

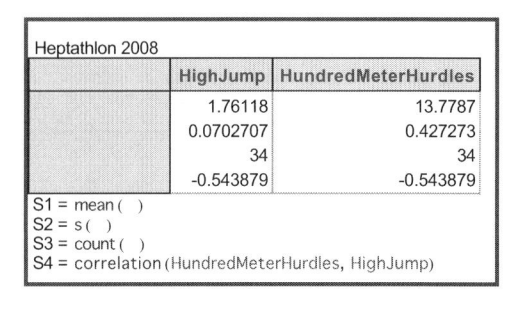

Heptathlon 2008	HighJump	HundredMeterHurdles
	1.76118	13.7787
	0.0702707	0.427273
	34	34
	-0.543879	-0.543879

S1 = mean ( )
S2 = s ( )
S3 = count ( )
S4 = correlation (HundredMeterHurdles, HighJump)

- Get a *Summary Table* showing the means, standard deviations, counts, and correlation coefficient *r* for the variables *HighJump* and *HundredMeterHurdles* shown here.
    c. Using information in the *Summary Table* and the formulas for the slope *b* and *y*-intercept *a* given in the **Notes,** confirm that $\hat{y} = 19.6 - 3.31x$. (You may have a rounding error, however.)
- With the *Graph* (the scatterplot) selected, go to *Graph>Plot Value* and type in *mean(HighJump)*.
- With the *Graph* (the scatterplot) selected, go to *Graph>Plot Function* and type in *mean(HundredMeterHurdles)*. You should see the plot divided into four quadrants.
    d. In which quadrant are the best athletes? Give a short reason for your answer.
    e. [Review of correlation] In which quadrants are the contributions to the correlation coefficient negative and in which quadrants are the contributions positive?
    f. Find an athlete who makes a positive (instead of negative) contribution to the correlation coefficient *r*. [By selecting a dot on the graph (so that it becomes red), that case will be highlighted in the *Case Table*.]
    g. Notice that the *Least Squares Line* goes through the means of the two variables. What in our formulas guarantees that the best-fitting line goes through the means?
- In the *Case Table,* select the number-two heptathlete, Hyleas Fountain, and note where she is on plot.
    h. Judging from where she is on the plot, will the residual predicting time in the *HundredMeterHurdles* from her *HighJump* performance be positive or negative for Hyleas Fountain? Is the direction of the residual good or bad for Ms. Fountain?
    i. From the *Case Table* find the values for Ms. Fountain's *HighJump* performance and her time in the *HundredMeterHurdles* and calculate the residual. (Two calculations are involved here.)
    j. Think about what direction the relationship between *HundredMeterHurdles* and the *JavelinThrow* should be. Should the relationship be positive, negative, or is there essentially no relationship? Give a reason.
- Get the scatterplot for the relationship between *HundredMeterHurdles* and the *JavelinThrow*. Make the *JavelinThrow* the response variable.
- On the scatterplot, get the *Least Squares Line* to predict distance in the *JavelinThrow* from the time in the *HundredMeterHurdles*. Note also on the plot $r^2 = 0.0043$. Take the square root to get *r*.
    k. Compare the correlation coefficient *r* with your ideas in part j and comment. Were you correct?

7. **Gearhead Cars: Power and the Quarter-Mile** [Fathom; put the answers, including graphs, into a Word document] For their road tests, *Road & Track*® records the maximum speed if they accelerate from a standing start to as fast as they can go in a distance of one-quarter mile. They also record how long it takes the car to reach the one-quarter-mile mark.

    In this exercise we will look at the relationships between these two measures and the power of the car, measured by *Horsepower*. The variables that measure the speed at the end of the one-quarter mile and the time taken to one-quarter mile are *QuarterMile_mph* and *QuarterMile_secs*.

a. Here are the means, standard deviations, and the correlation coefficient *r* for *Horsepower* and *QuarterMile_mph*. In the way that we usually think about cars, which variable of these two is most naturally the *explanatory* variable and which variable is most naturally thought of as the *response* variable?

	Horsepower	QuarterMile_mph
	356.474	105.032
	147.432	12.348
	133	133
	0.940495	0.940495

S1 = mean ( )
S2 = s ( )
S3 = count ( )
S4 = correlation (Horsepower, QuarterMile_mph)

Road&Track May 09 rev Cars

b. True or false, and explain: "Since the correlation coefficient *r* is the same whichever variable we choose as the explanatory and whichever variable we choose as the response, the *Least Squares Regression Equation* will also be the same."

c. Using the formulas (found in the **Notes**) for the slope and the *y*-intercept, calculate the *Least Squares Regression Equation* according to the choices that you made in part a.

- Open the Fathom file **RoadandTrackMay09C.ftm** and get a scatterplot of the relationship between *Horsepower* and *QuarterMile_mph* with the explanatory and response variables correctly assigned.
- Have Fathom get the *Least Squares Line* (**Graph>Least Squares Line**) and check your answer to your calculations for part c. Copy the graph into a Word document to submit with the answers to this exercise.

d. Identify the slope of the *Least Squares Regression Equation* and interpret the slope in the context of the data. (You may find it useful to consider an increase of ten horsepower rather than just one horsepower.)

- Highlight case 29 in your *Case Table*. The car is the Dodge Challenger R/T. When you highlight this row, its dot will show red on your scatterplot. Read off the *Horsepower* and *QuarterMile_mph* values for this car.

e. Use the information on *Horsepower* to find the *predicted* value of *QuarterMile_mph*.

f. From your calculations in part e, find the residual. Does the Dodge Challenger R/T have a speed that is better or worse than expected, given its horsepower?

g. Do you expect the relationship between *Horsepower* and *QuarterMile_secs* to be positive or negative? Give a reason for your answer.

- Get a scatterplot of the relationship between *Horsepower* and *QuarterMile_secs* values with the explanatory and response variables correctly assigned.
- Have Fathom get the *Least Squares Line* (**Graph>Least Squares Line**) to predict *QuarterMile_secs* from *Horsepower*. Copy the graph into a Word document to submit with the answers to this exercise.

h. Use the *Least Squares Regression Equation* and information from the *Case Table* to find the *predicted value* of the time taken for the Dodge Challenger R/T to cover a quarter-mile.

i. Find the residual for the *QuarterMile_secs* for the Dodge Challenger R/T. Is the result that you get in accord with the result you got for part f? Explain.

8. **Gearhead Cars: Power, Displacement, and MPG**

   Horsepower measures how powerful a car is, *MPG* (or *miles per gallon*) measures the amount of fuel a car uses (that is, the higher the *MPG*, the more miles the car travels on a gallon of fuel, and so less fuel is being used), and *Displacement* measures how big the engine is. We naturally expect that there is a tendency for cars with larger engines (the technical term is "displacement") to have more power and use more fuel, and therefore have lower *MPG*. (*Displacement* is expressed in measures of volume, and the metric measure is cubic centimeters, and that is why the variable name is *Displacementcc*.)

   a. With the description given just above (or from what you know about cars), determine which relationships should be positive, which should be negative.

      Displacementcc and *Horsepower*, *Displacementcc* and *MPG*, *Horsepower* and *MPG*

   b. Here is the *Summary Table* for the relationship between *Displacement* and *Horsepower*. Calculate, "by hand" the *Least Squares Regression Equation* to predict *Horsepower* from *Displacementcc*.

   Road&Track May 09 rev Cars

	Displacementcc	Horsepower
	3851.2	349.825
	1605.13	142.087
	126	126
	0.879342	0.879342

   S1 = mean ( )
   S2 = s ( )
   S3 = count ( )
   S4 = correlation (Displacementcc, Horsepower)
   ¬(Displacementcc = "missing")

   - Open the Fathom file **RoadandTrackMay09C.ftm** and get a scatterplot of the relationship between *Displacementcc* and *Horsepower* with the explanatory and response variables correctly assigned.

   - Have Fathom get the *Least Squares Line* (**Graph>Least Squares Line**) and check your answer to your calculations to part b. Copy the graph into a Word document to submit with the answers to this exercise.

   c. In the *Case Table*, select case 38, which happens to be the Ford Mustang Bullitt. (If you use the mouse to select the case, the case will also show in your scatterplot.) Use the values of the correct variable and the *Least Squares Regression Equation* to predict the *Horsepower* the Ford should have according to our equation.

   d. Find the residual for the Ford Mustang Bullitt. Is the residual positive or negative? Is the Ford above or below the *Least Squares Line*?

   e. Confused Conrad answers question b ("Find the residual...") by writing "$r = 0.879342$" from the Summary Table above. What is CC's confusion?

   f. Identify the slope of the *Least Squares Regression Equation* and interpret it in the context of the data.

   g. Identify the *y*-intercept and interpret this number if you think that it makes sense to interpret it; if you think it makes no sense, explain why it makes no sense.

   - Get a scatterplot of the relationship between *Horsepower* and *MPG* and get the *Least Squares Line*.

   h. For the Ford Mustang Bullitt, use the *Least Squares Regression Equation* to predict the *MPG* from the *Horsepower*.

   i. Choose any other car you think is interesting and use the two equations you have to predict the *Horsepower* from *Displacementcc* and *MPG* from the *Horsepower*.

§2.4 "Is the Model Good and Useful?"

A measure of how well a model fits: R^2

Recall the quote from George E. P. Box in §1.7: "All models are wrong, but some are useful." We have a way of assessing whether it is a good model or not. A general measure of how well the model fits is called the **coefficient of determination,** and its symbol is R^2. Here are some facts about R^2.

> **R^2: Facts and Interpretation**
> - For simple regression models, $R^2 = (r)^2$; that is, R^2 is the square of the correlation coefficient r.
> - $0 \leq R^2 \leq 1$
> - *Interpretation*: R^2 shows the proportion (or percentage) of variation in the *response* variable *explained* by the model being used. (See the examples just below for the wording.)

Here is the plot from the end of the last section on the relationship between the *Age* and the *Price* of used Lexus and Infiniti cars. (Note that the top-most line is the least squares line for the Lexus cars, and the second line is for the Infiniti cars, even though the equations are listed in alphabetical order in the Fathom output.) At the foot of the plot, you can read that for the Infiniti cars, $R^2 = r^2 = 0.81$ and for the Lexus cars, we see $R^2 = r^2 = 0.72$. For the regression model for the Infiniti cars, we would say:

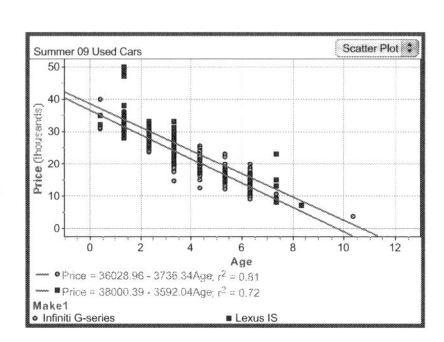

> 81 % of the variation in the *Price* of used Infiniti G-Series can be explained (or accounted for) by the linear model on the *Age* of the car.

For the Lexus cars, we would say:

> 72 % of the variation in the *Price* of used Lexus ISs can be explained (or accounted for) by the linear model on the *Age* of the car.

The linear model on the variable *Age* does a slightly better job of explaining the variation in the *Price* of used Infiniti cars than it does for the Lexus IS; however, to explain 70% or 80% of the variation in the response variable is doing pretty well. Notice also that there are some alternate verbs that can be used rather than "explained."

R^2 is also known as the **Coefficient of Determination,** but it is also convenient to refer to it as *"big R-squared"* to prevent confusion with the correlation coefficient r. It is easy to confuse R^2 (and its interpretation) with the correlation coefficient r or with residuals, $y_i - \hat{y}$, or with the least squares regression equation $\hat{y} = a + bx$ since all of these involve the letter "r" in their names or in their symbol.

Here is a further warning: be prepared to find the remainder of this section challenging since the ideas are new and different. However, the ideas are also important and are used widely in statistics.

Interactive Notes: How and why R^2 works. For this subsection, it will be helpful to have Fathom opened and follow the bulleted instructions as you read these **Notes** (that is why the subsection is called "Interactive Notes"). It will be helpful, although not strictly necessary, because all of the graphics will be shown, but seeing the graphics will give the argument a certain force; you will actually see. There is an exercise that follows the pattern of these "Interactive Notes" and so doing a "dry run" now when the answers are "given" will help. Our example will use the data for the women's heptathlon of the 2008 Summer Olympics in Beijing. (For a description of the data and the event, see Exercise 4 of §2.3.)

- Open the Fathom file **Heptathlon 2008 A.ftm** and get a scatterplot for the relationship between *HundredMeterHurdles* and *TwoHundredMeters*.

The scatterplot shows a positive linear relationship but with some scatter. We expect that a linear model that relates the time for the *HundredMeterHurdles* is appropriate since there does not appear to be a curved relationship. We also note scatter around the basic linear pattern. Here is the *Summary Table* showing the means, standard deviations, and the correlation coefficient *r* for this relationship. Notice that we have all the information here that we need to calculate the *Least Squares Regression Equation*.

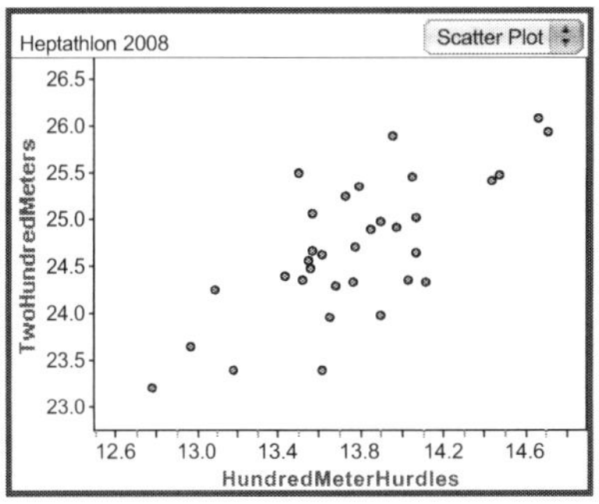

We *also* have all the information we need to calculate R^2, which is just $R^2 = r^2 = (0.723836)^2 = 0.5239$. We can interpret this R^2 and say that about 52% of the variation in the *TwoHundredMeters* times can be explained by the linear regression on the times for the *HundredMeterHurdles*. It does make some sense that the faster athletes should be faster in both events, the slower ones slower in both events. It also makes sense that there should be some other variation; one of the events is a purely running event, while the other event involves some jumping as well.

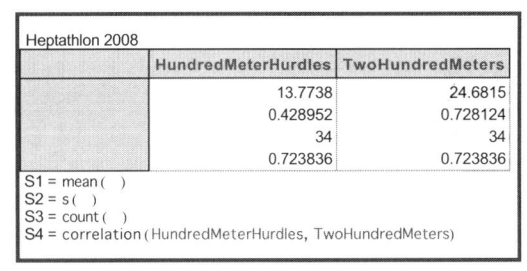

Now we shall proceed to see why we say that R^2 shows the "the proportion (or percentage) of variation in the *response* variable *explained* by the model being used"—in this instance, that 52% of the variation in the *TwoHundredMeters* times is explained by our model on the *HundredMeterHurdles*.

- Select the graph, and go to **Graph>Plot Function;** in the dialogue box, type in *mean(TwoHundredMeters)*.
- With the graph selected, go to **Graph>Show Squares.**

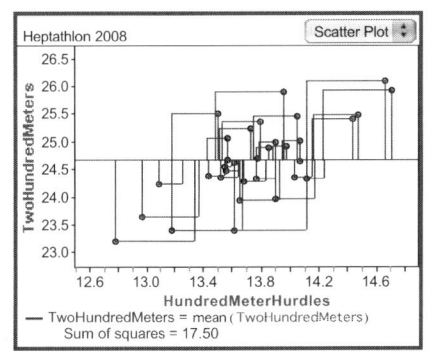

The plot at the right shows the mean of the response variable, *TwoHundredMeters* times, and many squares. Each square is a quantity $(y_i - \bar{y})^2$, where the y_i is the time for one of the athletes in the *TwoHundredMeters*, and $\bar{y}$ is the mean time for all the athletes. So the best "heptathlete" for both these events (the least time for both events) was Hyleas Fountain (of the USA), whose time was 23.21 seconds for the *TwoHundredMeters* (and 12.78 seconds for the *HundredMeterHurdles*, which means that she is the "lower-left-most" dot on the scatterplot). The mean for all $n = 34$ heptathletes was 24.68 s., so for Ms. Fountain the calculation of $(y_i - \bar{y})^2$ will be $(23.21 - 24.68)^2 = (-1.47)^2 = 2.161$. Ms. Fountain is 1.47 seconds below the mean time for the *TwoHundredMeters* (which is a good place to be!), and the square of this number is 2.16. For each heptathlete, there is a square like this. For the athletes whose times are near the mean time in the *TwoHundredMeters* of 24.68 seconds, the square is very small, and for the frontrunners and for the stragglers, the squares are big.

Total Sum of Squares What we do with these squares is add them to get a *Total Sum of Squares*, whose formula is $\sum_{i=1}^{n}(y_i-\bar{y})^2$. We should recognize $\sum_{i=1}^{n}(y_i-\bar{y})^2$ as part of the formula for the variance and the standard deviation, $s = \sqrt{\dfrac{\sum_{i=1}^{n}(y_i-\bar{y})^2}{n-1}}$. This sum tells us how much *variation* we have in the data as a whole. Usually we divide this by $n-1$ so that we can compare collections, but we will see we do not have to do that here.

For our example, the *Total Sum of Squares* $= \sum_{i=1}^{n}(y_i-\bar{y})^2 = 17.50$.

- With the graph selected, go to **Graph>Least-Squares Line.** You should see the plot shown here; if you see this plot on the computer, you should be able to see that there are two sets of squares shown in two different colors.

Sum of Squared Residuals The plot is looking really complicated because there is now an additional set of squares. If you are doing this with Fathom, you should just about be able to see that this new set is composed of the squares that come from the *residuals* between the data points and the *Least Squares Line.* Look at the dot for Hyleas Fountain (the best runner). Remember that her square for her distance from the *mean*, that is $(y_i-\bar{y})^2$ was large; now look closely and you will see a tiny square for her dot that represents her squared residual, that is $(y_i-\hat{y})^2$. Here is the calculation with comments.

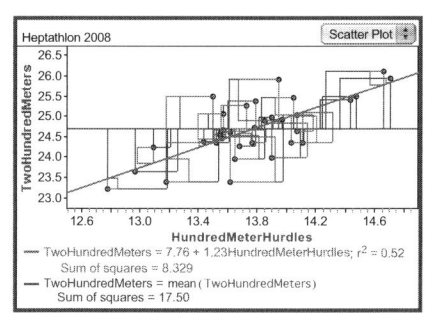

- Hyleas Fountain's time for the *HundredMeterHurdles* was 12.78 seconds. (You can select her on the graph if you are using Fathom, and her times for both races will show at the left foot of the window.)
- The *Least Squares Regression Equation* is $\hat{y} = 7.76 + 1.23x$, as can be checked from the graph.
- Therefore, Ms. Fountain's *predicted TwoHundredMeters* time should be
  $\hat{y} = 7.76 + 1.23(12.78) \approx 7.76 + 15.72 = 23.48$ seconds.
- Hence, Hyleas Fountain's residual is $y_{HyleasFountain} - \hat{y} = 23.21 - 23.48 = -0.27$, which means that she was even slightly faster than her predicted time.
- The squared residual for Ms. Fountain is $(y_i-\hat{y})^2 = (-0.27)^2 \approx 0.073$

What we have done and where this is going. Let us review what we have done. First, we calculated the sum of the squares from the mean of *y*, which in symbolic form is:

$$\text{Total Sum of Squares} = \sum_{i=1}^{n}(y_i-\bar{y})^2 = 17.50$$

We can think of this number as the *total amount of variation* in the data, or we can think of this number as a kind of *baseline* amount of variation. Then we calculated a similar—but different—sum of squares between the actual values and the predicted values, in symbolic form:

$$\text{Sum of Squared Residuals } \sum_{i=1}^{n}(y_i - \hat{y})^2 = 8.329$$

We can think of this as the variation still left over when we applied the least squares regression line. Notice that the *Sum of Squared Residuals* is smaller, and we would like to reduce the amount of variability, so smaller is good. That means that by applying the least squares line, we have disposed of, or gotten rid of, some of the original variation measured by the *Total Sum of Squares.* By applying the least squares line, we have accounted for (or, "explained") of some of the *baseline* amount of variation. If you look carefully at the plot again you should be able to see that the squares for the squared *residuals* $(y_i - \hat{y})^2$, the squares around the *Least Squares Line*, appear to be smaller than the squares around the mean, $(y_i - \bar{y})^2$. If the squares are smaller then, when we add them all up, we should get a smaller sum; and since Fathom reports the sum of squares, you can see that the sum is smaller.

The sum of squares for the least squares line, the *Sum of Squared Residuals* $\sum_{i=1}^{n}(y_i - \hat{y})^2 = 8.329$, whereas the *Total Sum of Squares* $= \sum_{i=1}^{n}(y_i - \bar{y})^2 = 17.50$. We have reduced the amount of variation by applying the linear model in the form of the *Least Squares Line*. The next step is to calculate—in percentage terms—how much we have reduced the original variation.

Calculating the proportion of explained and unexplained variation. Of course, we have not completely reduced the baseline variation to zero; the *Sum of Squared Residuals* is not zero. (If there were no variation, there would be no scatter around the least squares line.) There is variation that is still *unexplained* (or as yet *unaccounted for*) by the model. How much *unexplained variation* remains compared with what we started with? We calculate the **Proportion of Unexplained Variation** by making a fraction of the *Sum of Squared Residuals* to the *Total Sum of Squares*. Here is how it works out in our example:

$$\textbf{Proportion of Unexplained Variation} = \frac{\text{Sum of Squared Residuals}}{\text{Total Sum of Squares}} = \frac{\sum_{i=1}^{n}(y_i - \hat{y})^2}{\sum_{i=1}^{n}(y_i - \bar{y})^2} = \frac{8.329}{17.500} \approx 0.476$$

Now comes what may appear at first to be a linguistic trick, but it is not. Recall the rule that says that the probability of "not A" is equal to 1 minus the probability of the event. In our notation, we write $P(\text{Not } A) = 1 - P(A)$. What we have calculated so far is the proportion of (still) unexplained variation. If we want the **Proportion of Explained Variation,** we will subtract what we have just calculated from the number 1 (the 1 stands for "all the variation") and get

Proportion of Explained Variation = 1 – Proportion of Unexplained Variation

$$R^2 = \text{Proportion of Explained Variation} = 1 - \frac{\sum_{i=1}^{n}(y_i - \hat{y})^2}{\sum_{i=1}^{n}(y_i - \bar{y})^2} = 1 - \frac{8.329}{17.500} \approx 1 - 0.476 = 0.524.$$

Compare this with what we calculated before; you will find it just below the first heptathlon plot. There we calculated $R^2 = r^2 = (0.723836)^2 = 0.5239$. We can get R^2 by squaring the correlation coefficient or by calculating sums of squares. If the second appears to you more cumbersome, you should know that it has the advantage of showing *why* we *interpret* R^2 in the way that we do.

R^2 The calculation from sums of squares

$$R^2 = \text{Proportion of Explained Variation} = 1 - \frac{\sum_{i=1}^{n}(y_i - \hat{y})^2}{\sum_{i=1}^{n}(y_i - \bar{y})^2}$$

Flawed Models: Ways the model may not be adequate

There are a number of ways that our linear model can be flawed. Our model is built out of the data at hand. There may a few data points that have an undue influence on the construction of the model. If this happens, we speak of "influential observations."

Flawed Models: Influential Observations One or two observations may have the potential to determine the slope and the *y*-intercept of the *Least Squares Equation*. Such observations are called **influential observations.** Here is one example. (The Fathom instructions to get these plots are below.)

In the 2008 Olympic heptathlon, one competitor (Yana Maksimova of Belarus) fouled out in the long jump and got zero points for that event. As a result, her overall score of 4,806 was very low. Here are two plots showing the relationship between the total score (which determines the rankings in the event) and performance on two events, the *HundredMeterHurdles* and the *HighJump,* with Ms. Maksimova highlighted.

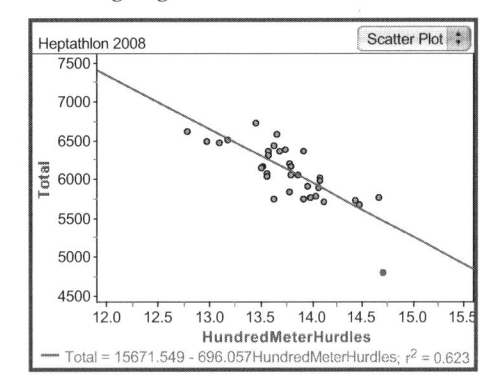

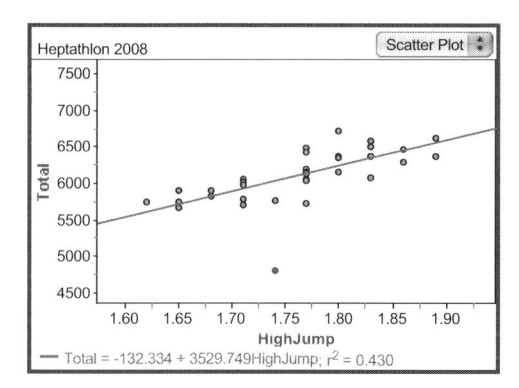

What if?   Now suppose that Ms. Maksimova had *not* fouled out in the long jump and so had a higher *TotalScore*. Then, if she had a *TotalScore* of about 6,000, the scatterplots might look like this.

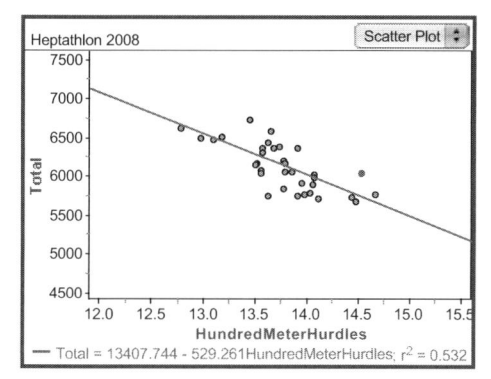

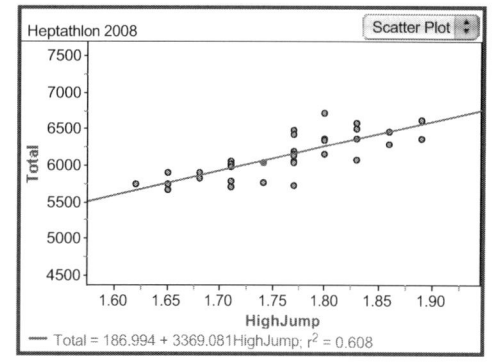

Here are the Fathom instructions for the plots shown above.

- Make the first set of scatterplots in the file **Heptathlon 2008 A.ftm**. You may want to adjust the scales in accordance with the plots shown and go to **Graph>Format Numbers** to *Fixed Decimal* with three decimal places.
- Select the dot that corresponds to Yana Maksimova and move the dot *up* to get something like the second set of plots.
- Move the dot anywhere you want to move it—you can make Ms. Maksimova faster or slower in the *HundedMeterHurdles*, and you can change her *TotalScore*, and watch how the least squares line is affected.
- To restore the data to its original state, go to **Edit>Undo Drag Point.** Another way to get back to "reality" is to select the collection icon and then go to **File>Revert Collection.**

So what has happened? Or not happened? If Ms. Maksimova's *TotalScore* had been higher then our least squares model looking at the effect that *HundedMeterHurdles* had on *TotalScore* would look much different. The slope would be different, and the R^2 is also different. Raising the *TotalScore* changes the slope of the *HighJump* least squares equation but much less so. The change for the *HundedMeterHurdles* is about 24% of its original value (about $\frac{167}{696} \approx 0.24$), but the change in the slope for the *HighJump* is only about 4.5% of the original value (about $\frac{160}{3529} \approx 0.045$). Notice that Ms. Maksimova is an *outlier* on the *response* variable (*TotalScore*) in both plots. However, her performance in the *HighJump* was neither an outlier or toward the "end" of the data, as it was for the *HundedMeterHurdles*. If an observation is also an outlier or near the end of the data on the *explanatory* variable then that observation has the potential to be influential. Whether a data point actually *is* an *influential observation* depends upon its relationship with the remainder of the data. With software (such as Fathom), it is a simple matter to run an analysis with and without a single data point to explore whether the data point is *influential*.

> **Influential Observations**
>
> A data point is an *influential observation* when small changes in the values of the data would substantially change the slope and y-intercept of the linear model fitted.

Flawed Models: Curves and Bends and Gearhead Cars Actually, "gearheads" tend to enjoy driving fast through curves and bends. However, when we use the correlation coefficient and the least squares line, we are using a *linear model*—we want a straight road. If our data is not linear, if it has curves and bends, then the conclusions that we draw from the linear model will not be useful. Here is an example from our *Road and Track* data. One of the most common measures with cars is the time in seconds that it takes for a car to accelerate from zero miles per hours to sixty miles per hour.

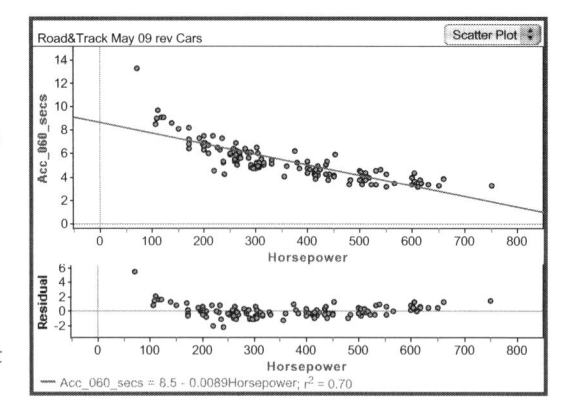

- Open the Fathom file **RoadandTrackMay09C** and get the scatterplot for the relationship between *Horsepower* and *Acc_060_secs* with the least squares regression line shown, as in our plot here.
- With the graph selected, go to **Graph>Make Residual Plot** to get the *residual plot* that is shown below the scatterplot.

The plot that you should get—showing how the acceleration time for cars depends upon horsepower—is displayed on the previous page. The relationship is negative—the more power a car has, the smaller the time it takes to accelerate to 60 mph—but there is also a certain amount of curvilinearity evident. Look at the cars with less than 150 horsepower; all of the residuals for these cars are positive. You can see that in the original scatterplot, but Fathom makes a plot of residual values that exhibits the same feature. Notice also that in all of the residuals for the cars that have more than 650 horsepower, the residuals are also positive. For the cars having 200 to 500 horsepower, there is a mix of positive and negative residuals, and the linear model is more appropriate. In general, you can detect **curvilinearity** by looking at the shape of the graph, but curves and bends can also be seen in the residuals; if the residuals exhibit a definite pattern, the linear model breaks down in some way. If the residuals show a random scatter then the linear model is a good model.

Here is another example of curves and bends in data (the Snook data); notice again how you can detect the curve in the scatterplot and also how the residuals form a pattern.

If the data are not linear, the data can still be analyzed. However, the techniques employed (such as **transforming** the data using logarithms) and using alternate models are for the next course in statistics. For this course, it is sufficient to *detect* curvilinearity in data.

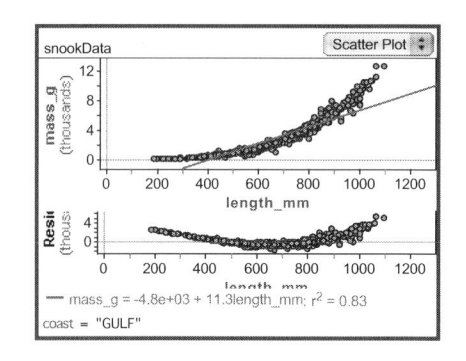

Curvilinearity

Curvilinearity (lack of a linear relationship) is reflected in a definite pattern to the residuals; residuals around the least squares regression line should exhibit random scatter if the linear model is appropriate.

Flawed Models; Extrapolation One of the purposes of the *Least Squares Regression Equation* is to make predictions. One of the mistakes that can be made is to make predictions for values of the *explanatory variable* outside of the range of the data we have. This mistake is called **extrapolation.** Here is an example that looks at the relationship between *Horsepower* and *Acc_060_secs* for just the cars with four cylinders.

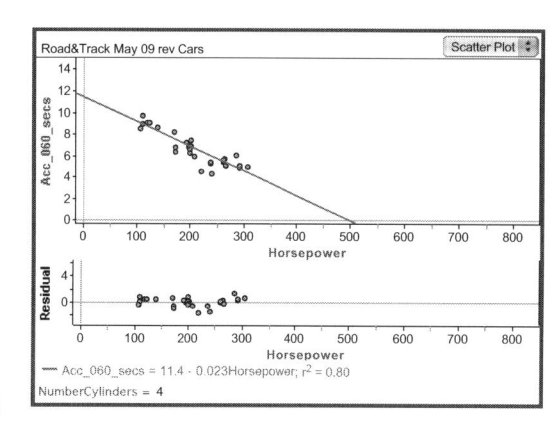

- With the graph selected, go to **Object>Add Filter** and type in *NumberCylinders = 4* to get the plot shown.

For just the four cylinder cars, the plot looks linear, and the least squares line has the equation $\hat{y} = 11.4 - 0.023x$, where x is *Horsepower* and y is the time to accelerate to 60 mph, or *Acc_060_secs*. It does not make sense at all to insert $x = 37$ into the equation and get $\hat{y} = 11.4 - 0.023(37) = 11.4 - 0.23 \approx 10.55$ because the data from which we derived the least squares line do not include cars with horsepower in the region of thirty-five to forty horsepower. (There were cars made with horsepower in this range, but their zero to sixty times tended to be in the neighborhood of twenty-five to thirty seconds!) For the same reason, we should not use this equation to predict the zero to sixty time for a four-cylinder car with 450 horsepower. There are cars with 450 horsepower, but these cars are not four-cylinder cars, and so a different linear model will be appropriate. In fact, with our data, it makes sense to look at linear models by the number of cylinders car have. Here is the plot. (Note: The variable *Cylinders* includes three-cylinder cars and five-cylinder cars in the categories "Four Cylinders" and "Six Cylinders" respectively.) As you can see from the plot with the lines shown, the linear relationship differs depending on the number of cylinders.

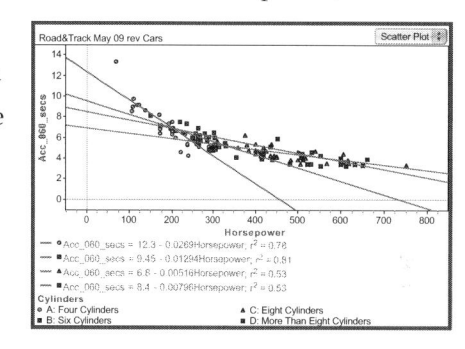

- To get the plot shown, go to **Object>Remove Filter** and then drag the variable *Cylinders* to the *body* of the scatterplot. You will still have the *Residual Plot*, which shows the residuals for the line selected.

Extrapolation

Extrapolation is the practice of using a linear model to predict outside the range of data from which the linear model was derived. *Extrapolation* is to be avoided.

Summary: Evaluating a Linear Model

Much data analysis involves fitting a model to data. But then the question immediately comes up as to whether our model—simple and beautiful as it may be, such a straight line—is any good. Our models attempt to make sense of this variability in our data—to "explain" it, to "account" for it. This section introduces a measure and perspectives for assessing how good linear models are.

- The **Coefficient of Determination,** whose symbol is R^2, is a measure to assess how good a linear model fits the data.
  - For simple regression models, $R^2 = (r)^2$; that is, R^2 is the square of the correlation coefficient r.
  - $0 \leq R^2 \leq 1$, so that the numbers can be interpreted as proportions;
- **Interpretation of R^2**: R^2 shows the proportion (or percentage) of variation in the *response* variable *explained* by the model being used.
- The **calculation of R^2** may also be done by considering:
  - The total variation in the response variable is taken as variation that is initially unexplained and measured by the *Total Sum of Squares* $= \sum_{i=1}^{n}(y_i - \bar{y})^2$ and,
  - The amount of variation still remaining after applying the model, measured by the *Sum of Squared Residuals* $\sum_{i=1}^{n}(y_i - \hat{y})^2$; then
  - $R^2 = 1 - \dfrac{\text{Sum of Squared Residuals}}{\text{Total Sum of Squares}} = 1 - \dfrac{\sum_{i=1}^{n}(y_i - \hat{y})^2}{\sum_{i=1}^{n}(y_i - \bar{y})^2}$
  - The *interpretation of R^2* follows from

    R^2 = Proportion of Explained Variation = 1 − Proportion of Unexplained Variation.

- **Flawed Models** It may be that our model is not good for a number of reasons.
  - There may be **Influential Observations,** observations whose position in the data would substantially change the slope and y-intercept of the linear model fitted if the data were changed or omitted. Hence, the model we are using may have a misleading slope or y-intercept.
  - The data may be **curvilinear** rather than linear. *Curvilinearity* is reflected in a definite pattern to the residuals around the least squares regression line rather than a random scatter about the line we expect to see if the linear model is appropriate. In this instance, the idea for the model itself is flawed.
  - We may be attempting to **extrapolate** prediction beyond the domain of data from which the model was constructed.

§2.4 Exercises on Evaluating Models

1. **Used Cars, Including BMWs** This exercise uses the same data set that was used for Exercise 1 in **Exercises §2.3,** and it is also the same data set that was used for the example in the **Notes §2.3** except that used BMWs have been added.

   - Open the Fathom file **Summer 09 BMWLexusInfiniti C.ftm** and get a scatterplot with *Price* as the response variable and *Age* as the explanatory variable. You did this in Exercise 1 of **Exercises §2.3**.
   - With the **Graph** selected, go in the menus to **Graph>Least Squares Lines.**
   - With the **Graph** selected, go in the menus to **Graph>Make Residual Plot.**
   - Drag the variable *Make1* to the *body* of the scatterplot so that the three different makes of cars are shown.

     You should see the plot like the one on the right. You may wish to go to **Graph>Format Numbers** to fixed decimal and two or three decimal places.

     a. Refer to the *Summary Table* in Exercise 1 of **Exercises §2.3** and the numbers to show that $R^2 = (r)^2$. Use all six decimal places and record your work for the three regression equations.

     b. Interpret the R^2 for the relationship between *Price* and *Age* in the context of the data.

     c. For which of the three cars does the linear model fit the best? Give a reason for your answer.

   - In the plot shown, the residual plot is for the BMW, and your plot may show that as well. However, you can get the residual plot for each of the other cars by selecting the least squares regression equation on your plot.

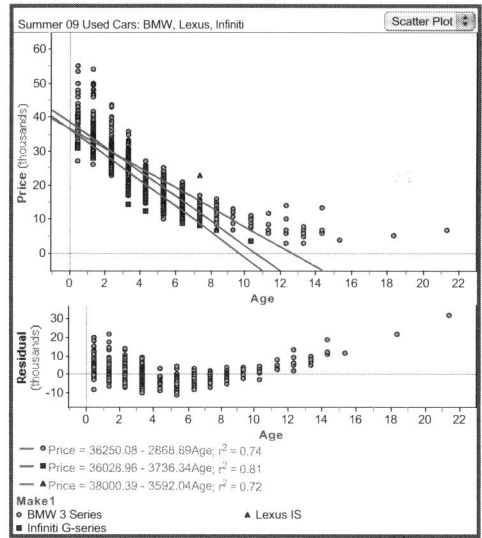

     d. If a linear model does not fit well, it may be that there is much scatter or that the data are not linear; that is the data curve. Is there evidence of curvilinearity for the BMW data? Give reasons for your answer.

     e. Get the residual plots (following the directions in the last bullet) for the Lexus ISs and the Infinitis. Is there evidence of curvilinearity for any of these two cars?

   - Select the **Graph** and go to **Object>Add Filter** and type *Age <11*. What we have done (as we did before) is to compare the relationship for the BMWs younger than eleven years old with the Lexus and Infinitis (where all the cars for sale were less than eleven years old).

     f. Compare the slope and the R^2 for the BMWs with age less than eleven years with the slope and the R^2 for *all* of the BMWs. What does this tell you about the linear model and also about how BMWs lose their value? (Think: at what ages do cars lose most of their value?)

     g. Are there any observations that appear to be *influential observations*? If so, why? If not, why?

2. Heptathlon for 1988 for the Eight Eastern Bloc Competitors: Calculating R^2 from the Total Sum of Squares and the Sum of Squared Residuals

This exercise follows the argument of the first part of **§2.4** but for a very small set of data. The data are just the eight competitors in the 1988 heptathlon who were from what was known as the Eastern Bloc. Here (on the right) are the mean and standard deviations for the times in the *HundredMeterHurdles* race and the *TwoHundredMeters*. The last number is *r*, the correlation coefficient.

Heptathlon 1988	HundredMeterHurdles	TwoHundredMeters
	13.38	23.9113
	0.297369	0.582983
	8	8
	0.730266	0.730266

S1 = mean ( )
S2 = s ( )
S3 = count ( )
S4 = correlation (HundredMeterHurdles, TwoHundredMeters)

$$R^2 = 1 - \frac{\sum_{i=1}^{n}(y_i - \hat{y})^2}{\sum_{i=1}^{n}(y_i - \bar{y})^2}$$ is the formula for the coefficient of determination. The formula is made up of two summations. The sum in the numerator is the **sum of squared residuals** $\sum_{i=1}^{n}(y_i - \hat{y})^2$ and the one in the denominator is the **total sum of squares,** $\sum_{i=1}^{n}(y_i - \bar{y})^2$. We will calculate

$$R^2 = 1 - \frac{\sum_{i=1}^{n}(y_i - \hat{y})^2}{\sum_{i=1}^{n}(y_i - \bar{y})^2}$$ for our small data set, after getting $R^2 = r^2$.

a. The easiest way to calculate R^2 is to simply square the correlation coefficient *r*. Find the R^2 for these data by doing that. At the end of the exercise you will compare the two calculations.

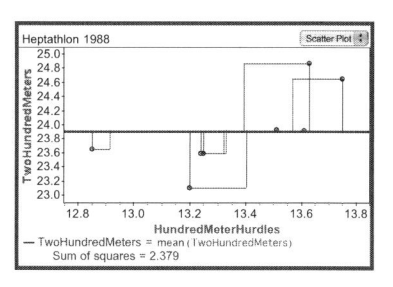

- Open the Fathom file **Heptathlon 1988 EastBloc.ftm** and make a scatterplot with the *TwoHundredMeters* as the response variable (*y*-axis) and with *HundredMeterHurdles* as the explanatory variable (*x*-axis).
- Select the graph, go to **Graph>Plot Function,** and type *mean(TwoHundredMeters)*.
- With the graph selected, go to **Graph>Show Squares.** You should see the plot shown above.

b. Which sum does the *Sum of Squares = 2.379* represent? Does it represent the **Sum of Squared Residuals** $\sum_{i=1}^{n}(y_i - \hat{y})^2$ or the **Total Sum of Squares** $\sum_{i=1}^{n}(y_i - \bar{y})^2$? Give a reason for your answer.

Num	Athlete	Team	Hundred MeterHurdles	TwoHundred Meters	y – y-bar	(y – y-bar)2
1	Sabine John	GDR	12.85	23.65	-0.2613	0.0683
2	Anke Behmer	GDR	13.2	23.10		
3	Nataliya Shubenkova	URS	13.61	23.92	0.0087	0.0001
4	Remigija Sablovskaité-N	URS	13.51	23.93	0.0187	0.0003
5	Ines Schulz	GDR	13.75	24.65	0.7387	0.5457
6	Zuzana Lajbnerová	TCH	13.63	24.86		
7	Svetlana Buraga	URS	13.25	23.59	-0.3213	0.1032
8	Svetla Dimitrova	BUL	13.24	23.59	-0.3213	0.1032
						2.3791

c. The calculation for the *Sum of Squares = 2.379* is shown at the bottom of the table (on the previous page) for our small data set. This sum of squares is the Total Sum of Squares. In the table, fill in the values for $(y_i - \bar{y})$ and $(y_i - \bar{y})^2$ for Anke Behmer and for Zuzanna Lajbnerová. Find the mean that you need in the Summary Table at the top of the previous page.

d. Add the last column to confirm that the Total Sum of squares is 2.379.

- Select the graph and go to **Graph>Least Squares Line.** The graph that you get should look like the one here.

e. The *Sum of Squares = 1.110* must represent the **Sum of Squared Residuals** $\sum_{i=1}^{n}(y_i - \hat{y})^2$. Print out this graph (or use an available handout) and indicate the residual for the fastest *TwoHundredMeters* runner (Anke Behmer, the one with the shortest time for that race). Is the residual for this runner positive or negative? Give a reason for your answer.

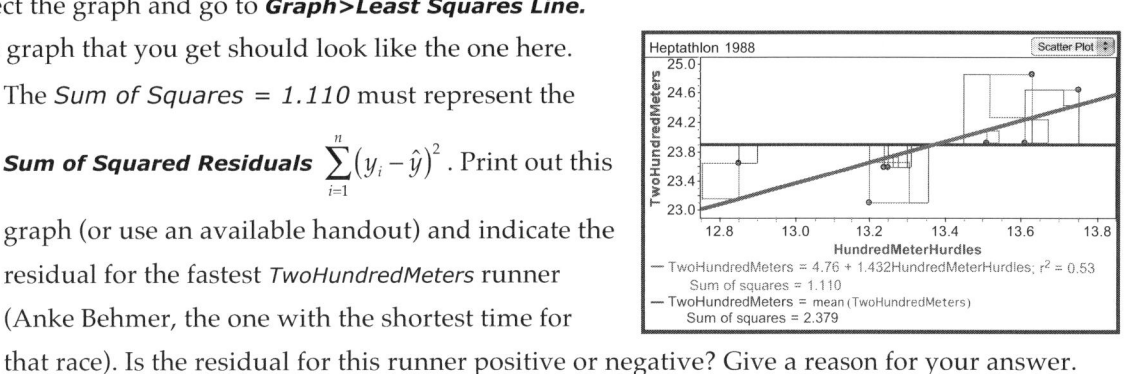

f. The second table shows the calculation for the *Sum of Squares = 1.110*, that is, the sum of squared residuals. Fill in the values for $(y_i - \hat{y})$ and $(y_i - \hat{y})^2$ for Anke Behmer and for Zuzanna Lajbnerová. The plot shows that the equation of the least squares regression line is $\hat{y} = 4.76 + 1.432x$; you will need this.

g. Add the last column to confirm that the value for the **Sum of Squared Residuals** $\sum_{i=1}^{n}(y_i - \hat{y})^2$ is 1.110.

Num	Athlete	HundredMeterHurdles	TwoHundredMeters	y-hat	y – y-hat	(y – y-hat)2
1	Sabine John	12.85	23.65	23.1612	0.4888	0.2389
2	Anke Behmer	13.2	23.10			
3	Nataliya Shubenkova	13.61	23.92	24.2495	-0.3295	0.1086
4	Remigija Sablovskaité	13.51	23.93	24.1063	-0.1763	0.0311
5	Ines Schulz	13.75	24.65	24.4500	0.2000	0.0400
6	Zuzana Lajbnerová	13.63	24.86			
7	Svetlana Buraga	13.25	23.59	23.7340	-0.1440	0.0207
8	Svetla Dimitrova	13.24	23.59	23.7197	-0.1297	0.0168
						1.1110

h. You now have the values of $\sum_{i=1}^{n}(y_i - \bar{y})^2$ and $\sum_{i=1}^{n}(y_i - \hat{y})^2$. Use these to calculate the value of $R^2 = 1 - \dfrac{\sum_{i=1}^{n}(y_i - \hat{y})^2}{\sum_{i=1}^{n}(y_i - \bar{y})^2}$.

You should get $R^2 = 0.53$. Show your work (there is not much). Does your result agree with what you calculated for part a?

i. Give a good interpretation of the R^2 (that you have calculated) in the context of the data.

j. The Coefficient of Determination $R^2 = 1 - \dfrac{\sum_{i=1}^{n}(y_i - \hat{y})^2}{\sum_{i=1}^{n}(y_i - \bar{y})^2}$ is known as the "proportion of explained variation in the response variable." What would be a good name for the fraction $\dfrac{\sum_{i=1}^{n}(y_i - \hat{y})^2}{\sum_{i=1}^{n}(y_i - \bar{y})^2}$?

3. **Playing with roller coasters** To do this exercise, you should have a Word document opened in so that you can paste in the Fathom graphics and answer the questions. Our *statistical* questions are:
   – We will examine the relationship between *Height* and *Speed*. Ultimately, we will ask whether the relationship differs for the wooden and steel roller coasters, and, if it does, how it differs.

- Open the Fathom file **Roller Coasters Summer 09 A.ftm** and make the scatterplot for the variables *Height* and *Speed*. Think about which variable should be the explanatory variable and which variable should be the response variable.
- With the graph selected, get **Graph>Least-Square Line.**
- With the graph selected, get **Graph>Show Residual Plot.**
- Drag the variable *Type* to the *body* of the graph to show the least squares lines for the two types of roller coasters.
   a. Judge, from the appearance of the scatterplot and the residual plot that you have, whether there is any evidence of non-linearity (curvilinearity). Give reasons for your answer. If you think there is at least some non-linearity present, indicate the roller coasters that are affected.
   b. Whatever your answer is to part *a*, explain what the value of $R^2 = r^2 = 0.83$ tells you about the relationship between the variables *Height* and *Speed*.
- Test whether the two highest roller coasters (the two over four hundred feet) are potential *influential observations* by selecting them and going to **Edit>Cut Cases.**
- To reverse cutting the cases, go to **Edit>Undo Cut Cases.** (You can toggle back and between *including* and *excluding* the two extreme cases in this fashion. If you seriously mess up the data, you can select the collection icon, and go to **File>Revert Collection** and the collection will revert to its most recent saved state.)
   c. Compare the slopes and the R^2 for the least squares regression line with and without the two outliers. Is there some change? Do you think that it is a big change or not a very big change?
- Toggle back so that you see the two outliers; select them (they should turn red) and drag them together so that they are both either very slow (something like 10–20 mph) or very fast. You can undo all this dragging at the end by again going to **Edit>Undo Drag Points.**
   d. Describe what happens to the slope of the least squares line and the R^2 when you drag.
   e. You can now have a lively discussion as to whether we are seeing real *influential observations* or not. Some of the issues may be whether you would ever actually see very high roller coasters that would have a very slow speed.
- Do **Edit>Undo Drag Points** or select the collection icon and go to **File>Revert Collection.**
- Drag the variable *Type* to the *body* of the scatterplot.
   f. Compare the slopes and R^2 of the wooden and the steel roller coasters for the relationship between *Height* and *Speed*. Are there differences between the wooden and steel coasters, or are the results basically similar?
   g. Do the *y*-intercepts have a meaning? If so, what meaning? If not, why not?

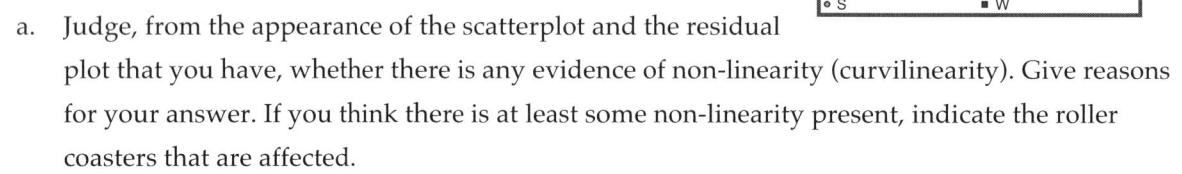

4. ***Australian High School Students' Body Measurements*** [Fathom] This exercise is about the relationship between three body measurements for Australian high school students.

- Open the Fathom file **CAS Australia B.ftm** and read just below about "clean data."

    Clean data: Read this, please. The measurements (hence, variables) are *HeightC*, *ArmSpanC*, and *RightFootC* (the length of the student's right foot), all measured in centimeters. [The "C" in these variables means "cleaned"; some of the Australian high school students reported being one thousand centimeters tall or said that they had feet one hundred centimeters long. All "measurements" clearly impossible for humans (even in Australia) were made into "missing data."] Our statistical questions are:
    - Can the relationship between height and length of arm span be modeled by a linear model?
    - If so, is the relationship different for male and female students?
    - Can the relationship between height and length of right foot be modeled well by a linear model?
    - If so, is the relationship different for male and female students?

- Open a **Word** or other word-processing program to paste in your Fathom graphics.
- **Arm Span Length and Height** Drag down a **graph** from the shelf and make a scatterplot using *ArmSpanC* as the response variable and *HeightC* as the explanatory variable.
- Select the **graph**; go to the menu **Graph>Least Squares line** to get the best-fitting line.
- With the graph still selected, go to **Graph>Make Residual Plot** to get a residual plot.
    a. Identify and interpret the slope for the relationship between arm span and height.
    b. Identify and interpret the R^2 ($= r^2$) for the relationship between arm span and height.
    c. The R^2 ($= r^2$) is not 1, so the linear model does not fit perfectly. Judging by the plot itself and by the residual plot, do you think that the fit is not perfect (a) because we should be using some kind of curvilinear model, or (b) because there is scatter (variability) around what seems to be a linear relationship? Give at least one reason for your answer from the graphic.
- Drag the variable *Sex* to the body of the **graph** so as to get least squares lines for the male and the female students separately. (Note: part d is extra credit and is shown at the foot of the exercise.)
    e. What differences do you notice in the slopes of the least squares lines for males and females?
    f. What differences do you notice in the R^2 ($= r^2$) for the least squares lines for males and females? Does the linear model appear to be a good fit? Give a reason for your answer.
- **Right Foot Length and Height.** Select the **graph** and go to the menu **Object>Duplicate Graph** then drag the variable *RightFootC* to the *y*-axis. This should give you the least squares lines for males and females for the relationship between height and the length of right foot.
    g. Identify and interpret the slopes of the lines for males and for females.
    h. Identify and interpret the R^2 ($= r^2$) of the lines for males and for females.
    i. Is height a good predictor of foot length for either males or females? Give a reason.
    [d. **Extra Credit:** Download an image of Leonardo da Vinci's *Vitruvian Man*. (A good site to visit may be http://leonardodavinci.stanford.edu/submissions/clabaugh/history/leonardo.html.) Explain in a concise paragraph or two how your linear model predicting arm span from height agrees or disagrees with da Vinci's picture.]

§3.1 Can We Trust Data (to Reflect Reality)?

Example: the Presidential Approval Rating

If you go to http://www.gallup.com/poll/113980/Gallup-Daily-Obama-Job-Approval.aspx, you will find something very much like:

> Gallup tracks daily the percentage of Americans who approve or disapprove of the job Barack Obama is doing as president. Daily results are based on telephone interviews with approximately 1,500 national adults; margin of error is ±3 percentage points.

There are approximately 217,000,000 adults in the USA who are eligible to vote. How can we trust findings that are based upon fifteen hundred out of 217,000,000 of the people involved? In the following sections, we will see:

- We *can* trust results based upon a miniscule sample of the voting population, if the data are collected and analyzed carefully. The next sections cover what "collected and analyzed carefully" means.

- We *cannot* trust results if data are collected badly, even if the number of cases being studied is large.

- The meaning of the term "margin of error" mentioned above

Some terminology We start with some essential terminology, making a distinction between a population and sample. By a population we mean whatever it is that we are ultimately interested in describing and understanding. A sample is a representative part of that population. In the example of the Presidential Approval Rating, the population of interest is the adults in the USA who are eligible to vote.

Population and Sample

Population: the collection that a statistical analysis of data seeks to describe or understand

Sample: a collection that is a representative part or subset of a population

Census if a "sample" includes all of the population, it is called a census

---

By a population we mean a *description* of something and not a number. The number would be the size of the population. The Gallup organization wants to know something about the adult voters of the USA, and not of Canada or Mexico. They could be interested in knowing something (incomes, for example) about those who work in the construction industry in California. Some other organization could be interested in knowing something about fir tree production in Oregon. Voters in the USA, construction workers in California, fir trees in Oregon are all different populations, with different descriptions.

Typically the thing we want to know about a population is either impossible (or very hard) to find out about for the entire population, but *is* possible for a sample of the population. It is possible for Gallup to contact fifteen hundred people and ask them about the president's job performance; it would not be feasible to ask *all* the voters.

Generalizing from a sample to a population: an example. We make the distinction between population and sample because we want to be clear about what we are searching for. If we do not clearly know what we want, we probably have a good chance of not finding it! However, sometimes we have data, and it is not easy to define a population to which the results can be reasonably generalized. An example is the student data from statistics classes that we have been analyzing. It comes from students taking statistics at one community college in the San Francisco Bay Area. Here are some possible descriptions of populations to which the results of the data might possibly be generalized but with an increasing amount of risk of being wrong.

- Students taking statistics at one community college
- Students at one community college
- Students at community colleges in the San Francisco Bay Area
- Students at all colleges and universities in the San Francisco Bay Area
- Students at all community colleges in Northern California
- Students at all colleges and universities in California
- Students at colleges and universities in North America

Moreover, the risk of being wrong as we generalize to a "wider" population may depend on what we are measuring. If we are looking at the proportion of students who speak more than one language, we may guess that this proportion may differ for students outside the San Francisco Bay Area (but perhaps not), and so it would be risky to generalize to all colleges and universities in Northern California. What about the heights of students? Can we generalize data on the heights of students from one college to students in all of North America?

Best practice. The best practice will be to start with the idea of a population and devise a good method to get a sample. That is what we consider next. However, as "consumers" of statistical analyses, we have to do the harder thing of thinking how a sample that has already been analyzed can be generalized or not.

How to get bad samples (aka non-random samples)

It may seem strange to start negatively, but bad samples are common, and it is well to deal with the bad before the good. There are a huge number of ways for data to come from bad samples, and as "consumers" of statistics, it is our job to look critically at how data are got. We start with some common examples of bad data.

Voluntary Response (or Self-Selected) Samples and Non-response. Whenever a web site or a publication asks for votes or "likes" or asks for opinions, that is a good example of a voluntary response or self-selected sample. A self-selected sample is one in which the data come *solely* from people voluntarily responding. Voluntary response is therefore not the same as *non-response* from people who are chosen to participate. A sample of people may be chosen to answer a questionnaire, and always there are some who do not respond; this is called *non-response*. A large proportion of *non-response* is problematic, but that large proportion does not make the data collection itself a voluntary response sample. We will reserve the words "voluntary response" for data collection schemes that depend wholly or solely on the voluntary response in a population. For voluntary response data collection (or for large *non-response*), the questions to ask are: (1) "Who is likely (and unlikely) to respond to the invitation to answer questions?" and (2) "Does the answer to question (1) represent any population in which we have an interest?"

Convenience Samples. Here is an (extreme?) example of a convenience sample: A college student is asked to collect data on some topic where the population is meant to be first-year college students at that student's college. So, the college student pesters all his or her friends until the friends give the required information about the topic. There may be an element of non-response in this procedure (some of the friends may refuse), but if the student is persistent enough, the non-response may not be a problem, although the student may lose some friends. We must still ask the same two questions: (1) "Who (or what) is likely to be included in the sample?" and (2) "Does the answer to question (1) fairly represent any population in which we have an interest?" Unless the student's friends really are a representative sample of all first-year college students at that college (unlikely!), this sample is not a good sample; it is a bad sample because it is likely to be unrepresentative.

Convenience samples may not always be as useless as our example suggests. The data collected from statistics students at one college in Northern California is a convenience sample. The answer to question (1) is clear: the data come from students in a statistics course at one community college. The little exercise in the sub-section above *(Generalizing from a sample to a population: an example)* comparing this sample to various populations in which we may have an interest is an attempt to answer question (2). One treads warily here.

Judgment (or quota) samples. One response to the extreme example of a convenience sample may be to say: "I would not be so naïve to think that my friends represent all first-year students. I would make certain that my sample had equal numbers of males and females, for example, and I would try to make the ethnic composition of my sample to reflect what I know of the ethnic composition of the first-year students at the college." This is a good move, and the sample produced will be a better sample than the one chosen naïvely in the example above. However, it is still not a good sample. Although the one choosing the sample may specify that half the sample be females (or 60% or whatever percentage at that college is), the choice of the individuals is still up to the one choosing the sample. Whenever there is human choice of the elements (the individuals or the cases), the sample has a chance of being unrepresentative. Odd as it may seem, the best way to choose a sample is to eliminate human choice in the final selection of the elements. Samples chosen by chance and not human choice are called *random samples*. The three types of "bad samples" listed above are not the only types of bad sampling. Here is a summary; in all of these types of sampling, human choice is ultimately involved.

Types of Non-Random Sampling

- **Voluntary Response Sampling**: Samples in which the elements in the sample (the individuals or cases) depend solely upon the choice of the individuals themselves
- **Convenience Sampling** Samples in which the choice of the elements in the sample (the individuals or cases) is determined primarily by which elements happen to be easily accessible to the researcher
- **Judgment or Quota Sampling** Samples in which the representativeness of the sample is furthered by having the elements fulfill quotas judged by the researcher to characterize the population

How to get good samples (aka random samples)

There are two good reasons for employing random sampling. One: in the long run, it is the only type of sampling that will guarantee truly representative samples. The second is that random sampling is an essential requirement for the techniques statisticians use to generalize. The "margin of error" cannot be calculated if the samples involved are not random, for example. But what are random samples?

Random is not random; so what is random? In that sentence, obviously the word "random" is used in several senses. The very first "random" refers to what people think of as random: "haphazard," "without pattern," "lacking regularity," "arbitrary," or perhaps even "unbidden" or "unwanted." If you are asked to choose from a group of people or things apparently randomly, your choices will not be random in the mathematical sense, so that kind of "random" is not "random" in a mathematical sense. Choices made by humans will always follow some pattern. We may not be able to know or predict what that pattern will be—it will be different for people from different cultures—but there will always be some pattern.

So what is random? It is helpful to reduce everything to numbers. Here is an initial working definition: if we have a finite set of numbers, a **random** choice will give each number an equal probability of being chosen. Here are some examples: on a ten-sided die there are ten choices, and the action of rolling the die gives each choice an equal chance of appearing. The equal probability is 1/10 or 0.10. In a lottery bin containing sixty-four numbers where a number is chosen "blindly" (without looking at the numbers) after thoroughly mixing the numbers in the bin, each of the sixty-four numbers has an equal chance of being chosen, with probability 1/64 or 0.015625.

Working Definition of Randomness
Given a finite set of numbers, a **random** choice gives each number the same probability of being chosen.

How this works in practice is that each case in a population is assigned a number. A device choosing numbers randomly (called a **random-number generator**) that essentially works like the ten-sided die (or picking from a lottery bin) gives the result that each number chosen has the same probability as any number. Using a random process will guarantee a random sample from a population.

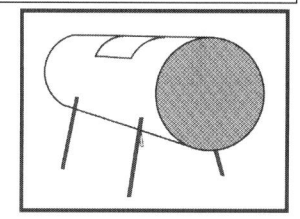

Simple random samples. The simplest kind of random sample involves two steps. First, there must be a list of the entire population, with a unique number assigned to each case in the population. Here is an example that comes from one of the exercises. We start with the collection of all of the houses that were sold in San Mateo County in 2005–2006. This is the population (a part of which is shown here), and in that population each of the houses is given a number: the assigned numbers are listed in the column labeled *Caseno*. Other variables are measured and recorded as well. A simple random sample from this population is chosen by getting a random-number generator (it could actually be a ten-sided die) choose however many numbers we want. Simple random

San Mateo Real Estate Y0506

Caseno	Address	City	Area	Beds	Sq_Ft	Age	List_Price
698	698 944 WALNUT ST	SCL	353	3	1570	80	795000
699	699 1026 ELM ST	SCL	353	3	1710	78	899000
700	700 1951 ELIZABETH ST	SCL	353	4	2060	47	999950
701	701 2806 SAN CARLOS AV	SCL	355	3	1210	99	829000
702	702 28 ARROYO VIEW CI	BEL	360	3	1420	8	889500
703	703 2603 CARLMONT DR	BEL	360	3	2060	9	1099800
704	704 2604 CARMELITA AV	BEL	361	2	740	59	669000
705	705 2700 CORONET	BEL	361	3	1160	53	749000
706	706 2841 SAN JUAN BL	BEL	361	3	1350	45	839000
707	707 2511 BUENA VISTA AV	BEL	361	4	1890	49	895000

sampling is the basis for all other sampling and involves two essential features, which are listed in the box below.

> ***Simple Random Sample (or SRS)***
> - To draw a **SRS** of sample size *n* from a population, you must have:
>   1. A list of all the cases in the population, and
>   2. A chance process that will choose randomly from the list.
> - In drawing a **SRS,** every possible sample of size *n* has an equal chance of being chosen.

Random Cluster Samples: an example Often it is not possible for researchers to make a list of the entire population. Or, even if it is possible to make a list, choosing a simple random sample may create too much work collecting the data. For example, suppose the population in which we have an interest is all college students in California. A simple random sample of students may include some students from Yreka (near the Oregon border) and other from Calexico (on the border with Mexico) and some everywhere in between. The travel costs to collect data on these students might be enormous. It would be better and cheaper if the sample data were clustered. The way to accomplish this is to randomly sample *clusters* of cases, and then use *all* the cases in the clusters. The example in the exercises with the real estate data is to draw a random sample of area numbers and then take *all* of the houses within the randomly areas. A common question that is asked about random cluster samples is: "What are the clusters?" The answer is that clusters can be *any* grouping of the cases where each case belongs to just one of the groups or clusters. In the real estate example in the exercises, the clusters are areas. In the student example above, it would make sense to group students by the college or university they attend, with provisions made for students who attend more than one school at the same time.

Stratified Random Samples: an example   Often researchers want to make certain that certain groups are represented in the sample in the same proportions as they think they exist in the population. One way to guarantee that this will be so is to randomly select from within groups. A researcher who is sampling students within one university may wish to randomly sample within the categories of the major administrative divisions of the university. Here are the "schools" that constitute Stanford University (see http://www.stanford.edu/academics/). The schools would become the groups, (or, as they are called, the "strata" in "stratified"), and from within each of these schools (business, earth sciences, etc.) a simple random sample of students would be chosen. It is the random selection makes it not merely a judgment sample. As with "clusters" in random cluster sampling, these "strata" are any groupings that make sense.

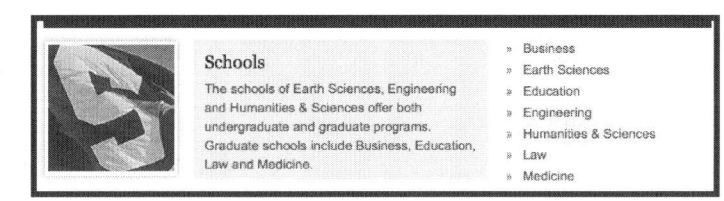

> **Random Cluster Sample**
> - To draw a ***random cluster sample*** from a population, you must have:
>   1. A list of the clusters to be sampled
>   2. A chance process that will choose randomly from the list of clusters
> - Then *all* of the cases within the chosen clusters constitute the random sample.

> **Stratified Random Sample**
> - To draw a **stratified random sample** from a population, you must have:
>   1. A list of all the possible cases in the population, grouped in the strata
>   2. A chance process that will choose randomly from the cases *within* each of the strata
> - Then the random sample is all of the cases randomly chosen from each of the strata (or groups).

Reality. Listed above are only three kinds of random samples. In reality, random samples are chosen in much more complicated ways. The sampling design may be a mixture of cluster and stratified sampling or a type of sampling that depends on the random arrangement of cases in the populations. Here is one example. (See: http://www.gallup.com/poll/101872/How-does-Gallup-Polling-work.aspx.)

> The majority of Gallup surveys in the U.S. are based on interviews conducted by landline and cellular telephones. Generally, Gallup refers to the target audience as "national adults," representing all adults, aged 18 and older, living in United States.
>
> The findings from Gallup's U.S. surveys are based on the organization's standard national telephone samples, consisting of directory-assisted random-digit-dial (RDD) telephone samples using a proportionate, stratified sampling design. A computer randomly generates the phone numbers Gallup calls from all working phone exchanges (the first three numbers of your local phone number) and not-listed phone numbers; thus, Gallup is as likely to call unlisted phone numbers as listed phone numbers.

In this example, what we have called the population Gallup calls the "target audience." The term "national adults" denotes the collection (of people, in this instance) that Gallup wants to describe or study. They do not have a list of the "national adults," but they do have a list of working telephone exchanges. They choose telephone numbers randomly by having a computer generate random telephone numbers within those exchanges. The important word here is "random": Gallup chose samples using a random process.

Experiments give data also, and randomization is still important

The examples in the first part of this section have all been from *observational studies*, but data also come from *experiments*. Here is an example of an experiment on the effect of negative political advertising.

Example: what effect does negative political advertising have? Do negative political ads turn voters against the political process, so that they are less likely to vote, or do negative political ads energize voters, so that they are actually more likely to vote? One political scientist from the Ohio State University conducted an experiment in a Florida mayoral election in 2003 to get some evidence[2]. Here is a brief description of the experiment (which can be found at http://www.jstor.org/stable/4148088, page 203):

> In brief, a random sample of voters was chosen for either the treatment (negative ads) or the control group (no ads) in a mayoral election. Subjects in the treatment received negative campaign ads (from an independent expenditure group) in the mail in the days immediately preceding the election. The resulting decision to vote was then measured by consulting official election records.

---

[2] Niven, David, "A Field Experiment on the Effects of Negative Campaign Mail on Voter Turnout in a Municipal Election," *Political Research Quarterly*, Vol. 59, No. 2 (Jun., 2006), pp. 203–210

We can make a diagram of this experiment that also shows some of the terminology associated with experiments.

Terminology of experiments: With this experiment, we can illustrate the vocabulary that goes with such experiments.

- **Factor**: In experiments, the explanatory variable or variables are called *factors*. Here the factor is whether or not the voter received negative campaign ads. The experimenter determined by random choice which voters in the sample of voters received negative political mailings and which voters did not. (Think of flipping a coin for each voter—heads: negative ads; tails, no ads.)
- **Treatment**: Here there are just two treatments, and they are the values of the explanatory variable or factor. The two treatment groups are: (1) received negative political advertising, and (2) did not receive negative political advertising. The word "treatment" reminds one of medical experiments, and often the default "treatment," representing either a standard procedure or no action at all, is labeled the "control group" or just "control." We will regard the control group as one of the treatments.
- **Response variable**: The response variable for this experiment is whether or not the voter actually voted: "the resulting decision to vote was then measured by consulting official election records." If negative ads depress voter turnout then the turnout should be lower in the treatment group that had negative ads; if negative ads encourage voter turnout then the voter turnout should be higher.
- **Experimental Units**: The cases for an experiment are typically either called "experimental units" or "subjects." The word "subjects" is typically used if the experiment involves humans. However, an experiment may have experimental units that are not humans, just as cases need not be humans.

Example: what effect does fertilizer have on begonias? Here is a diagram of an experiment. Some horticulturists were concerned that some gardeners were using too much fertilizer on their begonias. They bought thirty-six begonia plants and randomly allocated the plants to three treatment groups (no fertilizer, 100 ppm fertilizer, and 300 ppm fertilizer; ppm means "parts per million"), and then gave all the begonia plants the same amount of water, the same amount of sunlight, and the same amount of weeding for some weeks. At each week, they measured the growth of the begonia plants. So for this experiment, as shown in the diagram, the *experimental units* are begonia plants, the *factor* is the amount of fertilizer given to the plants, and the *response variable* is the amount of growth in plants. In case you are wondering, the researchers found that giving begonias big amounts of fertilizer stunted growth.

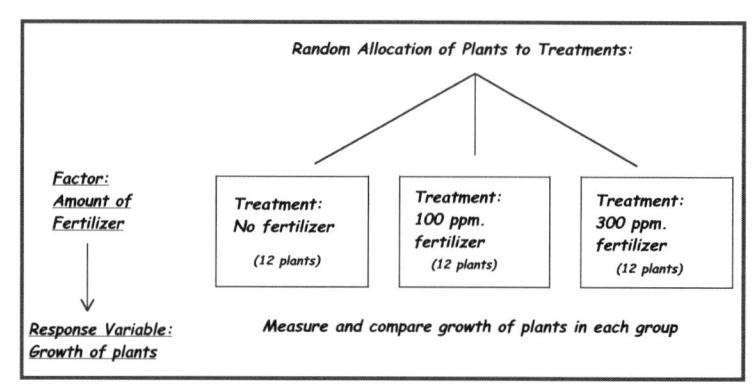

> ***Essential terminology about experiments:***
> - ***Factor:*** The explanatory variable in an experiment, whose value for each case is determined; there may be more than one factor.
> - ***Treatment:*** A value or combination of values of a factor or factors
> - ***Experimental units:*** The cases that are used in the experiment; they are sometimes also called "subjects."
> - ***Control group:*** A treatment against which the other treatments are compared; in medical experiments, the standard or no treatment

The role of randomization: The horticulturists who ran the begonia experiment were careful to treat all the begonia plants in the same way—the same amount of watering, etc. The important principle is that all of the cases should be as much alike as possible *except* for the value of the factor. The reason for this is logical: if the researchers *do* find a difference in the growth of the begonia plants after some weeks, they can legitimately conclude that the differences were caused by the amount of fertilizer given to the plants and not some other variable, such as the amount of water given, or the amount of sunlight, or the soil conditions. The researchers can confidently rule out some other cause.

On the other hand, if they find no difference in growth then they can legitimately conclude that the amount of fertilizer has no effect on growth, since all other things were kept equal. But can you be certain that the begonia plants were themselves "equal"? Perhaps some of them are initially healthier than others. What would happen if all the really healthy begonias got assigned to the "no fertilizer" treatment group? Then any greater growth of the no fertilizer treatment group that is found may have been because the plants in that group were healthier in the first place. That is where randomization comes in: using randomization to assign the begonias to the three treatment groups effectively makes it very unlikely that one particular treatment group gets all the healthy plants and another one unhealthy plants, since the probabilities of being assigned to each of the three groups should be equal. Where randomization is used to assign the experimental units to the treatment groups, the entire experiment is often called a ***randomized trial.*** This terminology is especially common in medical research.

The role of randomization for the negative political advertising experiment: The researcher first of all randomly selected 1400 voters from the list of voters. Then, he randomly assigned these sampled voters to the treatment groups, with 700 being randomly assigned to the group that did not receive any negative mailings. Randomly assigning the subjects (experimental units) made the two groups approximately equal in composition.

Blind, double blind, and the quest for equality. In the negative political ads experiment, neither the people who received the negative political ads nor those who did not realized that they were part of an experiment or that they were in one of the researcher's treatment groups. This is an example of a ***blind*** experiment, where the subjects are unaware ("blind") of the treatment group they are a part of. With people as experimental units (subjects) being aware of being in a "treatment" or "control" group may itself make a difference, which means that "not everything else is equal." In medical experiments that are blind, patients who are part of an experiment may be kept unaware of which treatment group they have been assigned. Sometimes this is not possible; one of the exercises concerns an experiment on comparing different diets, and, obviously, the subjects knew which diet they were following, so the experiment was *unblinded*. If in a medical experiment the medical personnel are also unaware of which treatment which patient has, that study is *double blinded*.

> ***Very elementary experimental design:***
> - In an experiment, the researcher is able to set the values of one explanatory variable (the *factor*) and keep all other variables as equal (or constant) as possible, so that the *treatment groups* differ only by the values of the *factor*.
> - In order to keep the treatment groups as equal as possible, studies sometimes keep human subjects from knowing which treatment group they are a part of; this is known as a *blind* or *blinded* experiment.
> - Where variables cannot be controlled to make the treatment groups equal then *randomization* is employed to assign experimental units to treatment groups, and the experiment is sometimes called a *randomized trial.*

Where the experimental units are not people (such as in the begonia experiment), the issue of the subjects' awareness of their treatment group as a threat to the equality of the treatments does not arise.

How experiments differ from observational studies and why.

In an experiment, the researchers deliberately set values to one variable vary and keep everything else the same. In the begonia experiment, the amount of fertilizer was set by the horticulturists at 0 ppm, 100 ppm, or 300 ppm. In the negative political mailings experiment, the researcher was able to determine that some people received negative mailings, and others did not. Since those who received and those who did not were randomly determined, the randomization works to make the two groups as equal as possible. In both examples, the researcher was able to set values of the explanatory variable (the factor) for the experimental units. It is this ability to set the values of the factor that makes an *experiment* distinct and different from an *observational study*. In a purely observational study, the researcher has no ability to set the values of any variable; the values of the variables are merely observed. The Gallup polls on the presidential approval rating are observational studies because the researchers do not have the ability to set any of the variables for the cases they select. One of the exercises draws samples from the population of houses sold in San Mateo County in 2005–2006. That is also an observational study since the data were simply observed, without any intervention by the researcher to set the values of any of the variables.

The advantage of an experiment over an observational study is that differences in a response variable can be directly linked to differences in the factor whose values have been set, since all other variables have been held constant or equal. If the begonia study had been an observational study where the growth of begonia plants was recorded (observed) as well as the amount of water, sunlight, fertilizer, weeding, etc. (all explanatory variables), it would be unclear whether any variation in growth (the response variable) was because of the fertilizer, the watering, the differences in sunlight, or some other variable.

Experiments and Observational Studies:
- In an **observational study**, the researchers simply collect the data as it is, without the ability to intervene to set the values of one of the explanatory variables.
- In an **experiment**, the researcher is able to intervene to set the values of one or more of the explanatory variables, often employing randomization to allocate or assign cases to treatment groups.
- **Experiments** offer a more direct basis for inferring effects (variation in response variables) from causes (variation in explanatory variables) than do **observational studies**, since variation in response variables can logically be attributed to many combinations of variables.

Trusting data: what we mean by it and the first steps toward it

By trusting data, we mean: "Do the data say anything beyond the data themselves?" If Gallup find that 42% of the fifteen hundred people in their sample approve of the president's performance, must they say that all we know is that 42% from that sample approve, or can Gallup generalize beyond those fifteen hundred to all voters? No and yes.

The answer is no if we (or Gallup or anyone else) are sloppy about how the fifteen hundred people are chosen. If they use a non-random sample, whether it is a voluntary response sample or a convenience sample or a judgment sample, then the results cannot be trusted. Generalizations from sloppy sampling cannot be trusted, even if the sample size is large. If we (or Gallup or anyone else) have not thought through about what the intended population is—the target audience, as Gallup terms it—then generalizing is risky. The first steps to being able to generalize are to be clear about the population to which we are generalizing and to use more sophisticated sampling. More sophisticated sampling means randomization.

We also saw in this section that there is numerical data that comes from experiments as well as from observational studies, and with these kinds of data, randomization also has a role to play.

However, getting data using randomization are just the first steps in what statisticians do to have confidence in their data to be able to generalize beyond as sample. The technical term for generalization that statisticians use is inference. In the next sections we tackle inference head-on.

Inference

Inference is the term that statisticians use to refer to generalizing from a sample to a population they seek to understand or describe.

Summary: Trusting Data

Population and Sample

Population: the collection that a statistical analysis of data seeks to describe or understand

Sample: a collection that is a representative part or subset of a population

Census: if a "sample" includes all of the population, it is called a census.

Types of Non-Random Sampling

Voluntary Response Sampling: Samples in which the elements in the sample (the individuals or cases) depend upon the choice of the individuals themselves, wholly or in part

Convenience Sampling Samples in which the choice of the elements in the sample (the individuals or cases) is determined primarily by which elements happen to be easily accessible to the researcher

Judgment or Quota Sampling Samples in which the representativeness of the sample is furthered by having the elements fulfill quotas judged by the researcher to characterize the population

Working Definition of Randomness:

Given a finite set of numbers, a *random* choice gives each number the same probability of being chosen.

Simple Random Sample (or SRS):

To draw a **SRS** of sample size n from a population, you must have:

1. A list of all the cases in the population, and
2. A chance process that will choose randomly from the list

In drawing a **SRS,** every possible sample of size n has an equal chance of being chosen.

Random Cluster Sample:

To draw a **random cluster sample** from a population, you must have:

1. A list of the clusters to be sampled
2. A chance process that will choose randomly from the list of clusters

Then all of the cases within the chosen clusters constitute the random sample.

Stratified Random Sample:

To draw a **stratified random sample** from a population, you must have:

1. A list of all the possible cases in the population, grouped in the strata
2. A chance process that will choose randomly from the cases *within* each of the strata

Then the random sample is all of the cases randomly chosen from each of the strata (or groups).

Essential terminology about experiments:

Factor: The explanatory variable in an experiment, whose value for each case is determined; there may be more than one factor.

Treatment: A value or combination of values of a factor or factors

Experimental units: The cases that are used in the experiment; they are sometimes also called "subjects."

Control group: A treatment against which the other treatments are compared; in medical experiments, the standard or no treatment.

Very elementary experimental design:

In an experiment, the researcher is able to set the values of one explanatory variable (the **factor**) and keep all other variables as equal as possible, so that the **treatment groups** differ only by the values of the **factor**.

When humans are used as the subjects in a study, the subjects are sometimes kept unaware of whether they are part of a treatment or control group. If this is done, the study is known as a **blind** or **blinded** experiment. If, in addition, the administrators of the experiment can be kept unaware of which subjects are in which treatment group, that study is called double blind.

Where variables cannot be controlled to make the treatment groups equal then **randomization** is employed to assign experimental units to treatment groups; the experiment is sometimes called a **randomized trial.**

Experiments and Observational Studies:

In an **observational study**, the researchers simply collect the data as it is, without the ability to intervene to set the values of one of the explanatory variables.

In an **experiment**, the researcher is able to intervene to set the values of one or more of the explanatory variables, often employing randomization to allocate or assign cases to treatment groups.

Experiments offer a more direct basis for inferring effects (variation in response variables) from causes (variation in explanatory variables) than do **observational studies**, since variation in response variables can logically be attributed to many combinations of variables.

Inference is the term that statisticians use to refer to generalizing from a sample to a population they seek to understand or describe.

§3.1 Exercises on Trusting Data

1. Opening of Johnston Road Fraser Suites McDonald's in Hong Kong

The file **Johnston Road/Fraser Suites McDonalds.pdf** shows a small crowd at the opening of the McDonald's on Johnston Road at the Fraser Suites Hotel in Hong Kong (the WikipediaCommons reference is: File:HK Wan Chai 莊士敦道 Johnston Road 輝盛閣 Fraser Suites McDonalds grand open 2010-May 泰昌餅家 Tai Cheong.jpg). Here it is reproduced:

We want to see what happens when we, using personal choice, choose individuals from this crowd. We actually do not know what will happen; it may be that people are able to choose representative samples in some sense. The only way to find out is to collect some data!

Your task is to choose $n = 5$ people from the "crowd" of 52 people. The five people will be interviewed about the "grand opening." Follow these directions.

- Open **Johnston Road/Fraser Suites McDonalds.pdf** so that you can see the people fairly clearly.
- Use any method you wish to choose five people and somehow mark your choices—perhaps on this paper. Use a method that you think will give you a representative, sample of five from the crowd.
- Open the file **NumberedJohnstonRoadCrowd.pdf** and record the numbers of the five people that you chose. (The numbers are not shown here on this small black and white picture, but they are on the file that you are to open.)

    a. How did you choose the five people? Did you attempt some kind of randomization? If so, how? Did you attempt some kind of judgment sample? If so, explain what categories you used. Write down what you did.

- Open Fathom, and get a new collection and a new case table.
- Go to **Collection>NewCases** and type in 5 to add five cases.
- Fill in the new variable with a suitable name (perhaps *Choice*), select the variable (it will turn blue), and go to **Edit>Edit Variable**.
- Either type in what you see here or go to **Functions>Random Numbers>randomInteger** and type in 1, 52 in the parenthesis. This will give you five random serial numbers between the numbers 1 and 52.

    b. Are you able to perceive any differences between your personal choice and the Fathom random sample? For example, did Fathom give you any people standing side by side? Hard-to-see people?

- Record your own personal choice on the Fathom file for this purpose on the website or in the classroom.

c. Record how the attempts of your class to choose a representative sample deviated from random samples chosen by Fathom.

d. If the choices are truly random, what should the probability of choosing person 14 be?

2. **Movie music on a classical music station.** In the San Francisco Bay Area, KDFC is the classical music station. Classical music stations have a difficult time because there is only a fraction of the population who enjoy classical music, and, at the time, KDFC was a commercial station. As a way to attract more listeners, the station wanted to play more movie music. However, they also did not want to alienate their "traditional" classical music listeners. The program manager of the station, in his blog, invited listeners to comment on whether the station should play more "movie music"—music written for films. Here is how the blog invited comments:

> **Movie Music** in small doses on KDFC was added almost 2 years ago. We've had a wide variety of feedback. In this survey we heard comments like "I would think that movie theme music is not quite for your targeted audiences. I personally detest it" to "I get a real kick out of hearing some of movie soundtrack music you throw in occasionally." Do movie soundtracks have a place on a classical station? Is it the new classical for a new audience, or should it be left out? If it's enjoyable, does it matter where it comes from? *What do you think?*

a. Which of the types of sampling discussed in the **Notes** will result from this invitation to comment on movie music? Give a reason for your answer.

b. You had to be a member of "Club KDFC" to actually participate in the survey; however, anyone with an e-mail address can sign up to be a Club KDFC member. (When you are a member of Club KDFC, you get e-mails and can participate in contests, etc.) Which of the following is the best description of the *population* from which the sample of responses is taken?

- (i) People who go to movies in the Bay Area and listen to KDFC
- (ii) All KDFC listeners
- (iii) Club KDFC members
- (iv) People in the Bay Area who like classical music
- (v) People who like movie music, whether or not they go to movies

c. Rank the following groups of people from the *least* likely to those *most* likely to respond.

- (i) Those who hate movie music and think KDFC should not play it at all
- (ii) Those who love movie music and want more of it played
- (iii) Those who think that the present amount of movie music is about right
- (iv) Those who have no strong opinions about movie music on the station

d. Rank the following groups of people from the *least* likely to those *most* likely to respond.

- (i) Those aged eighty-five years or above
- (ii) Those aged 60 to 84 years old
- (iii) Those aged 45 to 60 years old
- (iv) Those aged 30 to 45 years old
- (v) Those aged less than 30 years

e. The sample generated is not a random sample. Explain why it is not a random sample based upon what is said about random samples in the **Notes.**

f. Using the information in part b, devise a method for KDFC to get a random sample of Club KDFC members. (*Note:* KDFC are probably aware of the biased nature of their sampling design and probably do not need a random sample for their purposes, but they can't apply inferential statistics!)

3. **Mobile telephone use among Melbourne drivers**

   This exercise will use a study conducted by David McD Taylor and others about the incidence of mobile telephone use among drivers of vehicles in Melbourne, Australia, where there has been a law prohibiting the use of handheld mobile phones since 1999. (The study is in the pdf file: **Handheld mobile telephone use among Melbourne drivers** or on line at www.mja.com.au/journal/2007/187/8/handheld-mobile-telephone-use-among-melbourne-drivers.) The exercise will concentrate on the methods used to collect the data.

   We performed an observational study of motor vehicle drivers in metropolitan Melbourne, Australia, during October 2006… In brief, we used 12 sites to observe four major metropolitan roads, four central business district (CBD) roads, and four freeway exit ramps. Data were collected on three consecutive Tuesdays at exactly the same sites and by the same techniques as in 2002. Three observation sessions per day at each site (10:00–11:00, 14:00–15:00, 17:00–18:00) provided 36 hours of observation.

   a. The first sentence in the description identifies this as an observational study. If you did not have that identification, what is it about the study that indicates that it is *not* an experiment?
   b. The researchers chose to observe drivers rather than ask them about whether they had used their mobile phone while driving. List some advantages and disadvantages of this decision. (To answer, just think; there are no technicalities here and nothing in the **Notes** that will give you the "answer." Just think about people.)
   c. This is one of those situations in which we (the readers or consumers of statistical analysis) must make a decision about how we can generalize the results. What should we regard as the intended population? Choose from the following and give a reason for your choice as to what is the most reasonable choice of the intended population.
      i) Drivers who drive on the roads at the twelve places in Melbourne where the data were collected on Tuesdays at the hours when the data were collected
      ii) Drivers in Melbourne in the CBD (= Central Business District), on major metropolitan roads and at freeway exit ramps on Tuesdays at the hours the data were collected
      iii) Drivers in Melbourne in the CBD, on major metropolitan roads and at freeway exit ramps on weekdays at the hours the data were collected
      iv) Drivers in Melbourne in the CBD, major metropolitan roads, and freeway exit ramps on weekdays
      v) Drivers in Melbourne
      vi) Drivers in urban areas of Australia having a law banning mobile phone usage while driving
      vii) Drivers in urban areas where there is a law banning mobile phone usage while driving

   In this study, as with so many other studies with valuable data, the type of data collection does not clearly fall into any of the "textbook" categories, even though the way the authors collected the data makes sense. However, what they did does have connections with some of those categories, both negatively and positively.

d. For the sake of argument, let us say that the intended population is "Drivers in Melbourne." State in the context of the study why the methods used do *not* resemble the following types of sampling:

    i) Voluntary response sample   ii) Simple random sample

e. The researchers' methods resemble a "random cluster sample" even though what they did is not a true random cluster sample. In a true "cluster sample," the clusters would be chosen randomly. It is unlikely that the researchers chose the twelve locations randomly from a larger list; there must have been many potential locations that posed problems for data collection—such as having nowhere for the researchers to safely stand. But in what way does what they actually did resemble a "random cluster sample"?

f. The researchers' methods resemble a "stratified random sample" even though their sample is not a true stratified random sample. In a true stratified random sample, there would be random sampling within the strata. The researchers attempted to observe all cars and did not sample. What were the "strata" that the researchers were using?

g. [*Review*] Here are some of the results of the study for 2006. Because using a mobile phone is not common, the authors calculated the incidence in per thousands, that is "number of mobile phones in use/1000 drivers" rather than in percents. Calculate incidences in "per 1000"s in such a way as to compare the three kinds of locations as to incidence of mobile phone usage. Give an interpretation of your calculations.

Location	Not using mobile phone	Using Mobile Phone	Number of Drivers Screened
Major Metropolitan Road	4721	56	4777
Central Business District	6651	139	6790
Freeway Exit	8504	136	8640
	19876	331	20207

h. [*Review*] For the Melbourne data, here is the comparison of males and females for the "under 30" drivers for just the morning hours data collection.

Gender	Not using mobile phone	Using Mobile Phone	Number of Drivers Screened
Male	819	18	837
Female	430	11	441
	1249	29	1278

Do a calculation to analyze whether there is a difference in the incidence of mobile phone usage between males and females for drivers under thirty for the morning hours at all of the locations.

i. [*Review*] The same methodology was used to collect data on the campus of one community college in Northern California. That is, for about forty-five minutes in a time period that straddled the start and end of classes, drivers were observed at a stop sign coming in and going out of the campus. Here are some results for comparison with the Australian data above for this college campus.

California College Campus Data				
Gender	Not using mobile phone	Using Mobile Phone	Number of Drivers Screened
Male	1092	39	1131
Female	853	52	905
	1945	91	2036

These data are also for drivers under 30 years of age. From the table, calculate so as to compare incidence of mobile phone usage by gender. Interpret your calculations in the context of the data.

j. Compare the results for the California college campus with the results for the Melbourne drivers (who are of the same age group). What differences and similarities do you see?

k. The data for the California college campus were collected in 2011. Which would you assess is the most reasonable population from which these data were drawn?
   i) Drivers who drive on the college campus where the data were collected on the days and the hours when the data were collected
   ii) Drivers on college campuses in Northern California during the hours and days that classes are held
   iii) Drivers on college campuses in Northern California
   iv) Drivers in Northern California
   v) Drivers in California

4. Sampling Students at the University of Michigan

The University of Michigan is one of the biggest universities in North America, with about forty-two thousand students. On the UM news service on June 21, 2012, the university announced a 2.8% increase in tuition fees and housing costs for the next 2012–2013 academic year. (See: http://www.ns.umich.edu/new/releases/20608-aid-will-cover-total-increase-in-costs-for-students-with-financial-need as accessed on June 22, 2012.) Part of the news report reads as follows:

> ANN ARBOR, Mich.—The University of Michigan Board of Regents today approved a budget that will increase financial aid enough to cover the full rise in the cost of attendance for the typical Michigan resident student with financial need—and reduce the student's loan burden.
>
> This is the fourth consecutive year an increase in financial aid more than offsets the increase in tuition and room-and-board rates for most state resident students with financial need. Additionally, that increase will come in the form of grant aid —which does not need to be repaid—reducing the educational loan portion of those students' financial aid package.

It is evident from the news release that the university authorities know that a fee increase will not be popular with students. If the university authorities had wished to collect data on how the fee increase may affect students and whether students are able to take advantage of financial aid, the university's institutional research office could do it in several ways. For each of plans (A through D) described below, choose the best name for the type of sampling it appears to be from the list below and give a reason for your answer.

   i) Voluntary Response Sample   iv) Simple Random Sample
   ii) Convenience Sample   v) Random Cluster Sample
   iii) Judgment Sample   vi) Stratified Random Sample

a. Plan A: From the list of all students enrolled in the current semester who are not graduating, they draw a random sample. To these students they send a questionnaire (by e-mail or by posting a letter) containing questions about how the fee increase is likely to affect the student.

b. Plan B: The plan is the same as Plan A, except that a random sample is taken from the current first-year students, another random sample from the current second-year students, and a third from the current third year students. A fourth sample is drawn from the fourth-year students who are on a five-year degree program.

c. Plan C: From the schedule of classes, a random sample of classes being taught in the current semester is chosen. A short questionnaire on the effects of a fee increase is taken to the classes chosen for the students to fill in.

d. **Plan D:** The UM News Service have their own pages on several social networking sites. The responses to the news of the fee increase that are recorded on these pages are analyzed for comments about fees, etc.
e. Of the various plans, which is likely to have the lowest non-response rate?
f. Do any of these data collection plans constitute an experiment? If so, which one or ones, and why? If not (for the ones that are not), state why they are not experiments.

5. **Negative Political Advertisements** Read about the negative political ad experiment in the **Notes**. In the experiment, three different negative ads were used: A, B, and C. (You may be able to read the paper at *http://www.jstor.org/stable/4148088*.) The $n = 700$ subjects that received negative mailings were assigned (randomly) to seven groups, each having $n = 100$ cases. Three groups received just one of the negative mailings (A, B, or C) a week before the election. Three groups received two of the three mailings starting two weeks before the election, and one group received all three negative mailings, starting three weeks before the election. Using this information, diagram the experiment as it was actually done, showing the treatment groups and the response variable by making a diagram similar to the ones in the Notes.

6. **Sampling Houses from San Mateo County**

   This exercise will use data on all houses sold in San Mateo County in 2005–2006. We will regard as collection as a population and from this population draw various kinds of samples. We will also be able to compare what we get from a sample to the population. Researchers almost never have this privilege! Usually we know *nothing* about a population and have to make (albeit educated and well founded) guesses from samples. Be thankful!

- Open our population file **San Mateo Real Estate Y0506.ftm** and get a case table.
- Get a Summary Table and drag the variable *Region* to the down arrow of the Summary Table. Then drag the variable *RecentHouses* to the right-facing arrow. You should have a table that looks like the one shown here.

   a. Houses that were built in the ten years before 2006 (Age ≤ 10 years) are considered to be "Recently Built Houses." All other houses are older houses. Calculate proportions to determine which regions have the highest and lowest proportions of "Recently Built Houses" and state your conclusions.

   b. The table (here) and the analysis done in part "a" are done on the population as a whole. Select the best name for the data analysis from: i) Simple Random Sample; ii) Census; iii) Stratified Random Sample; iv) Convenience Sample.

San Mateo Real Estate Y0506 0

Region	RecentHouses Older Houses	Recently Built Houses	Row Summary
Central	1663	95	1758
Coast	417	76	493
North	1247	88	1335
South	1674	226	1900
Column Summary	5001	485	5486

S1 = count ( )

Getting a Simple Random Sample. We will get Fathom to draw a sample of $n = 120$.

- Select the **Collection** icon (so that it turns blue) and go to **Collection>Sample Cases**. A new collection entitled "**Sample of San Mateo Real Estate Y0506**" should appear with its Inspector panel.
- In the Inspector, *deselect* (by clicking on the x in the box) the options "Animation On" and "With Replacement." Change the sample size from $n = 10$ to $n = 120$ and click on "Sample More Cases."
- Select the **Sample of San Mateo Real Estate Y0506** icon and get a case table.
- Using the case table for the **Sample of San Mateo Real Estate Y0506,** get a Summary Table and drag the variable *Region* to the down arrow of the Summary Table. Then drag the variable *RecentHouses* to the right-facing arrow. (Your numbers will be different from those in the table here.)

		RecentHouses		Row
		Older Houses	Recently Built Houses	Summary
Region	Central	40	3	43
	Coast	10	2	12
	North	16	3	19
	South	38	8	46
Column Summary		104	16	120

    S1 = count ( )

    c. Calculate proportions to determine which regions have the highest and lowest proportions of "Recently Built Houses" for your sample and state your conclusions. Are the proportions similar to or very different from the population proportions? (Check in Fathom by adding a new formula and: **Special Measures>rowProportions**.)

Getting a Random Cluster Sample The idea of a random cluster sample is to randomly select "groups" (or "clusters") of cases and then take *all* of the cases in the clusters as the sample. In real estate, there is a system that lists the location of every house by an "area number." These area numbers are our "clusters." (Identifying a property by its area should make sense, as real estate people think that the location of property is one of its most important features. You can see a map of the areas for San Mateo County at http://www.mlslistings.com/Map-Search. Zoom in on the map enough so that you can hover on a specific place. For Half Moon Bay, it will say "Area Group 600-614.)

- Open the Fathom file **MLSAreaSanMateoCounty.ftm**. We will first randomly sample five area numbers.
- Select the **Collection** icon (so that it turns blue) and go to **Collection>Sample Cases**. A new collection "**Sample of MLSAreaSanMateoCounty**" should appear with its Inspector panel.
- In the Inspector, *deselect* (by clicking on the x in the box) the options "Animation On" and "With Replacement." Change $n = 10$ to $n = 5$ and click on "Sample More Cases."
- Select the **Sample of MLSAreaSanMateoCounty** icon and get a case table. It will be a good idea to copy down the five area numbers. Also add the numbers for the cases in each of the areas.
- Go back to file **San Mateo Real Estate Y0506.ftm** and select its collection icon, so it is blue.
- Go to **Object>Duplicate Collection**. A duplicate collection (with 1 after Y0506 instead of 0) will appear. Move this new duplicate collection to a free space on the screen and select it.
- Now, go to **Object>Add Filter** and in the dialogue box that appears, type in the area numbers that you have selected (*not* the ones shown here). Use the palette on the box to get the "or" between each number. The parentheses will come automatically. Take care! After typing in, click on OK.

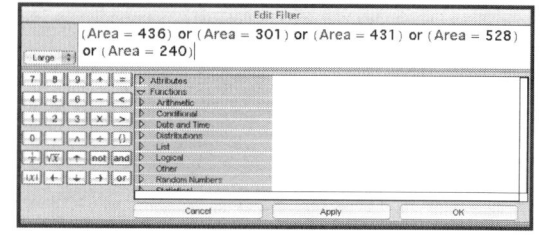

- Your cluster sample is the "filtered" duplicate collection. Get a case table for this collection and scroll down to see the sample size. It will probably not be $n = 120$. With a cluster sample, we cannot usually predict the sample size exactly, since the sample size depends upon the sizes of the clusters.

- Get a summary table showing the number of recent houses in the four regions—like the one for the **SRS** and for the population above. (You can repeat the instructions just above part c above; what you get may be quite different from this example, since you have a different cluster sample.)

  San Mateo Real Estate Y0506 1

		RecentHouses		Row Summary
		Older Houses	Recently Built Houses	
Region	Central	15	0	15
	North	16	0	16
	South	38	6	44
Column Summary		69	6	75

  S1 = count ( )

  d. The example cluster sample shown here did not select any houses from the region called "Coast." Explain (as though you are answering a test question) how this outcome is not surprising given that this sample is a cluster sample.

  e. Calculate the proportions of "Recently Built Houses" by Region for your sample and compare your results to the population proportions and the proportions from the **SRS**.

  f. [Extra credit]. The idea of a cluster sample is that clusters of cases are randomly selected and then *all* of the cases within those clusters are taken as the cases from the sample. Explain (as though you were answering a test question) how the procedure we used here—the directions represented by the bullets—accomplished this.

  Getting a Stratified Random Sample   The idea of a stratified random sample is to randomly sample *within* groups (the "strata"), usually with the idea of improving how well the sample represents the population. For our strata (our groups) we will use the four regions as our groups and sample with sample sizes as shown in the small table on the next page. There are many bullets here to get the stratified sample! You will have it by part g.

- Open a fresh copy of our population file **San Mateo Real Estate Y0506.ftm**. Or you can copy the collection (use **Edit>Copy Collection**) and the go to **File>New** and **Edit>Paste Collection**.

- Select the collection icon so that it turns blue and go to **Object>Duplicate Collection.** You should see a new collection with the last number in its "name" to be 1 instead of 0.

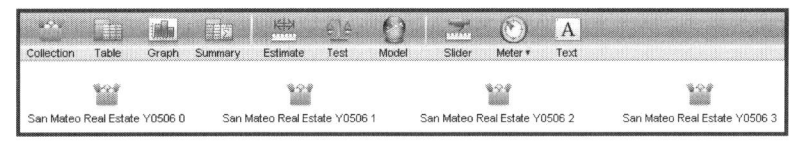

- Do **Object>Duplicate Collection** two more times, so that you have four duplicate collections displayed, like the picture here (on the previous page).

- Select the first collection and go to **Object>Add Filter.** In the dialogue box, type *Region = "Central"* as shown here. The first collection now has houses from only the Central Region. The Central Region is one of the groups or "strata."

- Repeat the bullet above for each of the other three collections, but using "Coast," "North," and "South" for each in turn in the dialogue box. Spread out the collections on the screen.

Getting the sample for the first group

- Go back to the first collection, the one for the Central Region, and select it so that a blue border appears. Go to **Collection>Sample Cases**. A new collection entitled "**Sample of San Mateo Real Estate Y0506 0**" should appear, with its Inspector panel.
- In the Inspector, *deselect* (by clicking on the x in the box) the options "Animation On" and "With Replacement." Change the sample size from $n = 10$ to $n = 36$ and click on "Sample More Cases."
- Select the **Sample of San Mateo Real Estate Y0506 0** icon and get a case table and check that you really do have $n = 36$ cases.

Getting the sample for the second, third, and fourth groups

Region	Number in stratified sample
Central	36
Coast	20
North	26
South	38
	120

- Repeat the three bullets just above (the ones following **Getting the sample for the first group**) for the three other collections, but use the sample sizes as in the table here instead of $n = 36$, so that for Coast, $n = 20$, for North $n = 26$, and for South $n = 38$. When you reach this point, the screen should look something like this.

Putting the samples for the groups together into one stratified sample

- Select the collection icon for **Sample of San Mateo Real Estate Y0506 0** (the sample for the Central Region), and go to **Object>Duplicate Collection**. A duplicate collection that will have "4" instead of "0" will appear.
- Get a case table for **Sample of San Mateo Real Estate Y0506 4** and scroll to the end of the data.
- Select the case table for the Coast Region go to **Edit>Select All Cases** then **Edit>Copy Cases**.
- Select the collection icon for **Sample of San Mateo Real Estate Y0506 4** and go to **Edit>Paste Cases**. The case table should now have $36 + 20 = 56$ cases.
- Repeat the last two bullets for the North Region and the South Region, adding them to the stratified sample **Sample of San Mateo Real Estate Y0506 4**. You should have $n = 120$.
- Get a summary table showing the number of recent houses in the four regions for the stratified sample.
    g. Calculate proportions to determine which regions have the highest and lowest proportions of "Recently Built Houses" for your sample and state your conclusions.
    h. Which of the types of random sampling (simple, cluster, or stratified) gave, as far as you can assess, the results closest to the population proportions?

7. A weight-loss study in Boston

This exercise is about a study carried out by researchers in Boston of weight-loss programs. The report can be found at http://jama.jamanetwork.com/article.aspx?articleid=200094, or:

Michael L. Dansinger, MD; Joi Augustin Gleason, MS, RD; John L. Griffith, PhD; Harry P. Selker, MD, MSPH; Ernst J. Schaefer, MD, *Comparison of the Atkins, Ornish, Weight Watchers, and Zone Diets for Weight Loss and Heart Disease Risk Reduction: A Randomized Trial,* Journal of the American Medical Association, 293(1), 2005.

Read the following, which is from the abstract of the paper, and answer the questions based upon this abstract. A few of the terms are explained below the abstract.

Objective *To assess adherence rates and the effectiveness of 4 popular diets (Atkins, Zone, Weight Watchers, and Ornish) for weight loss and cardiac risk factor reduction.*

Design, Setting, and Participants *A single-center randomized trial at an academic medical center in Boston, Mass., of overweight or obese (body mass index: mean, 35; range, 27-42) adults aged 22 to 72 years with known hypertension, dyslipidemia, or fasting hyperglycemia. Participants were enrolled starting July 18, 2000, and randomized to 4 popular diet groups until January 24, 2002.*

Intervention *A total of 160 participants were randomly assigned to either Atkins (carbohydrate restriction, n=40), Zone (macronutrient balance, n=40), Weight Watchers (calorie restriction, n=40), or Ornish (fat restriction, n=40) diet groups. After 2 months of maximum effort, participants selected their own levels of dietary adherence.*

Main Outcome Measures *One-year changes in baseline weight and cardiac risk factors, and self-selected dietary adherence rates per self-report.*

- "Single-center" in the context of medical studies simply means that the study was done at one place (in this instance, in a medical center in Boston) rather than at several different locations ("centers").
- "Body Mass Index" is a common measure used to assess whether people are overweight or obese. It takes account of the fact that taller people are naturally heavier. The formula is $BMI = \frac{mass}{(height)^2}$, where mass is measured in kilograms, and height is measured in meters. Adults having BMI in the range 25–29.9 are considered overweight, and those having BMI ≥ 30 are considered obese.
- The researchers were also interested in heart disease; that is why they chose adults having hypertension, dyslipidemia, or fasting hyperglycemia. The questions below do not depend on understanding the exact definitions of these terms.
- "Baseline weight" means weight at the beginning of the study. The researchers are interested in the *changes* in weight and cardiac (heart disease) risk factors from what they were at the beginning of the study.
    a. This is an experiment and not an observational study. What is it about the study that makes it an experiment and not an observational study? Answer in the context of the study.
    b. What are the experimental units? (One or more answers may be correct.)
        i) Diets (Atkins, Zone, Weight Watchers, Ornish)
        ii) Overweight or obese adults with a history of heart disease risk
        iii) Changes in weight and cardiac risk factors
        iv) Dietary adherence rates

c. What is the *factor*? (One or more answers may be correct.)
    i) Diets (Atkins, Zone, Weight Watchers, Ornish)
    ii) Overweight or obese adults with a history of heart disease risk
    iii) Changes in weight and cardiac risk factors
    iv) Dietary adherence rates
d. For this experiment, what are the treatment groups? Is there a "control" group?
e. Was random assignment to the treatment groups used?
f. What is (or are) the *response variable(s)*? (One or more answers may be correct.)
    i) Diets (Atkins, Zone, Weight Watchers, Ornish)
    ii) Overweight or obese adults with a history of heart disease risk
    iii) Changes in weight and cardiac risk factors
    iv) Dietary adherence rates

Here is a table of the baseline characteristics of the study participants. Baseline refers to characteristics at the beginning of the study. The next two questions are about this table. Here is some help in reading the table.

- Where the variable is quantitative, the mean is given with the standard deviation in parentheses.
- Where the variable is categorical, the number is given, and the percentage in that diet is in parentheses.
- For "Exercise," the measure is the number of people with more than "mild" exercise.
- "*P* value" will be introduced in §4.2; for now, know that *p* value ranges from 0 to 1, and that bigger values are an indication of "no difference" between the groups. A *p* value > 0.20 is big.

Table 1. Baseline Characteristics of Study Participants

Characteristics	Atkins Diet (n = 40)	Zone Diet (n = 40)	Weight Watchers Diet (n = 40)	Ornish Diet (n = 40)	All Diets (N = 160)	P Value
Demographics						
Age, mean (SD), y	47 (12)	51 (9)	49 (10)	49 (12)	49 (11)	.41
Women, No. (%)	21 (53)	20 (50)	23 (58)	17 (43)	81 (51)	.61
White race, No. (%)	32 (80)	26 (65)	30 (75)	32 (80)	120 (75)	.37
Risk factors, No. (%)						
Smoker*	3 (8)	5 (13)	1 (3)	4 (10)	13 (8)	.41
Hyperglycemia†	16 (40)	8 (20)	8 (20)	12 (30)	44 (28)	.14
Exercise‡	8 (20)	14 (35)	12 (30)	5 (13)	39 (24)	.09
Weight factors, mean (SD)						
BMI	35 (3.5)	34 (4.5)	35 (3.8)	35 (3.9)	35 (3.9)	.60
Body weight, kg	100 (14)	99 (18)	97 (14)	103 (15)	100 (15)	.43
Waist size, cm	109 (11)	108 (13)	108 (11)	111 (13)	109 (12)	.63

g. Look at the means and standard deviations for BMI, Body weight, and Waist size. Are these characteristics similar or markedly different in the four treatment groups?

h. In the context of an experiment, is it desirable that the baseline characteristics be almost the same for the treatment groups? Or should they be different? Or does it not matter? Give a reason for your answer. **PTO**

i. Here is some of the analysis of one of the response variables, the change in weight. (Ignore the daggers for now.) Summarize the figures in the table in the context of the study. What do they tell you about the four diets?

Table 3. Changes in Weight and Cardiac Risk Factors in an Analysis in Which Baseline Values Were Carried Forward in the Case of Missing Data*

Variable	Atkins (n = 40)	Zone (n = 40)	Weight Watchers (n = 40)	Ornish (n = 40)	P Value for Trend Across Diets
Weight, kg					
2 mo	−3.6 (3.3)†	−3.8 (3.6)†	−3.5 (3.8)†	−3.6 (3.4)†	.89
6 mo	−3.2 (4.9)†	−3.4 (5.7)†	−3.5 (5.6)†	−3.6 (6.7)†	.76
12 mo	−2.1 (4.8)†	−3.2 (6.0)†	−3.0 (4.9)†	−3.3 (7.3)†	.40

j. As explained above, a "big" *p* value indicates no difference between the groups. What do these *p* values suggest to you about the four diets and losing weight?

8. A weight-loss study in Israel. Read the description of the weight-loss experiment whose report was in *The New England Journal of Medicine* (**vol 359**: pp 229 – 241) entitled "Weight Loss with a Low-Carbohydrate, Mediterranean, or Low-Fat Diet" by I. Shai, D. Schwarzfuchs *et al*. The report was based upon an experimental comparison of the three types of diets. There were a total of 322 subjects in all, and these were randomly allocated so that 104 followed the low-fat diet, 109 followed the Mediterranean diet, and 109 followed the low-carbohydrate diet. Many response variables were measured, but one of these was change in weight (or weight change) from the baseline weight (the weight at the beginning of the experiment).

a. For the weight-loss experiment by Shai, Schwarzfuchs, *et al*, what are the treatments?
b. What is a good name for the *factor* for this experiment?
c. Diagram the experiment using a diagram as shown in the **Notes**.

- Open either the small Fathom file **TwelveObeseFemales.ftm** or the file **TwelveObeseMales.ftm**. You will randomly allocate these twelve to the three diets. (They are actual people from the NHANES data set, although the names in the Fathom file have been invented.)
- Get a ten-side die. For each person in the data set, you will roll the die once:
  – If the die comes up 0, roll again;
  – If the die comes up 1, 2, or 3, that person goes to the low-fat treatment;
  – If the die comes up 4, 5, or 6, that person goes to the Mediterranean treatment;
  – If the die comes up 7, 8, or 9, that person goes to the low-carbohydrate treatment.
- Do this for every person in the data set so as to assign each person to a group, but... (Read on.)
- Once you have four people in a diet group, do not add more. For example, if you get to the sixth person and you have four people in the LF group then keep rolling to allocate the remaining people in other groups.
- In the Fathom file assign LF, Med, or LC for each person for the variable *Treatment* according to your randomization.
- Use Fathom get the means and standard deviations of *weight* for the three treatment groups. Your **Summary Table** should look like this, but of course will be have different numbers.

Males for Experiment

	Treatment			Row Summary
	LC	LF	Med	
Weight	101.25	103.45	124.925	109.875
	9.32148	13.0857	26.2112	19.5498
	4	4	4	12

S1 = mean( )
S2 = s( )
S3 = count( )

d. In the Weight-Loss Study, the baseline mean weights, and standard deviations for the three groups were:

Low-Fat: $\bar{x}$: 91.3 kg   s: 12.3 kg.
Mediterranean: $\bar{x}$: 91.1 kg   s: 13.6 kg.
Low-Carbohydrate: $\bar{x}$: 91.8 kg   s: 14.3 kg.

Explain why for an experiment it is desirable that the baseline weights be almost equal and not desirable if the baseline weights are different.

e. We can allocate our twelve people, but we cannot make them diet. Here is a graphic from the results of the Shai, Schwarzfuchs, *et al* study. What can you say from this graphic about the three diets?

f. The study lasted twenty-four months (two years); what does the curved nature of the graphs for the three diets tell you about the dieting process?

9. **Australian Secondary Schools.** As part of the teaching curriculum, the Australian Bureau of Statistics (ABS) runs a data collection program in many schools called Census @ School (see http://www.abs.gov.au/censusatschool). Here, briefly, are the procedures for data collection:

– Teachers who wish to participate sign up for an account.
– Teachers are given Student Account Numbers (SANs).
– Each student uses his or her SAN to fill in the questionnaire.

The sampling plan as described above does not fit clearly in any one of our neat "textbook" categories, but it does resemble some of the categories. Consider how by answering the questions below.

a. Can this plan be a *simple random sample*? Look at the requirements for a SRS and answer.

b. Does this plan have features that make it like a *random cluster sample*? In what way does the plan resemble a random cluster sample and in what way is the plan *not* a random cluster sample?

c. Does this plan have features that make it like a *voluntary response sample*? In what way does the plan resemble a voluntary response sample and in what way may the plan *not* be a voluntary response sample? (It may be that we need more information. For example: it is likely that completing the questionnaire online is something like a required homework assignment. How does this affect the answer?)

d. Does this plan have features that make it like a *stratified random sample*? In what way does the plan resemble a stratified random sample and in what way is the plan *not* a stratified random sample?

When we have data that are not randomly drawn then we may want to assess if the data represents the population. Here is a bar graph provided by the ABS that compares the proportion of questionnaires by state and territory to the national population distribution. This kind of comparison does not guarantee that the data are representative of the population, but they give us a start.

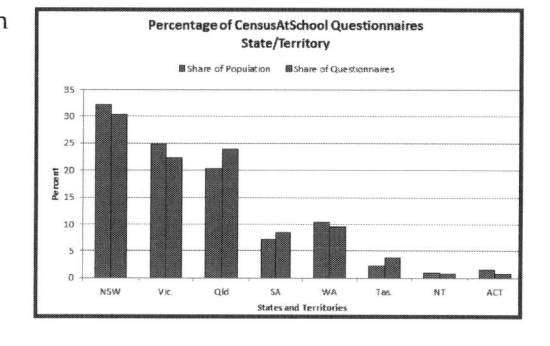

e. Are there parts of Australia that are seriously over- or underrepresented? Give a reason.

f. [*Review*] Here are data from the Australian C@S for high school students. Do some calculations to compare the proportion of students who "Never or Rarely" use the internet for social networking by region of Australia.

g. [*Review*] Here are some data for right foot lengths (in "cm") for male high school students in "Grade 10" or above. For the histogram shown here, what is the *bin width*?

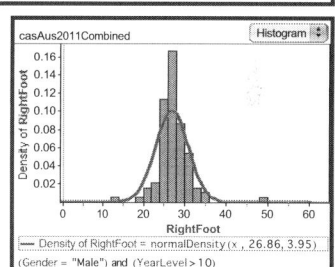

h. [*Review*] Use the histogram to estimate $P(RightFoot \geq 30)$ and then use the Normal distribution with $\mu = 26.86$ and $\sigma = 3.95$ to get $P(RightFoot \geq 30)$. (You may check your answers by getting **casAustralia2011Combined.ftm** or using **DistributionCalculator.ggb**.)

Special Exercise A on Sampling: How Big Were Ontario Farms in 1878?

Introduction: Here is a map from an atlas of one township in the province of Ontario in Canada. The map is from 1878, and it shows each farm, the name of the owner of the farm, and the size of the farm in acres. The map comes from the collection of maps on a site hosted by McGill University in Montreal (http://digital.library.mcgill.ca/countyatlas/). Our statistical question:

What was the mean size of farms in a Township in Ontario in 1878?

We could go through the map and record the size of each farm then get the mean of these numbers; in other words, we could look at *all* the data in our population.

1. a. What are the cases for these data?
   b. What is a name given to analyzing *all* of the data in a population rather than analyzing a sample of the population?

Collecting the data from all the farms in the township can probably be done, but doing that is tedious. Instead, we sample.

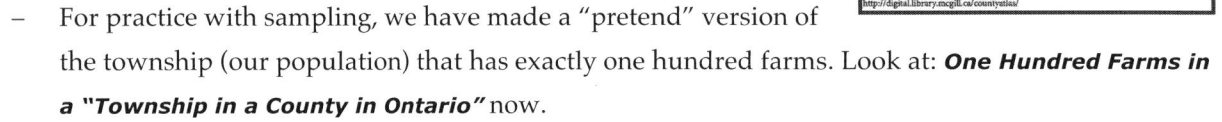

– For practice with sampling, we have made a "pretend" version of the township (our population) that has exactly one hundred farms. Look at: **One Hundred Farms in a "Township in a County in Ontario"** now.

– On the map **One Hundred Farms in a "Township in a County in Ontario,"** each farm is outlined with solid lines, and within each farm, the dashed lined squares are each ten acres, and each farm is numbered. So, for example, Farm number 1 (in Concession I) has 14 full squares, and so its size is $14 \times 10 = 140$ acres. Farm number 23 (also in Concession I) has two triangles and seven full squares, so $2 \times 5 + 7 \times 10 = 10 + 70 = 80$ acres.

I. Getting a Non-Random "Judgment" Sample

- After studying the map somewhat, use your *judgment* (not the *dice!*) to pick $n = 5$ farms from the entire township that you think represent the sizes of farms in the township. Record what you have chosen.
- Calculate the mean area of the rectangles (farms) in your sample and record it.

2. Let us think about what we are doing:
   a. For the sample that you have just taken, what is the sample size?
   b. Describe the population from which you chose the sample. Note: this is *not* asking for the size of the population but rather for a description of what the population is.
   c. Describe briefly how you got your sample and what you did (if anything) to try to make it representative. There is no "right" answer to this question; just describe what you did.

II. Getting A Simple Random Sample (SRS)

Read the box in the **Notes** about how to get a *simple random sample* and then answer question 3.

3. The first requirement for a *simple random sample* is to have a numbered list of the population. How does our map of the **One Hundred Farms in a "Township in a County in Ontario"** meet this requirement?

The second requirement for a *simple random sample* is to have a chance or random process to choose the units or elements of the sample. For this exercise, our random process uses *ten-sided* dice. To choose five farms randomly, you will roll the ten-sided dice to get numbers that have two digits, corresponding to the numbers of the farms on the map of the farms. Read the bulleted instructions.

- Roll the ten-sided die twice and note the outcome: this will give you the number of the first element in your sample. If 0 and then 9 appears, chose farm 9; if 34 (3 then 4) comes up, choose farm 34, etc. If you get 00, choose farm 100. Roll the dice again and again so as to get $n = 5$ farms randomly and record the farm number and the area of the farm. **Note:** If you get the same farm twice, record that farm twice.
- Calculate the mean area of the farms in your sample, and record it in the box with the thick border on your worksheet, if you have a worksheet for the **First Simple Random Sample (First SRS).**
- Repeat the entire process and record your results in the worksheet for a **Second Simple Random Sample.** You may get some of the same farms as you got for the first SRS or you may not; that is randomness!

II. Getting a Random Cluster Sample

The **Notes** tell you that in a random cluster sample, we randomly select "groups" (that is, "clusters") and then take *all* the cases in those clusters. We are choosing a very small sample, and so we will randomly choose just *one* cluster from the list on the back of the map and take *all* five farms in the cluster we choose.

- Roll the die twice and choose one farm. The cluster that you choose will be the one that has the farm you choose. Consult the list of clusters on the back of the map to see the cluster you chose. (For example, if you randomly choose farm 37, then your cluster is "H.")
- Record the numbers of the farms in your cluster and calculate the mean farm size for your sample.

III. Getting a Stratified Random Sample

For a **stratified random sample,** we first choose groups (or strata) and then randomly sample within the strata.

- *Read carefully:* For our exercise on *stratified random sampling* for the Ontario farms we will create just *two* strata or groups. **Stratum X**: the sixty farms in Concessions I and II—that is, those farms in the *southern* part of the township, called the *Southern Farms*. **Stratum Y**: the forty farms in Concessions III, IV, and V called the *Northern Farms*. (Have you noticed that the farms on the northern side of the township are generally larger?)
- To get the *stratified random sample,* roll the dice again, but this time do the rolling in two stages. First, roll the dice and *ignore* any farms whose numbers are 61 or greater; only choose farms from the numbers 01 to 60 until you get *three* farms among the *Southern Farms.*
- Then, after you have the three for the *Southern Farms,* get the *two* for the *Northern Farms* by rolling the dice, but now only paying attention to the numbers 61 through 00 (= farm 100).
- Record the numbers of the farms in your cluster and calculate the mean farm size for your sample.

Thinking about our samples The next two questions are best answered after a discussion.

4. a. If it is 1878 in Ontario (no phones, no cars, only horses), what are the advantages of selecting a *random cluster sample* to collect the data on farm sizes if you have to go from farm to farm (in winter)?

   b. Explain why it is (in the context of Ontario farms) that a *random cluster sample* could result in a very unrepresentative sample.

5. Think of the numbers in **Stratum X** and **Stratum Y** in the population (60 and 40 respectively) and the numbers we have in the sample of $n = 5$ for the two strata. Explain why it makes sense that we have three from the *Southern Farms* in the sample and two from the *Northern Farms* in the sample.

(• You may be asked to record your means for the various kinds of sampling so that there is collection of the means from many different samples. The mean farm size for the population is 138.65 acres. These questions are to be asked for the collection of the samples collected.)

One Hundred Farms in a "Township in Ontario"

North ---->

☐ = 10 acres   ◸ = 5 acres

Clusters of farms for the Ontario Farms Exercise

Cluster	Farms
A	1 – 5
B	6 – 10
C	11 – 15
D	16 – 20
E	21 – 25
F	26 – 30
G	30 – 35
H	36 – 40
I	41 – 45
J	46 – 50
K	51 – 55
L	56 – 60
M	61 – 65
N	66 – 70
O	71 – 75
P	76 – 80
Q	81 – 85
R	86 – 90
S	91 – 95
T	96 – 100

§3.2 Trusting Data, Part 2: Sampling Distributions

How tall are Hobbits?

Imagine being transported in both time and place to the Shire, land populated by Hobbits. You are surrounded by a group of Hobbits despite the fact that they are "shy of 'the Big Folk.'" But how tall are they on average? Can you trust what you see? Even if you were brave enough to collect measurements and use all the stats you have learnt to calculate means and standard deviations and make dot plots and box plots, how do you know that you have not stumbled upon an especially short or an especially tall sample of Hobbits? There is the story (pictured here)[3] of Bandobras Took (Bullroarer), son of Isengrim the Second, who was said to be four-foot-five (about 135 centimeters) tall and was able to ride a horse.[4] Was Bandobras an "outlier" in height, as implied by the story? Or, in his time, was he typical? How can we ever know the answer to this question?

The answer: The idea of a sampling distribution. There is a surprising answer to having doubts about the trustworthiness of the samples that we have: the answer is to have an idea of *all* the possible samples that we could possibly have. With this idea, we will be able to judge if "our" sample is trustworthy. How can this be done? If we do not trust our one sample, how can we have an idea of all of them? Read on. Height data for Hobbits is very difficult to find, and so we have to abandon this rather intriguing question. Instead, we will concentrate on something more accessible: the size of houses for sale.

Warning: Difficult Ideas Ahead Many students find the idea of a *sampling distribution* and the way that we use it in statistics quite difficult. This is correct: the ideas are not immediately intuitive and cannot be reduced to a number of steps to solve problems. Advice: ask many questions to try to picture what a sampling distribution is. (The exercises are meant to lead you through the kinds of questions to ask.)

The ideas will be developed in the context of observational studies, so we will look at samples from a population; however, the same ideas, with appropriate changes in terminology, also apply to randomization in experiments. In the context of population and samples, take care to distinguish between a **population distribution, a sample distribution,** and a **samp*ling* distribution.** (Here: "*ling*" is deliberately underlined.)

The Idea of a Sampling Distribution for Quantitative Variables

We are going to use an example where we know about the population; this is completely unrealistic because usually we do *not* know about the population. However, using an example where we *do* know everything about the population will help us better understand the concept.

---

[3] Bandobras Took during the fight with Golfimbul * Artist: Gregor Roffalski, alias <u>Sigismond</u>, * Copyright: Public Domain
[4] Tolkien, J. R. R., *The Fellowship of the Ring,* New York, Ballantine Books, 1965, p. 20

When we randomly sample from a population, we can never be completely certain how representative our sample is of the population. The plot here shows the distribution of the variable *Sq_Ft* (the number of square feet in the living area of the house) for the entire *population* of houses that were sold in San Mateo County in the years 2005–2006 (there were 1,890 houses sold in that year). The plot just below that shows the distribution of the same variable *Sq_Ft* for a random sample of $n = 40$. Since $n = 40$, there are far fewer dots in the dot plot. Is the sample like the population? One thing we can do is to compare the means, since we are fortunate to have the population. For the population, the mean for the variable *Sq_Ft* is $\mu = 1949.56$ ft^2 (we use the Greek letter μ because we are referring to the population mean). For the sample, the sample mean is $\bar{x} = 1965.08$ ft^2; it is not the same as the population mean but not that far away. How do we know that the sample mean is not far from the population mean? It is because we can compare our sample mean $\bar{x} = 1965.08$ ft^2 to the *collection of means from all possible random samples.* That little italicized phrase gives the idea of a *sampling distribution*. The idea is to collect together the sample means from *all* the possible random samples of size $n = 40$ drawn from the population of 1,890 houses in our population. Read the summary but do not stop there! The paragraphs below unpack the meaning.

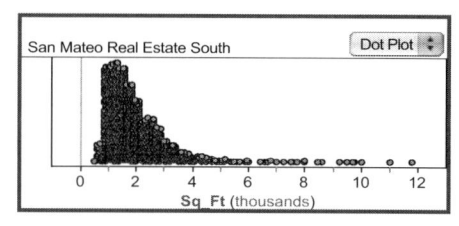

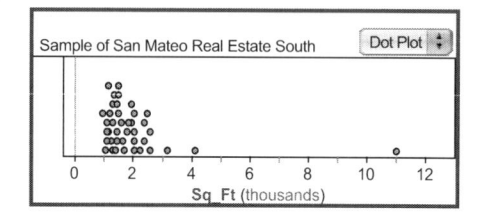

> **Idea of a Sampling Distribution of Sample Means**
> The **sampling distribution of sample means** is the distribution of the sample means for a specific variable of *all possible* random samples of a given size n drawn from a population.

Let us try to unpack the meaning of this. First, the number of possible different samples of size $n = 40$ is much greater than the size of the population. Our population has just 1,890 houses. The number of unique random samples of $n = 40$ (the count) that can be drawn from this population is approximately $9.254066615 \times 10^{82}$ if we each time we sample we sample *without replacement*. For our example, the population has 1,890 cases, a single sample has $n = 40$ cases, but the sampling distribution has a count of something close to that huge number, $9.254066615 \times 10^{82}$. Sampling distributions are huge.

Secondly, notice that the sampling distribution is a distribution of sample means, or in symbols, $\bar{x}$s. The cases for the population were houses, the cases for our single sample of $n = 40$ were also houses, but the cases for the sampling distribution are sample means of the variable *Sq_Ft*. With a *sampling distribution*, we are interested in looking at all the possible samples (all $9.254066615 \times 10^{82}$ of them). For each sample (each of the $9.254066615 \times 10^{82}$), we calculate some measure, such as a sample mean, or a sample proportion, or a sample median, and make the sampling distribution out of one of these measures calculated for the different samples. We can therefore see how we would go about getting a sampling distribution for our real estate example. Our process has a number of steps.

⇨ Step 1: Randomly sample a sample of $n = 40$ from the population of 1,890 houses.
⇨ Step 2: Calculate the sample mean $\bar{x}$ for the sample just collected; store the result in a safe place.
⇨ Step 3: Repeat steps 1 and 2 at least $9.254066615 \times 10^{82}$ times to include every possible sample.
⇨ Step 4: Collect together all the different sample means into a collection.

Sound impossible? Yes, it is impossible; actual sampling distributions are derived from mathematical theory, but the process gives you an idea of what goes into a sampling distribution. Moreover, we will *simulate* a sampling distribution, actually a part of a sampling distribution, using these steps.

Building a Sampling Distribution Using Fathom In §2.4, you worked "interactively." In this section it will be best to read the notes while (at the same time) following the bulleted Fathom instructions.

- Open the Fathom file **San Mateo Real Estate South.ftm** and if it is not already opened, open a **case table.** This file contains data on *all* the houses that were sold in the southern part of San Mateo County, California, from June 2005 to June 2006. It contains the *population* of houses sold.

- Get a **dot plot** of the variable *Sq_Ft* and also get a **Summary Table** showing the mean and the standard deviation and place all of these, including the **case table,** on the left side of the screen (we will put many other things on the remainder of the screen). What you should have should look like the (greatly reduced) picture on the next page.
  What you see is the **population distribution** of the variable *Sq_Ft*, for the population of *all* the houses that were sold in 2005–2006 in San Mateo County.

- Select the Collection Icon and go to the menu **Collection>Sample Cases.** A new collection should appear entitled *Sample of San Mateo Real Estate South* and its *"Inspector"* should also open. (If it does not, double click on the icon.)

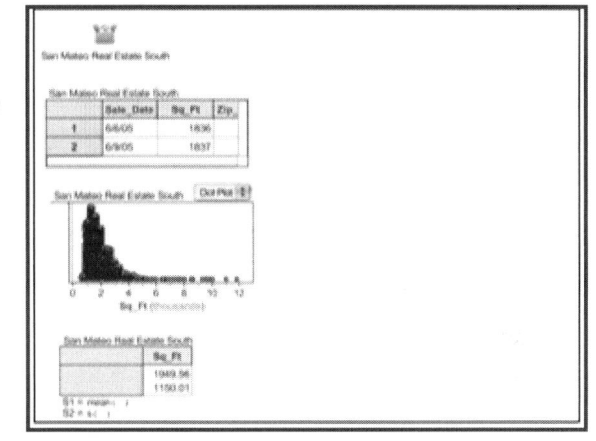

- In the *"Inspector"* three boxes should have an "x" in them; *de*select "with replacement" and *de*select "animation on" but leave the "x" in the box for "replace existing cases." Also change 10 cases to 40 cases and press the *Select More Cases* box.

- Select the *Sample of San Mateo Real Estate South* and get a **case table** for the sample. You may want to scroll down to see that you actually have $n = 40$ cases. We have completed Step 1 of our steps.

- For the *Sample of San Mateo Real Estate South* get a **dot plot** of the variable *Sq_Ft* and also get a **Summary Table** showing the mean and the standard deviation and place all of these, including the **case table,** in the center of the screen. What you should have should look like the (greatly reduced) picture below, but your sample will be different from the sample chosen here, and the mean and standard deviation will be different. Randomness at work!

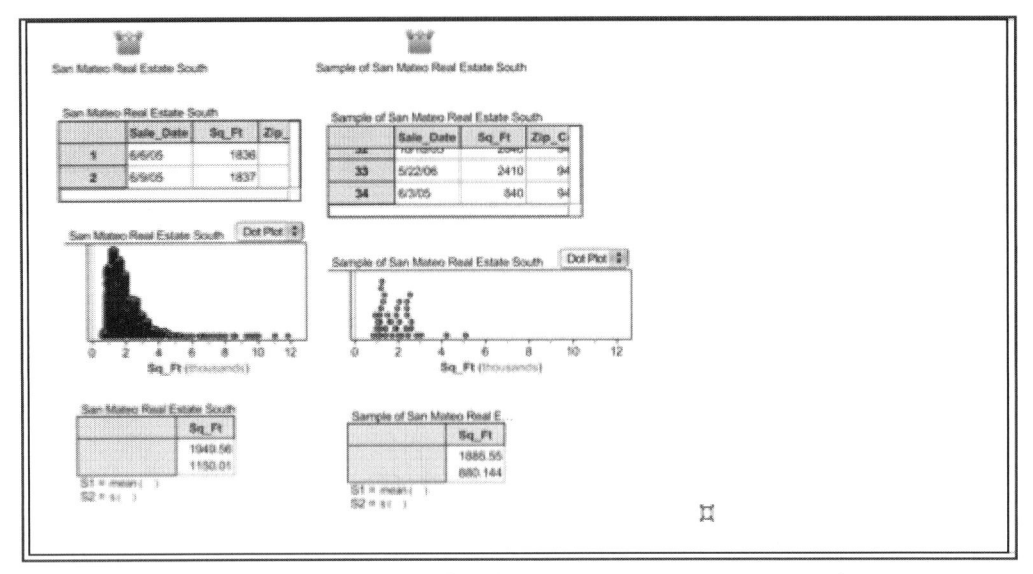

We have now completed (with the help of Fathom) Step 2 of our four steps; we have calculated the sample mean $\bar{x}$ for our sample. Now it is time to take Step 3: getting more and more samples, each time calculating the mean. We will have Fathom do the work, and we will start slowly.

- Get the *"Inspector"* for the *Sample of San Mateo Real Estate South* (it may be on your screen; if it is not, double click on the *Sample of San Mateo Real Estate South* collection icon). In the *Inspector,* select *Measures* (which may appear as just *"Meas..."*).
- In the left-hand box, where <new> appears, type in *xbar* and select the row so that it turns blue.
- Double click on the space under *Formula* (or go to **Edit>Edit Formula**) to open **Edit Formula**.
- In the dialogue box, type *mean(Sq_Ft)* and click on **OK**.

You should see the value of the sample mean of your sample appear in *value;* what we are doing is to prepare Fathom to collect *xbars* (sample means) for a huge number of different samples.

- You may now safely close the sample *Inspector*. It will just clutter the screen if it is open.
- With the *Sample of San Mateo Real Estate South* collection icon selected, go to **Collection>Collect measures.** When you do this, you will see your dot plot for your sample change five times and a new collection icon entitled *Measures from Sample of San Mateo Real Estate South* along with its *Inspector*.

The reason that you saw your sample data change is that Fathom is beginning to carry out Steps 3 and 4. Fathom has (quite quickly) randomly sampled five more samples of size $n = 40$, and for each one it has calculated the sample mean, *xbar*. Fathom has put these five *xbar* values in a new collection, which it has called *Measures from Sample of San Mateo Real Estate South*. However, we need many, many more than five values of *xbar;* ideally we would like *xbars* from $9.254066615 \times 10^{82}$ samples; we will have to be content with fewer than this.

- Select the *Measures from Sample of San Mateo Real Estate South* icon and get a **case table.** You should see the five *xbar*s that Fathom has calculated from the five random samples.
- For the *Measures from Sample of San Mateo Real Estate South*, get a **dot plot** of the five *xbar*s and also a **Summary Table** for the mean and standard deviation for *xbar*. Put these on your screen on the right.

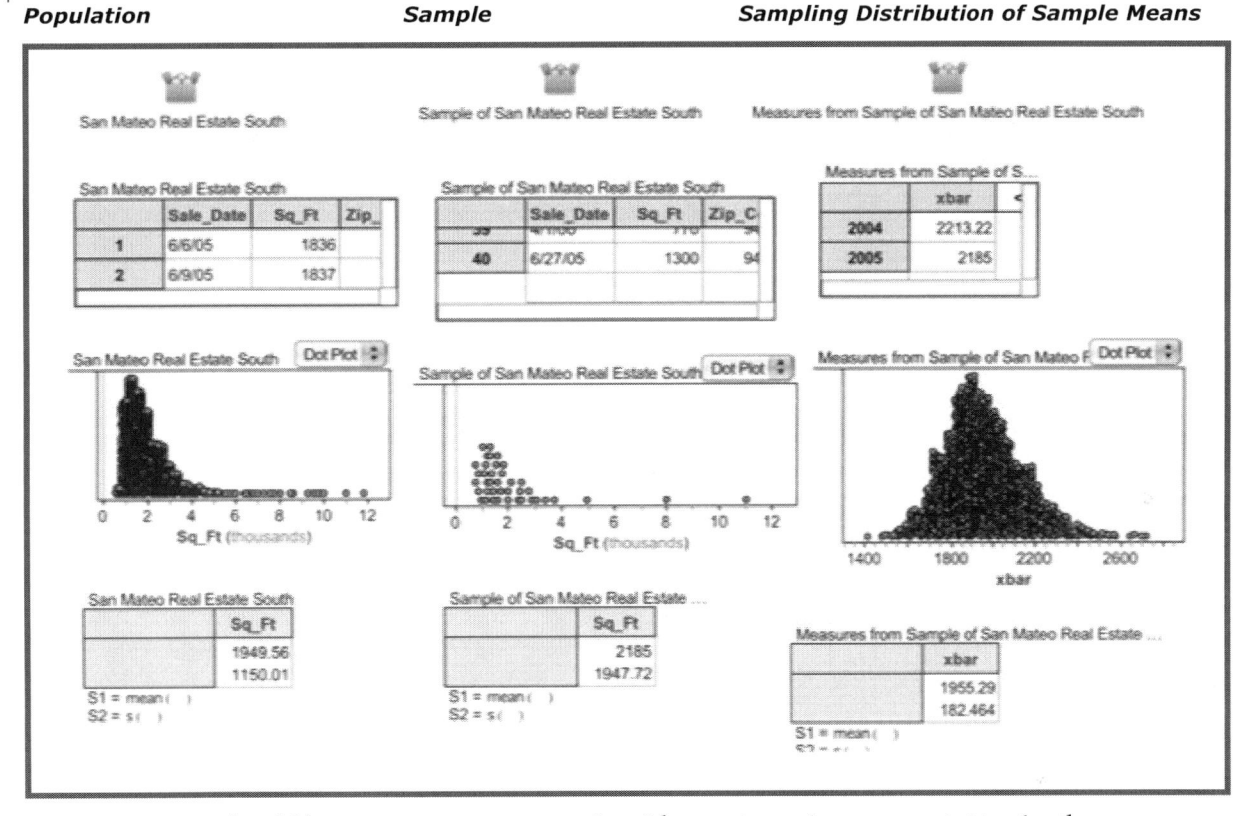

Population **Sample** **Sampling Distribution of Sample Means**

Your screen should have some symmetry to it, with *graphs* and *summary tables* for the population, for the sample, and now, on the right, for the **Sampling Distribution**. (Above is the graph of what your screen will look like when the process is complete; however, you will have just five dots in your graph instead of two thousand dots.)

- In the *Inspector* for the *Measures from Sample of San Mateo Real Estate South*, deselect the "*Animation on*" (if you do not, things will take a long time) and change the number of measures from 5 to 1000 (yes, 1000). Move the *Inspector* to the lower left side of the screen and click on *Collect More Measures*. Wait some time while Fathom works. Fathom is busy collecting samples and calculating.

After this process is done, you can check from your **case table** that there are 1,005 cases of *xbars*. Fathom has calculated 1,005 sample means and recorded them in this collection; this is the start of a sampling distribution.

- Go back to the *Inspector* and click on **Collect More Measures** (or with the *Measures from Sample...* . selected, go to **Collection>Collect More Measures**) and wait another four minutes or so. What you should see (if you then close the *Inspector*) should resemble—but not be the same as—the picture above.

Some Things to Notice about What We Have Done

We have not derived the complete *sampling distribution of sample means*. To get the complete sampling distribution, we would have to continue until we had got *all* possible random samples, which we said would be at about $9.254066615 \times 10^{82}$ different samples; Fathom would probably choke on this task, and we would also choke at the time it would take. We said that the facts about sampling distributions are derived mathematically. The simulation we have illustrates some facts about *sampling distributions of sample means*.

Compare the three distributions in your Fathom output (or the display above) and notice that:

- The population and the sample are both right skewed, but the sampling distribution looks symmetric. Indeed the sampling distribution looks like it may be a Normal distribution.
- The variable for the population and for the sample is the same, *Sq_Ft*, but for the sampling distribution the variable being measured is *xbar*. Look at the horizontal axes of the dot plots.
- The means for all three distributions are similar, but the standard deviation for the sampling distribution is much smaller than the standard deviation for the population and the sample.

There is another major, important difference between the population and sample on the one hand and the sampling distribution. Not only is the variable different, but the *cases* area not the same things; for the population and the sample, the cases are houses, but for the sampling distribution, the cases are *not* houses. The cases for the sampling distribution are *means from samples of n = 40*. Each dot in the dot plot for the *population* distribution is a house, and each of the $n = 40$ dots in the *sample* distribution is one of the houses in that sample, but each dot in the dot plot of the *sampling* distribution is a sample mean of some sample.

Since the cases for the sampling distribution are different from the cases for the sample, the means have to be interpreted in the correct context; for the *sample* (not *sampling*) distribution shown above, we can say that that the average house size for that sample was $\bar{x} = 2185$ ft^2. However, the mean that you see for the *sampling distribution* (which is 1955.29 ft^2) is the **mean of all the x-bars,** or we could say the **mean of all the sample means**. We assign a new symbol to the mean of all the *x*-bars; $\mu_{\bar{x}}$, which is read in short, "**mu of x-bar**" or "**mean of x-bar**." The standard deviation for a sampling distribution, since it is a standard deviation of all the sample means, is denoted $\sigma_{\bar{x}}$. Along with the three graphs, one for a population distribution, one for a sample distribution, and one for a sampling distribution, we have three sets of symbols for mean and standard deviation.

Notation for means and standard deviations			
	Population Distribution	Sample Distribution	Sampling Distribution
Mean	μ	$\bar{x}$	$\mu_{\bar{x}}$
Standard Deviation	σ	s	$\sigma_{\bar{x}}$

Be prepared to memorize these symbols and their meanings; there is a certain logic to the usage that will help you. Non-Greek symbols are used for what we can easily calculate (things from samples), whereas Greek symbols are used either for theoretical quantities (such as $\mu_{\bar{x}}$ and $\sigma_{\bar{x}}$) or quantities that are not accessible to researchers, such as the population mean μ and population standard deviation σ.

How Random Samples Behave: Facts about Sampling Distributions

Look back at the output for our *part-of-a-sampling-distribution* that we had Fathom simulate (it is only a part of a sampling distribution because we have only about two or three thousand samples, whereas the real sampling distribution would have $9.254066615 \times 10^{82}$ cases). Think, as always, about *shape, center,* and *spread*. The *shape* of our sampling distribution appears nearly Normal, despite the shape of the population distribution being very right-skewed. The center of the sampling distribution looks very close to the center of the population distribution; compare the means. However, the *standard deviation* looks much smaller than the standard deviation of the population distribution; compare the standard deviations.

Facts about Sampling Distributions of Sample Means Calculated from Random Samples

Shape: The shape of the sampling distribution of $\bar{x}$ will be approximately Normal if the sample size n is sufficiently large, even if the population distribution from which the samples were drawn is not Normal.

Center: The mean of the sampling distribution of $\bar{x}$ is equal to the mean of the population distribution. That is, $\mu_{\bar{x}} = \mu$.

Spread: The standard deviation of the sampling distribution of $\bar{x}$ is $\sigma_{\bar{x}} = \dfrac{\sigma}{\sqrt{n}}$, where σ is the population standard deviation.

Sampling Distributions are less variable than population distributions. It makes sense that the *spread* of sampling distribution is smaller than the spread of the population distribution. The standard deviation of the sampling distribution is the population standard deviation divided by the square root of the sample size $\sigma_{\bar{x}} = \dfrac{\sigma}{\sqrt{n}}$ and since $\sqrt{n}$ will be bigger than one, the standard deviation of the sampling distribution $\sigma_{\bar{x}} = \dfrac{\sigma}{\sqrt{n}}$ will be smaller than the standard deviation σ of the population. Hence the sampling distribution of sample means will be less variable than the distribution of the population since our sample size n will be bigger than one. Notice that the larger the sample size that we use, the less variation the sampling distribution has. This is important in practice.

The shape of sampling distributions may be Normal despite the population distribution being not Normal. The *shape* of the sampling distribution will be "approximately Normal" if the sample size n is sufficiently large, despite the *population distribution* being decidedly not at all Normal in shape. We can say more than this: we can say that the larger the sample we decide to draw, the closer the *sampling distribution* will be to a Normal distribution. Look at the shape of the distribution of our Fathom start-of-a-sampling-distribution (our simulation) with about two thousand different samples of $n = 40$. The dot plot looks Normal, but there is a hint of right skewness to it. If we increased the sample size to $n = 120$, the plot that we would get would conform more to a Normal distribution. The bigger the sample size, the more Normal is the *sampling distribution*. This characteristic of the shape of the sampling distribution is called the **Central Limit Theorem.**

Central Limit Theorem

If a simple random sample of size n is drawn from any shape population with population mean μ and finite standard deviation σ, then if n is sufficiently large, the **sampling distribution** of the sample mean $\bar{x}$ is approximately a Normal distribution with mean $\mu_{\bar{x}} = \mu$ and standard deviation $\sigma_{\bar{x}} = \frac{\sigma}{\sqrt{n}}$. The Normal approximation is better the bigger the sample size n and the closer the population distribution is to Normal.

Sampling distributions are useful: Reasonably Like and Rare. We can use the facts about sampling distributions to calculate what x-bars we are reasonably likely to see when we randomly sample. The discussion of the box below is on the next page.

Reasonably Likely and Rare

Sample means $\bar{x}$ that are *within* the interval $\mu - 1.96\frac{\sigma}{\sqrt{n}} < \bar{x} < \mu + 1.96\frac{\sigma}{\sqrt{n}}$ are called **reasonably likely**.

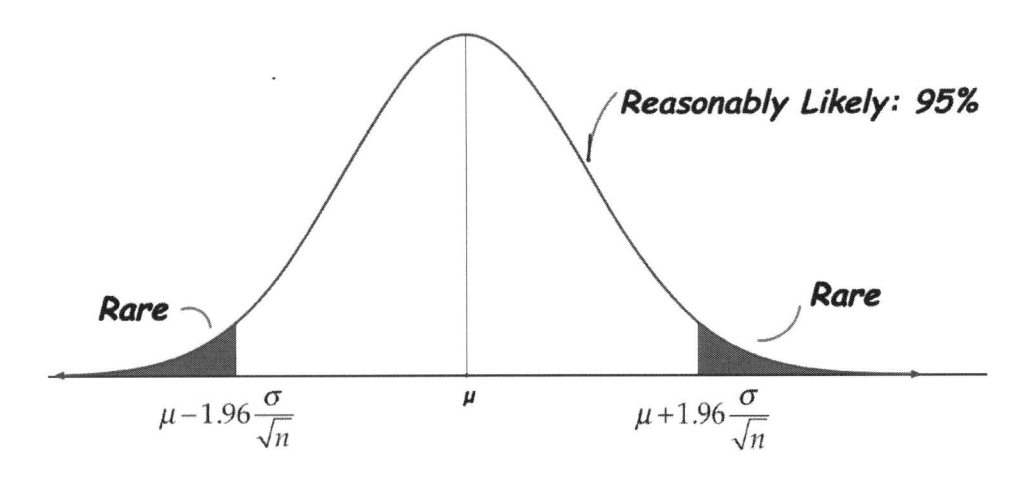

Reasonably Likely: 95%

Rare

Rare

$\mu - 1.96\frac{\sigma}{\sqrt{n}}$ μ $\mu + 1.96\frac{\sigma}{\sqrt{n}}$

Sample means $\bar{x}$ that are *outside* the interval $\mu - 1.96\frac{\sigma}{\sqrt{n}} < \bar{x} < \mu + 1.96\frac{\sigma}{\sqrt{n}}$ are called **Rare**.

In probability notation, $P\left(\mu - 1.96\frac{\sigma}{\sqrt{n}} < \bar{x} < \mu + 1.96\frac{\sigma}{\sqrt{n}}\right) = 0.95$.

We saw that with the standard Normal distribution, 95% of the distribution lay between 1.96 standard deviations less than the mean to 1.96 standard deviations more than the mean. (To check this again, remember that 95% in the middle of a Normal distribution leaves 5% for the tails, and hence 2.5% for each tail; look for .0250 in the *body* of the Standard Normal Chart and read off the z score.) We use this fact to define **reasonably likely** and **rare** sample means $\bar{x}$ from a random sample of size n.

Examples with the San Mateo Real Estate Data

What sample means are reasonably likely? We used the real estate data for the South Region of San Mateo County as our population. For the variable *Sq_Ft* we saw that the population mean μ = 1949.56 ft² and that the population standard deviation σ = 1150.81 ft². Using the facts in the box on the previous pages, we can specify the center, spread, and shape of the *Sampling Distribution* for sample means $\bar{x}$ for random samples of $n = 40$:

**Population Parameters
San Mateo South
Houses**

San Mateo Real Estate South

	Sq_Ft
	1949.56
	1150.01

S1 = mean ( )
S2 = s ( )

Center: The mean will be $\mu_{\bar{x}} = \mu = 1949.56$.

Spread: The standard deviation will be $\sigma_{\bar{x}} = \frac{\sigma}{\sqrt{n}} = \frac{1150.01}{\sqrt{40}} \approx 181.83$.

Shape: The sampling distribution will be approximately Normal.

Then we can calculate the **reasonably likely** sample means for sampling from this population with a sample size of $n = 40$ by calculating:

$$\mu - 1.96 \frac{\sigma}{\sqrt{n}} = 1949.56 - 1.96 \left(\frac{1150.01}{\sqrt{40}} \right) \approx 1949.56 - 1.96(181.83) = 1593.17 \text{ ft}^2, \text{ and}$$

$$\mu + 1.96 \frac{\sigma}{\sqrt{n}} = 1949.56 + 1.96 \left(\frac{1150.01}{\sqrt{40}} \right) \approx 1949.56 + 1.96(181.83) = 2305.95.$$ What this calculation means is that if we sample randomly from the population of San Mateo houses, we should expect to get sample means $\bar{x}$ between about 1593 ft² and 2306 ft² with probability 0.95. Another way of putting this is that if we sampled repeatedly—sampled over and over again—we expect 95% of the sample means to be between about 1593 ft² and 2306 ft². We expect to get sample means that are *outside* the interval $1593 < \bar{x} < 2306$ with probability 5%. Using probability notation, we can write $P(1593 < \bar{x} < 2306) \approx 0.95$. Having made this calculation, we can answer questions such as this: *Is getting a random sample of n = 40 with $\bar{x} = 1800$ ft² reasonably likely or rare?* Since $\bar{x} = 1800$ is *inside* the interval $1593 < \bar{x} < 2306$, getting such a sample is **reasonably likely.** Getting a sample with $\bar{x} = 2400$ would be **rare** since this sample mean is outside the interval $1593 < \bar{x} < 2306$.

What is the probability that we will get a sample mean greater than 2185 ft²? This was the last sample mean $\bar{x}$ that we collected in our simulation of two thousand sample means. (If you repeat the simulation, you will get some other sample mean as your "last one"; this is just a convenient one to focus on—there is nothing special about it.) Since we have a normal distribution for our sampling distribution, this question looks like a *"given a value, find a probability"* Normal distribution type of problem. We want to know the probability $P(\bar{x} > 2185)$. The first step is to make a sketch of the situation, which is shown in the graphic here. Notice that our $\bar{x} = 2185$ is in the *reasonably likely* region and *not* in the *rare* region,

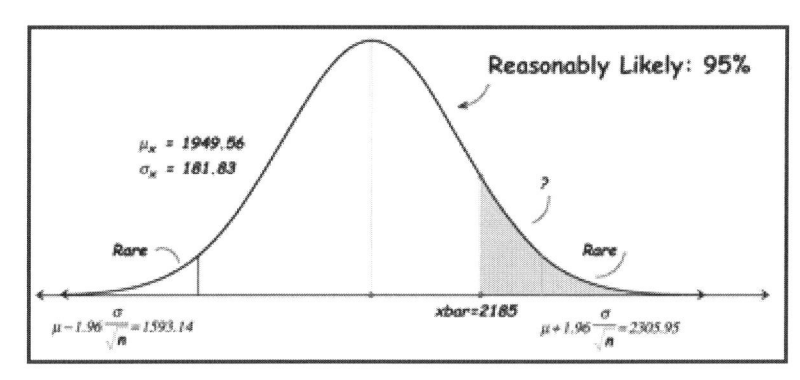

217

and so we should expect to get a z score that is smaller than 1.96. The z score is $z = \frac{2185 - 1949.56}{181.83} \approx 1.29$. Then $P(\bar{x} > 2185) = P(z > 1.29) = 1 - P(z < 1.29) = 1 - 0.9015 = 0.0985$. Hence we conclude that the probability that with random sampling, we would get a random sample with sample mean bigger than 2,185 square feet, or $\bar{x} > 2185$ is just under 10%.

What is the xbar for the lowest 5% of xbars that we could get with random sampling? Because the sampling distribution of the *xbars* is very approximately a Normal distribution, we can find such a value. This question looks like a *given a probability, find a value* type of Normal distribution problem. We want a value—in this situation a value for an *xbar*—that divides the lowest 5% of *xbars* from the remaining 95%.

Looking at the chart for the Normal distribution, we find that for the lowest 5%, we get a z score of $z = -1.645$, and we can use the formula for the z score to find the *xbar* that divides the lowest 5% from the rest. We calculate from: $z = \frac{\bar{x} - \mu_{\bar{x}}}{\sigma_{\bar{x}}}$. So, $-1.645 = \frac{\bar{x} - 1949.56}{181.83}$ and from this $(-1.645)(181.83) = \bar{x} - 1949.56$, and so solving for $\bar{x}$, we get $\bar{x} = 1949.56 - 181.83(1.645) \approx 1650.89$ ft^2. The lowest 5% of all *xbars* we expect to get will have $\bar{x} \leq 1650.89$.

Summary: Sampling Distributions for Quantitative Variables

The ideas in this section about sampling distributions are at the heart of how statisticians think about *inferring* from a sample to a population or from a specific random allocation in an experiment to all possible random allocations. These ideas are abstract; they are not "easy." Expect to be confused for a time—but hopefully not forever! Expect to have to work through the ideas. Avoid taking shortcuts; avoid learning just enough to answer the questions, parrot-like. Ask questions; draw pictures; get the idea.

- The **sampling distribution of sample means** is the distribution of the sample means for a specific variable of *all possible* random samples of a given size *n* drawn from a population.
    - A *sampling distribution* is *not* the same as a *sample distribution*, which is the distribution of a variable for a single sample drawn from a *population*.
    - A *sampling distribution* is *not* the same as a *population distribution*, which is the distribution of a variable for an entire population and which is usually unknown to researchers.
- A good way to understand *sampling distributions* is to think of how they could be constructed:
    ⇨ Step 1: Randomly sample a sample of a specific size *n*.
    ⇨ Step 2: Calculate the sample mean $\bar{x}$ for the sample just collected; store the result.
    ⇨ Step 3: Repeat steps 1 and 2 a huge number of times to include every possible different sample.
    ⇨ Step 4: Collect together all the different sample means into a collection.
- Another good way to keep straight the differences between *population distributions*, *sample distributions*, and *sampling distributions* is to recall the picture of the simulation done in this section.

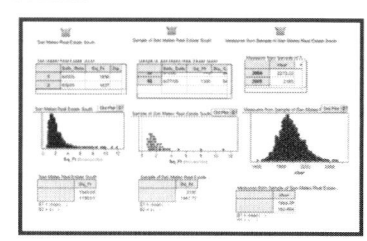

- **Notation: means and standard deviations:**

	Population Distribution	Sample Distribution	Sampling Distribution
Mean	μ	$\bar{x}$	$\mu_{\bar{x}}$
Standard Deviation	σ	s	$\sigma_{\bar{x}}$

- **Facts about Sampling Distributions of Sample Means Calculated from Random Samples**

  Shape: The shape of the sampling distribution of $\bar{x}$ will be approximately Normal if the sample size n is sufficiently large, even if the population distribution from which the samples were drawn is not Normal.

  Center: The mean of the sampling distribution of $\bar{x}$ is equal to the mean of the population distribution. That is, $\mu_{\bar{x}} = \mu$.

  Spread: The standard deviation of the sampling distribution of $\bar{x}$ is $\sigma_{\bar{x}} = \dfrac{\sigma}{\sqrt{n}}$, where σ is the population standard deviation.

- **Central Limit Theorem:** This theorem says that if a simple random sample of size n is drawn from any shape population with population mean μ and finite standard deviation σ, then if n is sufficiently large, the **sampling distribution** of the sample mean $\bar{x}$ is approximately a Normal distribution with mean $\mu_{\bar{x}} = \mu$ and standard deviation $\sigma_{\bar{x}} = \dfrac{\sigma}{\sqrt{n}}$. The Normal approximation is better the bigger the sample size n and the closer the population distribution is to Normal.

- **Reasonably Likely and Rare Sample Means**

  - **Reasonably likely** sample means $\bar{x}$ are *within* the interval $\mu - 1.96\dfrac{\sigma}{\sqrt{n}} < \bar{x} < \mu + 1.96\dfrac{\sigma}{\sqrt{n}}$.

  - **Rare** sample means $\bar{x}$ are *outside* the interval $\mu - 1.96\dfrac{\sigma}{\sqrt{n}} < \bar{x} < \mu + 1.96\dfrac{\sigma}{\sqrt{n}}$.

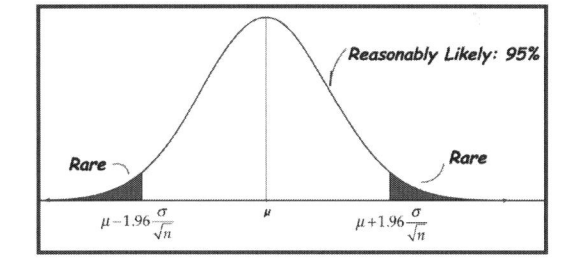

- With a sampling distribution for a specific variable and population, we can answer questions about:

  - The range of *reasonably likely* and *rare* sample means $\bar{x}$ s
  - Determine the likelihood that we see a sample mean $\bar{x}$ that is greater (or lesser) than a specific value
  - Calculate the sample mean that we would expect to see a given percentage of the time if we repeatedly sampled

§3.2 Exercises on Sampling Distributions

1. **What happens with a bigger sample size?** [Fathom] This exercise repeats the "interactive" part of the **Notes** but also reviews the conclusions about the *sampling distribution of the sample mean* $\bar{x}$. However, in this exercise, we use a sample size of $n = 120$ instead of a sample size of $n = 40$, but with the same population of houses sold in the South Region of San Mateo County. Hence the population mean for the variable *Sq_Ft* is $\mu = 1949.56$ ft^2, and the population standard deviation is $\sigma = 1150.01$ ft^2.

   a. Read the Summary Boxes **"Facts about Sampling Distributions..."** and **"Central Limit Theorem"** and consider the sampling distribution of sample means $\bar{x}$ for the variable *Sq_Ft* for samples of size $n = 120$ from the San Mateo County South real estate population. From the information in the boxes, state what value the *mean* $\mu_{\bar{x}}$ should have, what value the *standard deviation* $\sigma_{\bar{x}}$ should have, and what *shape* the sampling distribution should have. Use the correct notation for your answers. Notice that your overall answer must have three parts.

   b. Confused Conrad gives $\sigma_{\bar{x}} = 1150.01$ as part of his answer to the question in part a. What is Confused Conrad's confusion?

   c. Compare the value of the *standard deviation* $\sigma_{\bar{x}}$ that you have calculated (from the formula given in the boxes in the **Notes**) to the value we calculated for samples of size $n = 40$, which was $\sigma_{\bar{x}} = \dfrac{\sigma}{\sqrt{n}} = \dfrac{1150.01}{\sqrt{40}} \approx 181.83$. Is your value for samples of size $n = 120$ smaller or bigger?

   Explain from the formula why your answer differs from 181.83.

Follow the bullets to get the simulated sampling distribution using Fathom. These are the same instructions that are in the **Notes.** You may want to review the "four steps".

- Open the Fathom file **San Mateo Real Estate South.ftm** and if it is not already opened, open a **case table.** This file contains the *population* of houses sold.
- Get a **dot plot** of the variable *Sq_Ft*, and also get a **Summary Table** showing the mean and the standard deviation and place all of these including the **case table** on the left side of the screen
- Select the Collection Icon and go to the menu **Collection>Sample Cases.** A new collection should appear entitled *Sample of San Mateo Real Estate South,* and its *"Inspector"* should also open.
- In the *"Inspector"* three boxes should have an "x" in them; deselect "with replacement" and deselect "animation on" but leave the "x" in the box for "replace existing cases." Also change 10 cases to *120* cases and then press the *Select More Cases* box.
- Select the *Sample of San Mateo Real Estate South* and get a **case table** for the sample. You may want to scroll down to see that you actually have $n = 120$ cases. We have finished Step 1 of our steps in the **Notes.**
- For the *Sample of San Mateo Real Estate South* get a **dot plot** of the variable *Sq_Ft* and also get a **Summary Table** showing the mean and the standard deviation and place all of these, including the **case table,** in the center of the screen. What you should have should look like the picture in the **Notes,** but your sample will be different.

[We have now completed Step 2 of our steps; we have calculated (we had Fathom do this) the sample mean $\bar{x}$ for our sample. Now it is time to take Step 3: getting more and more samples, each time calculating the mean.]

- Get the *"Inspector"* for the *Sample of San Mateo Real Estate South*. In the *Inspector*, select *Measures* (which may appear as just *"Meas…"*).
- In the left hand box, where <new> appears, type in *xbar* and select the row so that it turns blue.
- Double click on the space under *Formula* (or go to **Edit>Edit Formula**) and **Edit Formula** should open.
- In the dialogue box, type *mean(Sq_Ft)* and click on **OK**.

You should see the value of the sample mean of your sample appear in *value*; what we are doing is preparing Fathom to collect *xbars* (sample means) for a huge number of different samples.

- You may now safely close the sample *Inspector*. It will just clutter the screen if it is open.
- With the *Sample of San Mateo Real Estate South* collection icon selected, go to **Collection>Collect measures.** When you do this, you will see your dot plot for your sample change five times and a new collection icon entitled *Measures from Sample of San Mateo Real Estate South* along with its *Inspector*.

The reason that you saw your sample data change is that Fathom is beginning to carry out Steps 3 and 4. Fathom has (quite quickly) randomly sampled five more samples of size $n = 120$, and for each one it has calculated the sample mean, *xbar*. It has put these five *xbar* values in a new collection, which it has called *Measures from Sample of San Mateo Real Estate South*. However, we need many, many more than five values of *xbar*.

- Select the *Measures from Sample of San Mateo Real Estate South* icon and get a **case table**. You should see the five *xbars* that Fathom has calculated from the five random samples.
- For the *Measures from Sample of San Mateo Real Estate South* get a **dot plot** of the five *xbars* and also a **Summary Table** for the mean and standard deviation for *xbar*. Put these on your screen on the right.

    Your screen should have *graphs* and *summary tables* for the population, for the sample, and now, on the right, for the *Sampling Distribution*. There is a picture in the **Notes**; however, at this point, you will have just five dots, not 2,005 dots.

- In the *Inspector* for the *Measures from Sample of San Mateo Real Estate South*, deselect the **"Animation on"** (if you do not, things will take a long time) and change the number of measures from 5 to 1,000 (yes, 1,000). Move the *Inspector* to the lower left side of the screen and click on **Collect More Measures.**

    After waiting some time for the collection to complete, you can check from your **case table** that there are 1,005 cases of *xbars*. Fathom has made calculated 1,005 samples means and recorded them in this collection; this is the start of a sampling distribution.

- Go back to the *Inspector* and click on **Collect More Measures** (or with the *Measures from Sample…* selected, go to **Collection>Collect More Measures**) and wait for Fathom to collect another thousand measures.
- Include a printed version of the screen you have with your submission of this question. (How you do this will depend on the kind of computer you have. What you want to do is to copy the screen, with *everything* on it, so that you can print the screen, or put it into a word processing document to be printed.)

    d. Look at the dot plot that you have created. What does each dot in that plot represent?

e. Confused Conrad answers question d with the statement: "Each dot represents the value of the variable *Sq_Ft* for a single house." CC is confused and his answer is wrong. Why is the answer wrong?

f. Look at your **Summary Table** with the mean and the standard deviation for the variable *xbar* as well as the **Graph.** Explain how what you got for the simulation agrees with your answers to question a, even though your numbers are slightly different from the "theoretical" answers.

g. Judging from comparing what you have done here and the information in the boxes **("Facts about Sampling Distributions...")**, what changes in the sampling distribution when you have a bigger sample size and how does it change? Does the shape change? Does the center of the sampling distribution change? Does the spread or the variability change? Be as specific as you can. The more specific you are in answering these questions, the more likely you are to understand and remember the facts.

2. **What happens with a bigger sample size?** This exercise is linked with Exercise 1. It also uses the *sampling distribution* of the *xbars* of the variable *Sq_Ft* for samples of $n = 120$ drawn from the population of houses sold in the South Region of San Mateo County. Since we have the same population, the population mean for the variable *Sq_Ft* is $\mu = 1949.56$ ft^2, and the population standard deviation is $\sigma = 1150.01$ ft^2.

   a. In the **Notes** there was a calculation of the interval of *Reasonably Likely* values for *xbars* of the variable *Sq_Ft* for samples of $n = 40$. Rounded to the nearest square foot, we found the interval to be $1593 < \bar{x} < 2306$. That is $P(1593 < \bar{x} < 2306) \approx 0.95$. Repeat this calculation for *xbars* of the variable *Sq_Ft* for samples of $n = 120$. You will get a different interval.

   b. Is the interval that you calculated narrower or wider than the interval found for samples of $n = 40$?

   c. What part of the formula $\mu - 1.96 \frac{\sigma}{\sqrt{n}} < \bar{x} < \mu + 1.96 \frac{\sigma}{\sqrt{n}}$ made your interval (which should have been $1744 < \bar{x} < 2155$) narrower than the interval calculated for samples of $n = 40$? Explain so that you are satisfied that you understand it how the formula made your interval narrower. (Your answer should make sense to a fellow stat student who it is just learning this material.)

   d. In the **Notes** we worked with an *xbar* equal to 2,185 ft^2. For samples of $n = 120$ (rather than for samples of $n = 40$), will an $\bar{x} = 2185$ ft^2 be a *reasonably likely* or a *rare* event (the event is "getting an *xbar*" in a single random sample of size $n = 120$)? Give a reason for your answer.

   e. We also calculated the probability of getting a sample mean *xbar* greater to 2,185 ft^2 by using the fact that the *sampling distribution* of the *xbars* is very nearly a Normal distribution. To calculate $P(\bar{x} > 2185)$ for a sample of $n = 120$, will we proceed as a "given a value—find a proportion" problem or "given a proportion—find a value" problem?

   f. Make a sketch of the problem and calculate $P(\bar{x} > 2185)$ for a sample of $n = 120$, using as a pattern the calculations done in the **Notes.**

   g. For $n = 40$, we found $z = \frac{2185 - 1949.56}{181.83} \approx 1.29$ and $P(\bar{x} > 2185) = P(z > 1.29) = 1 - 0.9015 = 0.0985$. For your sample of $n = 120$, did your probability come out larger or smaller?

h. Your probability of getting a sample mean greater than 2,185 sq. ft. should be smaller with a sample of n 120 compared with a sample of n = 40. Does this mean that samples of n = 120 are more (or less) accurate than samples of n = 40? Give a reason for your answer that convinces you but based upon your results.

i. For samples of n = 120, find a sample mean $\bar{x}$ of square feet for the smallest 5% of sample means we expect to see with random sampling. Draw a sketch and find a value for the sample mean $\bar{x}$. (Is this a "given a value—find a proportion" problem or "given a proportion—find a value" problem?)

j. For samples of size n = 40, we found that the value for the 5% of smallest means we would expect to see was 1,651 square feet. Compare this with your result from part i for samples of size n = 120. Is your result closer or farther away from the population mean μ = 1,949 square feet?

k. You should have found that your answer for the "5% smallest means" is *nearer* to the population mean μ for samples of size n = 120 compared with samples of size n = 40. Does your finding mean that samples of n = 120 are more accurate than samples of n = 40? Give a reason for your answer that convinces you but based upon your results. (Making a picture of the sampling distribution may help.)

3. **What happens with a Different Population** For this exercise, we will consider the same variable, *Sq_Ft*, but this time we will consider slightly different population. The population will be the houses that were sold in the **Central Region** of San Mateo County.

- Open the Fathom File **San Mateo Real Estate Central.ftm**.
- Get a **dot plot** of the variable *Sq_Ft* and also a **Summary Table** showing the mean and the standard deviation of the population. What you should get is displayed here.

  a. Use the correct symbols to show the mean and the standard deviation of the population.

  b. We will draw samples of size n = 120. For samples of this size, calculate and show, with the correct notation, the mean and the standard deviation of the *sampling distribution* of sample means $\bar{x}$ for the variable *Sq_Ft*.

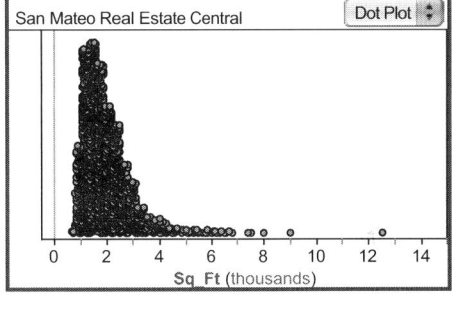

- Select the Collection Icon and go to the menu **Collection>Sample Cases.** A new collection should appear entitled *Sample of San Mateo Real Estate Central*.
- In the *"Inspector,"* three boxes should have an "x" in them; *de*select "with replacement" and *de*select "animation on" but leave the "x" in the box for "replace existing cases." Also change 10 cases to *120* cases, and then press the **Select More Cases** box.
- Select the **Sample of San Mateo Real Estate Central** and get a **case table** for the sample. You may want to scroll down to see that you actually have *n = 120* cases.
- For the *Sample of San Mateo Real Estate Central*, get a **dot plot** of the variable *Sq_Ft* and also get a **Summary Table** showing the mean and the standard deviation.

h

c. The population distribution of the variable *Sq_Ft* is clearly right-skewed. Do you expect that the random sample of size $n = 120$ that you draw will also be right-skewed or not? Give a reason for your answer.

d. The population distribution of the variable *Sq_Ft* is clearly right-skewed. Does this fact mean that the *sampling distribution* of sample means $\bar{x}$ for the variable *Sq_Ft* will also be right-skewed? Give a reason based on your experience with the same variable for the South Region.

e. For part b, you should have that $\mu_{\bar{x}} = 1984.92$ ft², and that $\sigma_{\bar{x}} = \dfrac{\sigma}{\sqrt{n}} = \dfrac{979.934}{\sqrt{120}} \approx 89.40$. Using this information, find the interval of *reasonably likely xbars* for this *sampling distribution* of sample means $\bar{x}$ for the variable *Sq_Ft*.

f. Is the sample mean $\bar{x}$ for the random sample of $n = 120$ that you have drawn *reasonably likely* or is it *rare*? Give a reason for your answer.

g. If your sample mean $\bar{x}$ is above the population mean then calculate the probability that you will get a mean greater than the one that you have gotten, using the fact that the sampling distribution is well approximated by a Normal distribution. If you sample mean is less than the population mean then find the probability that you would get a sample mean less than what you got.

h. The population mean for the Central Region for the variable *Sq_Ft* is bigger than the population mean for the *Sq_Ft* for the South Region. Is it possible that if we draw random samples of $n = 120$ from each of the populations, we would get a sample mean $\bar{x}$ for the *South* Region that was bigger than the sample mean $\bar{x}$ for the Central Region? Explain why or why not. (*Hint:* Think *reasonably likely*.)

§3.3 Trusting Data, Part 3: Categorical Data

The big ideas of this section

First off, recall from the last section the distinction between population and sample. A population consists of all the things we want to know about—perhaps people, but it does not have to be. A sample consists of the things about which we actually have data, hopefully a representative subset of the population.

- There are plenty of ways to get bad samples, and that even if we do all kinds of analyses on these bad samples, we cannot trust the results. Then, unlikely as it may seem, we saw that the best kinds of samples are those that are randomly chosen. In this section we will begin to see why that is so.
- The question remains, however: can we trust an analysis from a small sample when the population is so large, even if we have a good sample?
- Here is our strategy to answer our question about how to trust random samples—the same strategy as we employed in 3.2. First, we will pretend that we know all about the population. Then we will do two things: we will use the power of computing to get a picture of all the samples that we could possibly get—both truthful and misleading ones. And we will back that up with the power of mathematics to show that our computing picture is correct. In "real statistical life" we never know all about the population, but our pretending here will give us a handle on how to cope with our ignorance.

Our statistical question: Obesity amongst young adults

We need an example. Our question is: *What proportion of Americans between the ages of sixteen and thirty-five are obese?* First off, we need a definition of obese. A common definition depends upon the Body Mass Index, which is the mass of a person in kilograms divided by the square of the person's height, measured in meters, so in symbols: $BMI = \dfrac{mass}{(height)^2}$. Then the definition given by the Centers for Disease Control and Prevention is that adults with a BMI ≥ 30 are considered obese (see: http://www.cdc.gov/obesity/adult/defining.html). So, if we have a sample of adults and have the information to calculate their BMIs, we can classify each as either obese or not obese and then calculate the proportion classified as obese. But can we have to trust our results from a sample?

(What follows can be done without doing Exercise 1 or Exercise 6, but having done one of those exercises will help this section.)

To answer our statistical question, we randomly draw just one sample of adults from the population of all American adults and count the number that is obese in the sample; our idea is that this single sample will tell us about the population. But we know that a *single* sample may mislead us, especially if we choose a very small sample size, such as $n = 5$. Can we do something that will tell us the likelihood that our single sample will show us 20% obese (that is one obese out of five), or 40% (two obese of five), 60%, or perhaps 0%? What we can do, if we have a collection that we can "pretend" is the population, is to sample over and over again—with the same sample size, $n = 5$, every time we sample—and then we see the number of times our sample of $n = 5$ gives us 0 obese, 1 obese, etc. That is what is done in Exercises 1 and 6 where one thousand repeated small samples are used. Here we show the results of sampling $n = 5$ repeatedly ten thousand times from a pretend population. The "population" that we are drawing from has a proportion obese of about 20% (it is actually just under that figure), and the Summary Table shows the results of this repeated sampling.

The Summary Table shows that 4,132 of the ten thousand repeated samples of $n = 5$ (or 41.32%) gave us just one obese out of the five chosen people, and that 3,540 of the ten thousand repeated samples of $n = 5$ gave us no obese people. You may object that this shows that we need a bigger sample size to more accurately identify the proportion obese, but we can learn something about samples by using this small sample size. Later, to answer our statistical question, we will increase the sample size.

Measures from Sample of ...		
xObese	0	3540
	1	4132
	2	1855
	3	421
	4	51
	5	1
Column Summary		10000

S1 = count ( )

A mathematical model that fits what we see and why: the Binomial Model

Our statistical question and our strategy of repeatedly sampling with the same sample size fit what is called a Binomial Distribution Model. The Binomial Distribution Model is a mathematical model based on four conditions. Here in the box are the four essential conditions for the model.

Conditions for the Binomial Distribution Model

We can fit a Binomial Distribution Model if all of the following conditions are met:

B The outcomes are **binomial**; that is, there are just two outcomes. Often, the outcomes are called "success" and "failure."

I Each case is **independent** of the others. (The cases are often also called "trials.")

N There is a fixed total **number, n**, of cases (or "trials").

E We use a probability p of success that is the same ("**equal**") for each case, or "trial."

If the conditions are met, the Binomial Distribution Model will give us the probability that we see 0, 1,...n successes out of n cases or trials. See the next box for the formula that calculates the probabilities.

We can use a Binomial Model because in our situation all of the conditions are met; here is how:

B: We have just two outcomes: obese and not obese. We focused on "obese," which means that "obese" was our "success." (A "success" is just a choice of one of the two outcomes; it need not be something "desirable.") The CDC has four more categories besides "obese." Using these five categories would not meet the condition since we need just two ("bi"). Also, if our outcomes were waist circumference or BMI, our condition would not be met because these variables are quantitative variables. The **B** condition demands a categorical variable with just two outcomes.

I: Each case is independent of the others because our samples were randomly drawn. When you roll a die, the next roll is completely independent of the previous roll. The die does not say to itself: "Oh, my, I haven't come up '2' for such a long time; the next one better be '2.'" Also, when Fathom generates random numbers, independence is guaranteed. A random process is without memory and gives independent "trials."

N: Since our sample size was $n = 5$ each time, there is a fixed number of cases or "trials."

E: In our calculations we will fix the p at a specific number. In the example shown just below, we use $p = 0.20$ or 20%. The number that we use depends on out idea for the probability of a "success."

Using the model: P(X = k) and P(X > 1) If the conditions are met then the Binomial Model gives the probabilities of 0, 1, 2,…, up to *n* successes, using a mathematical formula based on the conditions. Here is an example connected with our statistical question. Here, we want the probabilities of the number *X* of successes (i.e., obese people) out of *n* = 5. Here is what the Binomial Model gives us. We have seen probabilities like this before. *k* = 1, which can be read "the probability that in a sample of *n* = 5, we see exactly one obese person," is 0.4096. In symbols, this is $P(X=1)=0.4096$. The height of the "spike" for $X=1$ is the proportion 0.4096. The hand calculation of this number is shown below; here we are content to use the already calculated values.

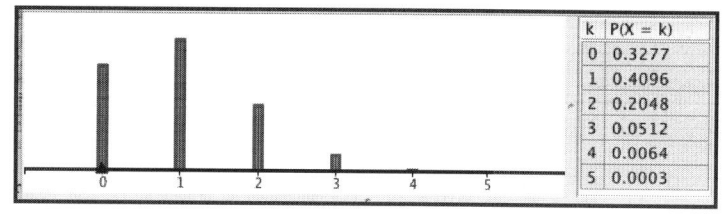

P(X > 1) By adding probabilities, we can also calculate the probability that we will see *more than one* obese person in a sample of five. That probability will be calculated as:

$$P(X>1) = P(X=2) + P(X=3) + P(X=4) + P(X=5)$$
$$= 0.2048 + 0.0512 + 0.0064 + 0.0003$$
$$= 0.2627$$

We can also see this graphically with by turning the spikes into a histogram, where the shaded in bars (each having $binwidth = 1$) show $P(X>1) = 0.2627$. (This graphic comes from **DistributionCalculator.ggb**.)

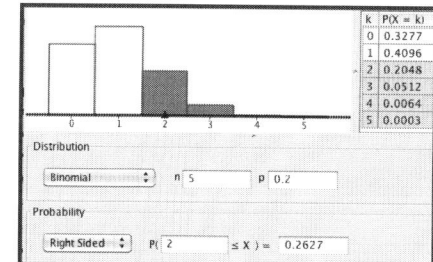

How the P(X = k) calculations are done. The formula (which actually has two parts) to calculate these probabilities is given in the box below.

Formula for the Binomial Model

To calculate the probability of X = k successes out of n cases (or "trials"), use:

$$P(X=k) = \binom{n}{k} p^k (1-p)^{n-k} \text{ where } \binom{n}{k} = \frac{n!}{k!(n-k)!}$$

and where $n! = n \cdot (n-1) \cdot (n-2) \cdots 2 \cdot 1$ and also $0! = 1$ by definition.

$\binom{n}{k}$ tells us the number of ways of placing k things into n boxes (also written $_nC_k$) and is "pronounced" "n choose k" or "the number of combinations of k things amongst n."

To see how the Binomial Model formula works, see the examples just below this box.

Example 1: P(X = 0) and the meanings of $\binom{n}{k}$ and n! We start with the easiest one; that is, we start with calculating the probability of getting no obese people in the sample of *n* = 5. Here, the value of $k = 0$, so plugging in the numbers, we have:

$$P(X=0) = \binom{5}{0} p^0 (1-p)^{5-0} = \left(\frac{5!}{0!5!}\right)(0.20)^0 (0.80)^5 \approx \frac{5 \cdot 4 \cdot 3 \cdot 2 \cdot 1}{1 \cdot 5 \cdot 4 \cdot 3 \cdot 2 \cdot 1}(1)(0.32768) = 0.32768$$

The most mysterious part of this for someone who has never seen it before is probably $\binom{n}{k}$. The formula says that it is $\binom{n}{k} = \frac{n!}{k!(n-k)!}$ and so here $\binom{5}{0} = \frac{5!}{0!(5-0)!} = \frac{5\cdot 4\cdot 3\cdot 2\cdot 1}{1\cdot 5\cdot 4\cdot 3\cdot 2\cdot 1} = 1$. First of all, the symbol $n!$ is pronounced "n factorial" (not n spoken with a loud voice!), and means the number n multiplied by one less multiplied by one less than the next one, and so on until you get to 1. There is a special case for 0! Which is defined to be one, so $0! = 1$.

n! is pronounced "n factorial" and means $n! = n\cdot(n-1)\cdot(n-2)\cdots 2\cdot 1$ and $0! = 1$

The calculation $\binom{n}{k} = \frac{n!}{k!(n-k)!}$ counts the number of ways that we can have k things if we have n to choose from, so, in our example $\binom{n}{k} = \binom{5}{0} = 1$ counts the number of ways we can have zero obese people in five choices. The answer is that there is just one way to have zero obese people in five people: all of the boxes (people) must be "not obese." In Exercise 1, the worksheet will look like the one shown here.

1	2	3	4	5		Number Obese
Not	Not	Not	Not	Not		0

Example 2: P(X = 1) and the meaning of $\binom{n}{k}$ continued Now, what about the calculation for the probability of getting exactly one obese person in the sample of $n = 5$? Here, $k = 1$, so plugging in the numbers, we have:

$$P(X=1) = \binom{5}{1}p^1(1-p)^{5-1} = \frac{5!}{1!(5-1)!}(0.20)^1(0.80)^{5-1} \approx \frac{5\cdot 4\cdot 3\cdot 2\cdot 1}{1\cdot(4\cdot 3\cdot 2\cdot 1)}(0.20)(0.4096) \approx 5\cdot(0.08192) = 0.4096$$

The part of the formula that shows $(0.20)^1(0.80)^{5-1} = (0.20)^1(0.80)^4 = (0.20)(0.4096) = 0.08192$ calculates the probability that the first of the five people we sample is obese and the rest are not. But there are five different places the one obese person could come, and that is why we multiply by

$\binom{n}{k} = \binom{5}{1} = \frac{5\cdot 4!}{1!4!} = 5$. The one obese person could be the first of the five, the second, the third, the fourth, or the fifth of the five. So the answer of "five different places" makes sense.

Example 3: P(X = 2) and the meaning of $\binom{n}{k}$ continued. The probability of getting exactly two obese people out of five sampled will have $k = 2$, so we have:

$$P(X=2) = \binom{5}{2}p^2(1-p)^{5-2} = \frac{5!}{2!(5-2)!}(0.20)^2(0.80)^{5-2} \approx \frac{5\cdot 4\cdot 3\cdot 2\cdot 1}{2\cdot 1\cdot(3\cdot 2\cdot 1)}(0.04)(0.512) \approx 0.2048$$

We calculate: $\binom{n}{k} = \binom{5}{2} = \frac{5\cdot 4\cdot 3!}{2\cdot 1\cdot 3!} = \frac{5\cdot 4}{2\cdot 1} = 10$. This calculation means that there are ten different ways that we could place the two obese people in the five boxes. The best way to see that is to tediously list them.

	A	B	C	D	E
1	O	O	N	N	N
2	O	N	O	N	N
3	O	N	N	O	N
4	O	N	N	N	O
5	N	O	O	N	N
6	N	O	N	O	N
7	N	O	N	N	O
8	N	N	O	O	N
9	N	N	O	N	O
10	N	N	N	O	O

With a much bigger sample size n and a much bigger number of successes k, this kind of listing obviously becomes very tiresome. Fortunately, all we need to know is how many different ways there are, and the formula $\binom{n}{k}$ gives that.

The Binomial Model with bigger sample sizes.

Here is what we have seen so far. If we specify a population proportion p (we used $p = 0.20$) and specify an n, which we are taking to be a sample size, then the Binomial Model will give us the probabilities of seeing all the possible outcomes of success, from 0, 1, 2,…, up to n if the conditions for using the model are met. By doing this, we should be convinced that $n = 5$ is too small a sample, since a wrong answer (number obese not 1) is more probable that a correct answer, if the correct answer is 20%. What happens is we increase the sample size?

Binomial Model with n = 20, and p = 0.20. We are using the same value for p, but we have increased the n. Compare this graph with the graph for the n = 5 Binomial Model. You will probably notice that the $n = 5$ graph appears to be much more right skewed than this graph, which seems almost symmetrical. In fact, as we increase the sample size n, the shape of the binomial becomes more like a Normal distribution. Since the Binomial Model is a distribution, we should be able to calculate the mean and standard deviation. The next box shows the mean and the standard deviation of a Binomial Model with specific n and p.

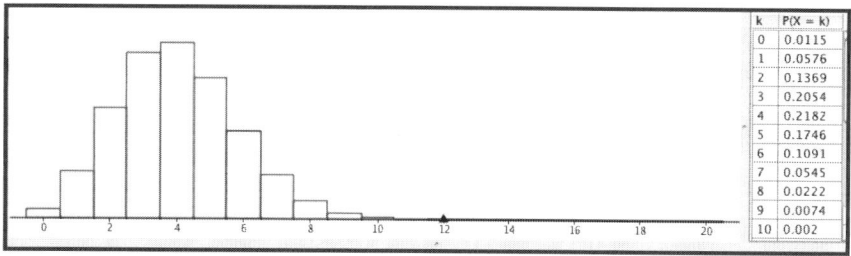

k	P(X = k)
0	0.0115
1	0.0576
2	0.1369
3	0.2054
4	0.2182
5	0.1746
6	0.1091
7	0.0545
8	0.0222
9	0.0074
10	0.002

The calculations become slightly more complicated with a bigger sample size. Look at the bar in the graph for $X = 4$ and also in the table. The table gives $P(X = 4) = 0.2182$, and this is the height of the bar in the graph. The meaning of this number is: "the probability that in a sample of $n = 20$, we get exactly four obese people is 0.2182". The calculation of this number is:

$$P(X=4) = \binom{20}{4} p^4 (1-p)^{20-4}$$

$$= \frac{20!}{4!(20-4)!}(0.20)^4 (0.80)^{20-4}$$

$$\approx \frac{20 \cdot 19 \cdot 18 \cdot 17 \cdot 16!}{4 \cdot 3 \cdot 2 \cdot 1 \cdot 16!}(0.0016)(0.028147)$$

$$\approx 5 \cdot 19 \cdot 3 \cdot 17 \cdot (0.000045036)$$

$$\approx (4845) \cdot (0.000045036)$$

$$\approx 0.2182$$

Calculating the $\binom{20}{4}$ takes a bit of arithmetic; the fraction $\frac{20 \cdot 19 \cdot 18 \cdot 17 \cdot 16!}{4 \cdot 3 \cdot 2 \cdot 1 \cdot 16!}$ reduces to $5 \cdot 19 \cdot 3 \cdot 17 = 4845$ by seeing that the 16! in the numerator cancels with the 16! in the denominator, and 20 divided by 4 is 5, etc. For large values of n, one must resort to a calculator or an online calculator. (On a TI–83 or TI–84, type 20 in the screen then go to **Math>PRB>3** and then type 4 in the screen. Or online: http://www.ohrt.com/odds/binomial.php.)

Mean, Standard Deviation, and Shape of the Binomial Model

A Binomial Model with n and p is a distribution that has mean: np

standard deviation: $\sqrt{np(1-p)}$

and a shape that becomes more nearly Normal as the sample size n increases.

If we apply these formulas to our example, we see that for the distribution with $n = 20$ and $p = 0.20$, we get the mean to be $np = 20 \cdot (0.20) = 4.0$. The mean does *not* necessary have to be a whole number, even though the number of successes can only be whole numbers. The standard deviation is:

$$\sqrt{np(1-p)} = \sqrt{20 \cdot (0.20)(1-0.20)} = \sqrt{20 \cdot (0.20)(0.80)} = \sqrt{3.2} \approx 1.789$$

See the graph above of this distribution and notice that the shape is more "Normal" and less skewed than the distribution for $n = 5$.

Binomial Model with n = 200, and p = 0.20. Let us see what happens when we increase the sample size to $n = 200$ but using the same value for p.

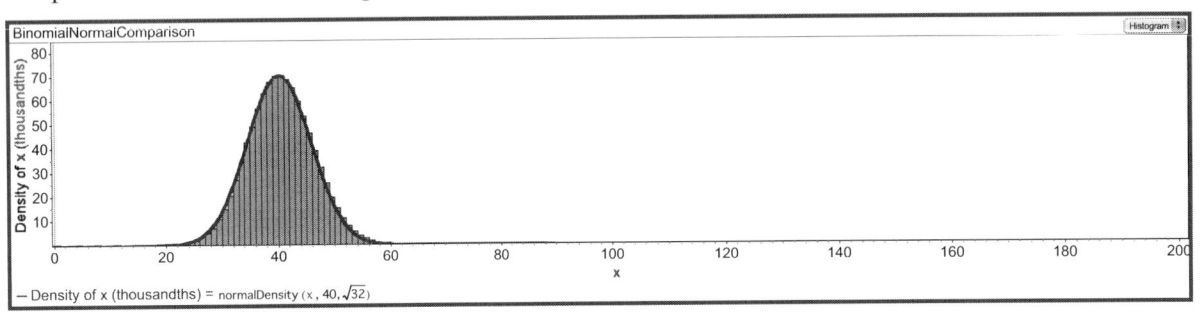

Here is a graphic showing the binomial distribution with $n = 200$ and $p = 0.20$, as the histogram, and the Normal distribution with mean $np = 200 \cdot (0.20) = 40$ and standard deviation

$$\sqrt{np(1-p)} = \sqrt{200 \cdot 0.20(1-0.020)} = \sqrt{32} \approx 5.657$$

Notice the following about the graphic: the binomial and the Normal are very close but not exactly the same. We should be able to use either the binomial or the Normal for calculations. The entire binomial distribution, although it has a Normal shape, is concentrated on the "left side" of the scale of possible outcomes of successes 0, 1, 2, 3,..., 200.

The meaning of the bars (again) and what we can do. The height of each bar represents $P(X = k)$, which is the probability of getting k successes out of the $n = 200$.

- Using our example, $P(X = 30) = 0.0147$ means that the probability is 0.0147 that we will have *exactly* thirty obese people in a random sample of $n = 200$.

- By adding the probabilities (or getting software to do it), we can calculate $P(X \le 30) = 0.043$, or the probability that we get thirty or fewer obese in a random sample of $n = 200$. The shaded-in area in the graphic is the number 0.043, or 4.3%.

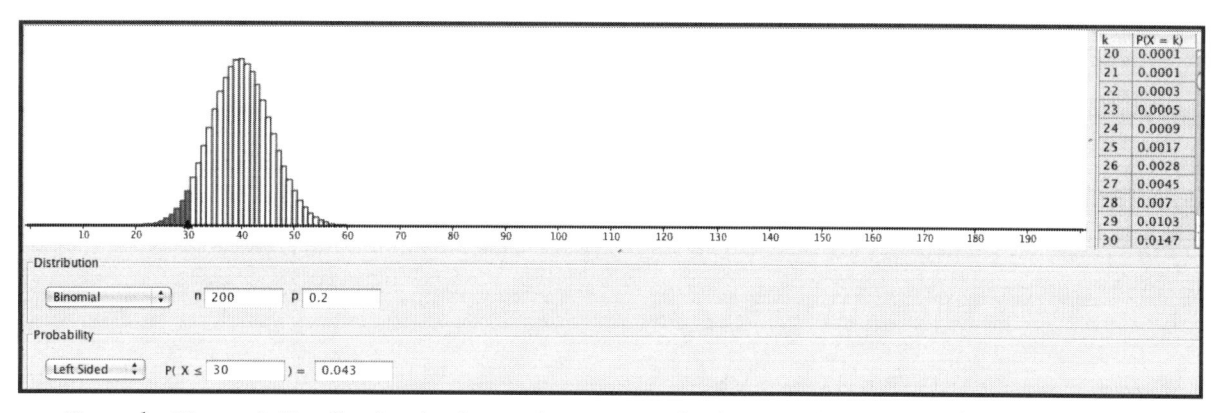

Since the Normal distribution is almost the same as the binomial, it appears that we could get a similar estimate using the Normal distribution, with mean 40 and standard deviation $\sqrt{32} \approx 5.657$. This is a "given a value, find a probability" problem, so we get the z score and consult the Normal Chart:

$$P(X \leq 30) \approx P\left(z \leq \frac{30-40}{5.657}\right) \approx P(z = -1.77) \approx 0.039$$

This is slightly different from the probability calculated from the binomial but close. The reason that it is different is that the binomial distribution only considers whole numbers (such as 29, 30, 32, etc.), but the Normal distribution is "smooth" and considers all the values between, say, 30 and 31 (such as 30.587675..., for example.) There are ways to correct for this difference (which are used in software), but the important thing is that the two results are close.

How accurate can we get? It depends... Introducing proportions and p-hat.

If the proportion obese is $p = 0.20$, and using the Binomial Model, we can see what happens with various sample sizes. If we add all the heights of the bars from $X = 29$ to $X = 51$, we get the probability that our sample has between twenty-nine and fifty-one obese people.

$$P(29 \leq X \leq 51) = P(X = 29) + P(X = 30) + \cdots + P(X = 50) + P(X = 51) = 0.9585$$

What this means is that we have more than a 95% chance of getting between 29 and 51 obese people if we have a sample of $n = 200$. The graph of this calculation is shown just below.

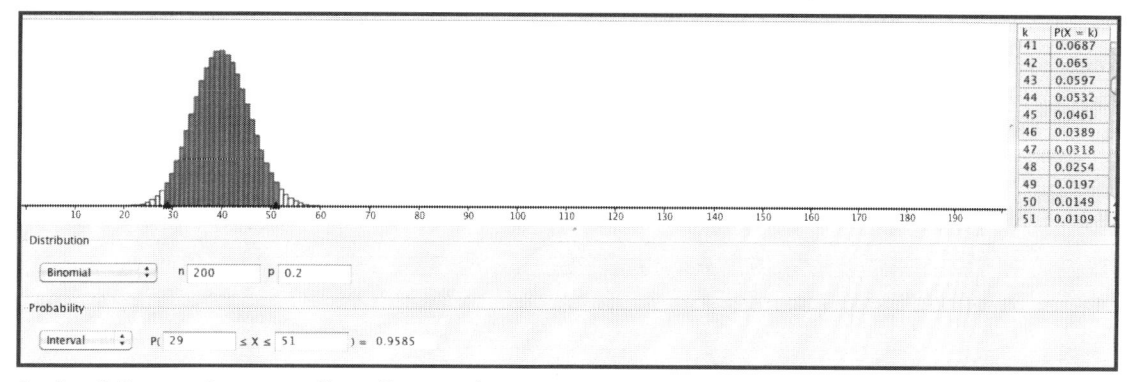

And so? Proportions usually tell more than "counts," so let us calculate proportions and even give them a new name. We will call the proportion $\hat{p}$, pronounced, "p-hat". We use the "hat" to distinguish a proportion calculated from a sample from the proportion for a population. If we have twenty-nine obese people out of two hundred people then $\hat{p} = \dfrac{29}{200} \approx 0.145$, and if we have fifty-one then $\hat{p} = \dfrac{51}{200} \approx 0.255$. (Notice that $\dfrac{1}{200} = 0.005$, and hence the third decimal.)

> **Definition of p-hat and p**
> - For a proportion that refers to—or is calculated from—a sample, we use the symbol $\hat{p}$.
> - For a proportion that refers to a population, we use the symbol p.
>
> Generally, it is not possible to actually calculate a population proportion p, but it deserves to have a symbol.

Our calculation $P(29 \leq X \leq 51) = P(X = 29) + P(X = 30) + \cdots + P(X = 50) + P(X = 51) = 0.9585$ can be written as: $P(0.145 \leq \hat{p} \leq 0.255) = P(\hat{p} = 0.145) + P(\hat{p} = 0.150) + \cdots + P(\hat{p} = 0.250) + P(\hat{p} = 0.255) = 0.9585$

We have a 95% chance of getting a sample proportion $\hat{p}$ between 0.145 and 0.255, if the population proportion is 0.20. Can we do better? Yes, by increasing the sample size. Here is the picture for $n = 1200$. Notice that the Binomial Distribution looks narrower and that $P(213 \leq X \leq 267) = 0.9529$. This translates into greater accuracy, so we can say $P(0.1775 \leq \hat{p} \leq 0.2225) = 0.9529$. That is we are 95% certain of getting a sample that gives us a proportion within 2.25% (0.2225 – 0.2000 and 0.2000 – 0.1775) from the "target" $p = 0.2000$.

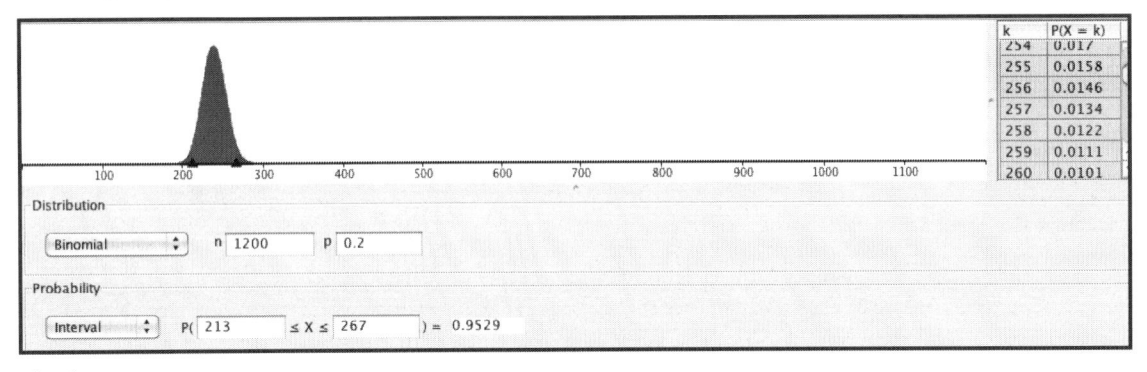

The lesson:
- The Binomial Model can tell us what to expect from our sample. For example, if the sample size is as large as $n = 1200$, it will say that the proportion of "successes" we will see in a sample $\hat{p}$ will be within 0.0225 of the population p.
- The bigger the sample size, the more accurately the sample will represent the population and therefore the more that we can trust our results.

However, notice that in all of this discussion, we never mentioned the size of the population! What we see is that we need a big sample size ($n = 1200$) to be accurate, but we need that sample size whether the size of our population is 100,000 or 1,000,000 or 10,000,000, or 100,000,000. What matters is the sample size. The Binomial Distribution Model has a life of its own as a theoretical (and important) distribution. We have been using it as a **sampling distribution,** which is the story of sections 3.2 and 3.4.

A mathematical joy ride (if you wish)

This sub-section is for those who are a little dissatisfied with just plugging in the numbers and who want to know why it is that the formula gives us what we want. The formula that we are riding for the joy ride is:

$$P(X=k) = \binom{n}{k} p^k (1-p)^{n-k}$$

This formula gives us the probability of getting k successes in n "trials" or cases. How does it manage to do that? It will help to have an example and one that is simple but not too simple. So, suppose that we have samples of size $n = 10$ and $p = 0.20$. In the context of out example of obese people, we are choosing ten people at random in a population in which 20% are obese. Suppose we want the probability that exactly three of these in the sample of ten are obese. That will work out to:

$$P(X=3) = \binom{10}{3} 0.2^3 (1-0.2)^7 = \frac{10!}{3!(7)!}(0.20)^3(0.80)^7 \approx \frac{10 \cdot 9 \cdot 8 \cdot 7!}{3 \cdot 2 \cdot 1(7!)}(0.008)(0.2097152) \approx 120 \cdot 0.0016778 \approx 0.2013$$

We start with: $p^k(1-p)^{n-k} = (0.2)^3(0.8)^7 \approx 0.0016778$. Think of the kind of thing that can be true if we have a sample of $n = 10$ in which there are exactly three successes—that is, three obese people. Think of the three successes as being placed in three positions in a sample worksheet, something like this:

N	O	N	O	O	N	N	N	N	N

What is the probability of this happening? This is an "and" probability and can be written:

$P(N$ and O and N and O and O and N and N and N and N and $N)$

It is here that the independence condition is used. Recall that two events A and B are independent if $P(A|B) = P(A)$. But since $P(A$ and $B) = P(A|B)P(B)$, the formula for an "and" probability has an especially simple form for independent events: $P(A$ and $B) = P(A)P(B)$. Now, since the trials must be independent, we can say that

$P(NONOONNNNN) = (0.8)(0.2)(0.8)(0.2)(0.2)(0.8)(0.8)(0.8)(0.8)(0.8) = (0.2)^3(0.8)^7$.

Also smuggled in here is the requirement that there are just two outcomes and that the probability remains the same. Now think: no matter where the three obese successes go in the ten boxes, the probability of that combination of three successes out of ten will work out the same: $(0.2)^3(0.8)^7$. But how many of these combinations are there when there are just three successes and seven failures? The answer to that question is in the $\binom{10}{3} = 120$, which says that there are 120 different combinations with this same probability. The formula for this, in general, is $\binom{n}{k} = \frac{n!}{k!(n-k)!}$.

Let us see how this works, although in our explanation, the formula will start out looking a bit different. Once again, think of the worksheet:

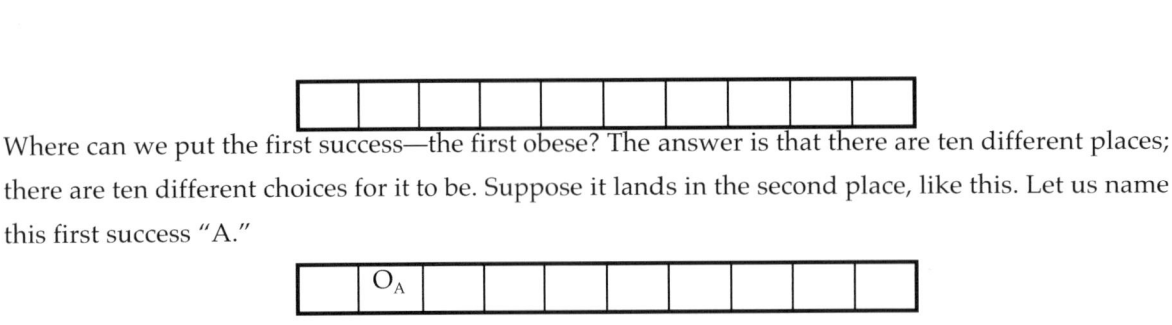

Where can we put the first success—the first obese? The answer is that there are ten different places; there are ten different choices for it to be. Suppose it lands in the second place, like this. Let us name this first success "A."

Now for the second success, there are only nine places left; one has been taken by the first success. Let us say that the second one (whose name happens to be "B") lands in the fourth position, like this:

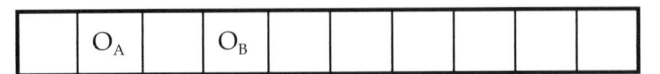

But this is just one possible placement for the two together; there are actually $10 \times 9 = 90$ different possible placements for the two together.. Now for the third (and last) success, there are just eight places left. Let us say the third success ("C") is found in the fifth position:

So, there are $10 \times 9 \times 8 = 720$ different ways we could have done this. However, this is actually too many, and the reason that it is too many is that many of these "landings" are exactly the same combination for us. The same three successes, A and B and C, could have come up like this:

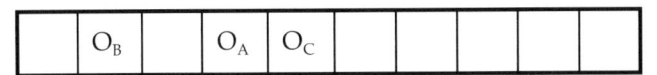

where "B" is the success in the second position and "A" in the fourth position. Or:

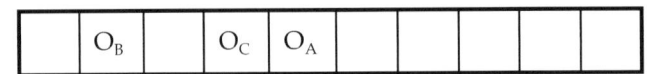

These two choices (the technical name is permutations) are both the same combination as far as we are concerned; having chosen A, B, and C, we are only concerned that they ended up in the second, fourth, and fifth place, not that specifically A got the second place, B got the fourth place, etc. In fact, with 720 we have listed six *times* too many because for the second place there are three of the A, B, and C available, for the fourth place two, and for the fifth one. Hence $\binom{10}{3} = \frac{10 \cdot 9 \cdot 8}{3 \cdot 2 \cdot 1} = \frac{720}{6} = 120$. The formula:

$\binom{n}{k} = \frac{n!}{k!(n-k)!} = \frac{10!}{3!7!} = \frac{10 \cdot 9 \cdot 8 \cdot 7!}{3 \cdot 2 \cdot 1 \cdot 7!} = \frac{720}{6} = 120$ is just a more convenient way of writing the same thing.

You should be able to work out that they are always equivalent.

One last thing: in the Fathom simulations, we sampled *with* replacement. That is, the sampled element was returned once chosen, able to be chosen again, and this was done to preserve independence. In real sampling, we usually sample *without* replacement. As long as the population size is more than ten times the size of the sample, independence is not compromised by using sampling *without* replacement.

Summary for the Binomial Distribution Model

The Basic Ideas The **Binomial Distribution Model** is a mathematical model that can be applied to many scenarios. In this section we have applied it to look at what we can expect from samples chosen randomly from a population. Specifically,

- If we choose samples of size n and look at a categorical variable with just two values or outcomes,
- Then the **Binomial Distribution Model** gives us the probabilities of getting 0, 1, 2,…, up to n "successes" in the sample,
- Where a "success" is one of the two values of the categorical variable.

Conditions for the Binomial Distribution Model We can fit a Binomial Model if *all* of the following conditions are met:

B The outcomes are binomial; that is, there are just *two* outcomes. Often, the outcomes are called "success" and "failure."

I Each case is independent of the others. (The cases are often also called "trials.")

N There is a fixed total number, n, of cases (or "trials").

E We use a probability p of success that is the same ("equal") for each case, or "trial."

Formula for the Binomial Model

To calculate the probability of $X = k$ successes out of n cases (or "trials"), use:

$$P(X=k) = \binom{n}{k} p^k (1-p)^{n-k} \text{ where } \binom{n}{k} = \frac{n!}{k!(n-k)!}$$

and where $n! = n \cdot (n-1) \cdot (n-2) \cdots 2 \cdot 1$ and also $0! = 1$ by definition.

$\binom{n}{k}$ tells us the number of ways of placing k things into n boxes (also written $_nC_k$) and is "pronounced" "n choose k" or "the number of combinations of k things amongst n."

n! is pronounced "n factorial" and means $n! = n \cdot (n-1) \cdot (n-2) \cdots 2 \cdot 1$ and $0! = 1$

Mean, Standard Deviation, and Shape of the Binomial Model

The **Binomial Distribution Model** with n and p is a distribution that has mean: np

standard deviation: $\sqrt{np(1-p)}$

and a shape that becomes more nearly Normal as the sample size n increases.

Looking Ahead: Definition of p-hat and p so that we can analyze proportions

- For a proportion that refers to—or is calculated from—a sample, we use the symbol $\hat{p}$.
- For a proportion that refers to a population, we use the symbol p.

§3.3 Exercises for Binomial Sampling Distributions

1. **Studying Obesity: Samples of Five.** To study obesity in a large population of people with a sample of only $n = 5$ does not make sense. However, such a small sample size *is* useful for learning about binomial distributions and how they work. We will graduate to bigger sample sizes.

   From studies by the Centers for Disease Control (CDC), the evidence is that approximately 20% of the adult population is obese. We will give this *population proportion* the symbol p, so $p = 0.20$.

   a. If we draw a random sample of $n = 5$, is it possible that we will get five obese people in our sample? Do you think it is likely? Why?

   b. If we have n = 5 in our sample, we could get zero, one, two, three, four, or five obese people in our sample. What number do you think is the most likely? Why?

   An imaginary population. We will imagine that we have a population of one million people. Now, imagine that we make the obese people—all 20 percent of the one million, or 200,000 of them—stand on our left so that we can enumerate them in order to take a SRS from the population. The very first obese person gets the number 000,000, the second one gets the number 000,001, and so on, until the last one, whose number is 199,999. On our right we put all the 800,000 non-obese people, and we number them, starting with number 200,000, 200,001, and so on, until the last one, whose number is 999,999. Now, using just one ten-sided die, we will choose five people and record whether each of the five is obese or not obese. Follow the directions.

   - Roll the ten-sided die; if it comes up "0" or "1," it means that we have chosen (randomly) one of the obese people (because their "serial numbers" start with 0 or 1), and an "O" should be recorded in the worksheet provided for the first sample. If the die comes up with a number 2 or more (serial numbers 200,000 plus), indicate "N" in the box indicating a "not obese" person.
   - Roll the die four more times to get the other four people in the random sample and record the number of obese people in your sample in the box provided on the worksheet.
   - Repeat the process three more times so that in all you get four samples, each of $n = 5$. Each time, record the number of obese people in your sample.
   - When you are finished, record the numbers of obese people in each of the four different samples either in the Fathom file in the classroom or in the Fathom survey online.

   c. The results from this exercise will be combined. What proportion of the samples of $n = 5$ had just one obese person? What proportion of samples had no obese people? What proportion of samples had five?

   Closer to reality: sampling from a random sample. We actually have a SRS with the data we want. It comes from the CDC and is part of the NHANES survey (http://www.cdc.gov/nchs/nhanes.htm), and we will use Fathom to randomly select very small samples ($n = 5$) so as to see what happens when we randomly chose just five people. Follow these directions:

   - Open the file **NHANESAge1635.ftm** and get a case table.
   - Get a **Summary Table** and drag the variable Obese to it.

nhanes.csv		
Obese	Not Obese	5783
	Obese	1340
	Column Summary	7123
S1 = count ( )		

d. **Getting bearings:** In these NHANES data we are using:
   i) What are the cases?
   ii) Is the variable *Obese* categorical or quantitative?
   iii) Are the data from an experiment or observational study?
   iv) What appears to be the population?
e. Calculate the proportion Obese from the numbers in the table. For this exercise, Fathom will consider this to be the *population proportion p*. Notice that it is close but not the same as the 0.20 we used in the first part of the exercise or in the **Notes**. The Binomial Model is a model.

- Put the case table and the Summary Table on the left-hand side of the screen.
- Select the collection icon (so that its border turns blue) and then go to the menu **Collection>Sample Cases**. You should see another collection icon (with Inspector) with the name **Sample of NHANESAge1635.ftm**.
- In the Inspector, *deselect* "Animation On" and change the "10" in the cases to the number "5." Leave the other boxes checked and click on "Sample More Cases."
- Select the **Sample of NHANESAge1635.ftm** icon and get a case table (to see that you have just five cases) and also get a **Summary Table** showing the number of obese and not obese people in your sample.
- Now we will command Fathom to repeat what we have done one thousand times. In the Inspector, select "Measures" (it may appear as "Meas…") and replace <new> with the variable name **xObese**. 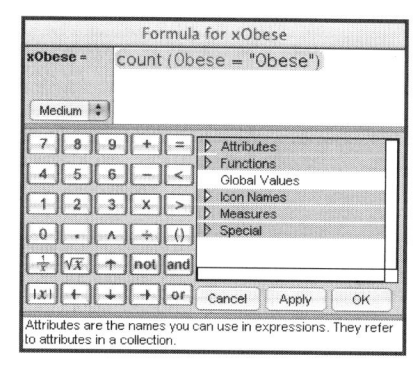
- Make the "Measure" row blue; double click on it so that you get a dialogue box like the one shown here, except that it will not have the formula part filled in. Fill it in with count(Obese = "Obese") as shown here. The number of obese in your sample should show.
- Select **Sample of NHANESAge1635.ftm** and go to menu **Collection>Collect Measures**. (You should see a new collection entitled *Measures of Sample of NHANESAge1635.ftm* and also an Inspector.)
- Select the icon of *Measures of Sample of NHANESAge1635.ftm* and get a case table and a dot plot of the variable *xObese*. Also, get a **Summary Table** and drag—with shift key depressed, so as to treat xObese as categorical—to the down arrow of the **Summary Table**.
- In the Inspector of **Measures of Sample of NHANESAge1635.ftm** deselect "Animation On" and change "5 measures to "995" and make the Inspector bigger until you see "Collect more measures" and click on it. Your screen should resemble this screen, with similar but different numbers.

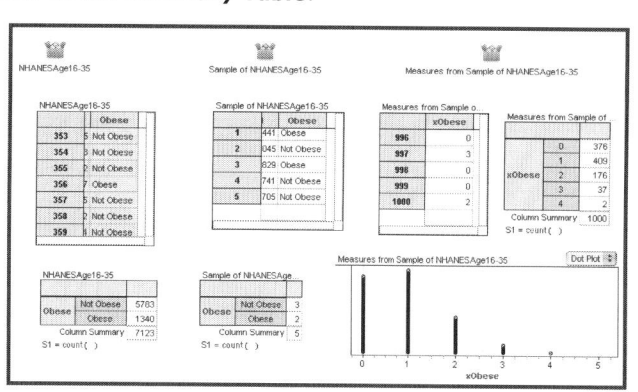

f. Think: In the dot plot that you got (like the one shown in the picture here), what does each *dot* in the dot plot represent? Choose the most accurate answer and give a reason for your choice.
   i) An obese person
   ii) A sample of size $n = 5$
   iii) A number of obese people in a sample of size $n = 5$
g. If the proportion of obese people in the entire population is $p = 0.20$, the number we would most expect to see in our very small sample of $n = 5$ is 1, the average, since $np = 5 \cdot 0.20 = 1$. What is the probability that getting a "1" from a random sample, according to your Fathom simulation work?

Conclusion: Small samples can give misleading results. Go back to the Notes and see how this exercise leads to further understanding of a model to predict what happens when we sample.

2. **Understanding the calculations: left-handed students in an ancient history class** In unit 4 of the **Notes**, we will look at the proportion of left-handed people in the general population and see that the proportion is about 12%, although there is much variation among cultures. This exercise focuses on the conditions and the calculations for the binomial distribution. Refer to the sections in the **Notes** where the conditions and the calculations are explained.

   An imaginary college and an imaginary course. In your college (say), all students are required to take a one-term course on Ancient Etruscan history (or AEH). (This is because the benefactor of the college's building program considers himself to be descended from the Etruscans.) Sections of the course are offered every hour from 7.00 am to 7.00 pm and are capped at a maximum of twenty-four students, and every section is full, since it is a college requirement. Our goal is to calculate the probabilities of 0, 1, 2, 3, 4, . . . , 24 left-handed students in a single section of twenty-four students, using the Binomial Model. We will make the assumption that the population proportion of left-handed is 0.12.

   a. For the scenario about left-handers in the section of AEH, identify the values for the symbols n and p.
   b. Read over the conditions **B, I, N, E** as well as the scenario presented here, and explain how each of the conditions for the binomial distribution are met if you think that they are met. If you think one or more of the conditions are problematical, clearly explain why you believe it is so.
   c. Random sampling or random allocation to the classes is not mentioned above. In the example in the Notes, random sampling is important because it guarantees the independence condition. However, independence of cases (or "trials") is also assured if students sign up for classes essentially randomly with respect to their handedness. From your experience as a student, do you think this is true?
   d. The meeting of "Left-Handers Action Front" (LHAF) has its weekly meetings at 10.00 am. Does this affect the conditions? If so, explain how; if not, explain why not.

- Use the **DistributionCalculator.ggb** to get the binomial distribution for this scenario. It should look like the graph and table here.
   e. In the histogram, what is the *bin width*?
   f. In the graph, do the heights of the bars represent: i) the number of left-handed students, or ii) the probability that a class will have a specific number of left-handed students?

g. Use the *table* and the correct notation to calculate $P(X<3)$ by adding probabilities.

h. Make a sketch of the graph, and indicate $P(X<3)$ by shading the graph. (You can check your answers to parts g and this question by using the *DistributionCalculator.ggb.*)

k	P(X = k)
0	0.0465
1	0.1522
2	0.2387
3	0.2387
4	0.1709
5	0.0932
6	0.0403
7	0.0141
8	0.0041
9	0.001
10	0.0002
11	0
12	0
13	0

Distribution: Binomial, n 24, p 0.12
Probability: Right Sided, P(15 ≤ X) = 0

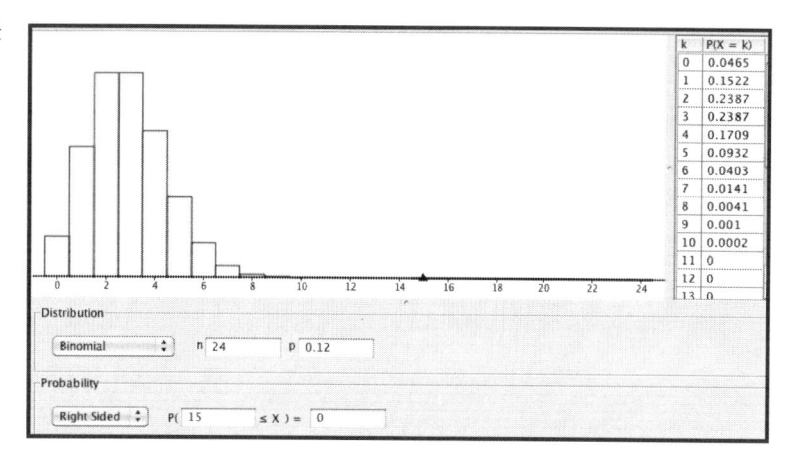

i. Give a meaning to $P(X<3)$ in terms of the number of left-handed students in one of the sections of AEH.

j. From the *table*, calculate the probability that a section will have six or more left-handed students, (by adding probabilities) and use the correct notation to display the result. (You can check the result of your calculations using *DistributionCalculator.ggb.*)

k. Make a sketch of the graph and show by shading the answer to part h. (If you checked your answer using *DistributionCalculator.ggb*, the shading should also show.)

l. Asked to calculate $P(X=2)$, Confused Conrad calculates $\binom{24}{2} = 276$. What is CC's confusion and how should he immediately know that 276 is not the answer to $P(X=2)$?

m. Using the information in part l, correctly use the formula to get $P(X=2)=0.2387$.

n. Do some simple arithmetic to confirm that $\binom{24}{2} = 276$. (See the example in the *Notes*.)

A binomial model for left-handers in seating arrangements.

o. Because of the generosity of the benefactor, all of the classrooms have tables with four chairs as pictured here. We are now interested in calculating the probabilities of zero, one, two, three, or four left-handers at a single table. For what we are doing now, what are the values of the symbols n and p?

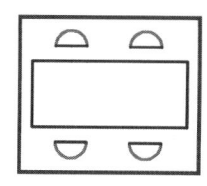

p. Explain how each of the conditions **B, I, N, E** are met if you determine they are met. If you have doubts about whether one of the conditions is met, explain which one and why.

q. What would happen if left-handers had a tendency to seek out their fellow "lefties" and tried to join a table where they saw another left-hander was sitting? Which condition or conditions would be affected?

r. By hand, calculate $P(X=k)$ for $k=0$, $k=1$, $k=2$, $k=3$ and $k=4$ using the formula for binomial probabilities. Show your work. Here are the answers you should get.

k	P(X = k)
0	0.5997
1	0.3271
2	0.0669
3	0.0061
4	0.0002

s. You should have found in the calculation of $P(X=2)$ that

$\binom{4}{2} = \frac{4!}{2!2!} = \frac{4 \cdot 3 \cdot 2!}{2 \cdot 1 \cdot 2!} = 6$ so there are six different placements of

the two left-handers at the four chairs at the table. Label the four chairs A, B, C, D and show the six different placements. Here is a start.

	A	B	C	D
1	L	L	R	R
2				
3				
4				
5				
6				

3. **Conditions: Is the Binomial Model applicable?** For each of the following scenarios, determine whether each of the conditions **B, I, N, E** are met. If they are, state why; if not, state why. You may well have doubts about some of the conditions and yet think others are fully met. You may also suggest ways to change the scenario so that the conditions are met. For each scenario that meets the conditions, identify the *n* and the *p*.

    a. In classes of thirty-six students, the students are asked whether they have a tattoo or not. We want to calculate the probability that we will see 0, 1, 2, . . ., 36 having a tattoo if we think that 20% of students have tattoos.

    b. The same classes of thirty-six students are also asked what kind of mobile telephone they have. (The choices are iPhone, Blackberry, Android, Other Smartphone, or no Smartphone; the proportions are about 0.50. 0.15, 0.20 and 0.05 and 0.10.) We want the probability of each type of device in a class of thirty-six.

    c. There is a law in California that prohibits talking on a mobile telephone. Data were collected by watching drivers as they came to a stop sign to see whether or not the driver was using a mobile 'phone. (Other variables were observed as well.) It is thought that about 5% of drivers talk or text while driving, and we are interested in modeling the number found talking or texting in an observation period of one hour.

    d. There is a random sample of mothers, *all* of whom have given birth to six children. Looking at all of their births, we are interested in knowing the probabilities of 0, 1, 2, 3, 4, 5, or 6 female children. (Additional question for those in medicine: is the **I** condition met?)

    e. For male college students in California, the mean height is about 178 centimeters and the standard deviation about eight centimeters. We want to model the probability that a male college student is shorter than 168 centimeters.

4. **Understanding the meaning: A bigger sample size, n = 120.** We have already concluded that samples of size $n = 5$ are far too small to estimate the proportion obese in a population. Here we will explore what happens if we randomly sample and get a larger sample size.

    - Use the **DistributionCalculator.ggb** to get the binomial distribution with for $n = 120$ and $p = 0.20$. Your picture should look like this one, but you must have the application open to answer some of the questions. First, let us choose just one of the bars (or bins), the bar for the number 18.

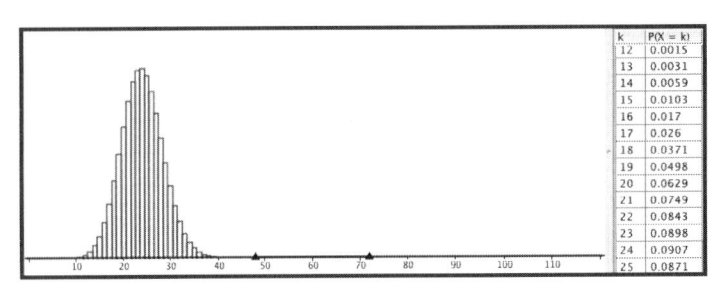

k	P(X = k)
12	0.0015
13	0.0031
14	0.0059
15	0.0103
16	0.017
17	0.026
18	0.0371
19	0.0498
20	0.0629
21	0.0749
22	0.0843
23	0.0898
24	0.0907
25	0.0871

a. What is a good name for the shape of this distribution?
b. What is the *best* answer for the meaning of this bar in the histogram? Choose one of the responses and give a reason for your answer.
   i) It shows that there are eighteen obese people in one sample of 120.
   ii) The bar shows the probability that in a sample of 120, we have at most eighteen obese people.
   iii) The height of the bar shows the probability that in sample of 120, we get exactly eighteen obese people.
c. [*Review*] By checking the **DistributionCalculator.ggb,** you will see that $P(X=18)=0.0371$. Show how 0.0371 was calculated by inserting the values of n and p in the proper places in the formula for the Binomial Model. Do not carry out the calculation at this point but show where the numbers go.
d. [*Review*] Part of the formula involves $\binom{120}{18} = \frac{120!}{18!(120-18)!} = 1,086,744,939,880,326,302,940$.

   The best answer for the meaning of this number in the formula for the probability in the Binomial Model is: (Choose one response and give a reason for your answer.)
   i) The number of different ways of placing eighteen obese people in a sample of 120 (Is one of them the second person of the 120 sampled, for example?)
   ii) The number of samples we would have to draw to get exactly eighteen obese people in a sample of 120
   iii) The probability of getting exactly eighteen obese people in a sample of 120
   iv) The number of eighteen-year-old obese people in the world
d. [*Review*] Question "b (ii)" above referred to "the probability that in a sample of 120, we have at most eighteen obese people." Find the probability that there are at most eighteen obese out of 120. Use the correct notation.
e. The Binomial Model gives us the probabilities of getting the various possible outcomes of the number of successes. In our example, these are the probabilities of getting 0, 1, 2, 3,..., up to 120 obese people in a sample of $n = 120$ if $p = 0.20$. According to the table in the **DistributionCalculator.ggb** what numbers of obese people in a sample of $n = 120$ have we a *zero* chance of seeing (at least with probabilities rounded to four decimal places)?
f. If someone said that in a sample of $n = 120$, there are fifty obese people, decide which of the following could be true and why, judging from our Binomial Model.
   i) The sample was not random. (Perhaps a voluntary response sample?)
   ii) The proportion of obese people in the population is bigger than $p = 0.20$.
   iii) The sample is very unusual. (Change the "rounding" in the **DistributionCalculator.ggb,** to ten decimal places.)
g. Use the table from the **DistributionCalculator.ggb** to show how (and to confirm that) $P(X \leq 15) = 0.0218$. (You will have to add some probabilities, but not fifteen of them.) Make a sketch of the graph of the binomial distribution to show the value 0.0218 by shading the graph. Indicate clearly what feature of the graph shows the 0.0218.

h. Now, go to the other side of the distribution and add probabilities so to confirm that $P(X \geq 33) = 0.0296$. On the same sketch you made for part d, show the value 0.0296 by shading the graph.

i. From the results of parts g and h, find the probability of getting a sample of 120 in which the number of obese people is between sixteen and thirty-two (including the end points). That is, find $P(16 \leq X \leq 32)$. (You may check your answer with the **DistributionCalculator.ggb** but show your calculations.)

j. On a sketch of the binomial distribution, show the answer to part "i" by shading.

k. The answer that you should have gotten for part "i" above is $P(16 \leq X \leq 32) = 0.9486$. The meaning of this probability is important. Choose the best answer (thinking about each) and give a reason for your choice.

   i) In a single sample of 120, we are certain to see between sixteen and thirty-two obese people.

   ii) The probability 0.9486 means that there is about a 95% chance that between sixteen and thirty-two people are obese out of every 120 in the population.

   iii) If the probability of being obese is 0.20 then the probability that we get a random sample of n = 120 with between sixteen and thirty-two obese people is 0.9486 (almost 95%).

l. For this binomial distribution with $n = 120$, and $p = 0.20$, find the mean and the standard deviation using the formulas given in the **Notes**. (There is a box for the mean and the standard deviation.)

m. [*Review*] Using the answers to part "l," make a sketch of a Normal distribution showing with mean 24 and standard deviation 4.382, showing the values for the mean ±1, ±2 and ±3 sd.

n. [*Review*] Using the Normal distribution with the mean and standard deviation from part "l," find the probability that we will see fifteen or fewer obese people. That is, calculate $P(X \leq 15)$.

o. [*Review*] Using the Normal distribution with the mean and standard deviation from part "l," find the probability that we will see thirty-three or more obese people. That is, calculate $P(X \geq 33)$.

p. [*Review*] Calculate, using the Normal distribution, $P(16 \leq X \leq 32)$.

q. Compare your answers for $P(X \leq 15)$, $P(X \geq 33)$ and $P(16 \leq X \leq 32)$ using the Normal distribution and the answers using the binomial distribution. Are they the same? Close but not the same? Show big differences?

You should find that the answers are close but not exactly the same. One reason for this is that the binomial distribution with p = 0.20 is still slightly right skewed compared with the Normal distribution with mean at 24, even though they have the same mean. Another reason is that the binomial distribution only takes on the whole numbers (16, 17, . . .31, 32, etc.) whereas the Normal distribution takes on any number—including 16.247, 31.678. Some people adjust the fit by using (in our example) 15.5, and 32.5 instead of 16 and 33.

5. **The shape of the Binomial Distribution Model.** Open the **DistributionCalculator.ggb** and try out different values of n—small and big—and values of $0 \le p \le 1$. Record what values you tried.

   a. Under what conditions (values of n and p) is the binomial distribution: i) right-skewed, ii) symmetric in shape, and iii) left-skewed?

   b. Under what conditions does the shape look most Normal?

6. **Preterm births and their incidence.** A premature or preterm birth is one that occurs before thirty-seven weeks of pregnancy. What proportion of all births is premature? There is quite a bit of variation among countries and regions of the world and even within North America. (See http://preemiehelp.com/about-preemies/preemie-facts-a-figures/general-preemie-statistics/incidence-of-preterm-birth-by-country and http://www.statehealthfacts.org/comparemaptable.jsp?ind=39&cat=2). For this exercise, we will use a guess for the population proportion of $p = 0.10$, or 10%.

   Samples of n = 10 "by hand" In the first part of this exercise, we will randomly choose samples "by hand"—that is, using a ten-sided die. To simulate drawing a simple random sample, imagine numbering our population of ten thousand births like this: we will gather together all of the premature births and give them serial numbers 0000 (the very first preterm birth), the second one 0001 up to 0999 (the vary last preterm birth), making one thousand preterm births, or 10% of the ten thousand births in all. Then the first full-term birth (not premature) is numbered 1000, the second one 1001, and on up to 9999. So we have a numbered list. So all the preterm births have serial numbers beginning with 0, and all the full-term births have serial numbers beginning with 1, 2,...up to 9. Number 0863 is a premature birth, but 7295 is a full-term birth.

   - Roll the ten-sided die; if "0" appears we have randomly chosen one of the preterm births (because their "serial numbers" start with 0), and a "P" should be recorded in the worksheet for the first sample. If the die shows 1, 2, 3,... (serial numbers 1000 plus), put "F" in the box indicating a full-term birth.
   - Roll the die nine more times to get the other nine births in the first random sample and record the number of preterm births in your sample in the box provided on the worksheet.
   - Repeat the process three more times so that in all you get four samples, each of $n = 10$. Each time record the number of preterm births in each sample.
   - When you are finished record the numbers of preterm births in each of the four different samples either in the Fathom file in the classroom or in the Fathom survey online.

     a. In this part of the exercise, the ten thousand births are "imaginary" (later we will look at real births). However, what we have imagined makes the sampling process a simple random sample. What about makes the process a simple random sample?

     b. Why is the sampling process not a voluntary response sample?

     c. The results from this exercise will be combined. Questions to answer: what proportion of the samples of $n = 10$ have just one preterm birth? What proportion of the samples of $n = 10$ have no preterm births?

   Sampling from real births. We have a collection of ten thousand actual births that we will treat as a population. The "population" is actually a random sample of all births in the USA in 2006.

- Open the file **DixMilleBirths4.ftm** and get a case table.
- Get a **Summary Table** and drag the variable *Premature* to it.
    d. **Getting bearings:** In these birth data:
        i) What are the cases?
        ii) Is the variable *Premature* categorical or quantitative?
        iii) Are the data from an experiment or observational study?
        iv) What appears to be the population?
    e. Calculate the proportion "Premature" from the numbers in the table. For this part of the exercise, Fathom will consider this to be the *population proportion p* even though it is calculated. Notice that it is close but not the same as the 0.10 we used in the first part of the exercise.
- Put the case table and the **Summary Table** on the left-hand side of the screen.
- Select the collection icon (so that its border turns blue) and then go to the menu **Collection>Sample Cases**. You should see another collection icon (with Inspector) with the name **Sample of DixMilleBirths4.ftm.**
- In the Inspector, *deselect* "Animation On." Leave the other boxes checked and click on "Sample More Cases."
- Select the **Sample of DixMilleBirths4.ftm** icon and get a case table (to see that you have just ten cases) and also a **Summary Table** showing the number of Premature and Full-term births in your sample.
- Now we will command Fathom to repeat what we have done one thousand times. In the Inspector, select "Measures" (it may appear as "Meas...") and replace <new> with the variable name **xPremature**.
- Color in the "Measure" row; double click on it so that you get a dialogue box like the one shown here, except that it will not have the formula part filled in. Fill it in with count(Premature = "Premature") as shown here. The number of premature births in your sample should show.
- Select **Sample of DixMilleBirths4.ftm** and go to menu **Collection>Collect Measures**. (You should see a new collection entitled **Measures of Sample of DixMilleBirths4.ftm** and also an Inspector.)
- Select the icon of **Measures of Sample of DixMilleBirths4.ftm** and get a case table and a dot plot of the variable *xPremature*. Also, get a **Summary Table** and drag—with the shift key depressed, so as to treat *xPremature* as categorical—to the down arrow of the **Summary Table**.
- In the Inspector of **Measures of Sample of DixMilleBirths4.ftm,** *deselect* "Animation On" and change "5 measures to "995" and make the Inspector bigger until you see "Collect more measures" and click on it. Your screen should resemble the screen shown on the next page, with similar but different numbers.

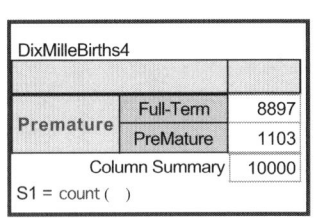

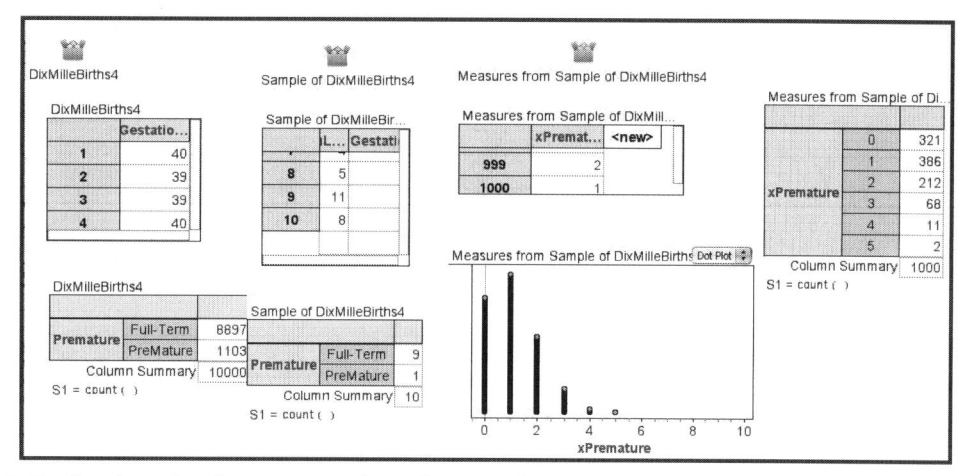

f. **Think:** In the dot plot that you got (like the one shown in the picture here), what does each *dot* in represent? Choose the *most* accurate answer and give a reason for your choice.
   i) A premature birth
   ii) A sample of size $n = 10$
   iii) A number between 0 and 10 of premature births in a sample of size $n = 10$

g. If the proportion of premature births in the entire population is $p = 0.10$, the number we would most like to see in our very small sample of $n = 10$ is 1, since $10 \times 0.10 = 1$. What is the probability that getting a "1" from a random sample, according to your Fathom simulation work? Use the correct notation.

h. Using the Fathom simulation results, calculate the probability of seeing less than or equal to one premature birth in a sample of $n = 10$.

7. **Preterm births and their incidence, continued.** (Note: It is helpful to have done Exercise 6 first. This exercise develops what was done there.) A premature or preterm birth is one that occurs before thirty-seven weeks of pregnancy. What proportion of all births is premature? There is quite a bit of variation among countries and regions of the world and even within North America. (See http://preemiehelp.com/about-preemies/preemie-facts-a-figures/general-preemie-statistics/incidence-of-preterm-birth-by-country and http://www.statehealthfacts.org/comparemaptable.jsp?ind=39&cat=2.) For this exercise, we will use a guess of $p = 0.10$ or 10% for the population proportion of premature births.

- Open the **DistributionCalculator.ggb** to get this graphic of the Binomial Distribution Model for $n = 10$ and $p = 0.10$.

   a. Confirm that the conditions **B, I, N, E** are met for the context of getting random samples of births and noting the number of premature births.

   b. Use the table labeled $P(X = k)$ to determine the probability that in a sample of $n = 10$, we get (by random sampling) just one premature birth. Use the correct notation.

245

c. (Answer this question only if you also worked Exercise 6.) Compare the probability from the Binomial Distribution Model (the answer to part "a") with the probability that you got from the Fathom simulation. Are they similar? [They should differ slightly, for two reasons: first, what Fathom did is a simulation (like the work you did with the ten-sided die), and so it does not reflect the Binomial Model exactly. Secondly, Fathom used the proportion $p = 0.1103$ rather than $p = 0.10$ in the simulation.]

d. Use the formula $P(X=k) = \binom{n}{k} p^k (1-p)^{n-k}$ to confirm that for $n = 10$ and $p = 0.10$, $P(X=1) = 0.3874$.

e. To calculate $P(X=2)$ for $n = 10$ and $p = 0.10$, Confused Conrad gives the answer $\binom{10}{2} = \frac{10 \cdot 9 \cdot 8!}{2 \cdot 1 \cdot 8!} = 45$. In what way is CC correct and in what way is he wrong?

f. Use the table from the **DistributionCalculator.ggb** to get this above to find the probability that our sample shows more than one premature birth. Use the correct notation. (You may use the calculator part of **DistributionCalculator.ggb** to check your answer.)

g. **Drawbacks of small samples** Even though our sample is random, its smallness can be misleading for the purpose of using the sample to estimate the population proportion p of premature births. Choose the answer or answers that give good reasons that small samples can be misleading.

   i) The sample is a convenience sample.

   ii) If the population proportion of premature births is actually $p = 0.12$, we would not be able to detect this using $n = 10$, since if we get $X = 1$, our estimate for the population proportion is 0.10, and if $X = 2$, our estimate for the population proportion is 0.20, and so on.

   iii) The Binomial Model tells us that even with random sampling, there is a probability of 0.2639 that the sample we draw may lead us to estimate the population proportion of premature births to be 0.20 or greater.

- **Getting a larger sample size** Let us see what happens if we increase the sample size to $n = 100$. We can explore in two ways. We can simulate using Fathom (as in Exercise 6) and we can get the Binomial Distribution Model with $n = 100$ and $p = 100$. We will do both. [To get your own simulation (rather than the one shown here just below, follow the directions in the bulleted parts of Exercise 6, except that in the second bullet, you must change the sample size from $n = 10$ to $n = 100$. To answer the questions, you can use your own simulation, or the one here. Remember that in the Fathom simulation, the population proportion is 0.1103.]

h. From the simulation shown (on the next page) $P(X=11) = \frac{133}{1000} = 0.133$, since there were one thousand different samples of sample size $n = 100$. In the context of getting random samples of $n = 100$ births, give a meaning to $P(X=11) = 0.133$.

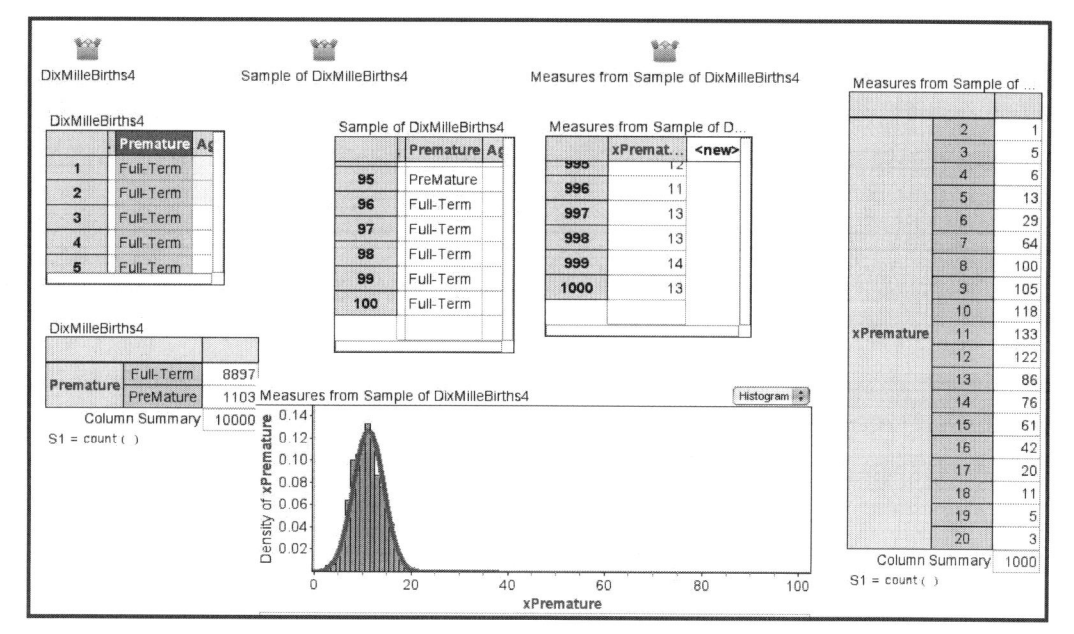

i. Change the *n* to 100 in the ***DistributionCalculator.ggb*** to get the binomial distribution shown here. (Notice the similarities to our simulation.) Use the ***DistributionCalculator.ggb*** to confirm that $P(X \geq 15) = 0.0726$. Interpret this probability in the context of selecting samples of births and recording the number of premature births.

j. Calculate the proportion of premature births implied by fifteen premature births out of one hundred. Since this proportion comes from a sample (here hypothetical), we give it the symbol $\hat{p}$, "p-hat."

k. The answer, of course, is $\hat{p} = 0.15$ or 15%. Choose the best interpretation of $P(\hat{p} \geq 0.15) = 0.0726$ and give a reason for your choice.

　　i) The proportion of premature births in the population from which the random sample of one hundred was drawn is 0.15.

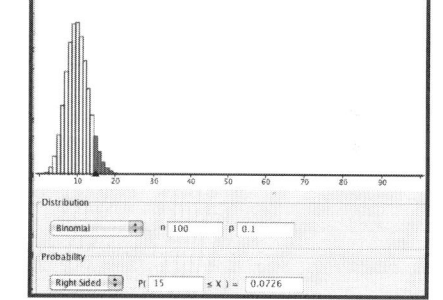

　　ii) Over 15% of the random samples of size $n = 100$ will have about 7.26% premature births in them.

　　iii) About 7.26% of the random samples of size $n = 100$ will have 15% or more premature births.

l. Calculate the mean and standard deviation of the Binomial Distribution Model with $n = 100$ and $p = 100$ using the formulas np and $\sqrt{np(1-p)}$. [In the next section, we will use the Normal distribution to calculate probabilities such as $P(X \geq 15)$.]

8. **A weight-loss experiment.** There have been comparatively few experimental studies of the various popular diet plans that have employed randomization of subjects to various diets. One that did was a study by Michael Dansinger, et al, whose results were published in the *Journal of the American Medical Association, 293:1, January, 2005,* and entitled *Comparison of the Atkins, Ornish, Weight Watchers, and Zone Diets for Weight Loss and Heart Disease Risk Reduction.* You may also find it at http://jama.jamanetwork.com/article.aspx?volume=293&issue=1&page=43. The design of the experiment was: there were 160 subjects (near Boston) who were randomly allocated the subjects to the four treatments, so that each treatment would have $n = 40$. The four diets, namely, Atkins, Ornish, Weight Watchers, and Zone, were the four treatments. They measured weight loss, persistence in the diets, and also changes in cholesterol levels.

   a. [*Review*] For this experiment, what is a good name for the factor or factors? (Choose all correct answers.)
      i) Dieters
      ii) Diets
      iii) Weight loss
      iv) Cholesterol levels

   b. [*Review*] What are the experimental units? (Choose all correct answers.)
      i) Dieters
      ii) Diets
      iii) Weight loss
      iv) Cholesterol levels

   c. [*Review*] Select the response variable or variables. (Choose all correct answers.)
      i) Dieters
      ii) Diets
      iii) Weight loss
      iv) Cholesterol levels

Part of experimental design is keeping everything in the treatment groups the same except for the differences in the factor. In this situation, it would not be good if one of the treatment groups was predominantly male and another predominantly female. Randomization of subjects to the groups does not guarantee that the groups each had exactly the same number of females, but the probability of big differences is minimized by randomization. As it happened, here is the number of females in each of the treatments:

Atkins: 21   Ornish: 17   Weight Watchers: 23   Zone: 20

§3.4 Trusting Data, Part 4: *Sampling distributions for proportions*

Introduction: Quantitative and categorical variables

Sections 3.2 and 3.3 introduced the idea of a sampling distribution. Our example in §3.2 was about the average area of a house, measured by the quantitative variable *Sq_Ft*. Since *Sq_Ft* is a quantitative variable, we used the sample mean $\bar{x}$ and we developed the idea of the sampling distribution of the sample mean.

In §3.3, we looked at a categorical variable—whether people are obese or not. There we measured the *number X* of obese people in a sample and saw that the Binomial Model served well to predict the number of "successes" in a sample with a given size. Then, toward the end of that section, we transitioned to thinking about proportions, which are simply the number of successes X, divided by the sample size *n*. We also saw that the Binomial Model started to look very much like the Normal distribution under some circumstances. The story was: the sampling distribution for proportions is essentially binomial, but sometimes the binomial looks a lot like a Normal distribution.

In this section, we look at the Normal sampling distribution as it is used in actual statistical work.

Example: More San Mateo Real Estate

When the San Mateo real estate data were first collected for 2005–2006, property values were high; there was a boom in the housing market. When there is a real estate boom, sellers sometimes find that they actually get more for their house than the listed price; the buyers actually compete with other buyers for the house. Here are data from a random sample of $n = 100$ of the real estate data for 2005–2006. Since the sample size is $n = 100$, it is easy to calculate the proportion (even in your head) of sellers who got more than they asked; using the notation that we saw in §1.2, we can write that $P(OLP) = 0.53$, if *OLP* stands for the event that the house sold for "Over List Price." We would interpret this by either saying that "53% of the houses in our sample sold for over the listed price" or by saying that "the probability that a house for sale in 2005–2006 sold for over its listed price was 0.53, at least according to our sample data."

Sample of San Mateo Real Estate Y0506		
SoldOverList	Not Over List Price	47
	Over List Price	53
	Column Summary	100
S1 = count ( )		

New notation for a new sampling distribution. Notice that our data come from a sample and not from a population, and that immediately raises our first question: "Can we generalize this 53% for 'Over List Price' to apply to the population from which the sample was drawn, since it is just from a sample of $n = 100$?" If we knew about the sampling distribution for a sample proportion, we might be able to answer the question of whether we have a sample result that is *rare*—far from the population value for the proportion of houses that were sold "Over List Price"—or a sample result that is *reasonably likely*—not far from the population value for *P(OLP)*, the proportion of houses that sold for over their listed price.

We need some new notation; we need to distinguish a proportion calculated from sample data from the corresponding proportion in the population. In §3.2, we used the symbol $\bar{x}$ for a sample mean, whose value we can calculate since we have the sample data, and μ for the population mean, whose value we generally do not know.

We have used *P(OLP)* to stand "the proportion of houses that sold for 'Over List Price.'" However, now we are aware that our calculation is for a sample that comes from a population. To distinguish between a sample proportion and the corresponding population proportion, we use the symbol $\hat{p}$ (pronounced "p-hat") for the proportion calculated from a *sample*. For our example, $\hat{p} = 0.53$; that is numerical value that we calculated as $P(OLP) = 0.53$ above. The population proportion for our example would be the proportion of *all* houses in the population that sold over list price, and for that we use the symbol p (p "without the hat.") Usually we are unable to calculate this population proportion; nevertheless, we need a symbol to refer to the population proportion even if we cannot calculate it. We use the $\hat{p}$ and p notation to keep it clear what we are referring to: the sample proportion that we can calculate and therefore know (the $\hat{p}$) or the population proportion that we cannot calculate and do not know but we really want to know (that is the p).

Notation

Sample proportion: $\hat{p}$ Population proportion: p

Building a Sampling Distribution Using Fathom

We can apply the idea of a sampling distribution to the sample proportion $\hat{p}$. We will build one using Fathom and then discuss the properties of the sampling distribution. Instead of the population mean μ and the sample mean $\bar{x}$ we will be interested in the population proportion p and sample proportion $\hat{p}$.

Idea of a Sampling Distribution of Sample Proportions $\hat{p}$

The **sampling distribution of sample proportions** is the distribution of the sample proportions for a specific variable of *all possible* random samples of a given size n drawn from a population.

Simulating a sampling distribution of sample proportions $\hat{p}$**.** Once again, we will use a specific example. The population is all of the houses sold in San Mateo County in 2005–2006 (not just from the South Region this time), and we will draw samples of size $n = 100$ (instead of $n = 40$ as we did in the sample mean example.) However, the four steps of the construction (or simulation) will be the same as before:

⇨ Step 1: Randomly sample a sample of $n = 100$ from the population of 5,486 houses.
⇨ Step 2: Calculate the sample proportion $\hat{p}$ for the sample just collected; store the result.
⇨ Step 3: Repeat steps 1 and 2 until 3.641×10^{216} unique samples are found.
⇨ Step 4: Collect together all the different sample proportions $\hat{p}$ into a collection.

To see the construction (or simulation) happen, follow the bulleted instructions.

- Open the Fathom file **San Mateo Real Estate Y0506.ftm** and open a **case table.** This file contains data on *all* the houses that were sold in San Mateo County, California, from June 2005 to June 2006. These data are the *population* of houses sold.
- Get a **Summary Table** and drag the variable *SoldOverList* to the **down-pointing arrow.**
- The **Summary Table** will show you just the counts for the two categories, since the variable *SoldOverList* is categorical. To get Fathom to show the proportions, select the **Summary Table**, go to **Summary>Add Formula** and in the dialogue box type: columnProportion.

- Drag down a **Graph** from the shelf and drag the variable *SoldOverList* to the *vertical side* of the graph. This will give a horizontal bar chart. Change *bar chart* to *ribbon chart*.

  What you see is the **population distribution** of the variable *SoldOverList* for the population of *all* the houses that were sold in 2005–2006 in San Mateo County. Notice that since our variable is categorical, we do not have a dot plot but rather just the ribbon chart. The summary table shows that the population proportion of houses that sold "Over List Price" is $P(OLP) = 0.590$. Using our notation, and knowing that in this instance the proportion is a population proportion, we will write $p = 0.590$. In the population of houses sold in 2005–2006, the proportion of houses that sold over list was 0.59. (Usually, we as researchers will not know the population proportion p; we have chosen an instance where we *do* know the population proportion to show how a sampling distribution is constructed.)

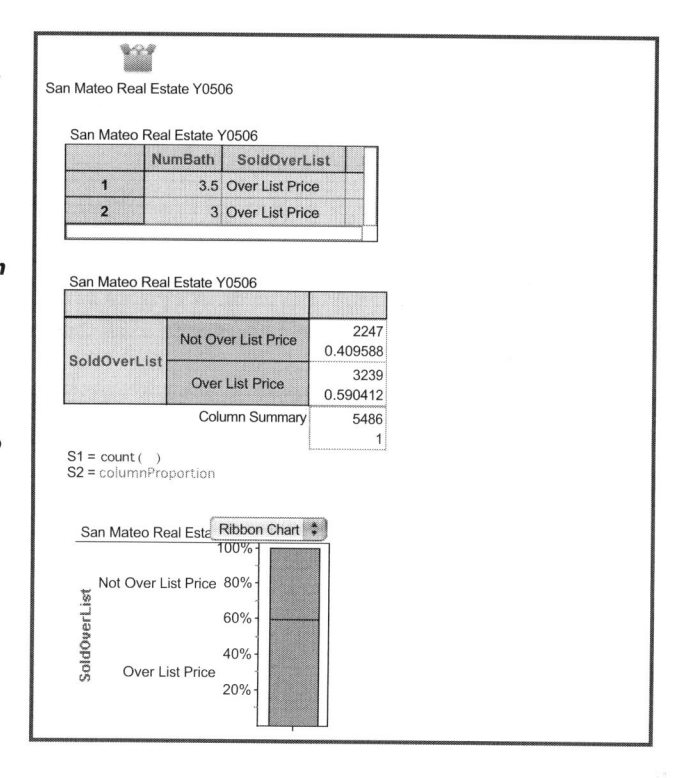

- Select the Collection Icon and go to the menu **Collection>Sample Cases.** A new collection should appear entitled *Sample of San Mateo Real Estate Y0506* and its *"Inspector"* should also open.
- In the *"Inspector,"* *de*select "with replacement" and *de*select "animation on" but leave the "x" in the box for "replace existing cases." Also change 10 cases to 100 cases and press: *Select More Cases*.
- Select the *Sample of San Mateo Real Estate Y0506* and get a **case table** for the sample.
- For the *Sample of San Mateo Real Estate Y0506*, repeat the bullets to get the *Summary Table* and the *ribbon chart* but for the sample rather than for the population. You should have two columns of Fathom output, similar to the graphic shown on the next page.

  Notice that the sample proportion shown in the example below is $P(OLP) = \dfrac{58}{100} = 0.58$; using our new notation, we would write $\hat{p} = 0.58$. This sample proportion is close to the population proportion.

- Get the *"Inspector"* for the *Sample of San Mateo Real Estate Y0506*. In the *Inspector*, select *Measures*.
- In the left-hand box, where <new> appears, type in *phat* and select the row so that it turns blue.
- Double click on the space under *Formula*, and **Edit Formula** should open.
- In the dialogue box, type: *proportion(SoldOverList = "Over List Price")* and click on *OK*.

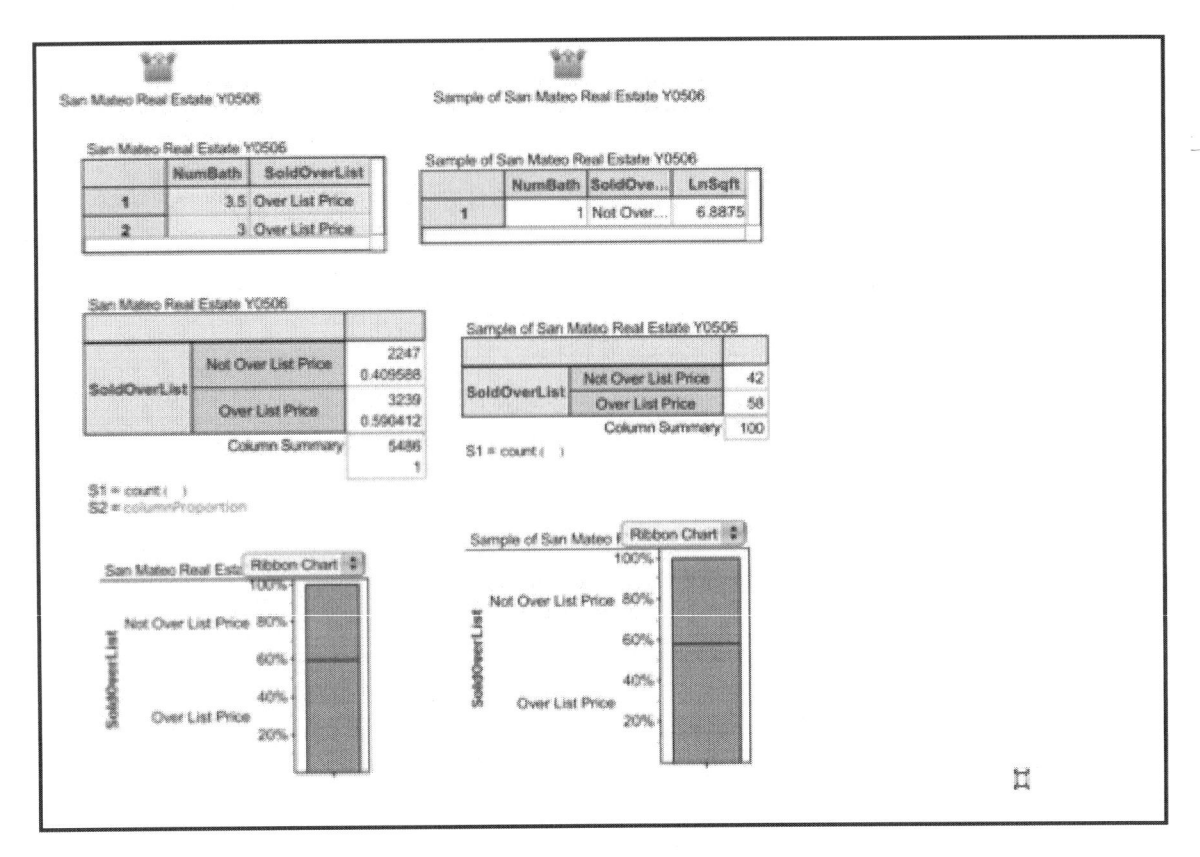

You should see the value of the sample proportion $\hat{p}$ of your sample appear in *value*; what we are doing is preparing Fathom to collect p-hats (that is, $\hat{p}$ s, sample proportions) for a huge number of different samples.

- You may now safely close the sample *Inspector*. It will just clutter the screen if it is open.
- With the *Sample of San Mateo Real Estate Y0506* collection icon selected, go to **Collection>Collect measures.** When you do this you will see your *ribbon chart* for your sample change five times (it will bounce) and a new collection icon entitled *Measures from Sample of San Mateo Real Estate Y0506*.

Fathom has put these five *phat* (that is, $\hat{p}$) values in a new collection, which it has called *Measures from Sample of San Mateo Real Estate Y0506*. However, we need a great many more than five values of *phat*.

- Select the *Measures from Sample of San Mateo Real Estate Y0506* icon and get a **case table**. You should see the five *phats* that Fathom has calculated from the five random samples.
- For the *Measures from Sample of San Mateo Real Estate South*, get a **dot plot** of the five *phats* and also a **Summary Table** for the mean and standard deviation for *phat*. Put these on your screen on the right.

For the sampling distribution, we have a dot plot and a mean and standard deviation. Why is this? It is because for the sampling distribution the cases are *phats* ($\hat{p}$ s) and these are quantitative *phats*; the variable for our population and for our sample is categorical, but for each sample we calculated a

252

quantitative measure, the sample proportion, $\hat{p}$. Our collection of $\hat{p}$ s is a distribution of quantitative things.

- In the *Inspector* for the *Measures from Sample of San Mateo Real Estate Y0506*, deselect the "Animation on" (if you do not, things will take a long time) and change the number of measures from 5 to 1000. Move the *Inspector* to the lower left side of the screen and click on *Collect More Measures*. Wait!

    After this process is done, you can check from your **case table** that there are 1,005 cases of *phats*.

- Go back to the *Inspector* and click on **Collect More Measures** (or with the *Measures from Sample...* selected, go to **Collection>Collect More Measures** and wait). What you should see (if you close the *Inspector*) should resemble but will not be the same as the picture below.

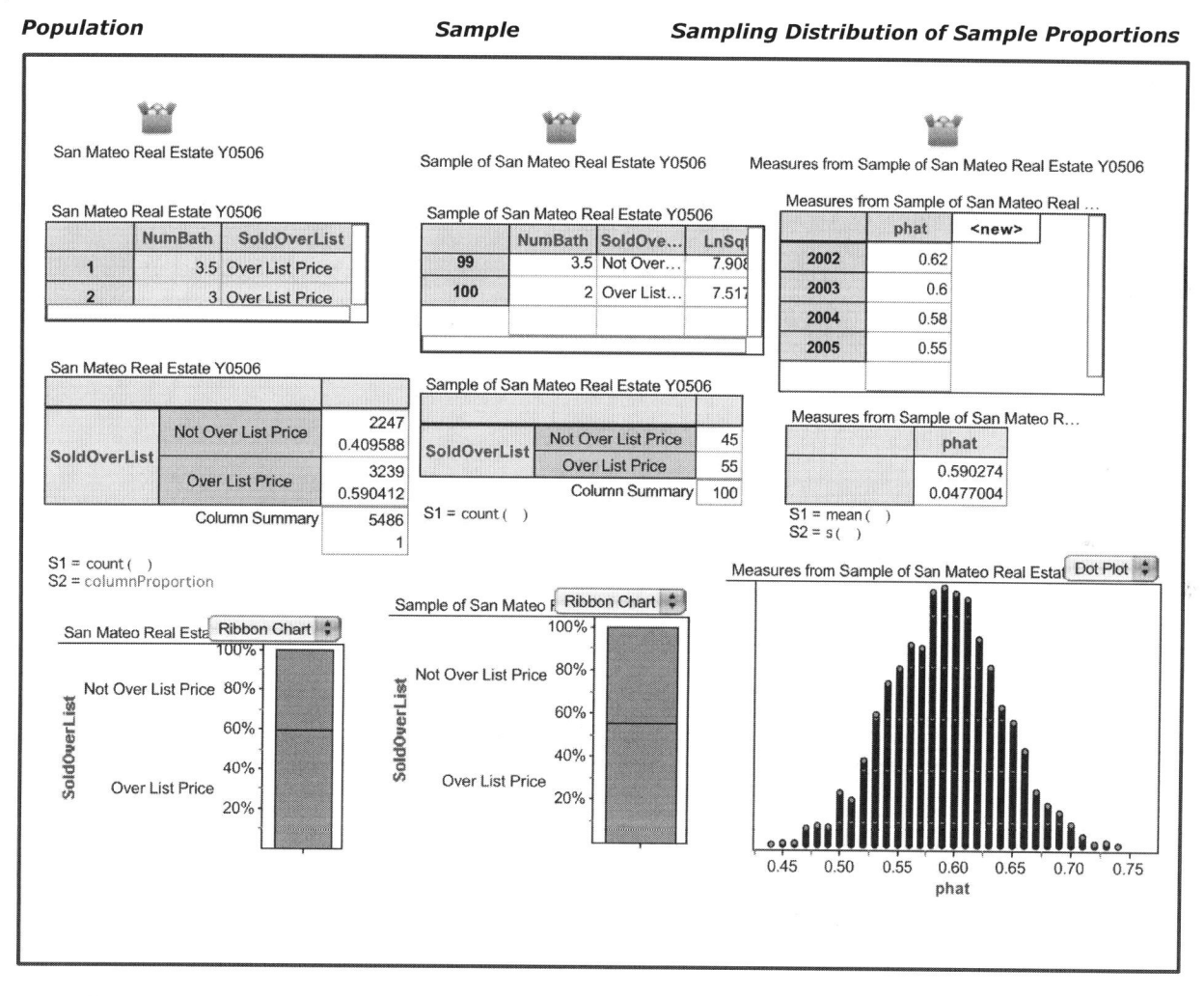

Why there are "spikes" in our sampling distribution. If you compare the plot on the right of the Fathom output to the similar output in §3.2 for the sampling distribution for quantitative, you see that the values of the $\hat{p}$ here are stacked up in "spikes" (or poles) whereas with the $\bar{x}$ s the values were not in neat vertical stacks. The picture for the sample means $\bar{x}$ s looked like a "normally shaped" collection of grapes rather than spikes. We see "spikes" for our sampling distribution of $\hat{p}$ because in a sample of $n - 100$, there can only be 0, 1, 2, 3,…, 96, 97, 98, 99, 100 houses that sold over list price. The number of houses that sold over list price conceivably could be any one of those numbers but cannot be

253

something like 41.275 houses that sold "Over List Price." Therefore, our *phats* can only take on certain values that come from dividing integers 0, 1, 2, 3,…96, 97, 98, 99, 100 by 100. So we only get certain values, and hence we get the appearance of stacked dots, or "spikes."

The shape of the sampling distribution looks Normal. In our calculations, we will usually use a Normal distribution to approximate the shape of a sampling distribution of a sample proportion. However, the actual sampling distribution is another family of theoretical (model) distributions called the **binomial distributions,** which we looked at in §3.3. Under certain conditions, the shapes of these binomial distributions are very close to the shape of a Normal distribution. For example, here is the plot for samples of size $n = 100$ drawn from a population in which the population proportion $p = 0.59$. The graphic is from **DistributionCalculator.ggb**.

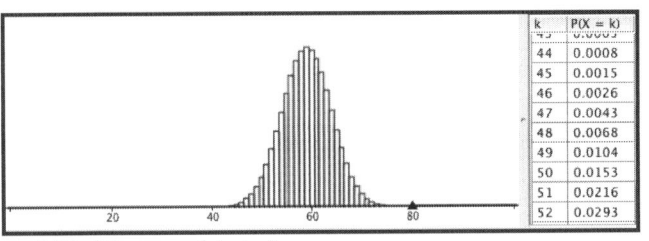

Having a sampling distribution that is Normal was good news because we can calculate what results from random sampling as *reasonably likely* or *rare.* The story is the same here: that the shape of the sampling distribution is approximately Normal allows us to work with the sampling distribution but only (as we shall see) under certain conditions. As with the sampling distribution for sample means $\bar{x}$, we need notation so that we will not lose our way. Here is a guide to the notation.

Notation for mean and standard deviation of the sampling distribution for $\hat{p}$

  Mean: $\mu_{\hat{p}}$ **Standard Deviation:** $\sigma_{\hat{p}}$

Now that we have some notation, we can compare the mean of the sampling distribution $\mu_{\hat{p}}$ with the population proportion p. From our simulation of 2005 different samples of $n = 100$, the two numbers look very, very close and exhibit what is actually a fact about the sampling distribution of sample proportions $\hat{p}$; the mean of the sampling distribution is equal to the population proportion: $\mu_{\hat{p}} = p$.

The standard deviation of the sampling distribution of sample proportions $\hat{p}$ is $\sigma_{\hat{p}} = \sqrt{\dfrac{p(1-p)}{n}}$. (This formula actually comes from the binomial distributions that we said are the actual forms the sampling distributions for $\hat{p}$ take). For our example where $p = 0.590$ and $n = 100$, this formula works out to be $\sigma_{\hat{p}} = \sqrt{\dfrac{p(1-p)}{n}} = \sqrt{\dfrac{(0.59)\cdot(1-0.59)}{100}} = \sqrt{\dfrac{0.59\cdot 0.41}{100}} = \sqrt{\dfrac{0.2419}{100}} = \sqrt{0.002419} \approx 0.0492$. When we compare this with our simulation, we see that the standard deviation of the $\hat{p}$ s that we collected is 0.0477, which is close to 0.0492. We can summarize these facts as follows.

> **Facts about Sampling Distributions of Sample Proportions $\hat{p}$ from Random Samples**
>
> **Shape:** The shape of the sampling distribution of $\hat{p}$ will be approximately Normal.
>
> **Center:** The mean of the sampling distribution of $\hat{p}$ is equal to the population proportion p. That is, $\mu_{\hat{p}} = p$.
>
> **Spread:** The standard deviation of the sampling distribution of $\hat{p}$ is $\sigma_{\hat{p}} = \sqrt{\dfrac{p(1-p)}{n}}$.
>
> **Conditions:** These facts hold only if *both* $np \geq 10$ and also $n(1-p) \geq 10$ are met and *also* the population is at least ten times the size of the sample being used.

The conditions $np \geq 10$ and also $n(1-p) \geq 10$ come from the fact that we are using a Normal distribution as an approximation to the binomial distributions. Binomial distributions can be quite skewed to the right or to the left if the p is small (near 0) or big (near 1) and the n also small. If the n is big then the Normal distribution is still a good approximation of the binomial distribution, even if the p is extreme; our guide as to whether the combination of sample size n and the population proportion will work are the rules: it must be true that $np \geq 10$ and also $n(1-p) \geq 10$.

Using Sampling Distributions for Sample Proportions $\hat{p}$ from Random Samples

Checking the conditions The first thing that we always need to do is to check the conditions $np \geq 10$ and also $n(1-p) \geq 10$. With our example of the San Mateo real estate, where $p = 0.59$, we can proceed, since we see that $np = 100 \cdot 0.59 = 59 > 10$ and $n(1-p) = 100 \cdot (1-0.59) = 100 \cdot 0.41 = 41 > 10$. However, if we had a sample size of only $n = 20$, we would *not* certain that the Normal approximation to the binomial distribution would be close enough; when we calculate $np = 20 \cdot 0.59 = 11.8 > 10$ and this "passes the test" (just barely), but when we calculate $n(1-p) = 20 \cdot (1-0.59) = 20 \cdot 0.41 = 8.2 < 10$, this does not pass the test. Both parts of the conditions must be true for us to proceed with confidence.

Reasonably Likely and Rare As we did with sample means $\bar{x}$, we can calculate an interval for what sample proportions $\hat{p}$ we can reasonably expect to see (the **reasonably likely** $\hat{p}$ s) and therefore what sample proportions would be **rare** if we have some knowledge of the population proportion p. The box below gives the facts, and then just below it there is a worked example.

> **Reasonably Likely and Rare**
>
> Sample proportions $\hat{p}$ that are *within* the interval $p - 1.96\sqrt{\frac{p \cdot (1-p)}{n}} < \hat{p} < p + 1.96\sqrt{\frac{p \cdot (1-p)}{n}}$ are called ***reasonably likely.***
>
> [Normal distribution curve showing Reasonably Likely: 95% in the center and Rare regions in the tails, bounded by $p - 1.96\sqrt{\frac{p \cdot (1-p)}{n}}$ and $p + 1.96\sqrt{\frac{p \cdot (1-p)}{n}}$]
>
> Sample proportions $\hat{p}$ that are *outside* the interval $p - 1.96\sqrt{\frac{p \cdot (1-p)}{n}} < \hat{p} < p + 1.96\sqrt{\frac{p \cdot (1-p)}{n}}$ are called ***rare.***
>
> In probability notation, $P\left(p - 1.96\sqrt{\frac{p \cdot (1-p)}{n}} < \hat{p} < p + 1.96\sqrt{\frac{p \cdot (1-p)}{n}} \right) = 0.95$

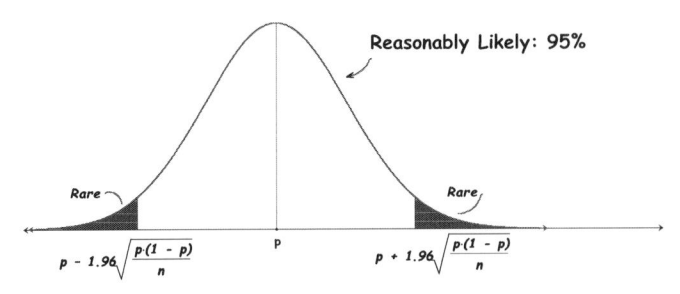

Reasonably Likely and Rare: San Mateo Real Estate Houses Over List Price To follow the argument, recall that the population proportion of houses that sold for more than the list price was $p = 0.59$. Therefore we know that the mean of the sampling distribution of $\hat{p}$s calculated from samples of $n = 100$ is $\mu_{\hat{p}} = p = 0.59$, and the standard deviation of the sampling distribution is

$$\sigma_{\hat{p}} = \sqrt{\frac{0.59 \cdot (1 - 0.59)}{100}} = \sqrt{\frac{0.59 \cdot 0.41}{100}} \approx 0.0492$$. We can then calculate the *reasonably likely* interval:

Lower limit: $p - 1.96\sqrt{\frac{p(1-p)}{n}} = 0.59 - 1.96\sqrt{\frac{0.59 \cdot 0.41}{100}} \approx 0.59 - 1.96(0.0492) = 0.59 - 0.096 = 0.494$ and

Upper limit: $p + 1.96\sqrt{\frac{p(1-p)}{n}} = 0.59 + 1.96\sqrt{\frac{0.59 \cdot 0.41}{100}} \approx 0.59 + 1.96(0.0492) = 0.59 + 0.096 = 0.686$.

So for this example, the interval of *reasonably likely* $\hat{p}$s is the interval $0.494 < \hat{p} < 0.686$. What does this mean? We can give two interpretations. We can say that if we sampled samples of $n = 100$ repeatedly (and randomly) from our population of all houses, 95% of those samples would have $\hat{p}$s in the interval $0.494 < \hat{p} < 0.686$. Alternately, we can say that if we draw a random sample of size $n = 100$ houses from our population of houses in San Mateo real estate, we are *95% certain* that our sample proportion $\hat{p}$ of houses that sold over their list price will be in the interval 49.4% to 68.6%. We might be "unlucky" with our random sample and get a $\hat{p}$ outside this interval, but the probability that we will get such a "wild" $\hat{p}$ is only 5%.

How likely is it that we get a sample with "this $\hat{p}$" or one even more extreme? What do we mean by this? The full context of exactly *why* we want to know this will be clearer in the next unit; but for now, know that we will be interested in knowing how likely it is that we actually see a $\hat{p}$ as far away from a population p as we have actually seen. We have everything we need to calculate this probability.

Here is an example. The last sample that we collected in our simulation had a $\hat{p} = 0.55$; here $\hat{p} = 0.55$ is the "this $\hat{p}$". This *sample* of just $n = 100$ houses told us that 55% of the houses have sold for over list price. We can ask the question: "If the population proportion is $p = 0.59$, what is the probability that we get a sample proportion this small or one even smaller?" We are asking the question: "What is $P(\hat{p} < 0.55)$?" This is a "given a value (the value $\hat{p} = 0.55$)— find a proportion" type of problem, and since we have an approximately Normal sampling distribution, we can solve this problem.

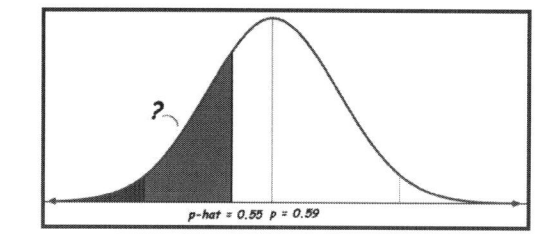

The first step is to draw a picture of the situation showing what we have and what we want—what we want by an arrow; we want a probability. Next calculate the z score so that we get

$$z = \frac{\hat{p} - \mu_{\hat{p}}}{\sigma_{\hat{p}}} = \frac{0.55 - 0.59}{\sqrt{\frac{0.59 \cdot (1 - 0.59)}{100}}} \approx \frac{-0.04}{0.049} \approx -0.813 \approx -0.81.$$

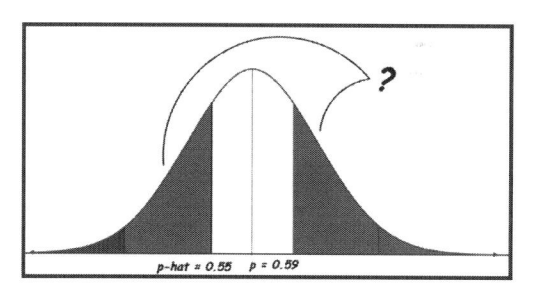

With the $z = -0.81$, we consult the Normal Probability Chart and write (using the correct notation) $P(\hat{p} < 0.55) = P(z < 0.81) = 0.2090$. The probability of getting a *p-hat* of 0.55 or smaller is about 21%.

The answer $P(\hat{p} < 0.55) = P(z < 0.81) = 0.2090$ is correct for the question "What is $P(\hat{p} < 0.55)$?" That calculation is correct to answer the question: "What is the probability that we get a $\hat{p}$ less than 0.55 if the population proportion is 0.59?" However, it does not do justice to the question: "What is the probability of getting a $\hat{p}$ that is *more extreme?*" Why? How far away from 0.59 is 0.55? Answer: –.04. So anything 0.04 farther away from 0.59 is more extreme; but this can happen (randomly) on the upper side of 0.59 as well as the lower side. "More extreme" could be a $\hat{p}$ greater than $0.59 + 0.04 = 0.63$ as well as a $\hat{p}$ less than 0.55. Our picture of "more extreme" should have two tails, rather than one tail like the picture shown.

To actually calculate this "more extreme" probability is very easy because of the symmetry of the Normal distribution. The probability for the two tails is simply twice the probability of one tail. The mathematical way to express this is to use the absolute value notation, since we want to embody both the left and the right tail in one expression. Therefore, we would write, as an expression of "more extreme than $\hat{p} = 0.55$," $P(\hat{p} > |0.55|) = 2P(z < 0.81) = 2(0.2090) = 0.4180 \approx 0.42$.

Summary: Normal Sampling Distributions for Categorical Variables

- The relevant measure for categorical variables is a **proportion,** so we examine the sampling distribution for sample proportions.

- **Notation:** Sample proportion: $\hat{p}$ Population proportion: p

- **Notation for mean and standard deviation of the sampling distribution for $\hat{p}$**

    Mean: $\mu_{\hat{p}}$ **Standard Deviation:** $\sigma_{\hat{p}}$

- **Facts about Sampling Distributions of Sample Proportions $\hat{p}$**

    - **Shape:** The shape of the sampling distribution of $\hat{p}$ will be approximately Normal.

    - **Center:** The mean of the sampling distribution of $\hat{p}$ is equal to the population proportion p. That is, $\mu_{\hat{p}} = p$.

    - **Spread:** The standard deviation of the sampling distribution of $\hat{p}$ is $\sigma_{\hat{p}} = \sqrt{\dfrac{p(1-p)}{n}}$.

    - **Conditions:** These facts hold only if *both* $np \geq 10$ and also $n(1-p) \geq 10$ are met and *also* the population is at least ten times the size of the sample being used.

- **Reasonably Likely and Rare**

    - **Reasonably likely** sample proportions $\hat{p}$ are those *within* the interval

    $$p - 1.96\sqrt{\dfrac{p \cdot (1-p)}{n}} < \hat{p} < p + 1.96\sqrt{\dfrac{p \cdot (1-p)}{n}}.$$

    - **Rare** sample proportions $\hat{p}$ are *outside* the interval $p - 1.96\sqrt{\dfrac{p \cdot (1-p)}{n}} < \hat{p} < p + 1.96\sqrt{\dfrac{p \cdot (1-p)}{n}}$.

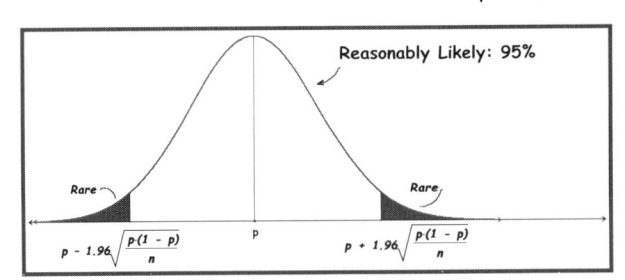

- With a sampling distribution for a specific categorical variable and population, we can:

    - Find the range of *reasonably likely* and *rare* sample proportions $\hat{p}$ s.

    - Determine the likelihood of a sample proportion $\hat{p}$ that is greater (or lesser) than a specific value.

    - Calculate the sample proportion $\hat{p}$ that we would expect to see a given percentage of the time if we repeatedly sampled.

    See the **Notes** above for examples of these kinds of calculations.

§3.4 Exercises: Sampling Distributions for Proportions

1. **Inference: Sample and Population** This exercise is about the proportion of students in colleges having a tattoo. Here we have shown the proportions of male and female students reporting having a tattoo for a sample at Penn State (in 1999) and samples for several semesters at one community college in Northern California.

Semester, Year and Place	Females Proportion having a tattoo	n	Males Proportion having a tattoo	n	Overall Proportion having a tattoo	n
Both, 1999, PA	0.131	137	0.191	68	0.151	205
Spring 2008. CA	0.255	47	0.160	50	0.206	97
Autumn 2008, CA	0.254	63	0.127	63	0.190	126
Spring 2009. CA	0.372	86	0.219	64	0.307	150
Autumn 2009, CA	0.258	89	0.115	78	0.192	167
Spring 2010. CA	0.266	94	0.309	81	0.286	175
Autumn 2010, CA	0.149	101	0.146	89	0.147	190
Spring 2011. CA	0.191	94	0.218	87	0.204	181

   a. **Sample** Study the numbers in the table and decide whether or not each of the following statements are supported by the data in the samples. Give reasons.
      (i) Tattoos are more popular among students in California than at Penn State.
      (ii) For the California community college students, it appears that tattoos were most popular in 2009.
      (iii) Among college students in California, females are more likely than males to have a tattoo.

   b. **Population** What you see in the samples may or may not be generalizable; although it is not often done, it is helpful to think about what population the samples represent. (If we decide that the sample is a good representation of a population, we feel safe generalizing the findings to that population.) The California sample data were collected from students who had signed up for a statistics class. For each option for a "population," decide whether you think the California samples (for the information in the table above) are likely to be a representative sample of that population.
      (i) all the students taking statistics at that college?
      (ii) all the students at that college?
      (iii) all community college students in Northern California?
      (iv) all college and university students in California?
      (v) all college and university students in North America?

   c. The sample data were collected from all the students who had signed up for some sections of statistics at one college. Why is this sample *not* a simple random sample of all the students taking statistics at that college? (In practice, we often have samples that may be representative but are not random samples.)

   The logic of inference For the remainder of this exercise, let us agree that the population we have in mind is "all college and university students in California." The next questions have to do with the proper use of notation. Then we tackle some logic.

d. Calculate the overall proportion of students having a tattoo in the Summary Table for the sample for spring 2009 and assign the symbol p or the symbol $\hat{p}$. Give a reason for your choice.

Combined Class Data Spr 09			
	Tattoo		Row
	n	y	Summary
Gender f	54	32	86
Gender m	50	14	64
Column Summary	104	46	150
S1 = count ( )			

e. Pete, who proclaims himself an expert on "student trends," is very happy to see the result of your calculation. He is happy because the result is in line with his idea that the *population* proportion of college and university students in California who have a tattoo is 30%, or 0.30. Suppose Pete is right; we do not know whether Pete is right, but let us suppose his idea is correct. *If* Pete really is right, what symbol should be assigned to Pete's idea, the symbol p or the symbol $\hat{p}$? Give a reason for your choice.

f. If Pete is correct that the population proportion $p = 0.30$ (so now you know the answers to questions d and e—but you still need to have the reasons right), describe the *sampling distribution* of sample proportions $\hat{p}$ for simple random samples of $n = 150$ drawn from the population of college and university students in California. The words "describe the sampling distribution" mean that you need to specify the *shape, center,* and *spread* of the sampling distribution.

g. Are the conditions involving $np \geq 10$ and $n(1-p) \geq 10$ met for the spring 2009 sample using Pete's idea?

h. Use the information you have put down in your answer to part f to calculate the interval of *reasonably likely* $\hat{p}$ s if Pete is right and $p = 0.30$.

i. Will the $\hat{p}$ we calculated be *reasonably likely* or *rare?* Give a reason for your answer.

j. Pete has an identical twin brother named Repete, but, amazingly, they disagree on nearly everything. Repete thinks that the population proportion of college and university students who have a tattoo is $p = 0.20$. Repete thinks that only 20% of all college and university students in California have a tattoo. Are the *conditions* for the sampling distribution of sample proportions $\hat{p}$ met if $n = 150$ and Repete is right and $p = 0.20$? Show that they are (or are not) by calculation.

k. Calculate the interval of *reasonably likely* $\hat{p}$ s if Repete is right and $p = 0.20$.

l. If $p = 0.20$ (Repete's idea) then is $\hat{p} = 0.3067$ *reasonably likely* or *rare?* Give a reason.

m. If we had a $\hat{p} = 0.3067$ from an SRS of $n = 150$ from the population of all California college and university students, would Pete have good evidence that his brother Repete is wrong? What should his argument be? (*Hint:* Think *reasonably likely* or *rare.* You may wish to discuss this question in class; this question actually introduces a way of arguing that we will use.)

n. Suppose instead of forty-six students having a tattoo (which leads to $\hat{p} = 0.3067$), we actually had just twenty-nine out of the $n = 150$ who had a tattoo. Calculate the $\hat{p}$ and then decide whether Pete or Repete would have the stronger evidence for his case. Explain why one or the other has the stronger evidence.

o. Now suppose there were thirty-eight students out of the $n = 150$ who had a tattoo. Pete and Repete's mother, Maria (who knows statistics), says to her sons: "This result will *not* settle your argument." Explain, using the ideas of *reasonably likely* and *rare* to show why Maria is right.

2. **Inference on tattoos, cont.** All of the questions in Ex. 1 used a sample size of $n = 150$. Suppose that our sample size is only $n = 50$. We still work with Pete's idea that $p = 0.30$. The sampling distribution of sample proportions may be different with a smaller sample size or not.

   a. Determine whether the conditions involving $np \geq 10$ and $n(1-p) \geq 10$ are met.
   b. For $n = 50$, find the mean of the sampling distribution $\mu_{\hat{p}}$ and what is the $\sigma_{\hat{p}}$.
   c. Make a sketch of the sampling distribution showing the mean and the standard deviation and the interval of reasonably likely $\hat{p}$ s. Use a sketch of a Normal distribution.
   d. Will a $\hat{p} = 0.20$ be reasonably likely or rare if $p = 0.30$ and $n = 50$? Give a reason.

3. **San Mateo Real Estate, Act II** In 2005–2006, the housing market was booming, and one result of the boom was (as we saw in the **Notes**) that nearly 60% of the houses sold for more than their list price. In contrast, 2007–2008 was a depressed market, and we should expect the proportion of homes that were sold over their list price to be much lower. This exercise takes you through the "simulation" of the sampling distribution and then asks some questions about the sampling distribution from this population of houses. Follow the bullets and answer the questions.

   a. What are the cases and what is the population for this exercise? (It may seem that the answers to "cases" and "population" should be the same, and they are almost the same.)

- Open the Fathom file **San Mateo Real Estate Y0708.ftm** and open a **case table.**
- Drag down a **Graph** from the shelf and drag the variable *SoldOverList* to the *vertical side* of the graph. This will give a horizontal bar chart. In the upper right-hand corner, change *bar chart* to *ribbon chart*.
- Get a **Summary Table** for the variable *SoldOverList*. Since the variable *SoldOverList* is categorical, the Summary Table will show you just the counts for the two categories. To show the proportions, select the *Summary Table*, go to **Summary>Add Formula** and in the dialogue box type: columnProportion.

   b. Your screen should resemble the picture here. Record, using the correct notation, the population proportion for the houses that were sold "Over List Price." **PTO**

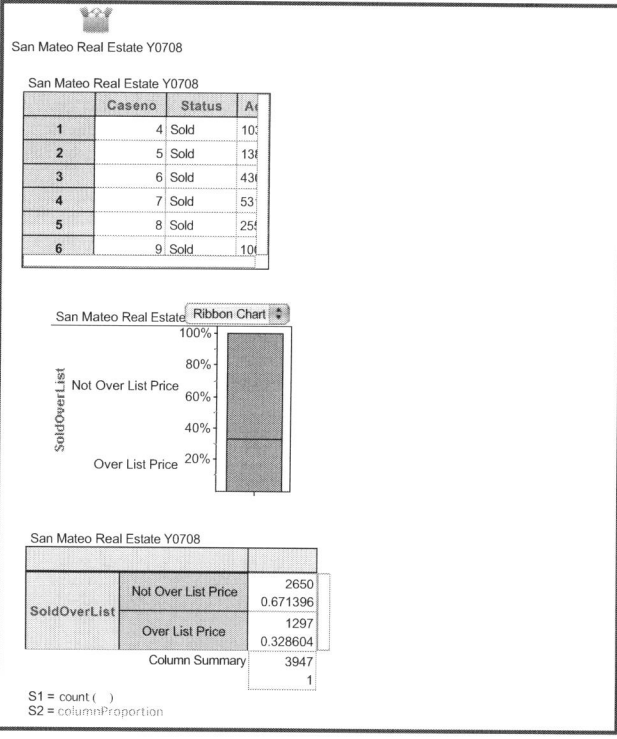

-
    c. Compare the population proportion of houses sold *"Over List Price"* for 2007–2008 with the population proportion for 2005–2006. Explain briefly how the comparison makes sense in the context of the changing housing market described in the introduction to the exercise.
- Select the Collection Icon and go to the menu **Collection>Sample Cases.** A new collection should appear entitled *Sample of San Mateo Real Estate Y0708,* and its *"Inspector"* should also open.
- In the *"Inspector,"* *de*select "with replacement" and *de*select "animation on" but leave the "x" in the box for "replace existing cases." Also change 10 cases to 100 cases and press the *Select More Cases* box.
- Select the *Sample of San Mateo Real Estate Y0708* and get a **case table** for the sample.
- For the *Sample of San Mateo Real Estate Y0708,* repeat the bullets to get the *Summary Table* and the *ribbon chart* but for the sample rather than for the population. You should have two columns of Fathom output, similar to the two columns shown in the **Notes** at this point.
    d. Record your sample proportion using the correct notation for the sample proportion.
    e. In the **Notes**, read the four steps for getting a sampling distribution. At which step are we now?
- Get the *"Inspector"* for the *Sample of San Mateo Real Estate Y0708.* In the *Inspector*, select *Measures*.
- In the left-hand box, where <new> appears, type in *phat* and select the row so that it turns blue.
- Double click on the space under *Formula,* and **Edit Formula** should open.
- In the dialogue box, type: *proportion(SoldOverList = "Over List Price")* and click on *OK.*
- With the *Sample of San Mateo Real Estate Y0708* collection icon selected, go to **Collection>Collect measures.** When you do this you will see your *ribbon chart* for your sample change five times (it will bounce) and a new collection icon entitled *Measures from Sample of San Mateo Real Estate Y0708.*
- Select the *Measures from Sample of San Mateo Real Estate Y0708* icon and get a **case table.** You should see the five *phats* that Fathom has calculated from the five random samples.
- For the *Measures from Sample of San Mateo Real Estate Y0708,* get a **dot plot** of the five *phats* and also a **Summary Table** for the mean and standard deviation for *phat.* Put these on your screen on the right.
    f. Notice that for the sampling distribution, we have got a dot plot and a mean and standard deviation. Why is this?
- In the *Inspector* for the *Measures from Sample of San Mateo Real Estate Y0708,* deselect the *"Animation on"* (if you do not, things will take a long time) and change the number of measures from 5 to 1000. Move the *Inspector* to the lower left side of the screen and click on *Collect More Measures.* Wait!

- Go back to the *Inspector* and click on **Collect More Measures** (or with the *Measures from Sample...* selected, go to **Collection>Collect More Measures**) and wait. What you should see (if you then close the *Inspector*) should resemble but not be the same as the picture here.

  g. Use the information in the box **Facts about Sampling Distributions of Sample Proportions...** to find what the mean $\mu_{\hat{p}}$ and standard deviation $\sigma_{\hat{p}}$ of the sampling distribution of $\hat{p}$ drawn from samples of $n = 100$ from the population of San Mateo Real Estate Y0708. You may round your value for p to three decimal places. Your simulated results should be fairly close to your calculations. Are they?

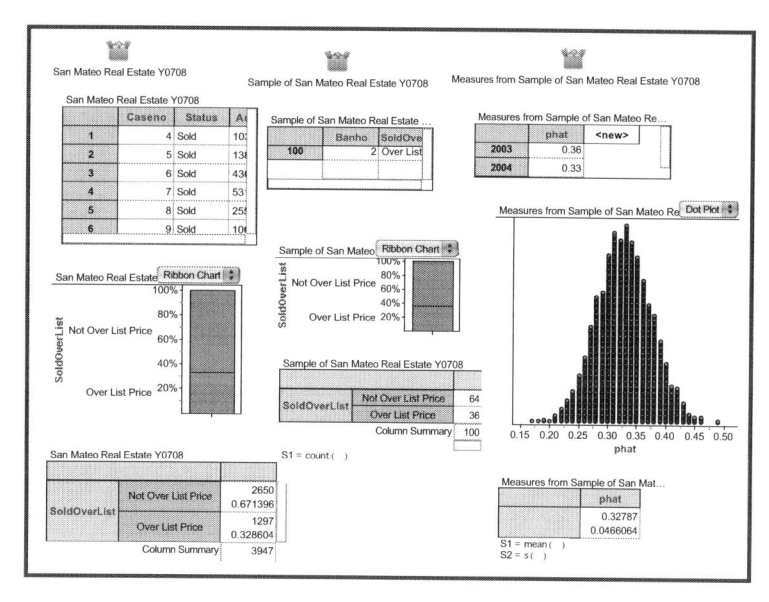

  h. Calculate the interval of reasonably likely $\hat{p}$ s from your answer to part g.

  i. Was the last sample that Fathom collected for you reasonably likely or rare? Give a reason for your answer.

  j. Find a value for $\hat{p}$ so that only 5% of the $\hat{p}$ s are smaller than this value. (*Hint:* Is this a "given a value—find a proportion" problem or a "given a proportion—find a value" problem?)

4. **Students and Languages: Proportions and means** In the combined class data for spring '09, eighty-two students of the 146 who answered the question said that they speak two or more languages. Another way of looking at the same data (the variable *Language*) is to calculate the *mean* number of languages spoken. For the spring '09 students, the mean number of languages is 1.70. This exercise looks at both the sampling distributions of proportions (§3.4) and the sampling distribution of sample means (§3.2).

   Note: parts a through h are about the sampling distribution of sample proportions.

   a. When we calculate a proportion of students who speak two or more languages, are we treating the variable *Language* as a *categorical* or a *quantitative* variable? When we calculate the mean number of languages spoken, are we treating the variable *Language* as a *categorical* or a *quantitative* variable? (Notice that although *Language* is a quantitative variable, it is helpful to analyze it using "categorical" techniques.)

   b. Calculate the proportion who speak two or more languages and express this proportion in two ways: first, using the P( ) notation, using *X* to indicate the number of languages spoken, and then secondly, using the notation for a sample proportion that was introduced in this section.

   c. We do not know the *population* proportion of California college and university students who speak two or more languages; we have to "guesstimate," although our sample data at least gives the ballpark in which to guess. So let us guess that one half of the population of all college and university students in California speak two or more languages. Express this guess for the population proportion using the correct symbol for population proportion.

d. Using the box in the **Notes** called **Facts about sampling distributions...** do some calculations to determine that $n = 146$ is big enough for the sampling distributions of $\hat{p}$ to be approximately Normal.

e. Again using the **Facts about sampling distributions...** and *also making certain to use the correct notation*, describe, with numbers and symbols and words, the center and the spread of the sampling distribution of $\hat{p}$ from samples of $n = 146$.

f. Use your answer to part e to find the interval of reasonably likely $\hat{p}$ s for samples of $n = 146$ and $p = 0.50$.

g. Make a sketch of the sampling distribution showing the mean and the standard deviation and the interval of reasonably likely $\hat{p}$ s. Use a sketch something like the one shown here but with the horizontal axis labeled with the units that are used for the proportions in the context here.

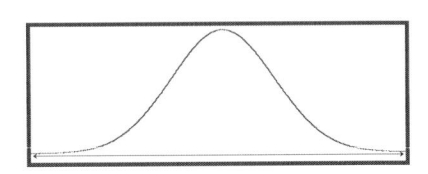

h. Is the sample proportion found from the spring '09 data a reasonably likely or rare $\hat{p}$ if $p = .50$?

Note: parts i through p are about the sampling distribution of sample means.

i. Now consider the *mean* number of languages spoken, which was given above as 1.7 for the spring '09 students. What symbol should be used for this number? Give a reason for your choice.

j. Again, we do not really know the population mean number of languages spoken or the population standard deviation for the variable *Language* for the population of all California college and university students. We can only guess, although our data gives some guidance. Let us guess that the population mean is 1.5 and that the population standard deviation is 1. Use the correct symbols for these numbers. (Notice: we can use the symbols even though we do not really know the values; the symbols are a way of *naming* what we are working with.)

k. Using the **Facts about sampling distributions of sample means...** in §3.2 and *also making certain to use the correct notation*, describe, with numbers and symbols and words, the center and the spread of the sampling distribution of $\bar{x}$ from samples of $n = 146$ from California college and university students.

l. Find the interval of reasonably likely $\bar{x}$ s for samples of $n = 146$ where $\mu = 1.5$ and $\sigma = 1$.

m. Make a sketch of the sampling distribution showing the mean and the standard deviation and the interval of reasonably likely $\bar{x}$ s. Use a sketch something like the one shown here but with the horizontal axis labeled with the units that are used for the proportions in the context here.

n. Is the $\bar{x} = 1.70$ from our sample of $n = 146$ reasonably likely or rare if $\mu = 1.5$ and $\sigma = 1$?

o. On your sketch of the sampling distribution, indicate our $\bar{x} = 1.70$ and calculate the probability of getting an $\bar{x}$ this large or larger with random sampling. ("With random sampling" means: if our Normal sampling distribution applies.) **PTO**

p. You should have found that the probability that we would see an $\bar{x} = 1.70$ *if* $\mu = 1.5$ and $\sigma = 1$. You should have $P(\bar{x} \geq 1.7) = 0.0078$. Because we have a *rare* outcome, there are several possibilities:

- $\mu = 1.5$ and $\sigma = 1$ are wrong. Perhaps μ is actually bigger than 1.5.
- We were unlucky in our sampling; we just got one of the weird samples.
- We have a biased sample, since it is not a simple random sample of all California colleges and university students. Perhaps $\mu = 1.5$ is correct, but our college has a bigger mean.

Choose one or more of these possibilities and make your case.

5. **Cell Phone Usage by Drivers** It is now illegal in California to talk on a cell telephone while driving. Our question is: what proportion of drivers breaks the law at a given time? The state of Victoria in Australia has a similar law to the one in California. A small team of researchers in Melbourne, Australia, collected data by *observing* whether drivers were talking on their cell phones [D. Taylor, C. MacBean, A. Das and R. Rosli, "Handheld mobile telephone use among Melbourne drivers," *Medical Journal of Australia* (2007) 187: 432-434].

Here is what they did: Four observers with just a clipboard stood on a corner of a road or a freeway ramp exit and observed drivers. There were three kinds of locations: one freeway ramp exit, one city street, and one urban area road. They categorized drivers in age/gender categories and, most importantly, as to whether or not the driver was using a handheld cell phone. They observed on three consecutive Tuesdays in October 2006 for one-hour periods (10.00–11.00, 14.00–15.00, and 17.00–18.00; see Exercise 3 of §3.1).

They had done a previous study and were able to guess that about 1.5% of the drivers would be using cell phones. So their guess was that $p = 0.015$.

a. Look again at the *conditions* for the sampling distribution of the sample proportion $\hat{p}$ which are that $np \geq 10$ and $n(1 - p) \geq 10$. For the freeway exit they observed $n = 8{,}640$ drivers. Was this enough?

b. Find the mean and the standard deviation of the *sampling distribution* of $\hat{p}$ if $p = 0.015$ and $n = 8640$.

c. Calculate the interval of reasonably likely $\hat{p}$ s if $p = 0.015$ and $n = 8640$.

d. Is the sample used by Taylor, et. al. a random sample? Give a reason for your answer. (Here is a situation in which getting a random sample is probably impossible, but getting some kind of sample is better than none.)

Special Exercises on Sampling Distributions: Sample Means and Sample Proportions

These exercises use the *population* of houses being sold in the *North* part of San Mateo County from June 2005 to June 2006. These houses were sold during the housing boom. It therefore makes sense that a large proportion of the houses were sold for something over the list price. In fact, the population proportion for the houses that were sold for more than their list price was 0.693.

		x	P(x)
Over List?	No	0	0.307
	Yes	1	0.693

Special Exercise A. Sampling Distribution for a Sample Proportion

1. What is the symbol that we should use for the 0.693? (*Hint:* Is this the population proportion or is this a sample proportion?)
2. You are going to describe the sampling distribution of sample proportions calculated from samples of size $n = 50$ drawn from the population of houses sold in the North region.
   a. What symbol should be used for a sample proportion?
   b. What will be the mean of the sampling distribution of sample proportions calculated from samples of size $n = 50$ drawn from the population of houses sold in the North region? Refer to the box in the **Notes** entitled **Facts about Sampling Distributions of Sample Proportions...**
   c. What will be the standard deviation of this *sampling distribution*? Do a calculation.
   d. Will the shape of the sampling distribution be approximately Normal? Check to see if the conditions are met for the sampling distribution to be approximately Normal.
   e. Make a sketch of the sampling distribution showing the mean and the standard deviation. Use a sketch something like the one shown here but with the horizontal axis labeled with the units that are used for the proportions in the context here.
3. To work with the sampling distribution you have just described, answer these questions.
   a. If you draw a random sample of $n = 50$ from the population, use your sampling distribution to find the probability that your sample proportion shows three-quarters or more of the houses in the sample sold for more than the list price. Translate the question (including three-quarters to decimals) using the correct probability and Normal distribution symbols, show your calculations, and express your answer in the context of the question here.
   b. If you draw a random sample of $n = 50$ from the population, use your sampling distribution to find the probability that your sample proportion shows that one-half or less of the houses in the sample sold for more than the list price. Translate the question using the correct probability and Normal distribution symbols, show your calculations, and express your answer in the context of the question here.
   c. Use your sampling distribution to find the interval of values of the sample proportion that are **reasonably likely**. Refer to the box in the **Notes** entitled **Reasonably Likely and Rare...**
   d. Which of the results in parts a and b are reasonably likely and which are rare: both, neither, one of each? If so, which?

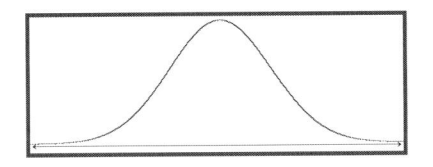

Special Exercise B. Sampling Distribution for Sample Mean

About 70% of the houses sold for more than their list price; so how much more? Here is the population distribution of the *difference* between the sale price and the list price. As you can see, some of the houses actually sold for less than their list price (so the value is negative), and many houses sold for more than the list price.

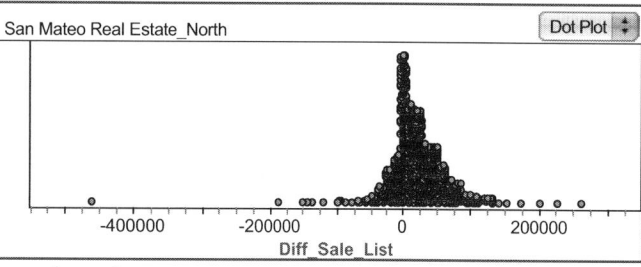

The population mean and standard deviation are given in the Fathom Summary Table below.

1. a. What is the symbol that we should use for the mean, and what is the symbol we should use for the standard deviation shown in the Fathom Summary Table? (*Hint:* Are the numbers in the table to be regarded as population figures or sample figures?)

   b. The meaning of the numbers. Which of these interpretations (I, II, or III below) is most in accord with the numbers? The mean list price was about $780,000.

      (I) Most houses sold for way more than their list price, but there was some variation.

      (II) On average the amount that a house sold for was not very much over the list, but there was considerable variability.

      (III) On average, houses were selling for less than what people asked for them.

   c. For each of the two interpretations that you discarded, state briefly (but in a sentence) why you discarded that interpretation.

2. You are going to describe the **sampling distribution of sample means** calculated from samples of size $n = 50$ drawn from the population of houses sold in the North region.

   a. What symbol should be used for the sample mean?

   b. What will be the mean of this sampling distribution of sample means calculated from samples of size $n = 50$ drawn from the population of houses sold in the North region?

   c. What will be the standard deviation of this sampling distribution? You will have to do a small calculation.

   d. Will the shape of the sampling distribution be approximately Normal? Give a reason for your answer.

   e. Make a beautiful sketch of the sampling distribution showing the mean and the standard deviation.

3. Answer these questions by working with the **sampling distribution** you have just described.

   a. If you draw a random sample of $n = 50$ from the population, use your sampling distribution to find the probability you get a sample mean for the amount over list of forty thousand dollars or more of the houses in the sample sold for more than the list price. Translate the question using the correct probability and Normal distribution symbols, show your calculations, and express your answer in the context of the question here.

   b. Find the range of values of the sample mean that are *reasonably likely.*

   c. Is your result in part a rare or reasonable likely? Give a reason.

§4.1 Politics and Confidence: Estimating a Proportion

Approving the President

For decades the Gallup® organization has been asking people about whether they "approve the job the president is doing." There was an interesting website (http://www.usatoday.com/news/washington/presidential-approval-tracker.htm) that that had tracked these approval ratings over time. Here is a graph showing the comparison of the approval ratings for Clinton, Bush (the younger), and Obama (up to July 2012). For Bush and Clinton, the rating is tracked to the end of their second terms.

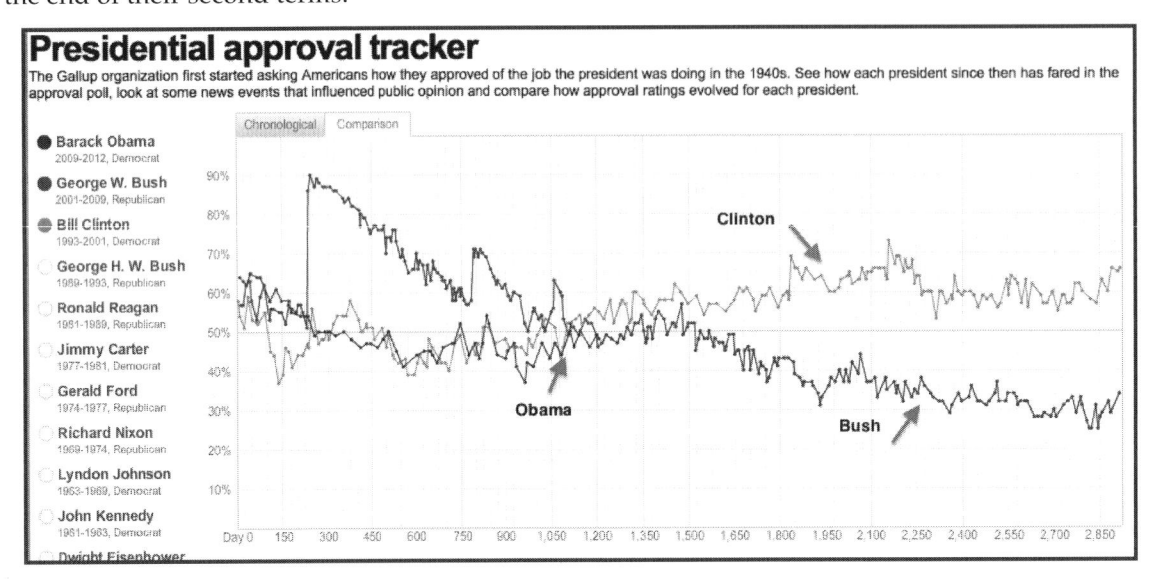

This approval rating is an interesting population proportion p because the population is well defined as "all Americans," but the actual population proportion p is always changing, even by the minute, and no one can know what it "actually" is. The best that can be done is to estimate the p with results from a sample of people. So how is this done? Just after Barack Obama received the Nobel Prize for Peace, the Gallup® organization reported Obama's approval rating was 56% (see http://www.gallup.com/poll/123629/Obama-Job-Approval-56-After-Nobel-Win.aspx). At the foot of that press release, there was the paragraph below that informs the reader how they did the survey. (Gallup® press releases typically give this kind of information, although the details vary.)

Survey Methods

Results are based on telephone interviews with 1,532 national adults, aged 18 and older, conducted Oct. 9-11, 2009, as part of Gallup Daily tracking. For results based on the total sample of national adults, one can say with 95% confidence that the maximum margin of sampling error is ±3 percentage points.

Interviews are conducted with respondents on land-line telephones and cellular phones.

In addition to sampling error, question wording and practical difficulties in conducting surveys can introduce error or bias into the findings of public opinion polls.

This tells us that the sample size was $n = 1532$, that the population included adults eighteen years and older, and that they interviewed the respondents on either land line or cell telephones. Note the sentence: "For results based on the total sample of national adults, one can say with 95% confidence that the maximum margin of sampling error is ±3 percentage points." What Gallup® has calculated is an *interval estimate* of the population proportion p of Americans who approve of Obama's job performance. The technical name for the *interval estimate* that they have calculated is a **confidence interval.** In this section, we will see:

- How **confidence intervals** are calculated
- The meaning of the phrase: "One can say with 95% confidence…"
- The meaning of the term: "margin of sampling error"
- How it is that Gallup® can use a sample of only $n = 1532$ to make their estimate for the entire population
- How what Gallup® has done is connected with the *reasonably likely interval* calculated from a sampling distribution of $\hat{p}$ s, that we looked at in §3.4

What Gallup® has done is to calculate a **95% confidence interval** for the population proportion p of Americans who approve of the president's job performance. They can say "with 95% confidence" that the percent of Americans who approved of Obama's performance (for the time they took their survey) was 56%±3%. The 3% is called the **margin of sampling error** (or just **margin of error),** and by subtracting and adding this *margin of sampling error* we get an interval for the population proportion p: the *interval* in this situation is $53\% \leq p \leq 59\%$. That is, it is possible that only 53% of Americans approve of Obama's performance or it is possible that the proportion is as high as 59% or any percentage between 53% and 59%. Gallup® can say this with 95% confidence.

How 95% Confidence Intervals Come from Reasonably Likely Intervals

Imagine that we try to repeat what Gallup® did on a college campus. Instead of a sample size of $n = 1532$, we will manage (for simplicity) with a smaller sample size of $n = 100$. However, our sample *must* still be a random sample; everything depends upon the sample being random. One thing we can do even before we collect the data is to calculate intervals of *reasonably likely* $\hat{p}$ s for various *population p s*. Since a President's Approval Rating could be any p between 0.00 and 1.00, we have calculated many reasonably likely intervals by steps of 0.05. For example, if the population proportion $p = 0.60$ then the lower end of the reasonably likely interval is calculated by

p	$p - 1.96\sqrt{\frac{p(1-p)}{n}}$	$p + 1.96\sqrt{\frac{p(1-p)}{n}}$
0.950	0.907	0.993
0.900	0.841	0.959
0.850	0.780	0.920
0.800	0.722	0.878
0.750	0.665	0.835
0.700	0.610	0.790
0.650	0.557	0.743
0.600	0.504	0.696
0.550	0.452	0.648
0.500	0.402	0.598
0.450	0.352	0.548
0.400	0.304	0.496
0.350	0.257	0.443
0.300	0.210	0.390
0.250	0.165	0.335
0.200	0.122	0.278
0.150	0.080	0.220
0.100	0.041	0.159
0.050	0.007	0.093

$p - 1.96\sqrt{\frac{p \cdot (1-p)}{n}}$, so $0.60 - 1.96\sqrt{\frac{0.60 \cdot (1-0.60)}{100}} \approx 0.60 - 1.96(0.049) = 0.60 - 0.096 = 0.504$. The upper end of the interval will be $0.60 + 1.96\sqrt{\frac{0.60 \cdot (1-0.60)}{100}} \approx 0.60 + 1.96(0.049) = 0.60 + 0.096 = 0.696$. That is, the interval $0.504 \leq \hat{p} \leq 0.696$ contains the reasonably likely sample proportions $\hat{p}$ for samples of size $n = 100$ if the population proportion $p = 0.60$.

Now suppose from our *sample* of n = 100, we get fifty-six people who approve of the job the president is doing. This means that we have a $\hat{p} = 0.56$, which falls in the interval $0.504 \le \hat{p} \le 0.696$. Here is the question: how can we use this sample proportion $\hat{p} = 0.56$ to get an *estimate* of the population proportion? In other words, how can we *infer* (or generalize to) the population proportion p using our sample proportion $\hat{p} = 0.56$? To see the logic for this inference, we will use a **Swift Diagram**, as shown here.

The horizontal lines on the **Swift Diagram** show the intervals of reasonably likely $\hat{p}$ s shown in the table on the previous page. You should look at the reasonably likely interval for $p = 0.60$ and see that it does appear to reach from $\hat{p} = 0.504$ to $\hat{p} = 0.696$. Actually, although we have only put the horizontal reasonably likely intervals for $p = 0.05$, $p = 0.10$, $p = 0.15$... up to $p = 0.95$, you should imagine that the spaces between these horizontal lines are filled with other reasonably likely intervals or the values of p between.

In the Swift Diagram, the *vertical line* shows the $\hat{p} = 0.56$, which is the *phat* that we actually got.

Using the Swift Diagram, here is how we reason. If the population proportion of people approving the job the president is doing were (only) $p = 0.40$ then would we be likely to see the $\hat{p} = 0.56$ we actually got? No, we would not; a $\hat{p} = 0.56$ is far *outside* the reasonably likely interval for $p = 0.40$. If the population proportion were $p = 0.70$, would we be likely to see the $\hat{p} = 0.56$ that we actually got? Again, no, for the same reason; the $\hat{p} = 0.56$ is far *outside* the reasonably likely interval for the population proportion $p = 0.70$. However, the population proportion could be $p = 0.50$, or $p = 0.55$, or $p = 0.60$ or $p = 0.65$ because for all of *these* values for p, our value of $\hat{p} = 0.56$ is *inside* the reasonably likely intervals.

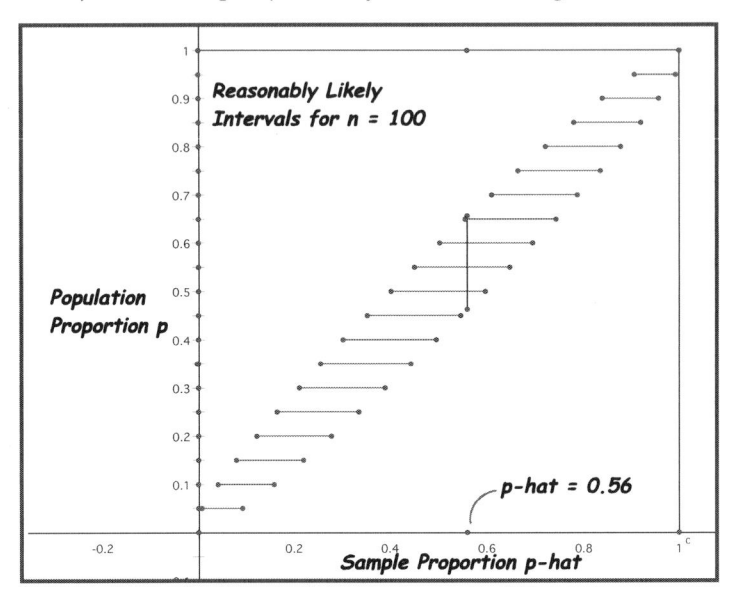

We can be even more precise. By looking at the bolded part of the vertical-line Swift Diagram, we can see that the population proportion should be somewhere between $p = 0.46$ and $p = 0.66$. We can say that we are **95% confident** that the population proportion is in the interval $0.46 < p < 0.66$ because those are the values of the population proportion p that have $\hat{p} = 0.56$ inside their 95% reasonably likely intervals for $\hat{p}$. The values of p outside this interval $0.46 < p < 0.66$ have $\hat{p} = 0.56$ *outside* their reasonably likely intervals. The bolded part of the vertical line in the Swift Diagram is the **95% confidence interval for a population proportion p.**

If you measure the length of the bolded part of the vertical line, you will find that its length was exactly the same as the length of the reasonably likely interval for $p = 0.56$. This gives a way of being very precise in our calculation of the bolded part of the vertical line in the Swift Diagram. The calculation for our example will be

$$\hat{p} \pm 1.96\sqrt{\frac{\hat{p}\cdot(1-\hat{p})}{n}} = 0.56 \pm 1.96\sqrt{\frac{0.56\cdot(1-0.56)}{100}} \approx 0.56 \pm 1.96(0.0496) \approx 0.56 \pm 0.097$$, and this results in an interval whose lower end is 0.56 − 0.097 = 0.463 and upper end is 0.56 + 0.097 = 0.657. Our 95% confidence interval is $0.463 < p < 0.657$. With our small random sample of $n = 100$, we can be 95% confident that the proportion of people who approve of the president's job performance is between 46.3% and 65.7%.

Calculation of a Confidence Interval for a Population Proportion p

$$\hat{p} \pm z^*\sqrt{\frac{\hat{p}\cdot(1-\hat{p})}{n}}, \text{ where}$$

$z^* = 1.645$ for a 90% confidence interval

$z^* = 1.96$ for a 95% confidence interval

$z^* = 2.576$ for a 99% confidence interval

This formula works under the **conditions** that:

1. The sample of size n from which $\hat{p}$ was calculated is a simple random sample.
2. Both $n\hat{p} \geq 10$ and $n(1-\hat{p}) \geq 10$.
3. The size of the population is at least ten times the size of the sample.

In the box above, there is a z^* in the formula in the position where we used 1.96 in our calculations. We have z^* instead of 1.96 because we can have confidence intervals that estimate the population proportion p with 90% or 99% confidence, and then the value for z^* will be different, although the basic reasoning is the same. (Actually, we could calculate confidence intervals with any level of confidence we want, but 90%, 95%, and 99% are the most common.)

The Margin of Error for a Confidence Interval

The formula $\hat{p} \pm z^*\sqrt{\frac{\hat{p}\cdot(1-\hat{p})}{n}}$ can be thought of as having two parts: the part on the left side of the "±", that is, our sample proportion $\hat{p}$, and the part on the right-hand side of the "±". The part on the left of the ± is sometimes called a **point estimate** or just **estimate.** This estimate is the number that we derive from calculating a proportion in our sample. The part of the right-hand side is called the **margin of error** (or **margin of sampling error**). The *margin of error* reflects what we know from the sampling distribution theory about how much variability there can be in sample proportions when we use randomization. Putting these together, we can say that a confidence interval is constructed by an **estimate ± margin of error.**

Margin of Sampling Error (or Margin of Error) for a Confidence Interval

$$ME = z^*\sqrt{\frac{\hat{p}\cdot(1-\hat{p})}{n}}$$

For the sample proportion $\hat{p} = 0.56$ for the president's approval proportion, we can calculate

$$ME = z^* \sqrt{\frac{\hat{p}\cdot(1-\hat{p})}{n}} = 1.96\sqrt{\frac{0.56\cdot(1-0.56)}{100}} \approx 1.96(0.0496) \approx 0.097.$$

When we calculated the confidence interval, we actually calculated the margin of error; it was the part that we subtracted from and added to our $\hat{p} = 0.56$. The **margin of error** for our 95% confidence interval is 0.097 or about 9.7%. If we interpret the confidence interval in the way the Gallup® organization has, we would say: "We estimate Obama's approval rating to be 56% with a margin of sampling error of 9.7%." It may be that a 9.7% margin of error is not that impressive; recall that the Gallup® organization said that their margin of error was at most 3%. Why? The answer is that they had a much bigger sample size. We can calculate the margin of error with their sample size and the same $\hat{p} = 0.56$ as:

$$ME = z^* \sqrt{\frac{\hat{p}\cdot(1-\hat{p})}{n}} = 1.96\sqrt{\frac{0.56\cdot(1-0.56)}{1532}} \approx 1.96(0.0127) \approx 0.025$$

The margin of error is 2.5 (less than what Gallup® said; this is because they advertised a "maximum" margin of error of 3%). If we look at the formula for the margin of error, we can easily see that as the sample size increases, the margin of error decreases (if the other numbers are the same) since we are dividing by a larger number for n.

The bigger the sample size, the smaller the margin of error. If we decide that we want a specific margin of error (say, 3%), we can do the algebra and calculate what sample size we need for that margin of error. The algebra is not too difficult. The next section shows a detailed example.

Sample Size for a Given Margin of Error What sample size would be necessary for a 3% (or 0.03) margin of error? We calculate by setting 3% or 0.03 to be bigger than or equal to ME. In symbols, we set $0.03 \geq ME$ in the ME formula $z^*\sqrt{\frac{\hat{p}\cdot(1-\hat{p})}{n}}$ so we have $0.03 \geq z^*\sqrt{\frac{\hat{p}\cdot(1-\hat{p})}{n}}$ and then solve for n using algebra.

$ME = z^*\sqrt{\frac{\hat{p}(1-\hat{p})}{n}}$	
$0.03 \geq 1.96\sqrt{\frac{0.56\cdot(1-0.56)}{n}}$	The "$\geq$" translates ME $\leq$ 0.03 as 0.03 $\geq$ ME
$\frac{0.03}{1.96} \geq \sqrt{\frac{0.56\cdot(1-0.56)}{n}}$	Divide both sides by 1.96
$\left(\frac{0.03}{1.96}\right)^2 \geq \frac{0.56\cdot(1-0.56)}{n}$	Square both sides
$n \geq \frac{0.56\cdot(1-0.56)}{\left(\frac{0.03}{1.96}\right)^2}$	Multiply both sides by n and divide by $\left(\frac{0.03}{1.96}\right)^2$
$n \geq \left(\frac{1.96}{0.03}\right)^2 [0.56\cdot(1-0.56)]$	Dividing by $\left(\frac{0.03}{1.96}\right)^2$ is the same as multiplying by $\left(\frac{1.96}{0.03}\right)^2$
$n \geq (4268.444)\cdot(0.2464)$	
$n \geq 1051.74$	
$n \geq 1052$	Sample sizes must be whole numbers, so round up.

For a 3% margin of error, the sample size that is needed is just over $n = 1000$. This means that we can estimate a population proportion p, whatever the size of the population, if we have a sample size of about $n = 1000$. The margin of error does *not* depend on the size of the population; the margin of error depends on the level of confidence (so the z^*), the $\hat{p}$, and the sample size n. This answers the question of how organizations such as Gallup® can estimate population proportions with what appear to be very small sample sizes, such as $n = 1532$.

It is sometimes simpler to have done the algebra in general and to have the formula already solved for n.

Calculation of the Sample Size for a Given Margin of Error

$$n \geq \left(\frac{z^*}{ME}\right)^2 \left[\hat{p} \cdot (1-\hat{p})\right]$$

Note: If the $\hat{p}$ is not known (as, for example, before data are collected), use $\hat{p} = 0.50$ since this gives the biggest possible value for $\hat{p} \cdot (1-\hat{p})$ and gives a slightly larger sample size than is actually needed.

Interpretation of a Confidence Interval

There are a number of ways to correctly interpret a confidence interval. There are also a number of ways to go wrong and give an interpretation that makes no sense. Gallup® and other opinion polls report the estimate—the $\hat{p}$—along with the margin of error, with a note stating the confidence level. For our presidential approval example, Gallup® would report that the rating was 56% and then note that "one can say with confidence that the maximum margin of error is ± 3 percentage points," as in the quote at the beginning of this section. For our example of the students on the campus and the proportion who approve of the president's job performance, we could say (following this pattern): "With 95% confidence, we can say that 56% of students on the campus approve of how the president is doing his job, with a margin of sampling error of 9.7%"

A standard interpretation in textbooks reports the confidence interval, the confidence level, and the population proportion one is estimating. So, for the example we have been using, where the population is the collection of all students on one campus, we would say: "We can be 95% confident that the proportion of all students on the campus who approve of the president's job performance is between 46.3% and 65.7%"

Here is a kind of template for these two ways of interpreting a confidence interval.

Interpretation of a Confidence Interval I

"With 95% (or 90%, or 99%, whatever the level is) confidence, we can say that the proportion for [variable] for [the population from which the sample was drawn] is [the sample proportion] with a margin of error of [margin of error]."

Interpretation of a Confidence Interval II

"We can be 95% (or 90%, or 99%, whatever the level is) confident that the proportion for [variable] for [the population from which the sample was drawn] is between [the lower limit] and [the upper limit]. The margin of error is [margin of error]."

In both of these interpretations, what is made clear is the (i) confidence level, (ii) the variable being measured, (iii) the population to which we are inferring, and (iv) the results.

Reading technical reports In technical research reports, the interpretation is typically left to the reader, who is assumed to have taken a course in statistics and to understand what these intervals are.

Often, the confidence intervals are given (without interpretation), or sample proportions $\hat{p}$ and the margins of error are given, especially when there are many estimates to be reported. However, in this situation, somewhere in the report, researchers make clear what confidence level (90%, 95%, 99%, or some other) is being used and what the population is. In the exercises, some of the common mistakes that are made in interpretation will be pointed out.

Repeated sampling process interpretation There is still another way to interpret confidence intervals. Here is the Fathom output for a confidence interval for our Presidential Approval Rating example. The last sentence states that if the process were performed

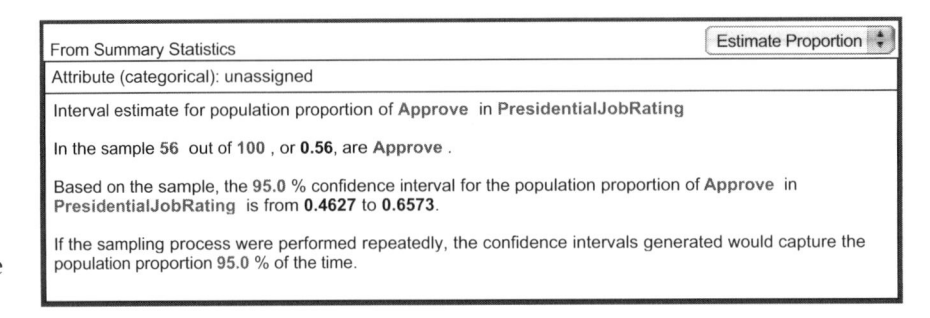

From Summary Statistics	Estimate Proportion
Attribute (categorical): unassigned	
Interval estimate for population proportion of Approve in PresidentialJobRating	
In the sample 56 out of 100, or **0.56**, are Approve.	
Based on the sample, the **95.0** % confidence interval for the population proportion of Approve in PresidentialJobRating is from **0.4627** to **0.6573**.	
If the sampling process were performed repeatedly, the confidence intervals generated would capture the population proportion 95.0 % of the time.	

repeatedly then 95% of the confidence intervals that would be generated would capture the population proportion. This sounds a bit like the process for making a sampling distribution and is in fact related to it. What is meant by "95% of the confidence Intervals would capture the population proportion" can be illustrated by using a Swift Diagram.

How the Swift Diagram works; many lines that we usually do not draw. In this diagram, we have specified the "true population proportion" to be 0.54, and that is shown by the thick horizontal line. The thin, solid, vertical lines are possible confidence intervals that we could calculate depending on what our $\hat{p}$ happened to be. We cannot show all of them, of course, but we can show what would happen with a few; 95% of the confidence intervals will be based upon the $\hat{p}$ s that are reasonably likely. Confidence intervals calculated from any of these reasonably likely $\hat{p}$ s will "capture" or cross the true population proportion. The only confidence intervals that will miss will be those based on the 5% *rare* $\hat{p}$ s. This diagram shows two of them, shown as the thicker vertical confidence interval lines. For those rare 5% of the confidence intervals, the confidence interval will *not* cross the true population proportion. We can be 95% confident that our confidence interval is one of those based on a reasonably likely $\hat{p}$ and captures the *p*.

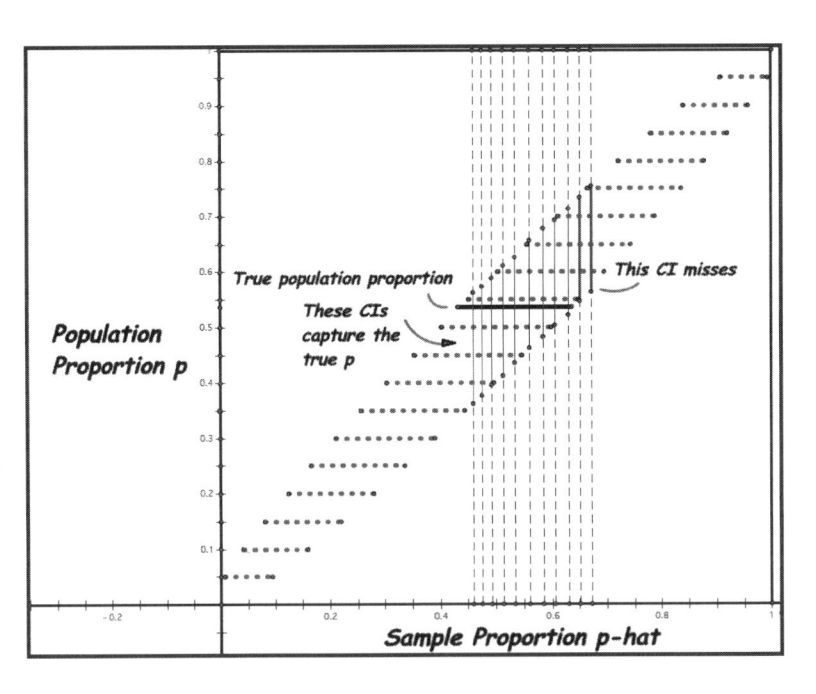

Summary: Confidence Intervals for Proportions

- One formal type of statistical inference is to calculate an *interval estimate* of the population proportion p based upon a sample proportion $\hat{p}$, where the sample is randomly drawn from the population.
    - This *interval estimate* is called a **confidence interval.**
    - The connection between the calculation of a *confidence interval* and the *reasonably likely* intervals based upon the sampling distribution of $\hat{p}$ can be seen in a Swift Diagram, in which the reasonably likely intervals are the horizontal lines, and the confidence intervals are the vertical lines. (See the example above.)

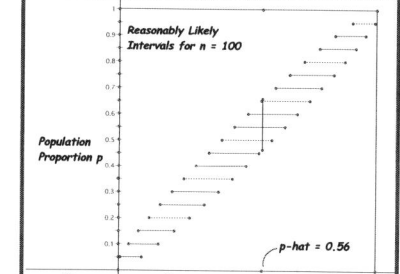

- **Formula for a Confidence Interval for a Population Proportion p:** $\hat{p} \pm z^* \sqrt{\dfrac{\hat{p} \cdot (1-\hat{p})}{n}}$, where $z^* = 1.645$ for a 90%, $z^* = 1.96$ for a 95% and $z^* = 2.576$ for a 99% confidence interval,

- The formula given above for a confidence interval works under the **conditions** that:
    - The sample of size n from which $\hat{p}$ was calculated is a simple random sample.
    - Both $n\hat{p} \geq 10$ and $n(1-\hat{p}) \geq 10$.
    - The size of the population is at least ten times the size of the sample.

- Another way of expressing the general form of a confidence interval is: **estimate ± margin of error.**

- **Margin of Error for a Confidence Interval:** $ME = z^* \sqrt{\dfrac{\hat{p} \cdot (1-\hat{p})}{n}}$; the margin of error is the part of the formula that is subtracted from and added to the point estimate.
    - The larger the sample size, the smaller the margin of error, and hence the narrower the confidence interval.
    - It is possible, using algebra, to calculate the sample size needed to achieve a specific margin of error; the following formula may also be used: $n \geq \left(\dfrac{z^*}{ME}\right)^2 [\hat{p} \cdot (1-\hat{p})]$.

- **Interpretation of a Confidence Interval** There is a number of correct interpretations of a confidence interval and also many incorrect interpretations. Here are two templates for correct interpretations:
    - "With 95% (or 90%, or 99%, whatever the level is) confidence, we can say that the proportion for [variable] for [the population from which the sample was drawn] is [the sample proportion] with a margin of error of [margin of error]."
    - "We can be 95% (or 90%, or 99%, whatever the level is) confident that the proportion for [variable] for [the population from which the sample was drawn] is between [the lower limit] and [the upper limit]. The margin of error is [margin of error]."

§4.1 Exercises on Confidence Intervals for Proportions

1. **What is the probability of a delayed flight to LA?** The government agency Bureau of Transportation Statistics (see www.bts.gov) collects data on every commercial flight in the United States. The Fathom file **OntimeCombLAXSample.ftm** is a random sample of flights from the San Francisco Bay Area to Los Angeles International (LAX). For this exercise, we will be analyzing the variable *DelayOver15* whose categories are: "Departure Delay 15 min. or less" and "Departure Delay Over 15 min." Our goal will be to get an *estimate* of the probability that a passenger will be delayed more than fifteen minutes on a flight from the San Francisco Bay Area to LAX.

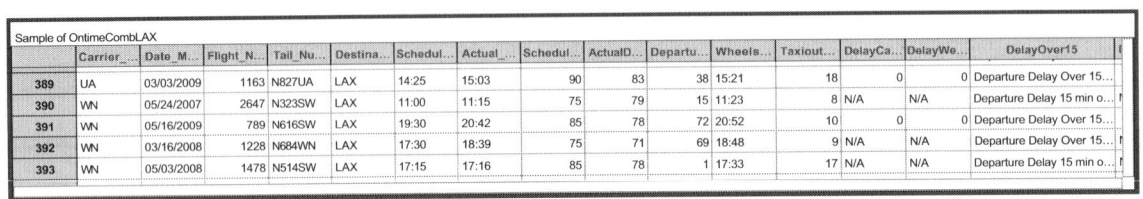

	Carrier	Date_M...	Flight_N...	Tail_Nu...	Destina...	Schedul...	Actual_...	Schedul...	ActualD...	Departu...	Wheels...	Taxiout...	DelayCa...	DelayWe...	DelayOver15
389	UA	03/03/2009	1163	N827UA	LAX	14:25	15:03	90	83	38	15:21	18	0	0	Departure Delay Over 15...
390	WN	05/24/2007	2647	N323SW	LAX	11:00	11:15	75	79	15	11:23	8	N/A	N/A	Departure Delay 15 min o...
391	WN	05/16/2009	789	N616SW	LAX	19:30	20:42	85	78	72	20:52	10	0	0	Departure Delay Over 15...
392	WN	03/16/2008	1228	N684WN	LAX	17:30	18:39	75	71	69	18:48	9	N/A	N/A	Departure Delay Over 15...
393	WN	05/03/2008	1478	N514SW	LAX	17:15	17:16	85	78	1	17:33	17	N/A	N/A	Departure Delay 15 min o...

    - Use the Fathom file **OntimeCombLAXSample.ftm** and get the case table.
    - Get a summary table and drag the variable *DelayOver15* to the downward-pointing arrow.

Sample of OntimeCombLAX		
DelayOver15	Departure Delay 15 min or Less	332
	Departure Delay Over 15 min	93
	Column Summary	425
S1 = count ( )		

        a. What are the cases for this analysis? Are they airlines, airplanes, airports, delays, flights, passengers, or times?

        b. Is the variable *DelayOver15* categorical or quantitative?

        c. [Review] Use the probability notation of §1.2 with the "symbol" *Over* to show the probability *from this sample* that a flight will be delayed over fifteen minutes.

        d. In calculating an estimate of the population proportion, what symbol should be used for the proportion of *all* flights from the San Francisco Bay Area to LAX, p or $\hat{p}$? Give a reason for your answer.

        e. In calculating an estimate of the population proportion, what symbol should be used for the proportion that you calculated in part c, p or $\hat{p}$? Give a reason for your answer.

        f. Calculate a 95% confidence interval for the proportion (or probability) of flights delayed over fifteen minutes.

    - You will check your calculated answer to part f by having Fathom do the calculation. Drag an **Estimate** box (a "ruler") from the shelf and change the box from an "Empty Estimate" to "Estimate Proportion."
    - Drag the variable *DelayOver15* to the "Attribute (categorical) unassigned" space.
    - You will find that Fathom has calculated the confidence interval for the category "Departure Delay 15 min or less" rather than for "Delay Over 15 min," but since those are in blue, you can change those to what you want.

g. Fathom gives a correct interpretation that is different from the standard interpretations shown in the *Notes*, although Fathom's interpretation is also discussed:

> If the sampling process were performed repeatedly, the confidence intervals generated would capture the population proportion 95% of the time.

To get an idea of what this means, think of getting two thousand different random samples of $n = 425$ from the flights from the Bay Area to LAX, and each time you calculate a confidence interval. How many of the two thousand confidence intervals will include (or capture) the population proportion p of "Delay Over 15 min"?

h. Give an interpretation (using the pattern in the *Notes*) of the confidence interval you have calculated in the context of the problem. **PTO**

i. Think about your results; if you flew to LAX from the SF Bay Area five times, would you be surprised if all five flights were delayed over fifteen minutes? Would you be surprised if one of them was delayed? Explain in the light of your estimate.

j. What would happen to your confidence interval if you increased the *level of confidence* from 95% to 99%? (You can actually see by changing the level of confidence in the Fathom output.) Explain in terms of the formula for the confidence interval.

2. The President's Approval Rating

- Log on to http://www.gallup.com/poll/124922/Presidential-Approval-Center.aspx, which gives a running account of the president's approval rating. The president's approval rating for December 7, 2009, was 47%. Gallup® used a sample size of $n = 1529$ for this survey (in the link, look under "Survey Methods").

    a. What symbol should be used for the 47% (or 0.47), p or $\hat{p}$? Give a reason for your answer.

    b. From the information given, calculate the margin of error for a 95% confidence interval for the proportion of Americans who gave the president their approval.

    c. Use what you have calculated in part b to find the 95% confidence interval for the population proportion of Americans who, on December 7, 2009, approved of the president's handling of his job.

    d. Confused Conrad has exactly the same answer to parts b and c. Explain to CC exactly what his error is.

    e. Express your confidence interval using mathematical notation; that is, $0.445 < < 0.495$. What symbol should be placed between the inequality symbols, p or $\hat{p}$? Give a reason for your answer.

    f. Give a valid interpretation of the 95% confidence interval that you have calculated.

    g. Will a 99% confidence interval be wider or narrower? Give a reason for your answer. (You can answer by actually doing a calculation or by reasoning.)

    h. What is the term by which the numbers 90%, 95%, 99% is known in the context of confidence intervals?

- Gallup® also tracks the approval rating on a weekly basis. The results for the weekly assessment can be found at http://www.gallup.com/poll/121199/Obama-Weekly-Job-Approval-Demographic-Groups.aspx.

i. For these figures, calculate the margin of error for a 95% confidence interval. You will find that the margin of error for these numbers is considerably smaller than the answer you got for part b. What in the formula for margin of error makes this answer smaller?

j. Complete this sentence: "In general, if the sample proportion is the same and the level of confidence is the same then with a bigger sample size, we will have a _____ margin of error."

k. Calculate the 95% confidence interval for the approval rating for the week of December 14–20, 2009. Is this confidence interval wider or narrower than the one that you calculated in part c?

l. For the week of December 14–20, 2009, for a sample size of $n = 3646$, 7% of those sampled had "no opinion." Is the condition about sample size for the calculation of a 95% confidence interval met with this small a sample proportion? Do a calculation to determine whether the condition is met and come to a conclusion.

Special Exercise 3: Confidence Intervals on Real Estate

- Open the Fathom file **San Mateo Real Estate Y0506.**
- Select the Collection Icon and go to **Collection>Sample Cases.**
- In the Inspector (which will appear), deselect "With Replacement," change the sample size from 10 to 64, and click on **Sample More Cases.**
- With the **Sample of San Mateo...** selected, get a **summary table.**
- In the **case table,** scroll over to find the variable *Style_2*. Drag the variable to the **summary table.** You should have a table that looks something like this, but, of course, since your sample is a different sample, the numbers will be different.

We are interested in the proportion (or the probability) that a house is a *Ranch* house.

1. If we were in the "world of probability calculations" (§1.2) and if we use *R* for "ranch house," calculate the probability that a house is a ranch house" and express the probability using the correct notation. (Remember to use your own sample data and not the data shown in the table above.)

2. However, we are not in the world of probability calculations (though that world is just next door); we are in the world of making confidence intervals, so we will use either p or $\hat{p}$. Which one should we use? (*Hint:* Are the numbers from a sample or a population?) Give a reason for your choice.

3. Using the confidence interval formula, calculate a 95% confidence interval for p using your $\hat{p}$. (The Fathom instructions just below will allow you to check your results.)

- With the collection icon for the **Sample of San Mateo...** selected, drag down from the shelf an **Estimate.**
- Choose the kind of estimate we want from the pop-up menu on the box. (We want to estimate a Proportion.)
- Drag the variable *Style_2* to the space labeled "Attribute (categorical) Unassigned." What you should see should be similar to this picture.
- We do not want the proportion of "Contemporary Houses"; we want the proportion of "Ranch" houses. The selection is in blue so is open to modification. Click on it and choose Ranch.

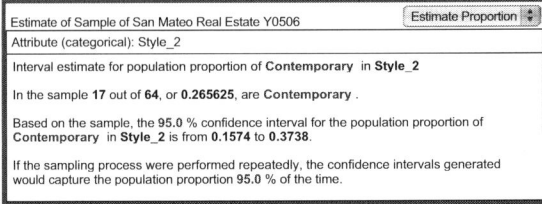

Fathom has calculated the confidence interval. (You can now check whether your calculations for your confidence interval were correct.)

4.
   a. In your calculation, what is the value of the "margin of error"?
   b. Subtract the CI lower limit *from* the CI upper limit and divide by 2; does this number agree with your answer to a?
   c. Add the CI lower limit to the CI upper limit and divide by 2; what is the symbol for the number you get?
   d. True or false, and explain: "The margin of error is just half the width of the confidence interval."

5. Give an interpretation of your confidence interval as outlined in the boxes in the **Notes,** making certain that what you say is the context of the data, which are "houses sold in San Mateo County in 2005–2006."

6. Compare your confidence interval with the confidence interval calculated by someone else with a different sample. Do you expect their confidence interval to be exactly the same as yours (that is, assuming that both of you did everything correctly)? Similar to yours? Why or why not?

- You will find **Swift Diagram** with the reasonably likely intervals shown on the last page of these exercises. (You may also be able to print one out from the course website, or you may be given one — free.)

•7. a. On the *x* axis of the **Swift Diagram**, plot your $\hat{p}$. Draw a light vertical line for your $\hat{p}$. (That is why there is a scale at the top of the plot.)
   b. On this light vertical line, plot the lower limit and the upper limit for your estimate of *p* as it is expressed in the confidence interval. (See the **Swift Diagram** in the **Notes.**)
   c. Darken in the part of the line that is your confidence interval.

- Select the collection icon for the **San Mateo Real Estate Y0506** (not the sample) and get a **case table.**
- Get a **summary table** for the variable *Style_2*.

8. a. The **summary table** you have just found gives you the population number for different styles of houses. Calculate the population proportion of "Ranch" houses and give it the correct symbol. Also draw a horizontal line on the **Swift Diagram**.
   b. Did your confidence interval capture (or include) the true population proportion *p*?

A Bigger Sample Size We will see the effects of getting a bigger sample size on the confidence interval. Instead of a sample size of $n = 64$, we will have a sample size of $n = 144$, but we are interested in the proportion of ranch houses in the population *p*, and we want to estimate this *p* using our $\hat{p}$ of ranch houses.

9. Look at the formula for the confidence interval.
   a. If you increase the sample size, what will happen to the size of the margin of error? Explain your answer by referring to what the formula does.
   b. If you increase the sample size, what should happen to the width of the confidence interval? Explain your answer.

- Select the Collection Icon and go to **Collection>Sample Cases.**

- In the Inspector, deselect the "With Replacement," change the sample size from 10 to 144, and click on *Sample More Cases*. We are getting a sample size of $n = 144$ instead of $n = 64$.
- With **Sample of San Mateo...** selected, open a **case table** and get a **summary table**.
- In the **case table**, scroll over to find the variable *Style_2*. Drag the variable to the **summary table**. You should have a table that looks something like this, but, of course, your sample is a different sample, so the numbers will be different.

Sample of San Mateo Real Estate...		
Style_2	Contemporary	38
	Other	39
	Ranch	43
	Traditional	24
	Column Summary	144
S1 = count ( )		

10. a. Use a calculator to calculate the 95% confidence interval for the population proportion of ranch houses using the sample proportion $\hat{p}$ found from your Summary Table, and express your answer using the proper notation.
    b. What is the margin of error for your confidence interval for p using $n = 144$?
    c. Compare the size of your margin of error for the CI based on $n = 144$ with the margin of error for the CI based on $n = 64$. Does the comparison follow what you predicted in answering question 9a?
    d. Give a good interpretation of your new CI.

More Confident What happens when we increase the level of confidence from 95% to 99%?

11. a. What number in the formula will be different for the calculation of the 99% CI compared with the 95%?
    b. Will the margin of error be bigger or smaller with 99%? Explain.
    c. Do the calculation and express the answer using proper notation.
    d. Compare the size of your margin of error for the 99% CI with the 95% CI. Does the comparison follow what you predicted in answering question 11b?

- Get the Fathom output for your sample of $n = 144$ and check your calculations in questions 10 and 11. Follow the instructions after question 3 above. Notice that the confidence level is in blue, so it can be changed from 95% to 99%.

Special Exercise 4: Internet in Australia?

Australian High School Students The data for this exercise are from a sample of high school students in the Australian states of Queensland and Victoria.
- The sample size is $n = 200$.
- Here is a part of the Case Table.

1. Based on the description just above, write down what the cases are and what population you could reasonably say the sample represents. (It is reasonable to ask questions about how the data for the Censuss @ School were collected.)

CAS Australia Vic Qld								
	Qtime	Sex	BrthMnth	BrthYear	BrthPlce	BedRms	PeoplHme	InetAxs
154		19 Male	July	1990	Victoria	5	6	Yes - br...
155		42 Male	May	1989	Victoria	4	4	Yes - br...
156		17 Male	July	1992	Queensl...	3	4	Yes - br...
157		23 Male	January	1992	Queensl...	4	4	Yes - br...
158		20 Female	September	1990	Victoria	3	3	Yes - dia...
159		18 Male	March	1989	South Au...	4	2	Yes - dia...
160		16 Male	January	1992	Victoria	5	5	Yes - br...

Here are the results by year of school for the type of Internet connection the students reported.

CAS Australia Vic Qld		First Year	Second Year	Third or Fourth Year	Row Summary
	No - Internet connection	15	9	1	25
InetAxs	Yes - broadband connection	69	37	14	120
	Yes - dial-up connection	27	15	10	52
	Yes - other (include Internet access through mobile phone etc)	1	1	1	3
	Column Summary	112	62	26	200

S1 = count ( )

2. From the table above, you can see that sixty-nine out of the 112 first-year students said that they had a broadband Internet connection at home.
   a. Calculate a proportion from the "69 out of the 112" and give this proportion its correct symbol that we will use in calculating a confidence interval. Notice that now for the sample of just first-year students, n is smaller than it was for the entire sample. (So, now, $n = 112$; this is the n you will use for the other parts of the question.)
   b. Determine whether the conditions necessary for calculating a confidence interval are met for the proportion of first-year students who have a broadband Internet connection at home. (See the box in the ***Notes.***)
   c. Using the formula for a confidence interval, calculate a 95% confidence interval for the proportion of first-year students in Queensland and Victoria who have a broadband Internet connection. (Ans: 0.616 ± 0.090 or $0.526 < p < 0.706$)
   d. One part of the formula is $\sqrt{\dfrac{\hat{p}(1-\hat{p})}{n}}$. Confused Conrad puts this into his calculator: $\sqrt{\dfrac{69(1-69)}{112}}$ and he finds that his calculator refuses to give him an answer. Tell Conrad what his mistake is, and why his calculator is being troublesome.
   e. Give a correct interpretation for your confidence interval.
3. For the second-year students, the confidence interval for the proportion of students in the population of students who have broadband Internet connection comes out as 0.597 ± 0.122, which translates into the interval $0.475 < p < 0.719$.
   a. The margin of error (the number to the right of the $\pm$) is 0.122. Show that you get this number by getting half the width of the confidence interval.

281

b. What happens to the margin of error when you increase the sample size? Explain by referring to the formula.

c. Suppose you wanted the margin of error to be as small as 0.06 (you want that amount of accuracy) for the second-year students. Calculate the sample size of second-year students you would have to have if the sample proportion $\hat{p}$ is the same.

d. Here are some bad interpretations for the confidence interval in this question, which is $0.475 < p < 0.719$. For each bad interpretation, say why it is incorrect.
   > "The confidence interval is 0.475 to 0.719."
   > "This confidence interval shows that 95% of the students have Internet connection between 47.5% and 71.9% of the time."
   > "We can be 95% confident that the sample percentage of second-year students who have broadband Internet connection in Victoria and Queensland states is between 47.5% and 71.9%." (*Hint:* The problem is *not* that the interpretation uses percentages; that is acceptable: percentages are just another way of expressing proportions.)

e. Give a correct interpretation for the confidence interval $0.475 < p < 0.719$ or 0.597 ± 0.122.

4. **Thinking about Confidence Intervals.** You have all the information you need to answer the questions below; you just need to think and perhaps think in reverse.

   Large parts of Australia are dry. One of the questions all two hundred high school students were asked was: "Do you agree or disagree that Australia will always have plenty of water?' The confidence interval for the "disagree" answer to this question (so that the students showed some concern about the supply of water) was $0.652 < p < 0.778$.

   a. What was the $\hat{p}$? (Think of how the interval was calculated.)
   b. What was the margin of error?
   c. Approximately what number of students disagreed?

5. Suppose we wanted confidence intervals for the proportion of Australian students who do *not* have Internet connection.
   a. For the first-year students, are the conditions met for calculating a confidence interval? Explain.
   b. For the second-year students, are the conditions met for calculating a confidence interval? Explain.

6. The Australian Bureau of Statistics administers the Census @ School Project, and a great many schools in Australia participate. When a school participates, all of the students in the school answer the questionnaire, which is administered online. However, school participation is voluntary, so not all schools in Australia are part of the project.
   a. Are the data from a simple random sample of Australian high school students? Give reasons for your answer. If you think that the data area not SRS, what label would you give the sampling process? (Check back to §3.1 for the various types of sampling.)
   b. If the data are not from an SRS, what implication does that have for the calculation of confidence intervals?

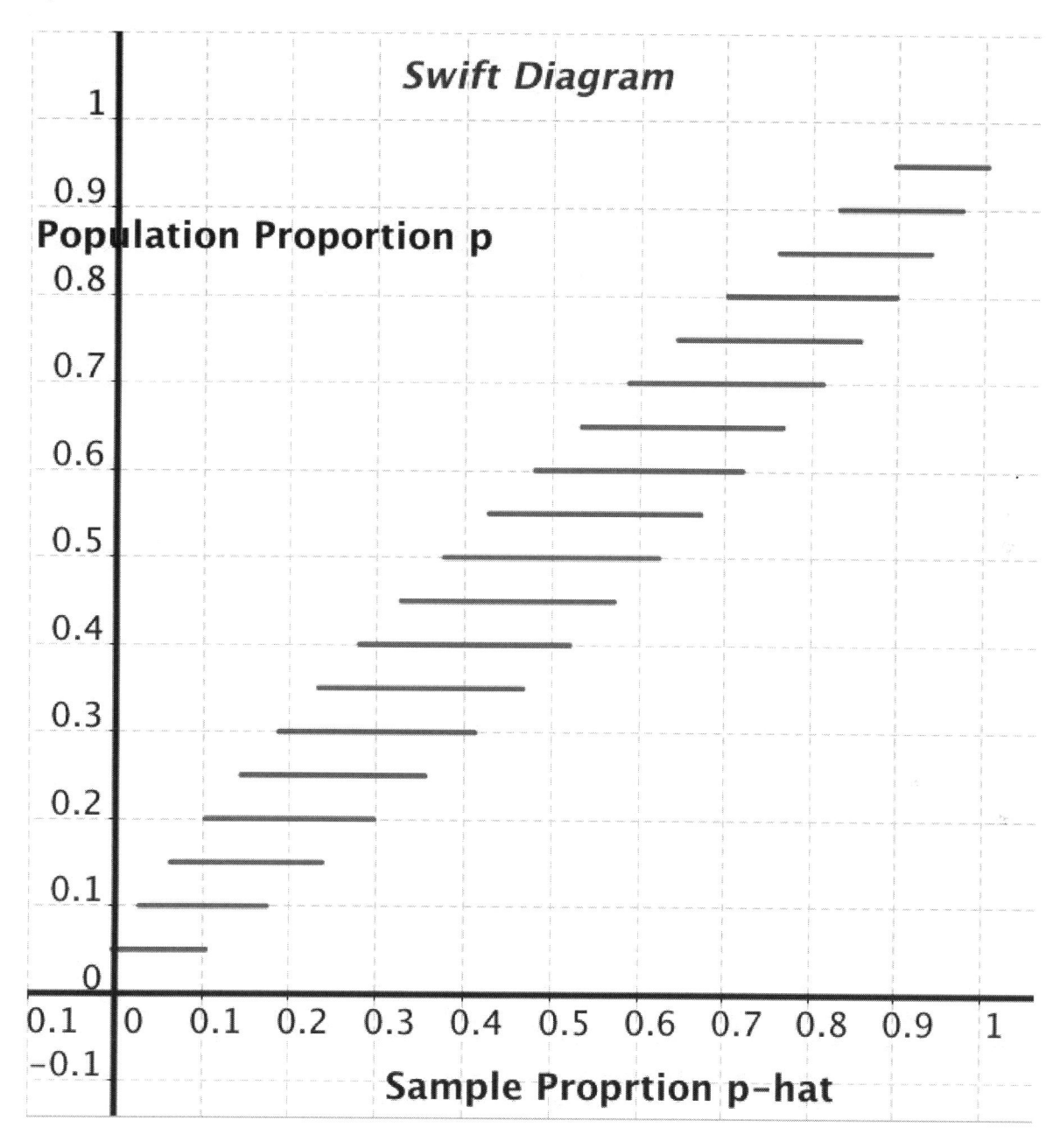

Special Exercise 3

Swift Diagram

Population Proportion p

Sample Proprtion p-hat

§4.2 Hypothesis Testing: Handedness

Background: What is the percentage of left-handed people?

The usual answer to this question is that about 10% to 12% of the population is left-handed. However, the proportion has been found to vary across cultures. A web page by the Australian Broadcasting Corporation "News in Science" (www.abc.net.au/science/news/stories/s1196384.htm, Sept. 13, 2004) reports on research done by a team of Australian and American researchers.[5]

> There may be more left-handed people than we realise, an international study has found. If we include the number of people who throw a ball, strike a match or use a pair of scissors with their left hand, the researchers say the world looks more of a left-handed place.
>
> Australian researcher Sarah Medland of the Queensland Institute of Medical Research in Brisbane and team publish the research in the current issue of the journal *Laterality*.
>
> Left-handed people face problems in a world where most things, from scissors to can-openers and computers to power tools, are designed for right-handers.
>
> In the past left-handers faced even greater problems. Some schoolchildren were forced, under threat of the strap, to write with their right hand, regardless of their natural tendency. Medland and team hoped to shed light on the contribution of cultural factors like this on the distribution of handedness.

The researchers studied a sample of $n = 8528$ people from various parts of the world and measured "handedness" in various ways. A previous larger study had found that the proportion of left-handers varied in different countries and in different cultures. For example, they found that the proportion was 2.5% for Mexico and 12.8% for Canada.[6] Other researchers have found very low proportions of left-handedness in Chinese (3.5%) and in Japanese (0.7%) schoolchildren.

These researchers think that the differences in the proportion left-handed are probably cultural rather than genetic. That is, in many cultures there is great pressure to conform to the "norm" of right-handedness, but in some cultures left-handed or ambidextrous people are "allowed" to be left-handed, although they still have to live in a right-handed world and have to make accommodations for computer mice, keyboards, and can openers.

Our statistical question is:

> *Does the percentage of left-handers in the population of all students at one college differ from the "standard" 11 percent that is found generally?*

This college has a very diverse student body, and so our reasoning is that the proportion left-handed may be lower or perhaps higher than the accepted 11%. We are unwilling to say which way, but we want to know if we have evidence that the proportion is different.

Language and Logic of Hypothesis Tests

Estimation and Testing: the Difference We have been doing *inference* (or generalization) from a sample $\hat{p}$ to a population p by calculating a *confidence interval estimate*. We can calculate a confidence interval even when we have no idea at all about what the population proportion should be;

---

[5] Medland, S. E., Perelle, I, De Monte, V., and Ehrman, L. (2004) "Effects of culture, sex, and age on the distribution of handedness: An evaluation of three measures of handedness." *Laterality: Asymmetries of Body, Brain and Cognition*, **9**(3): 287–297

[6] Medland, et. al (2004), page 288

all we need is a sample proportion $\hat{p}$ drawn from a sample of size n, and we can calculate a confidence interval to get an estimate of the population proportion p.

Hypothesis testing is different: it starts with a preconceived *idea* about the population proportion p and uses the sample proportion $\hat{p}$ to test whether our sample information (that is, the $\hat{p}$) is in agreement with our preconceived idea or (on the other hand) the $\hat{p}$ disagrees with our preconceived idea. In our example, our preconceived idea is that the proportion of left-handers in the population is $p = 0.11$. Our sample data is for the **CombinedClassData** for 2009. The proportion of students in our sample of who reported being left-handed was

$\hat{p} = \dfrac{27}{317} = 0.0852$, about 8.5%. This 8.5% is somewhat lower than the 11% that we were expecting. Does this mean, necessarily, that the entire population of CSM has a lower percentage of

Combined Class Data 09				
	DominantHand			Row Summary
	Right	Left	Ambidextrous	
	274	27	16	317
	0.864353	0.0851735	0.0504732	1

S1 = count ( )
S2 = rowProportion

left-handers than the 11% that is reported generally? No, it does not; remember that our sample proportion $\hat{p}$ varies from the population proportion p just because it is a calculated from a sample and not from the entire population. We need to use the idea of a *sampling distribution*. So how do we do this? There is some standard terminology and structure to hypothesis tests.

Null and Alternate Hypotheses. The first thing we do is to formalize our statistical question by expressing our preconceived idea in the form of two competing hypotheses about the population proportion p. One hypothesis is called the **null hypothesis** and usually represents what is commonly accepted; in our example, the null hypothesis is that the "population proportion of left-handers among CSM students is $p = 0.11$." A second hypothesis, which is counter (or contrary) to the *null hypothesis* and can be thought of as a *challenge* or a *claim against* the null hypothesis, is called the **alternate hypothesis.** For our statistical question, the *alternate hypothesis* is that the "population proportion of left-handers among CSM students *differs* from $p = 0.11$, that is $p \neq 0.11$." There is notation for these two hypotheses, and that notation is shown in the box on the next page. In that box, the symbol p_0 (pronounced "p naught" or "p zero") denotes the basic value that we have for the null hypothesis. In our example, p_0 has the value 0.11.

Notice that both the *null hypothesis* and the *alternate hypothesis* use the same value for p_0, but say different things about the p_0; the null hypothesis says that the population proportion p is equal to the p_0, and the alternate hypothesis says that the population proportion p does *not* equal p_0.

Notation for Null and Alternate Hypotheses for Population Proportions

The *null hypothesis* that the population proportion p is equal to p_0, written: **H₀: p = p₀**

The *alternate hypothesis* that the population proportion p is not equal to p_0, written: **Hₐ: p ≠ p₀**

[**Note:** This alternate hypothesis is known as **two-sided** because we do not specify in which way the population proportion differs from the p_0—whether p is larger or smaller than p_0. **One-sided** tests are discussed later.]

For our example, we would write: $H_0 : p = 0.11$
                                     $H_a : p \neq 0.11$

Set-up and Calculations for a Hypothesis Test: First Four Steps

We can think of a conducting a hypothesis test as a having **five steps**; the fifth and last step is interpretation, and in these **Notes** that step is so important that it has its own sub-section.

The **first step** in any hypothesis test is to specify clearly and correctly the null and alternate hypotheses. Then where do we go from there? Checking the conditions is the **second step**. We check the conditions so that we can be sure that we can use the *sampling distribution implied by the null hypothesis*. For our example, the sampling distribution is the sampling distribution for $\hat{p}$ s for random samples of $n = 317$ drawn from a population in which $p_0 = 0.11$. What is that sampling distribution? We can describe it:

Shape: The shape is approximately Normal because the **conditions** for Normality
$np_0 = 317(0.11) = 34.87 > 10$ and $n(1 - p_0) = 317(1 - 0.11) = 317(0.89) = 282.13 > 10$ are met.

Center: $\mu_{\hat{p}} = p_0 = 0.11$

Spread: $\sigma_{\hat{p}} = \sqrt{\dfrac{p_0(1 - p_0)}{n}} = \sqrt{\dfrac{0.11 \cdot (1 - 0.11)}{317}} \approx 0.01757$

From this we can work out the interval of reasonably likely $\hat{p}$ s using

$p_0 \pm 1.96 \sqrt{\dfrac{p_0(1 - p_0)}{n}} = 0.11 \pm 1.96(0.01757) = 0.11 \pm 0.03444$, which implies an interval:

$0.0756 < \hat{p} < 0.1444$. The picture of the sampling distribution (shown below) for our example looks like this. There is a general principle involved here; that general principle is that a hypothesis test is *always* based upon the sampling distribution implied by the null hypothesis.

Logic of a Hypothesis Test: This picture gives us an idea of how the testing will proceed. If the $\hat{p}$ that we get from our sample is one of the *rare* $\hat{p}$ s according to the sampling distribution, we will consider getting this rare $\hat{p}$ as evidence *against the null hypothesis*. But if (on the other hand) the $\hat{p}$ that we get from our sample is one of the **reasonably likely** $\hat{p}$ s according to the sampling distribution then we will consider that our test has *not* provided us with evidence against the null hypothesis. The logic of the testing is that the sampling distribution of $\hat{p}$ s shows us which $\hat{p}$ s are reasonably likely *if the null hypothesis is true* (that is, if $p = p_0$) and also what $\hat{p}$ s are rare *if the null hypothesis is true* (that is, if $p = p_0$). So, if we get one of the *rare* $\hat{p}$ s, we count that as evidence against the truth of the null hypothesis. If we get a reasonably likely $\hat{p}$ then we cannot count that $\hat{p}$ as evidence against the H_0. If we get a *reasonably likely* $\hat{p}$, it does not necessarily mean that the null hypothesis is true; it only means that we have no evidence that it is *not* true.

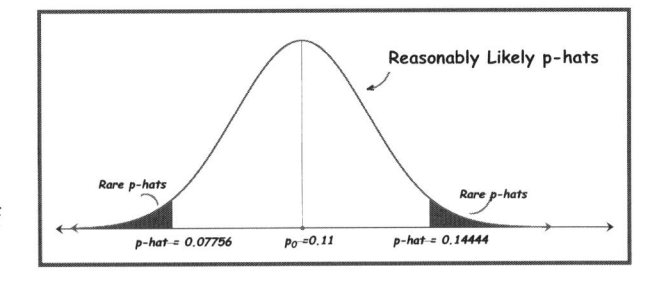

It is the *null hypothesis* that is the focus of our attention; the *alternate hypothesis* serves as a challenge to the null hypothesis. Our data shows us whether or not the challenge will be successful. For our example, we got $\hat{p} = 0.0852$. This value happens to be within the interval of reasonably likely $\hat{p}$ s, which is $0.0756 < \hat{p} < 0.1444$. Therefore, for our example, we do *not* have sufficient evidence from our sample against the null hypothesis. Since our $\hat{p} = 0.0852$ is a reasonably likely $\hat{p}$, we cannot challenge the accepted standard of $p_0 = 0.11$ for the proportion of left-handers in the population of students at CSM.

Calculating the Test Statistic. We have come to our conclusion by noting that our $\hat{p} = 0.0852$ is in the interval of reasonably likely $\hat{p}$ s. However, hypothesis testing has a certain structure to it, and the **third step** in carrying out a hypothesis test is to calculate the **test statistic.** A test statistic is something that is *always* calculated for any hypothesis test, but the nature of what is calculated differs depending on what is being tested. For tests that involve one $\hat{p}$, the test statistic is the z score based upon the sampling distribution for the null hypothesis. The general form of a z score is

$$z = \frac{\text{observed sample value} - \text{mean}}{\text{standard deviation}}$$; we have used it with the Normal model to tell us how far and in what direction a value is—in standard deviation units—from the mean. Now, we are using sampling distributions, so the values are $\hat{p}$ s and the mean of our sampling distribution is $\mu_{\hat{p}} = p_0$ and the standard deviation is $\sigma_{\hat{p}} = \sqrt{\frac{p_0(1-p_0)}{n}}$. Our z score takes the form shown just below.

Test Statistic for Testing a Single Population Proportion p

$$z = \frac{\hat{p} - \mu_{\hat{p}}}{\sigma_{\hat{p}}} = \frac{\hat{p} - p_0}{\sqrt{\frac{p_0(1-p_0)}{n}}}$$ where p_0 is the value used in the null hypothesis H_0: $p = p_0$.

Below, the *test statistic* for our example is worked out and a graph shows what we have calculated and how it relates to the reasonably likely and rare $\hat{p}$ s.

For our example, where $\hat{p} = 0.0852$ and $p_0 = 0.11$, and $n = 317$, we would calculate:

$$z = \frac{\hat{p} - p_0}{\sqrt{\frac{p_0(1-p_0)}{n}}}$$

$$= \frac{0.0852 - 0.11}{\sqrt{\frac{0.11 \cdot (1-0.11)}{317}}}$$

$$= \frac{-0.0248}{0.01757}$$

$$\approx -1.41$$

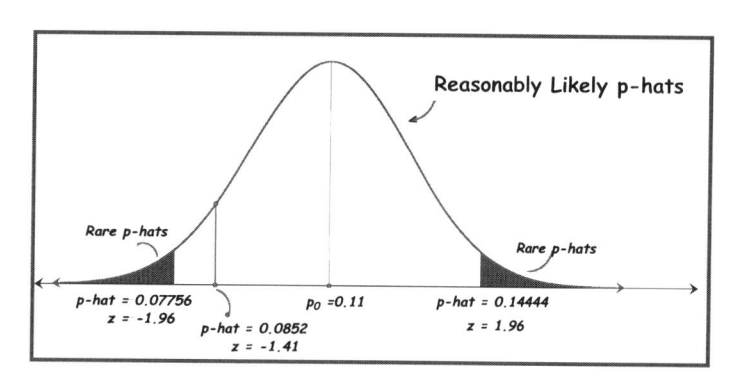

For $\hat{p} = 0.0852$ (or 8.52% left-handers) we get $z = -1.41$; this tells us that our $\hat{p} = 0.0852$ is to the left of the mean of the sampling distribution of $p_0 = 0.11$, and it is 1.41 standard deviations to the left. This also tells us (just from the z score) that our $\hat{p} = 0.0852$ is within the interval of reasonably likely $\hat{p}$ s, since our $z = -1.41$ is between $z = -1.96$ and $z = 1.96$. We knew that our $\hat{p}$ was reasonably likely by seeing that our $\hat{p} = 0.0852$ was inside the reasonably likely interval. The calculation of the test statistic puts our conclusion in the world of the Normal Distribution Chart. A test statistic that is greater than z

=1.96 or less than z = -1.96 would be rare, but a test statistic between z = – 1.96 and z = 1.96 will be reasonably likely.

Reasonably Likely p-hats and the Value of the Test Statistic

For a given $\hat{p}$, if the *test statistic* calculated by $z = \dfrac{\hat{p} - p_0}{\sqrt{\dfrac{p_0(1-p_0)}{n}}}$ is within the interval $-1.96 \leq z \leq 1.96$

then the $\hat{p}$ is *reasonably likely* for the p_0 and the n used in the calculation. If the test statistic z is outside the interval $-1.96 \leq z \leq 1.96$ then the $\hat{p}$ is *rare*.

Warnings. In Unit 3, there were many things that either had the symbol *r* or names that began with *r*, and you needed to keep them straight. Now we are about to have the same problem with the letter *p*. We have encountered P( ), $\hat{p}$, *p*, and p_0—all of them referring to some kind of probability or proportion (hence the letter *p*) but used in slightly different ways. Now comes still another one; we are about to discuss the **p-value,** which is easily confused with all the other "*p*"s. All of these are used commonly; again, your job is to keep them straight in your thinking. Second warning: the next paragraphs will probably have to be read more than once! These paragraphs concern the calculation of a **p-value** and its meaning. It is not simple.

Calculation of a p-value. This is the **fourth step.** We have calculated similar probabilities before. The *p-value* is the probability of getting a random sample with a $\hat{p}$ *as extreme or more extreme than the one that we actually got*. The calculation is another "given a value—find a proportion" type of calculation that uses the Normal distribution, but we will have to see how the "*as extreme or more extreme* "comes in. Using our example of the percentage of left-handed students in our sample 8.52% or $\hat{p} = 0.0852$, we note that this $\hat{p}$ is less than the $p_0 = 0.11$ and ask (first of all): what is the probability that, if the null hypothesis H_0 is true so $p_0 = 0.11$, we end up seeing a $\hat{p} = 0.0852$ *or less*? Anything less than $\hat{p} = 0.0852$ would be "*more extreme*" or farther away from $p_0 = 0.11$ than what we got.

> We should draw a sketch; here is the Fathom drawing for this part of the problem.
> Then we should calculate a z score remembering that we are dealing with a sampling distribution, so we get

$$z = \dfrac{\hat{p} - p_0}{\sqrt{\dfrac{p_0(1-p_0)}{n}}} = \dfrac{0.0852 - 0.11}{\sqrt{\dfrac{0.11 \cdot (1-0.11)}{317}}} = \dfrac{-0.0248}{0.01757} \approx -1.41,$$

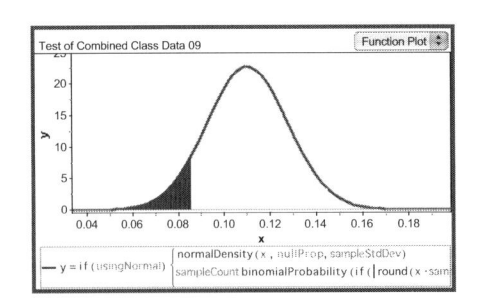

> which we recognize as our *test statistic*.
> Then we consult the Normal Distribution Chart and find that $P(z < -1.41) = 0.0793$. That is the area shown shaded here.

More extreme and its meaning. However, since we are using random sampling, "*more extreme*" could also mean that we get a $\hat{p}$ that is more extreme on the positive side of $p_0 = 0.11$ than the one that we got on the negative side. So our picture for "*more extreme*" should look like the Fathom picture on the right. This is not hard to calculate because of the symmetry of the Normal distribution. We multiply the area in the left shaded area by two. In probability notation, we have

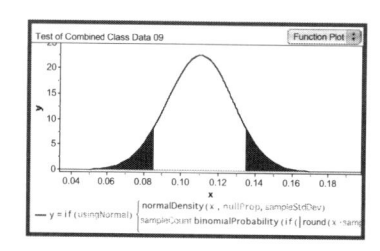

$P(\hat{p} \text{ that is more extreme}) = P(Z > |-1.41|) = 2P(Z > 1.41) = 2 \times 0.0793 = 0.1586$. The *capital* Z in this notation refers to the horizontal scale for the standard Normal distribution and not our z score; *our z score* in this example is -1.41. The **p-value** for our example is 0.159, that is: *p*-value = 0.159, or approximately 0.16.

Many calculations but the same result. It appears that we are calculating several things and that they are all related. This is true. We can see whether our $\hat{p}$ is reasonably likely or rare; we can calculate the test statistic (getting the $\hat{p}$ in terms of standard deviation units); we can calculate the *p*-value to see the probability that we get this $\hat{p}$ or one more extreme. All of these calculations are related, and the relationships are spelled out in the arrowed bullets below.

➤ A sample proportion $\hat{p}$ that is **rare** will have a **test statistic** whose absolute value is relatively **large** and therefore a **p-value** that is **small**. (The *p*-value will be *smaller* than 0.05.)

➤ A sample proportion $\hat{p}$ that is **reasonably likely** will have a **test statistic** whose absolute value is relatively **small** and therefore a **p-value** that is **large**. (The *p*-value will be *larger* than 0.05.)

In any particular hypothesis test, all these calculations should agree: either they should show evidence consistent (or in accord) with the null hypothesis or evidence against (or contrary to) the null hypothesis.

Here is the reasoning behind the relationships shown above: we will think about how the *p-value* is going to work; if we get a sample proportion $\hat{p}$ that is *rare* then the test statistic will either be bigger than $z = 1.96$ or smaller than $z = -1.96$. Then the picture like the one just above will have very small shaded areas on either side, and the *p*-value will be very small. In fact, if we have $\hat{p}$ that is rare, we know that the *p*-value must be less than 0.05, since the tails amount to just 5%. On the other hand, if we have (as we do here in our example) a $\hat{p}$ that is *reasonably likely* then we know that the area in the shaded part of the picture must be bigger than 0.05, and so the *p*-value must bigger than 0.05. There are some relationships between *p*-values, test statistics, and reasonably likely or rare $\hat{p}$ s.

Since we will use a *p-value* in most of our calculations with hypothesis tests, , and because the idea of a *p*-value is extremely important, we give a general definition, which is found on the next page. It will be a good thing to think carefully about this idea; it is one of the most difficult concepts in the course.

> **Definition of p-value for a hypothesis test**
>
> The *p*-value for a hypothesis test is the probability of getting a *test statistic as extreme as or more extreme than* the one observed from the null hypothesized value for the population when randomization is used.

This definition is constructed so that it will apply to all the hypothesis testing we will do. In our example here, the test statistic is a *z* score, and (for the moment) we are thinking of a sample of students as being a simple random sample. We want our definition to apply to data that come from an experiment, where randomization would mean that the experimental units are randomly allocated to the treatments.

Fathom Output and Review of the First Four Steps. Below is the Fathom output and a review of the first four steps of our hypothesis test for the statistical question: *Does the percentage of left-handers in the population of all students at one college differ from the "standard" 11 percent that is found generally?* You need to understand how to carry each step by hand, and you need to understand each step to use Fathom, but the Fathom output provides a nice summary. It will be a good idea to follow along by using Fathom at this point, since using the computer will show the choices within Fathom.

Fathom Instructions for the Percentage of Left-handed Students Example

- Open the Fathom file **Combined ClassData Y09.ftm** and get a **CaseTable**.
- Get the **Summary Table** showing the row proportion like the one shown here.
- Pull down a **Test** icon from the shelf; it should say "**Empty Test.**"
- Change **Empty Test** to **Test Proportion**.
- Drag the variable *DominantHand* to the "Attribute (categorical) Unassigned." You will notice that there are things in blue; anything in blue can be changed. You may notice that the hypothesis test is set up to test the idea that 50% of people are right-handed.
- Put your cursor on **Right** (meaning right-handed) and change it to **Left**.
- Put your cursor on **0.50** and change it to **0.11**, which is our p_0.
- With the **Test** selected, go to the menu **Test** and deselect **Verbose**.

You should see the Fathom output below, which effectively shows the four steps we have done.

Step 1: The null and alternate hypotheses are specified, with $p_0 = 0.11$.

Step 2: Checking conditions; Fathom *cannot* know whether your sample is a random sample. You have to check that condition. If either $np \geq 10$ or $n(1-p) \geq 10$ fails to be true, Fathom does not use the Normal distribution.

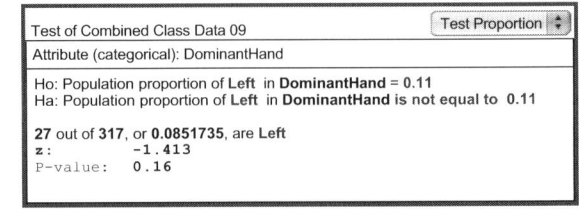

```
Test of Combined Class Data 09              Test Proportion
Attribute (categorical): DominantHand

Ho: Population proportion of Left in DominantHand = 0.11
Ha: Population proportion of Left in DominantHand is not equal to 0.11

27 out of 317, or 0.0851735, are Left
z:         -1.413
P-value:    0.16
```

Step 3: The $\hat{p} = 0.0852$ is shown as well as the *test statistic* $z = -1.41$.

Step 4: The *p*-value of 0.16 is shown.

The **fifth step** is about what all this means. See below.

The Fifth Step: Interpreting Hypothesis Tests

Logic of the hypothesis test, again. It may be wise to reread the section above with this same title because the way we interpret a hypothesis test depends on this logic. Recall that the data in the form of a sample proportion $\hat{p}$ and the alternate hypothesis form a kind of challenge to the null hypothesis; the challenge can be successful—we have evidence against the null hypothesized p_0—or it can fail: we have evidence, but it is not strong enough, considering sampling variation. If the $\hat{p}$ is rare, and *therefore* the test statistic is relatively large in absolute value, the *p*-value will be small: *then* we have sufficient evidence *against* the null hypothesis, and the challenge is successful. On the other hand, if the sample proportion $\hat{p}$ is reasonably likely, the test statistic will be relatively small in absolute value, and the *p*-value is relatively large: *then* we do *not* have sufficient evidence against the null hypothesis over and above sampling variation, and the challenge to the null hypothesis fails. We live with the null hypothesis.

An interpretation of our example. We found that our sample proportion $\hat{p} = 0.0852$ was reasonably likely if $p_0 = 0.11$ is true, and our test statistic $z = -1.41$ is not large in absolute value (it was *not* beyond $z = -1.96$) and the *p*-value = 0.16 was relatively large (that is, it was larger than 0.05). Therefore, our evidence does *not* successfully challenge the null hypothesized value of $p_0 = 0.11$. We should add that our challenge with the sample size we used ($n = 317$) was not successful.

In the context of our data, with a $\hat{p} = 0.0852$, which is reasonably likely if $p_0 = 0.11$, a test statistic of $z = -1.41$, and a *p*-value of 0.16, we have *insufficient* evidence to say that the percentage of CSM students who are left-handed is different from the usual standard of 11%. This is a good interpretation because it makes clear the hypothesis ("the percentage of CSM students who are left-handed is different from the usual standard of 11%"), the evidence ("a $\hat{p} = 0.0852$, which is reasonably likely if $p_0 = 0.11$, a test statistic of $z = -1.41$, and a *p*-value of 0.16"), and our conclusion ("we have insufficient evidence to say…").

However, there are certain conventions of language that are used with hypothesis tests. One has to do with "rejecting" the null hypothesis, and the other has to do with the term "statistical significance."

Rejecting the Null Hypothesis—Or Not A very common way of speaking about the conclusion to a hypothesis test speaks about "rejecting the null hypothesis." If we have evidence against the null hypothesis—a very small *p-value*, for example—then we could say that we *reject* the null hypothesis H_0. If, instead of finding twenty-seven left-handers in our sample, we had found just twenty-one then our results would be as the Fathom output shown here. In this scenario, we

```
From Summary Statistics                        Test Proportion
Attribute (categorical): unassigned
Ho: Population proportion of Left in DominantHand = 0.11
Ha: Population proportion of Left in DominantHand is not equal to 0.11
21 out of 317, or 0.0662461, are Left
z:            -2.49
P-value:       0.013
```

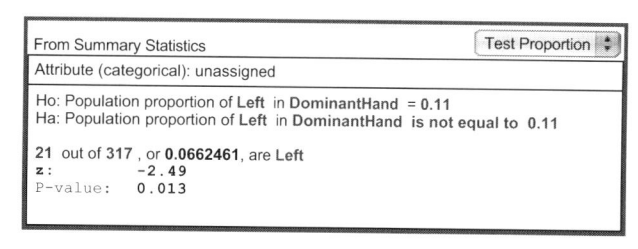

would have evidence *against* the null hypothesis that the population proportion of left-handers is 11%. Notice how all the evidence is consistent: a test statistic whose absolute value $|-2.49| = 2.49$ is larger than 1.96 and a *p*-value that is smaller than 0.05. The $\hat{p}$ we got cannot be the result of sampling variation.

Our example did not in fact look like this. We did not have sufficient evidence against the null hypothesis. We would say that we *fail* to reject the null hypothesis. The percentage of left-handers that we saw can easily be explained as a reasonably likely result when we sample randomly from a population. Notice that we do not say that we "accept" the null hypothesis; the accepted terminology is that we "fail to reject H_0." We brought a challenge to the H_0 that did not work.

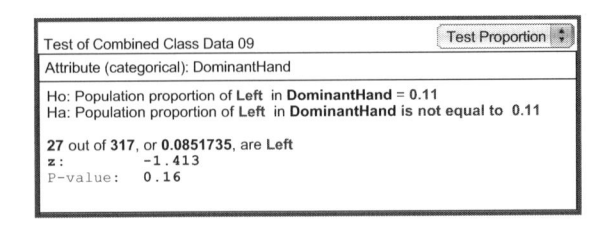

```
Test of Combined Class Data 09          Test Proportion
Attribute (categorical): DominantHand
Ho: Population proportion of Left in DominantHand = 0.11
Ha: Population proportion of Left in DominantHand is not equal to 0.11

27 out of 317, or 0.0851735, are Left
z:         -1.413
P-value:    0.16
```

Statistically Significant—Or Not. The term **statistically significant** has a very specific meaning. Once again, we think of the null hypothesis: if the null hypothesis is ($p = 0.11$), there is a reasonably likely range of sample proportions $\hat{p}$ that we expect just by random variation. However, if our $\hat{p}$ is outside that range then we believe that something more than sampling variation going on. So, in the scenario above, if we had $\hat{p} = \frac{21}{317} \approx 0.066$, we would say that this test was "statistically significant." In our actual test with the sample proportion of $\hat{p} = 0.0852$ we would say that the test was *not* statistically significant. The word "significant" should be modified with "statistically"; moreover, the term does *not* mean necessarily that a result is important or profound. It merely means that what we saw could not have been explained by sampling variation from random sampling or allocation. Some publications try not to use the word "significant" and instead use the term **statistically detectable**.

How small is small? How big is big? We have said that a small *p*-value and a large test statistic means we have a *rare* sample proportion $\hat{p}$; these three go together and are evidence against the null hypothesis. However, this raises the question: how small does a *p-value* have to be, and how large does the test statistic need to be? We have already hinted at the standard above. Notice that in the box below, both the test statistic and the *p-value* are mentioned. These two go together; if you get a small *z* and a big *p*-value, you have made some mistake. The box below gives some standards that depend upon taking 5% as rare.

Standards: How small is small, how big is big? If *rare* means the 5% most extreme $\hat{p}$s, then:

- If the *p*-value < 0.05 and the test statistic *z* is *outside* the interval $-1.96 < z < 1.96$ then we *have* evidence against the null hypothesis, and
  - the test *is statistically significant* and
  - we can *reject* the null hypothesized p_0.
- If the *p*-value > 0.05 and the test statistic *z* is *inside* the interval $-1.96 < z < 1.96$ then we do *not* have sufficient evidence against the null hypothesis, and
  - the test is *not statistically significant* and
  - we *fail* to *reject* the null hypothesized p_0.

Interpreting our example, again, and with caution. We must be clear on four points:
- What the population is
- The variable in the null hypothesis and the hypothesized value
- The evidence: the *p*-value can be mentioned, or the value of the test statistic can be mentioned, or both
- Whether the evidence against the null hypothesis is sufficient or not, and here we can say:
    Whether the test is statistically significant or not and
    Whether we reject the null hypothesis or not

Therefore, a good interpretation of our example could go something like this:

"We found that 8.52% of our sample of $n = 317$ CSM students were left-handed. This 8.52% is not significantly different from the hypothesized proportion of 11%. The calculations give a *p*-value of 0.16, and so 8.52% is a reasonably likely sample result if 11% of all CSM students are left-handed. So we fail to reject the standard that 11% of CSM are left-handed."

There is one more important thing that should be added to this interpretation, however. One of the conditions for doing these tests is that the sample be a simple random sample (a SRS). We know that our sample of students is *not* a simple random sample of CSM students, and so we should add a cautionary note to our interpretation. We should say that our results must be treated with caution. Now it may be that with respect to handedness, our sample is as good as a SRS, but we actually do not know whether it is or not.

Other Standards

We have been taking the definition of **rare** to be the most extreme 5% of $\hat{p}$ s, leaving 95% as **reasonably likely** as in the picture to the right. However, although this 5% is very common (even something like an "industry standard"), it is not the only definition of *rare* possible. We could define *rare* to be the most extreme 2% or the most extreme 1%. The choice is left to the researcher doing the analysis. Because that choice is available, this choice has a name and symbol. It is called the **level of significance,** and its symbol is the Greek letter α. If *rare* means the most extreme 5% of $\hat{p}$ s then we have chosen $\alpha = 0.05$.

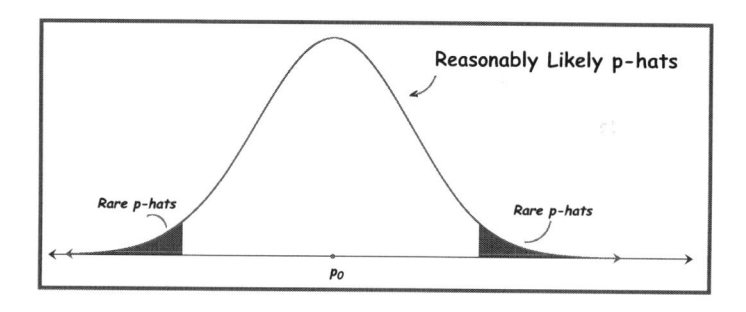

In research publications, the phrase "significant at the 1%" means that the hypothesis test would be statistically significant if $\alpha = 0.01$. Since the choice of standard is left to researchers, it is quite common now to report *p*-values rather than significance levels, leaving the choice of the level of significance to the reader.

We can put these considerations together and revise the "How small is small" box.

> ***Other standards:*** If we define *rare* as the 100α percent most extreme $\hat{p}$ s then:
>
> - If the *p*-value $< \alpha$ and the test statistic z is *outside* the interval $-z^*_{\alpha/2} < z < z^*_{\alpha/2}$ then we *have* evidence against the null hypothesis, and the hypothesis test *is statistically significant* and we can *reject* the null hypothesized p_0.
>
> - If the *p*-value $> \alpha$ and the test statistic z is *inside* the interval $-z^*_{\alpha/2} < z < z^*_{\alpha/2}$ then we do *not* have sufficient evidence against the null hypothesis, and the hypothesis test is *not statistically significant* and we *fail* to *reject* the null hypothesized p_0.
>
> Where $z^*_{\alpha/2}$ refers to the z^* from the Normal Distribution Chart, that cuts off the lowest $\alpha/2$ and highest $\alpha/2$ of the standard Normal distribution. If $\alpha = 0.01$, for example, $z^*_{\alpha/2} = 2.576$.

Summary: Hypothesis Tests

- The basic notion of a hypothesis test is to see if sample data (in this section, a sample proportion $\hat{p}$) agrees with or disagrees with a preconceived idea (in this section, an idea about a population value p), taking account of the fact that the sample values may vary by random sampling variation.
  - The preconceived idea is formalized into the **null hypothesis** and the **alternate hypothesis**.
  - The test is based upon the sampling distribution for the **test statistic** assuming the null hypothesis true.
- **Definition of p-value for a hypothesis test.** The *p*-value for a hypothesis test is the probability of getting a *test statistic as extreme as or more extreme than* the one observed from the null hypothesized value for the population when randomization is used.
  - Small *p*-values indicate that, given the null hypothesis, the sample result is relatively *rare*.
  - Large *p*-values indicate that, given the null hypothesis, the sample result is reasonably *likely*.
- **Standards.** Researchers or readers of research are the ones who decide how small a *p*-value is necessary to declare that a sample result is *rare* (as opposed to *reasonably likely*). These standards are known as **levels of significance**, expressed as proportions and denoted by the Greek letter α. The most common standard is $\alpha = 0.05$. This means that the **test statistic** will be rare with a probability of 5% when the null hypothesis is true if the hypothesis test were repeatedly done.
- When the evidence from a hypothesis test is *rare* enough according to the significance level chosen, the hypothesis test is deemed **statistically significant.** A statistically significant result means that the sample results must have been rare enough to provide evidence against the null hypothesis.
- **Interpretation of hypothesis tests.** A good interpretation should include the following:
  - A clear description of the *population*, the *variable* in the null hypothesis, and the hypothesized *values*
  - The evidence: the *p*-value can be mentioned, or the value of the test statistic can be mentioned, or both
  - Whether the evidence against the null hypothesis is sufficient or not, and here we can say: Whether the test is statistically significant or not and whether we reject the null hypothesis or not.

Summary: Five-Step Hypothesis Testing

Step 1:	Set up the null and alternate hypotheses using an idea for the population proportion p_0. H0: $p = p_0$ Ha: $p \neq p_0$
Step 2:	Check the conditions for a trustworthy hypothesis test. The sample must be a simple random sample from a population ten times n. Both $n\, p_0 \geq 10$ and also $n(1 - p_0) \geq 10$.
Step 3:	Calculate the test statistic $z = \dfrac{\hat{p} - p_0}{\sqrt{\dfrac{p_0(1-p_0)}{n}}}$ using the $\hat{p}$ from the sample and the hypothesized p_0.
Step 4:	Calculate the p-value, by getting $P(\hat{p}\text{ more extreme}) = 2P(Z \geq |z|)$ where $|z|$ is the absolute value of the test statistic you calculated in Step 3.
Step 5:	Evaluate the evidence that the p-value and the test statistic give you to determine whether your test successfully challenges the null hypothesized population proportion p0 or not. Give an interpretation in the context of the data using the terminology of "statistical significance" and "rejecting the null hypothesis."

§4.2   Exercises on Hypothesis Testing for Proportions

1. **Shoes on Beaches in the Netherlands.** Often hypothesis testing is used just to test the idea of whether there is something more than random variability happening. Here is an example.

   There is a light-hearted discussion among biologists who study seabirds in Northern Europe about the proportion of left and right shoes found on beaches. Martin Heubeck, who works in the Shetland Islands in Scotland, relates that at a seabird conference in Glasgow,

   > …the conversation drifted around to the amazing variety and number of shoes found during beached bird surveys. Mardik Leopold, normally an amiable and respected seabird biologist, claimed that it was a little known fact that due to some physical process or other, more left than right shoes wash ashore on beaches, at least in the Netherlands.
   > *[Shetland Bird Club Newsletter 107 (Spring 1997)]*

   Here are Leopold's sample data for the Dutch island of Texel.

Collection 1		
Foot	Left	66
	Right	38
Column Summary		104
S1 = count ( )		

   Our statistical question is: *The proportion of left shoes should be near 50%; so does the proportion of left shoes deviate from 50% so much that we think something more than random variability accounts for the high proportion of left shoes on Texel Island?* We can analyze these data using a hypothesis test.

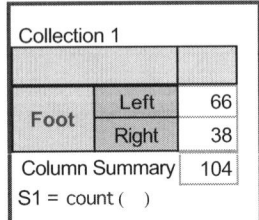

   a. First, use the data for Texel to calculate the sample proportion of *left* shoes. (You should get 0.635, rounded to three decimal places.) Assign the correct symbol to this proportion; should the symbol be $\hat{p}$, p, or p_0? Give a reason for your choice of symbol.

   b. What is the sample size? Assign the correct symbol to the sample size.

   Step 1: Setting Up the Null and Alternate Hypotheses. Our interest will be in the population proportion p of *left* shoes on beaches. (That, everything will refer to *left* shoes.)

   c. What is the logical proportion of left shoes in the world? That is, what is a good choice for p_0?

   d. Write the null hypothesis using the notation shown in the **Notes.**

   e. The alternate hypothesis should be written as though you have no expectation that the number of left shoes should outnumber the right shoes or right outnumber the left, only that they *may* be different. Write the alternate hypothesis using the notation shown in the **Notes.**

   In our presidential approval example, we are able to clearly identify the population: it was all American voters. In a situation like this, things are not so clear because we do not know how the shoes ended up on the beaches. Were they left there by beach-goers or lost from ships passing by? In a sense, it does not matter since whatever it is, the p_0 is probably 0.50.

   Step 2: Checking the Conditions. There are three conditions to be checked, some of them difficult.

   f. Simple Random Sample. State why the shoes found on beaches do *not* conform to the definition of a deliberately chosen simple random sample. (That leaves unclear whether how the shoes got there is a "random process" or not; that is what is being tested.)

   g. Check the conditions $np_0 \geq 10$ and $n(1-p_0) \geq 10$. There is an easy way of doing this, which you may only discover after you have gone through the calculations. If you see the easy way, state why it works—perhaps using fractions.

h. The population is required to be ten times the size of the sample. Whatever the source of the shoes on the beaches (beach-goers or washed-overboard shoes), do you think that there are more than 1,040 shoes in the population? If so, the condition is satisfied.

Step 3: Calculating the Test Statistic

i. Calculate the test statistic using the formula given in the *Notes*.

j. Draw a sketch of a Normal distribution (see below) and show on it the p_0. In your calculation of the test statistics, the denominator should have $\sqrt{\dfrac{p_0(1-p_0)}{n}} = \sqrt{\dfrac{0.50(1-.50)}{104}} \approx 0.049$. Use this to find the locations of 0.500 ± 0.049, $0.500 \pm 2*0.049$ and $0.500 \pm 3*0.049$ on a sketch of the sampling distribution we are using in this hypothesis test. Locate $\hat{p} = 0.635$ on your sketch.

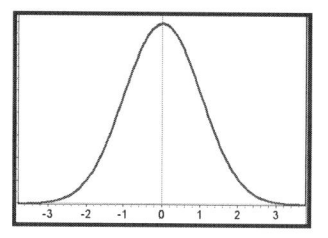

k. From the value of the *test statistic,* is the $\hat{p}$ *rare* or *reasonably likely* given the p_0 we are using? Give a reason for your answer. (You should also be able to see this from your sketch.)

Step 4: Calculating the p-value. (We will do this in two sub-steps: the answer to part *l* is half the p-value.)

l. Your test statistic (your answer to part *i*) should be a positive number. Use the Normal Distribution Chart to get the area (i.e., the probability) in the right tail. (Will you subtract a probability from 1?)

m. Your answer to part *l* (just above) is only one-half the *p*-value. Use your result to get the *p*-value.

n. On your sketch of the Normal distribution, show the *p*-value by shading tails.

Check your calculations using Fathom.

- Open the Fathom file **Texel Shoes 1997.ftm** and get a *case table.*
- Get the **Summary Table** showing number of right and left shoes using the variable *Foot.*
- Pull down a **Test** icon from the shelf; it should say "**Empty Test.**"
- Change **Empty Test** to **Test Proportion.**
- Drag the variable *Foot* to the "Attribute (categorical) Unassigned." There are things in blue; anything in blue can be changed.
- With the **Test** selected, go to the menu **Test** and deselect **Verbose.**
- With the **Test** selected, go to the menu and get **Test>Show Test Statistic Distribution.** The plot shown, with its tiny sliver of shading, should resemble the picture you made in part j.

o. Check that your calculated values for the $\hat{p}$, for the *z*, and for the *p*-value are in accord with Fathom's.

p. How is the very small value of the *p*-value shown on the sketch? Get this correct! (The *p*-value is *not* shown by a point on the horizontal axis! Check your answers for this; many get it wrong!)

Step 5: Interpreting the result.

q. Is your test statistically significant? Give a reason for your answer.

r. With this test, can you reject the null hypothesis? Give a reason for your answer.

s. Give an interpretation in the context of the statistical question posed at the beginning. That is, do the results of our hypothesis test show that "something more" than random variability "accounts for the high proportion of left shoes" on Texel Island beaches?

(**Note:** Your answers to parts q and r should be "yes." No one knows why there should be a significantly higher proportion than 50% of left shoes on Texel beaches. Do we have just a "weird" sample? It could be, or perhaps the shoe collection process favored left shoes for some reason.)

2. **Shoes on Beaches in Scotland.** Dr. Martin Heubeck (of Scotland) responded to his Dutch colleagues by collecting some data on shoes on Scottish beaches. Here is how he relates it:

> The gauntlet having been flung, I decided to record the same on 'my' beaches during the end of the February beached bird survey. I've done some crazy things in my time, but picking up shoes and wellingtons in Force 9 sleet takes some beating and I just hope few people saw what I was doing!
> [*Shetland Bird Club Newsletter 107 (Spring 1997)*]

Actually, Dr. Heubeck managed to collect sample data for both February and March. Here is the Fathom summary table for all the shoes he and a colleague collected.

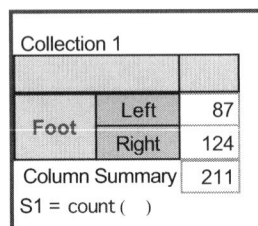

Collection 1		
Foot	Left	87
	Right	124
Column Summary		211
S1 = count ( )		

a. Calculate the proportion of *left* shoes collected and assign the correct symbol (the one that we will use in hypothesis testing) to it.

b. Compare the *sample* results for Scotland to the *sample* results for Texel Island. What are the differences? Where do right shoes "go"? Left shoes?

c. To do the same kind of hypothesis test as we did for the Texel Island data, we need to set up the null and alternate hypotheses. Set up the hypotheses using the correct notation.

d. Check the sample size conditions for the test (i.e., $np_0 \geq 10$ and $n(1-p_0) \geq 10$).

e. Calculate the test statistic for this test, showing your work.

f. Judging from your test statistic, is the sample proportion $\hat{p}$ reasonably likely or rare given the p_0 we are using? Give a reason for your answer. (Consult the **Notes.**)

g. Before calculating the *p*-value, you should be able to predict whether the *p*-value will be smaller or bigger than 0.05. Predict and give a reason for your prediction.

h. Make a sketch of the sampling distribution, showing the location and value of p_0 and also the approximate location and value of $\hat{p}$. [Make this sketch precise by using $\sqrt{\dfrac{p_0(1-p_0)}{n}}$].

- Open the Fathom file **Scottish Shoes.ftm** and get a **case table.**
- Get the **Summary Table** showing number of right and left shoes using the variable *Foot.*
- Pull down a **Test** icon from the shelf; it should say "**Empty Test.**"
- Change **Empty Test** to **Test Proportion.**
- Drag the variable *Foot* to the "Attribute (categorical) unassigned."
- With the **Test** selected, go to the menu **Test** and deselect **Verbose.** Check your answers.

- With the **Test** selected, go to the menu and get **Test>Show Test Statistic Distribution.** Adjust the scales so that your plot resembles the one here. Compare the plot you made in part h to the plot that you got from Fathom. (Have you shown the *p*-value as small shaded in tails to the graph?)
    i. Is your test statistically significant? Give a reason for your answer.
    j. Can you reject the null hypothesis? Give a reason for your answer.
    k. Give a good interpretation of your test in the context of the data.
    l. Confused Conrad's answer to part k reads like this: "The test is not statistically significant because the *p*-value is so small. Only 1.1% of shoes left on Scottish beaches are left shoes." Poor Conrad; he is really confused! Identify as many of his errors as you can.

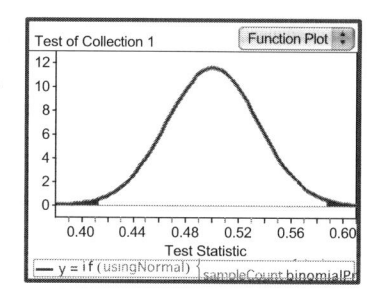

3. **Right-Handers** The traditional figure given for the percentage of left-handers is from 10%–12%, which is the reason we have used the figure 11%. The paper cited in the **Notes** (Medland, et.al, 2004) and other research suggests that another 4% or 5% of people are "partially" left-handed. That is, they do some things with their left hands and some things with their right hands. In our sample data (shown here again), about 5% of the students identified themselves as "ambidextrous." These students may be among those who do some things left-handed and other things right-handed. If about 15% of the population is either left-handed or ambidextrous, that leaves 85% who are "purely" right-handed. We will use our data to test whether the idea that 85% of the population of CSM students is right-handed or whether it is some other number.

    a. Using the table calculate $P(R)$, where R stands for "right-handed," confirm that the formula from §1.2 $P(\text{Not } A) = 1 - P(A)$ can be used to give you the proportion of left-handed or ambidextrous.
    b. Using the *notation for hypothesis tests,* what symbol should be used for the fraction 274/317: $P(R)$, $\hat{p}$, p or p_0? Give a reason for your answer and say briefly why the others are wrong.
    c. What number are we using for the p_0 for this test? Why is it not 0.864353?
    d. **Step 1:** Using the information in the previous answers, set up the null and alternate hypotheses for this test. Use the correct notation.
    e. **Step 2:** We have already discussed in the **Notes** that our sample is not a simple random sample. Are the other conditions for a hypothesis test met? Show simple calculations to make your point.
    f. Describe the sampling distribution of $\hat{p}$ s for the p_0 we have chosen. Your description should include shape, center, and spread.
    g. Use the answers to part f to make a sketch of the sampling distribution, showing the mean and the standard deviation.
    h. **Step 3:** Calculate the test statistic using the correct formula and using the correct notation.

i. Judging from the test statistic, is the $\hat{p}$ reasonably likely or rare? Give a reason for your answer.

j. Judging from the test statistic, will the *p*-value be smaller than 0.05 or bigger than 0.05? Give a reason for your answer, perhaps accompanied by a sketch.

k. **Step 4**: Find the *p*-value, using the correct notation. Use the Normal chart.

- Open the Fathom file **Combined ClassData Y09.ftm** and get a **case table.**
- Get the **Summary Table** showing the row proportion like the one shown here.
- Pull down a **Test** icon from the shelf; it should say "**Empty Test.**"
- Change **Empty Test** to **Test Proportion.**
- Drag the variable *DominantHand* to the "Attribute (categorical) Unassigned." You will notice that there are things in blue in the dialogue box; anything in blue can be changed.
- Put your cursor on **0.50** and change it to our p_0 for this test. (Neglecting this will direct Fathom to test that the proportion of right-handers is 50%.)
- With the **Test** selected, go to the menu **Test** and deselect **Verbose.** Check your answers.
- With the **Test** selected, go to the menu and get **Test>Show Test Statistic Distribution.** Adjust the scales so that your plot resembles the one shown below. Compare your plot to the one shown below.

l. Is your test statistically significant? Give a reason for your answer.

m. Can you reject the null hypothesis? Give a reason for your answer.

n. **Step 5**: Give a good interpretation of your test in the context of the data.

o. What additional information do you need to add to your interpretation given that we know that sample is not random? Explain.

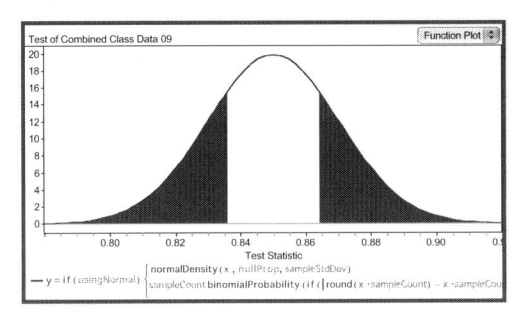

p. Using the $\hat{p}$ from the sample and using the formulas from §4.1, calculate the 95% confidence interval for the true population proportion *p* of right-handers among CSM students.

q. Does your confidence interval include the p_0 we have been testing?

r. True or false, and explain why. If a hypothesis test (such as we have been doing) is *not* statistically significant then the confidence interval based upon a $\hat{p}$ will capture the p_0 for the test.

s. True or false, and explain why. If a hypothesis test (such as we have been doing) *is* statistically significant then the confidence interval based upon a $\hat{p}$ will *not* capture the p_0 for the test.

4. **The King of Rufutania's Seventieth Birthday and Palace Intrigue.** The king of Rufutania is a megalomaniac, jealously guarding his standing and power and always on the lookout for unrest in his kingdom. His picture appears everywhere as well as on the coins of his realm, which all bear his image on the head side. To celebrate his seventieth birthday, the king directs the director of the Royal Mint to make coins that come up heads exactly 70% of the time when flipped. The questions are on the next page. **PTO**

a. The king, always suspicious, suspects that the director of the Royal Mint has not followed his directions. The king summons the director to bring twenty coins, chosen at random, which the king will flip. When he was a youth, the king had taken statistics (some of which he has forgotten, however) and agrees to abide by a hypothesis test based on the flipping of the twenty coins. What are the null and alternate hypotheses? What is the p_0?
b. Is the sample size large enough for the test to be valid?
c. At the answer to part b, the king is furious, but the director of the Royal Mint, scrambling for his life, suggests they do a test with forty coins. Now is the sample size large enough? Explain.
d. They do the coin flipping, and 25 of the 40 flips come up heads. Will the director of the Royal Mint lose his head? (He will lose his head if the king is convinced that the coins come up heads less than 70% of the time.) Explain with a hypothesis test. Follow the five steps. Be complete.
e. The enemy of the director of the Royal Mint arranges to collect $n = 400$ coins, and, surprisingly, 250 of them turn up heads when flipped, yielding the same sample proportion $\hat{p}$ as with $n = 40$. Should the director of the Royal Mint be very worried? Explain why or why not. You may well have to do the hypothesis test again.
f. What is the (statistical) moral of this story?

5. Pete Repete Reprise Recall the question about Pete and Repete and their argument about the proportions of college students having tattoos. Recall that:

Pete's idea was that the population proportion is $p = 0.30$, and
Repete's idea was that the population proportion is $p = 0.20$.

We now have data for many semesters, and we can use these data to evaluate their ideas:

Semester, Year and Place	Females Proportion having a tattoo	Females Number having a tattoo	n	Males Proportion having a tattoo	Males Number having a tattoo	n	Overall Proportion having a tattoo	Overall Number having a tattoo	n
Both, 1999, PA	0.131	18	137	0.191	13	68	0.151	31	205
Spring 2008, CA	0.255	12	47	0.160	8	50	0.206	20	97
Autumn 2008, CA	0.254	16	63	0.127	8	63	0.190	24	126
Spring 2009, CA	0.372	32	86	0.219	14	64	0.307	46	150
Autumn 2009, CA	0.258	23	89	0.115	9	78	0.192	32	167
Spring 2010, CA	0.266	25	94	0.309	25	81	0.286	50	175
Autumn 2010, CA	0.149	15	101	0.146	13	89	0.147	28	190
Spring 2011, CA	0.191	18	94	0.218	19	87	0.204	37	181

You may use Fathom to do this exercise, or you may do the exercise completely by hand. If you use Fathom, follow the directions just below.

- Open the Fathom file **PeteRepeteSetUp.ftm.** You will see no data, but rather the Test display. Anything in blue can be changed. **PTO**

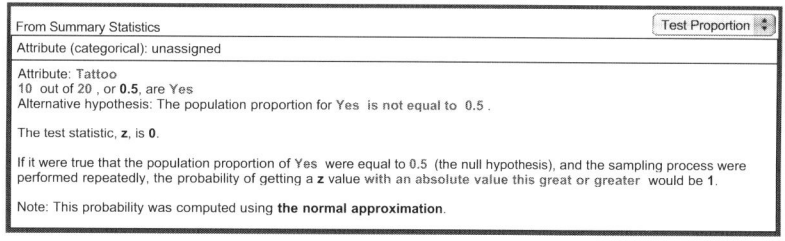

From Summary Statistics — Test Proportion
Attribute (categorical): unassigned
Attribute: Tattoo
10 out of 20, or 0.5, are Yes
Alternative hypothesis: The population proportion for Yes is not equal to 0.5.
The test statistic, z, is 0.
If it were true that the population proportion of Yes were equal to 0.5 (the null hypothesis), and the sampling process were performed repeatedly, the probability of getting a z value with an absolute value this great or greater would be 1.
Note: This probability was computed using **the normal approximation**.

- To get the test using spring 2008 overall data, and testing *Pete's idea*, change **"10 out of 20"** to **"20 out of 97."** Also change **"Yes is not equal to 0.5"** to **"Yes is not equal to 0.3"** by changing just the 0.5 to 0.3.
    a. Is the test statistic in the reasonably likely region or the rare region?
    b. What is the *p*-value? (Notice the wording that Fathom uses.)
    c. So does Pete lose this argument? Why?
- With the **Test** selected, go to **Object>Duplicate Hypothesis Test**. This will allow you to see the old test, as well as the current one.
    d. Repeat (!!) the test but using the overall data for spring 2010. What is the test statistic and what is the *p-value*? Does Pete lose this argument? Express the outcome in terms of the null hypothesis.
    e. With these same data (spring 2010), test Repete's idea. (What do you have to change?) Does Repete lose? Explain why or why not.
- In **PeteRepeteReprise.ftm,** you will also see a set-up for a confidence interval. Change the **"10 out of 20"** to the number for spring 2010.
    f. The confidence interval gives plausible values for *p*. Explain how it supports Pete's idea and does not support Repete's idea.
    g. Get the confidence interval for the autumn 2010 overall data. What can you say about Pete's and Repete's ideas? Do your results support either one? Why?

6. **Born on Sunday?** Are babies more likely or less likely to be born on some days of the week rather than others? Are they less likely to be born on weekends? Or more likely? Specifically, are babies less or more likely to be born on *Sundays*? We can answer this question, because we have a random sample of all births in the USA for 2006. For this exercise we will work with a small random sample ($n = 210$) of births. Our statistical question is:

    "Do we have evidence that for the population of babies born in the USA, the proportion of Sunday births is different from what we expect if births for each day of the week are equally likely?"

    - Open the Fathom file: **TwoTenBirthSample.ftm** and get a case table.
    - From the Shelf, drag down a **Summary Table**, and drag the variable SundayBirth to get the table shown here. [SundayBirth records whether the birth was on a Sunday.]

TwoTenBirthSample		
SundayBirth	Not Sunday	191
	Sunday	19
Column Summary		210
S1 = count ( )		

    a. First, use the data shown here to calculate the sample proportion of *Sunday* births in our small sample of $n = 210$. Assign the correct symbol to this proportion; should the symbol be $\hat{p}$, p or p_0? Give a reason for your choice of symbol.
    b. Thinking about the statistical question: If babies are *equally likely* to be born on any specific day of the week, then in a sample of $n = 210$, how many babies should be born on Sunday?
    c. If babies are *equally likely* to be born on any specific day of the week, then what *proportion* of babies should be born on Sunday? Do a calculation, and in this instance, give the answer rounded to *six* decimal places.

d. The proportion that was calculated in part c is the proportion of Sunday births, if the proportions by day are equally likely, and it will be a part of both our null and alternate hypotheses. What is the best symbol for the proportion calculated in part c? Should the symbol be $\hat{p}$, p or p_0?

e. Give a reason that the answer to part d should *not* be p rather than the symbol you chose.

Step 1: Setting up the Null and Alternate Hypotheses

f. So far, we have seen that we have a sample proportion $\hat{p} = \dfrac{19}{210} \approx 0.0905$ and we have determined that our idea for the population proportion is $p_0 = 1/7 \approx 0.142857$. Using the H_0 format and symbols shown in the **Notes**, write the Null Hypothesis.

g. Write the Alternate Hypothesis using the H_a format shown in the Notes. So which symbol of the three: $>$, $<$, or $\neq$ is the most appropriate for the alternate hypothesis? (To decide which of the symbols of $>$, $<$, or $\neq$ should be used in the alternate hypothesis, read the statistical question. Our statistical question does not state that the proportion born on Sunday is to be greater than the "equally likely" proportion nor does it predict that the population proportion will be less than the "equally likely" proportion. It allows for *either* greater or less with the word "different." Note: the H_a is based upon our statistical question, not our sample data!)

Step 2: Checking the Conditions There are three conditions to be checked.

h. *Simple Random Sample*: Our sample of $n = 210$ is a Simple Random Sample of all of the births recorded in the year 2006. So, if our population of interest is *all* births in the USA, is the condition that our sample is a SRS met? (We must always check for randomness.)

i. Check the (second) condition that $np_0 \geq 10$ and also $n(1-p_0) \geq 10$. (This is the condition that must be met if we use a Normal Distribution rather than a Binomial Distribution.)

j. A third condition is that the size of the population should be at least ten times the size of the sample? Is that condition met if the population is all births in the USA in 2006? Give a very brief reason.

Step 3: Calculating the Test Statistic

k. Calculate the test statistic using the formula for the test statistic.

l. Does the value of the test statistic indicate that the $\hat{p}$ is *rare* or *reasonably likely*? Give a reason for your answer.

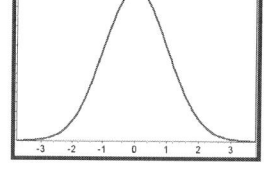

m. Draw a sketch of a Normal Distribution, (as shown) and put $p_0 \approx 0.143$ as the mean of the distribution. In your calculation of the test statistic, the denominator should have been $\sqrt{\dfrac{p_0(1-p_0)}{n}} = \sqrt{\dfrac{0.142857(1-0.142857)}{210}} \approx 0.024$. This number is the standard deviation of the sampling distribution we are using. Use the numbers for the mean and standard deviation to determine the location of p-hats that are one and two standard deviations from the mean; that is 0.143 ± 0.024, $0.143 \pm 2(0.024)$. Locate the $\hat{p} \approx 0.0905$ on the sketch. [Your answer to part k should be $z \approx -2.17$. Does the p-hat shown in your sketch come out to be just beyond two standard deviations from the mean? Your sketch and your conclusion to part l (i.e. part "el") should agree.]

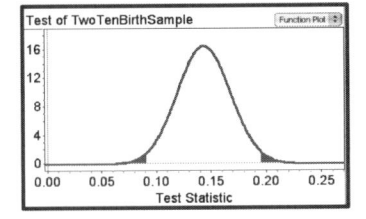

Step 4: Calculating the p-value

We will calculate the p-value in two sub-steps;

part "n" is just the first step.

n. Using the test statistic calculated in part k, and the Normal Distribution Chart, find the probability

$P(\hat{p} < 0.0905) = P(z < -2.17)$. On your sketch show this probability by appropriate shading and an arrow.

o. Because our alternate hypothesis is two sided (H_a: $p \neq 0.142857$), the p-value is twice what was found in part n. Show the p-value by shading the graph made to answer part m.

Check your calculations using Fathom.

- Open the Fathom file **TwoTenBirthSample.ftm** and get a **case table**.
- Get the **Summary Table** showing Sunday and "not Sunday" births using the variable *SundayBirth*.
- Pull down a **Test** icon from the shelf; it should say "**Empty Test**."
- Change **Empty Test** to **Test Proportion**
- Drag the variable *SundayBirth* to the "Attribute (categorical) Unassigned." There are things in blue, and anything in blue can be changed. Change "Not Sunday" to "Sunday" and 0.5 to 0.142857.
- With the **Test** selected go to the menu **Test** and deselect **Verbose**. Check your calculations with the Fathom output.
- With the **Test** selected, get (from the menu) **Test>Show Test Statistic Distribution**. [You may have to move the scales to see this picture.]

p. On the **Show Test Statistic Distribution** how is the p-value = 0.03 shown? Is it (i) on the "x-axis" with 0.03 just to the right of zero, OR is it (ii) by the slivers of shading to the left of 0.10 and to the right of 0.20. Annotate your sketch to illustrate the p-value.

Step 5 Interpreting the Result

q. Is the test statistically significant? Give a reason for the answer.

r. Do we have evidence against the Null Hypothesis, or is our test consistent with the H_0?

s. Give an interpretation of the result in the context of the statistical question. That is, do our results indicate that babies are just as likely to be born on Sunday as any other day, or not?

t. Using $\hat{p} \approx 0.0905$ and n = 210 find (either by calculating or by using Fathom) a 95% Confidence Interval for the proportion of Sunday births *p*.

7. **Births on Monday?** The exercise just above showed evidence that the proportion of births on Sunday is different from the proportion one would expect if each day's births were equally likely. Our small sample of n = 210 shows evidence that in the population of all births, Sunday births are less numerous than expected. What about Monday? We can do the same kind of test for that day of the week.

TwoTenBirthSample		
MondayBirth	Monday	33
	Not Monday	177
	Column Summary	210
S1 = count ( )		

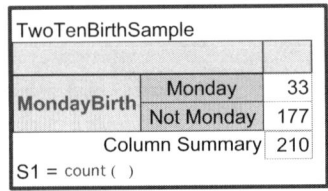

Having seen that the proportion for Sunday is lower, perhaps we may think that the proportion for Monday is higher than the $p_0 = 1/7 \approx 0.142857$ expected if the proportions for each day of the week are equal. Our statistical question is:

Do we have evidence that the proportion of births on Monday is higher than one-seventh of all births?

- Use the Fathom file **TwoTenBirthSample.ftm**.
- Get the **Summary Table** showing Monday and "not Monday" births using the variable *MondayBirth*.
  a. Calculate the proportion of Monday births in the sample and assign the correct symbol. Is the proportion higher than the expected proportion?
  b. Set up the Null and Alternate Hypotheses for the statistical question written in italics just above. [Should the alternate hypothesis be one-sided or two sided? Be prepared to give a reason for the choice.]
  c. Although we checked the conditions for these data in Exercise 6, give reasons why each of the three conditions [(i) Random sampling, (ii) $np_0 \geq 10$ and $n(1-p_0) \geq 10$ and (iii) Population large compared with the sample.] are met.
  d. Calculate the test statistic for the hypothesis test.
  e. Judging from the test statistic, do you think that this hypothesis test will be statistically significant? Give a reason for your answer.
  f. Use the Normal Distribution Chart to find the *p*-value.
  g. Give an interpretation of the hypothesis test in the context of the statistical question asked.
  h. Make a sketch of the sampling distribution used, and show (i) the mean of the sampling distribution, (ii) the *p*-hat on the *x*-axis, and (iii) by shading, and labeled, the *p*-value.

  Check your calculations using Fathom.
- Pull down a **Test** icon from the shelf. Change **Empty Test** to **Test Proportion**
- Drag the variable *MondayBirth* to the "Attribute (categorical) Unassigned." Change what needs to be changed.
- With the **Test** selected go to the menu **Test** and deselect **Verbose**. Check your calculations.
- With the **Test** selected, get **Test>Show Test Statistic Distribution**. [You may have to move the scales.]
  i. Your sampling distribution sketch should resemble the Fathom "Show Test Statistic Distribution" shown here. Explain why the shading is only on the right side, and not on both sides.
  j. The *p*-value for this test is 0.28. Confused Conrad, to attempting to interpret the *p*-value, writes: "The test is statistically significant because this is a big *p*-value and shows that 28% of the children are born on Mondays, which is a lot" Correct all of CC's mistakes.
  k. We have failed to reject the Null hypothesis. Our test statistic shows that given our H_0 our sample proportion is reasonably likely; our *p*-value indicates that there is a fairly high probability that we would get our *p*-hat or one greater if the H_0 is true. Have we therefore proved the Null Hypothesis to be true? Why or why not? [This question may be discussed in class in general.]

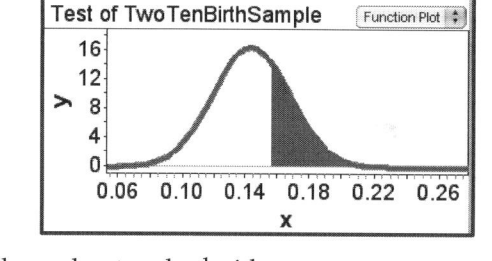

§4.3 Comparing Proportions, or What Is the Difference?

Tattoos Again

The first statistical question we asked was: *"Are male or female students more likely to have a tattoo?"* We have looked at various samples (Penn State and from California), but here are the data for the **CombinedClassDataY09.ftm**. To answer our statistical question, one of the first things we did was to compare the proportions of males and females having a tattoo. Using the notation from §1.2, we would write $P(T \mid M) = \frac{23}{142} \approx 0.162$, and $P(T \mid F) = \frac{55}{175} \approx 0.314$. So it appears that females (in a college in California in 2009) are more likely to have a tattoo than males.

Combined Class Data 09		Tattoo		Row Summary
		N	Y	
Gender	F	120 / 0.685714	55 / 0.314286	175 / 1
	M	119 / 0.838028	23 / 0.161972	142 / 1
Column Summary		239 / 0.753943	78 / 0.246057	317 / 1

S1 = count ( )
S2 = rowProportion

But are we certain? We have since learned that to *generalize* from a sample—to *infer* from a sample to a population—takes some special techniques; we cannot just say that because we saw it in a sample, it *must* be true for a population that we have in mind. We can generalize but only after doing some work and only under certain conditions. Now we come to two very common questions.

Two Questions: The first question is:

<u>Is there</u> any difference between two groups in the population from which the samples were drawn?

For our example, we are asking whether there really is a difference in the percentage of females and the percentage of males who have tattoos among all CSM students. For this question, the "Is there any difference?" question, we will use a hypothesis test.

The second question is:

<u>How big</u> is the difference between two groups and in what direction?

In our sample, it looks like the difference is about 15%, but we would like to have an estimate for the size and direction of the difference in the population. For this question we will *estimate* using a *confidence interval*.

More new notation: Since we have two groups instead of just one, we need some new notation. We need to distinguish between the proportions for each of the groups, both for our sample data and also for the population. For our example, there are different sample sizes for the males and the female (and the sample sizes do *not* have to be the same!), so in our calculations, we want to distinguish between them. It will be convenient to use subscripts and write $n_M = 142$ and $n_F = 175$. For the sample proportions, rather than having just one $\hat{p}$, we now have two, so we would write $\hat{p}_M \approx 0.162$, and $\hat{p}_F \approx 0.314$. And we must not forget the population proportion symbols; remember that even though we never know the exact values for the population proportions, we must still have symbols for them so that we can talk about them. For our example, we would write p_M and p_F—these not wearing hats because they are population proportions.

Advice: you will find it best to assign letters to the subscripts of the two groups to help keep track of which group is which. In the general formulation, since we do not know what we will use them for, we will have to use numerical subscripts. Try to use meaningful letters in any actual application.

Notation for Comparing Groups:			
	Sample Size	Sample Proportion	Population Proportion
Group 1:	n_1	$\hat{p}_1$	p_1
Group 2:	n_2	$\hat{p}_2$	p_2

A sampling distribution for differences of sample proportions. Everything that we did with confidence intervals (in §4.1) and hypothesis testing (in §4.2) depended on having a sampling distribution. Now, we will be interested in the sampling distribution for differences of sample proportions, or $\hat{p}_1 - \hat{p}_2$. What does this mean? In our example above, $\hat{p}_F - \hat{p}_M \approx 0.314 - 0.162 = 0.152$. These kinds of differences are what go into our sampling distribution.

(You may wonder whether it matters which proportion comes first. It does not matter—if you reversed the order you would get a negative number in this case but the same numerical difference. But you *do* have to be consistent; if you start with $\hat{p}_F - \hat{p}_M$ then everything must be in terms of $\hat{p}_F - \hat{p}_M$ and not $\hat{p}_M - \hat{p}_F$.)

Sampling Distribution for the Difference of Sample Proportions

The sampling distribution of $\hat{p}_1 - \hat{p}_2$ for samples of size n_1 and n_2 drawn randomly from a population with population proportions p_1 and p_2 will have shape, center, and spread according to:

Shape: The shape is approximately Normal under the **conditions** that $n_1 p_1$, $n_2 p_2$, $n_1(1-p_1)$, $n_2(1-p_2)$ are all 5 or more.

Center: The mean is $\mu_{\hat{p}_1 - \hat{p}_2} = p_1 - p_2$.

Spread: The standard deviation is $\sigma_{\hat{p}_1 - \hat{p}_2} = \sqrt{\dfrac{p_1(1-p_1)}{n_1} + \dfrac{p_2(1-p_2)}{n_2}}$.

The next thing is to see how we actually use this sampling distribution. We will take up our first question: "Is there actually a difference of proportions in the population?" Perhaps the sample difference we have seen is actually just the result of random sampling variation that we naturally have when we randomly sample. (Notice that, for the moment, we are thinking of our sample as a SRS.) We answer the question with a hypothesis test, and we will use the same five steps but with different formulas.

Example: Is there any difference?

Our specific question for the tattoo data is "Is there really a difference in the population proportions of males compared with female students at CSM who have a tattoo?" Realistically, there probably is *some* difference in the tattooed population proportion of males and females. We would be very surprised to find *exactly* the same proportions in any collection. For simplicity in doing the test, however, we will take as our preconceived idea that there is *no difference*; that is our standard. (When we get to confidence intervals then we can estimate the size of any difference.) What does it mean that there is no difference in the population? "No difference" would mean $p_F = p_M$, that the two population proportions are equal. In fact, if it is true that the proportion of females having is the same as the proportion of males having a tattoo then it must be the same proportion, and we could write $p_F = p_M = p$, where p is the common proportion.

The idea that the population proportions of the two groups are the same will form the **null hypothesis** for all of the hypothesis tests that we will do with two groups. (It is possible to have other null hypotheses; we could test whether the population proportions differ by 0.12, for example. The most common null hypothesis is one of "no difference.") Having made this decision, we are now ready to go through our **five steps** of a hypothesis test with our example, after one preliminary step that we already made.

We have to decide whether what proportions we are comparing. Here we could compare the proportions of males and females that *do* have a tattoo or the proportion of males and females that *do not* have a tattoo. The results would be the same, but we must make a choice. The one that we choose ("Yes"—tattooed, in our example) is called a **success** and defines what counts go into the numerators of our proportions.

Step 1: <u>Setting Up the Hypotheses</u>: There are two equivalent ways of writing the null and alternate hypotheses:

$$H_0 : p_F = p_M \quad \text{or} \quad H_0 : p_F - p_M = 0$$
$$H_a : p_F \neq p_M \qquad\qquad H_a : p_F - p_M \neq 0$$

(where the proportions refer to the proportions having a tattoo)

Step 2: <u>Checking the Conditions</u>: To check the conditions, we must calculate one more thing. The sampling distribution is based on the null hypothesis, which says that $p_F - p_M = 0$. If it is true that $p_F - p_M = 0$ then $p_1 = p_2 = p$. Of course we do not know this p, so our best estimate is the *overall proportion of students who have a tattoo* in our sample. This estimate of p is designated $\hat{p}$, written without any subscript, since it refers to both groups. For our example, $\hat{p} = \dfrac{\text{total number of students with a tattoo}}{\text{Total number of students}} = \dfrac{78}{317} \approx 0.246$.

Combined Class Data 09

		Tattoo N	Tattoo Y	Row Summary
Gender	F	120 0.685714	55 0.314286	175 1
	M	119 0.838028	23 0.161972	142 1
Column Summary		239 0.753943	78 0.246057	317 1

S1 = count ( )
S2 = rowProportion

One condition for our test is that $n_1\hat{p}$, $n_2\hat{p}$, $n_1(1-\hat{p})$, $n_2(1-\hat{p})$ are five or greater. We can calculate these: for example: $n_M\hat{p} = (142)\cdot(0.246) \approx 34.9 > 5$. In practice, it suffices to calculate only those that may appear problematic. Here the calculation we have chosen is for the cell with the smallest value. However, notice that we use an "expected" value for p of $\hat{p} = 0.246$ for these calculations; the $\hat{p} = 0.246$ is based on our best estimate of the overall proportion of students who have a tattoo.

We can also now calculate the standard deviation of the sampling distribution using the formula $\sqrt{\hat{p}(1-\hat{p})\left[\dfrac{1}{n_1} + \dfrac{1}{n_2}\right]}$; this is really the same formula as in the box above but simplified because $p_1 = p_2 = p$, and p is estimated with $\hat{p}$.

For our example, where $\hat{p} \approx 0.246$, we calculate

$$\sqrt{\hat{p}(1-\hat{p})\left[\frac{1}{n_1}+\frac{1}{n_2}\right]} = \sqrt{0.246(1-0.246)\left[\frac{1}{175}+\frac{1}{142}\right]} \approx 0.0486.$$

Then the sampling distribution looks like this Normal distribution, where the interval of reasonably likely differences is $0-1.96(0.0486) < \hat{p}_F - \hat{p}_M < 0+1.96(0.0486)$ or $-0.0953 < \hat{p}_F - \hat{p}_M < 0.0953$. For this example, the shaded (rare) region on the left is to the left of $\hat{p}_F - \hat{p}_M = -0.0953$ and the shaded (rare) region on the right is to the right of $\hat{p}_F - \hat{p}_M = 0.0953$. These are *differences of sample proportions*. Our actual sample difference happened to be $\hat{p}_F - \hat{p}_M \approx 0.314 - 0.162 = 0.152$. You should easily be able to determine whether *our* difference of proportions is reasonably likely or rare. (Is 0.152 inside or outside the reasonably likely interval?)

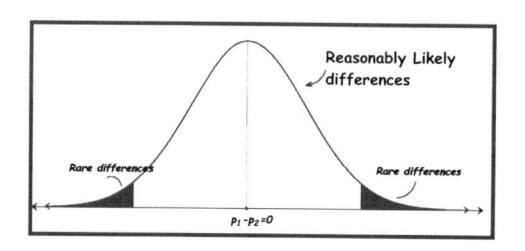

Step 3: <u>Calculating the Test Statistic:</u> The test statistic is just a z score, and its formula comes from the fact that our null hypothesis is $p_F - p_M = 0$, and that therefore (under the null hypothesis) $p_1 = p_2 = p$, which we are estimating from our sample with $\hat{p} \approx 0.246$. The test statistic for this test is as shown in the box.

Test statistic for a hypothesis test for comparison of proportions.

$$z = \frac{(\hat{p}_1 - \hat{p}_2) - 0}{\sqrt{\hat{p}(1-\hat{p})\left[\frac{1}{n_1}+\frac{1}{n_2}\right]}} \quad \text{where } \hat{p} = \frac{\text{Total number of cases that are "successes"}}{\text{Total number of cases}}$$

Here is the calculation for our example.

$$z = \frac{(\hat{p}_1 - \hat{p}_2) - 0}{\sqrt{\hat{p}(1-\hat{p})\left[\frac{1}{n_1}+\frac{1}{n_2}\right]}} = \frac{0.314 - 0.162}{\sqrt{0.246(1-0.246)\left[\frac{1}{175}+\frac{1}{142}\right]}} \approx \frac{0.152}{0.0486} \approx 3.13$$

Since $z = 3.13 > 1.96$, we can see immediately that our observed sample difference is *rare* and that we have evidence *against* the null hypothesis that $p_F - p_M = 0$. Since the z is big, it is very unlikely that we would see as big a difference (or one bigger) that we have seen in our sample, if it were really true that $p_F - p_M = 0$. We can calculate the probability of getting a difference of proportions this big or more by calculating the *p*-value.

Step 4: <u>Calculating the p-value:</u> We first find the probability of getting $z = 3.13$ or greater by consulting the Normal Distribution Chart and getting $P(z > 3.13) = 1 - P(z < 3.13) = 1 - 0.9991 = 0.0009$. The *p*-value will be double this so that *p*-value = $2P(z > 3.13) = 2(0.0009) = 0.0018$. Since this *p*-value is far *less* than $\alpha = 0.05$, we see again that our result for the difference of proportions is rare if the null hypothesis is true (that is, if $p_F - p_M = 0$).

Step 5: <u>Interpretation:</u> The test statistic and the *p*-value point in the same direction (as they must always!), and that is that we have observed a rare difference of proportions if it were true that $p_F = p_M$. Since our result is so rare, we are led to reject the null hypothesis that the proportion of female students (in the population) who have tattoos is the same as the proportion of male students in the population of all CSM students who have tattoos. Our result is *statistically significant*. We think that we have evidence that the proportions of male and females who have tattoos really do differ; we seem to have successfully challenged the null hypothesis.

However, remember that our sample is not random; it is just possible that the tattooed males (as opposed to the tattooed females) avoid taking statistics. It may be that the non-tattooed males have a greater likelihood of taking statistics. For the test to be trustworthy, we need a random sample.

Two-Sided and One-Sided. All of the hypothesis tests that we have done have had null and alternate hypotheses of the form $\begin{array}{l} H_0: p = p_0 \\ H_a: p \neq p_0 \end{array}$ (the ones we did in §4.2) or of the form $\begin{array}{l} H_0: p_1 = p_2 \\ H_a: p_1 \neq p_2 \end{array}$ (in this section). We have been challenging the null hypothesis with the idea that the proportion *differs* or is *not equal* to the standard or that the proportion for one group *differs* or is *not equal* to the proportion for the second group. All of these tests are **two-sided tests**. In our example of the proportions of males and females who have tattoos, we tested $\begin{array}{l} H_0: p_F = p_M \\ H_a: p_F \neq p_M \end{array}$. However, we may be able to bring a more refined "preconceived idea" to the test before we actually do it. We may have reason to believe (*before* we look at the data) that the proportion of tattooed females is bigger than the proportion of tattooed males. We can make this refined preconceived idea to be the challenge to the null hypothesis that $H_0: p_F = p_M$. If so, then what we are doing is a **one-sided test**. In a one-sided test, the null stays the same, but the alternate is different. The hypotheses would be: $\begin{array}{l} H_0: p_F = p_M \\ H_a: p_F > p_M \end{array}$.

Differences between two-sided tests and one-sided tests. Some calculations remain the same whether the test is one- or two-sided. The test statistic will be the same. However, since the hypotheses are different, the calculation of the *p*-value will be different. We are making a hypothesis in only one direction, so in this situation "more extreme" does not take account of random variation in both directions. Hence, the *p*-value will be calculated on only one shaded area. The test statistic is the same, but the *p*-value will be different. In our interpretation, we will mention that we hypothesized that one proportion would be greater than the other.

We can see these differences by looking at the Fathom output for the two-sided and one-sided tests for our example. Inspect the output carefully in the graphics on the next page to see the differences. (Note: the graphs shown are from another test to show the typical differences in the pictures; the shaded areas for *our* test would be so small that you would not be able to see them.)

A two-sided test:

```
Test of Combined Class Data 09                      Compare Proportions
Attribute (categorical): Tattoo
Attribute (categorical or grouping): Gender

Ho: Population proportion of Y in Tattoo where Gender is F  equals that of Y in Tattoo where Gender is M
Ha: Population proportion of Y in Tattoo where Gender is F  is not equal to that of Y in Tattoo where Gender is M

55 out of 175, or 0.314286, in Tattoo where Gender is F  are Y
23 out of 142, or 0.161972, in Tattoo where Gender is M  are Y
z:          3.131
P-value:    0.0017
```

A one-sided test:

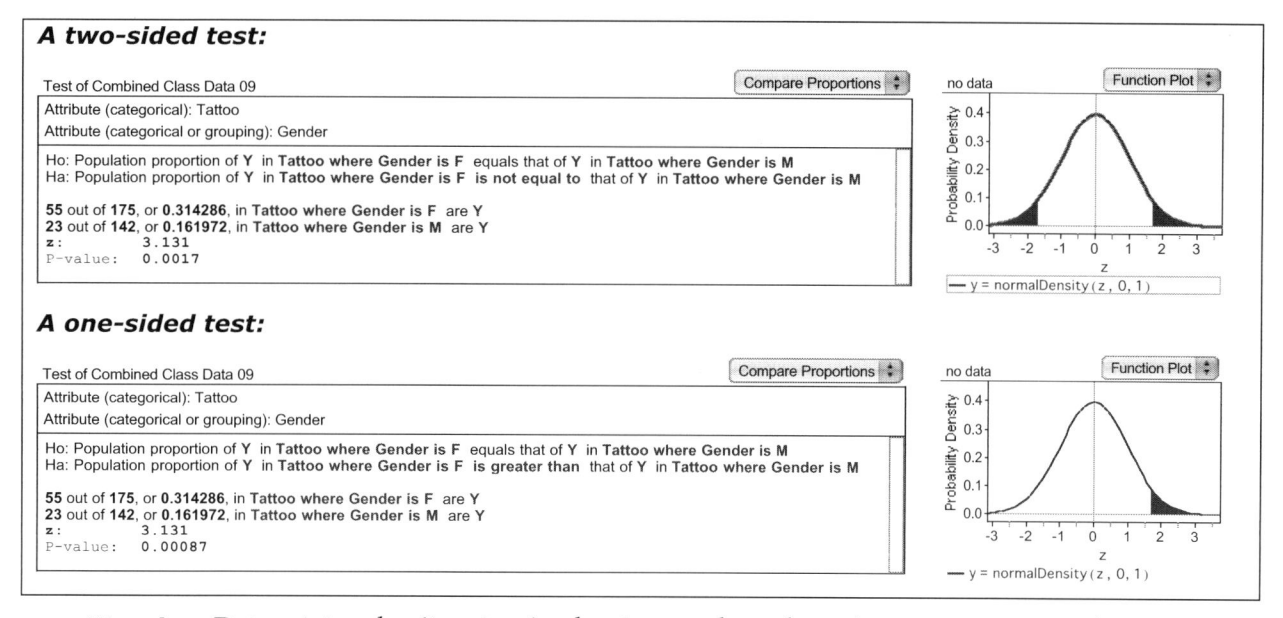

```
Test of Combined Class Data 09                      Compare Proportions
Attribute (categorical): Tattoo
Attribute (categorical or grouping): Gender

Ho: Population proportion of Y in Tattoo where Gender is F  equals that of Y in Tattoo where Gender is M
Ha: Population proportion of Y in Tattoo where Gender is F  is greater than that of Y in Tattoo where Gender is M

55 out of 175, or 0.314286, in Tattoo where Gender is F  are Y
23 out of 142, or 0.161972, in Tattoo where Gender is M  are Y
z:          3.131
P-value:    0.00087
```

Warning. Determining the direction for the alternate hypothesis for a *one-sided test* should not be based on the data that is used *for the test*. In our example, it would be "cheating" to say $H_a: p_F > p_M$ based on the data that we have. The idea for $H_a: p_F > p_M$ must come from ideas that the tester had before seeing the data.

Example: Estimating how big the difference is—the second question

The second question that we had concerned the size of the difference of the proportions. We want to estimate how big the difference is between the proportions in our population. *Estimation* means using confidence intervals. In this case what we are estimating is the difference of the proportions of tattooed females and tattooed males among *all the students*, our population. The formula is given in the box below.

Formula for the Confidence Interval for the Difference of Two Proportions

$$(\hat{p}_1 - \hat{p}_2) \pm z^* \sqrt{\frac{\hat{p}_1(1-\hat{p}_1)}{n_1} + \frac{\hat{p}_2(1-\hat{p}_2)}{n_2}}$$

where $z^* = 1.645$ for a 90% confidence interval
 $z^* = 1.96$ for a 95% confidence interval
 $z^* = 2.576$ for a 99% confidence interval

From the summary table shown again below, we see as before that $\hat{p}_F = 0.314$ and $\hat{p}_M = 0.162$, and so we can calculate a 95% confidence interval for the difference in proportions having a tattoo between female and male students using the formula shown in the box above:

$$(\hat{p}_F - \hat{p}_M) \pm z^* \sqrt{\frac{\hat{p}_F(1-\hat{p}_F)}{n_F} + \frac{\hat{p}_M(1-\hat{p}_M)}{n_M}} = (0.314 - 0.162) \pm 1.96 \sqrt{\frac{0.314(1-0.314)}{175} + \frac{0.162(1-0.162)}{142}}$$

$$= 0.152 \pm 1.96\sqrt{0.001231 + 0.000956}$$

$$\approx 0.152 \pm 1.96(0.0468)$$

$$= 0.152 \pm 0.09166$$

and this implies a confidence interval of $0.0603 < p_F - p_M < 0.2437$. We can interpret the interval by saying:

> "With 95% confidence, 15.2% (with a margin of sampling error of 9.2%) more female students than male students have a tattoo among the entire student body," or

> "We are 95% confident that the difference in the proportions of female and male students having a tattoo among all of the students at the college is between 6.0% and 24.4%," or

> "With 95% confidence, we can say that the likelihood that a female student at the college has a tattoo is 15.2% greater than the likelihood that a male student has a tattoo, with a margin of error of 9.2%."

Another example, showing Fathom output

We can ask another similar question comparing male and female students. The Summary Table shows the numbers and the proportions of male and female students who have visited "More than Seven States" against those who have visited "Seven or Fewer States." If a student has visited more than seven states then we call that student "Well-Travelled" or WT. So, using our notation from §1.2, we can see that

$P(WT \mid F) = \dfrac{56}{172} \approx 0.326$, and $P(WT \mid M) = \dfrac{59}{137} \approx 0.431$. Using the notation for this section, we would write $\hat{p}_F = \dfrac{56}{172} \approx 0.326$ and $\hat{p}_M = \dfrac{59}{137} \approx 0.431$. We have two questions:

> In the college as a whole, is there a difference in the proportions of "well-traveled" students by gender, or is the proportion essentially the same for male and female students?

> If there is a difference, what are the size and the direction of that difference?

The first question will be answered using a hypothesis test, since it is testing an idea: "Is there a difference by gender?" The second question calls for an estimate, so we will use a confidence interval.

For the first question, we can set up the null and alternate hypotheses (*Step 1*) in symbolic form as

$$H_0 : p_F - p_M = 0$$
$$H_a : p_F - p_M \neq 0$$

We can do a quick check on the conditions of a Normal sampling distribution by calculating

$n_M \hat{p} \approx (137) \cdot (0.372) \approx 51.0$ where $\hat{p} = \dfrac{115}{309} \approx 0.372$ (*Step 2*). However, we still have our usual concerns about the sample not being a random sample from all students at the college. The results of the next two steps (*Step 3: Calculating the test statistic* and *Step 4: Finding the p-value*) are shown in the Fathom output.

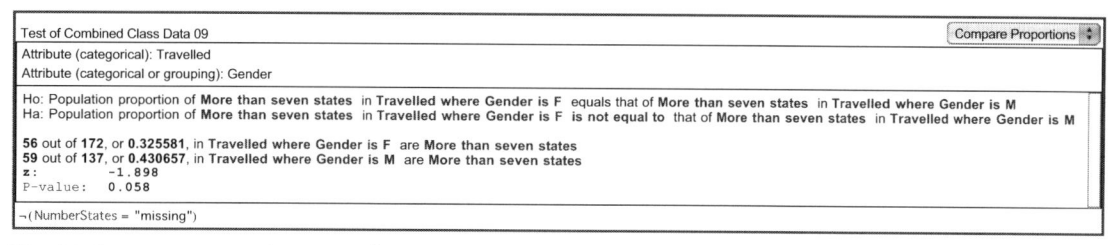

```
Test of Combined Class Data 09                                    Compare Proportions
Attribute (categorical): Travelled
Attribute (categorical or grouping): Gender

Ho: Population proportion of More than seven states in Travelled where Gender is F equals that of More than seven states in Travelled where Gender is M
Ha: Population proportion of More than seven states in Travelled where Gender is F is not equal to that of More than seven states in Travelled where Gender is M

56 out of 172, or 0.325581, in Travelled where Gender is F are More than seven states
59 out of 137, or 0.430657, in Travelled where Gender is M are More than seven states
z:         -1.898
P-value:    0.058

¬(NumberStates = "missing")
```

The Fathom commands to get this output are:

- From the shelf, drag down a **Test** icon and change "Empty Test" to "Compare Proportions."
- To the (first) attribute, drag the categorical variable whose success are to be compared.
- To the (second) attribute, drag the variable that contains the groups or categories in the comparison. Fathom automatically considers as a "success" whichever category is first in the alphabet. This may have to be changed.

How do we interpret these calculations? Is the test statistically significant? Can we reject the null hypothesis that the proportions of well-travelled are equal for male and female students? No, and no. The test is *not statistically significant* at α=0.05 because the *p*-value of 0.058 is bigger than 0.05. However, notice that it is very close, so we have *some evidence* of a gender difference. Also, the test statistic $z = -1.898$ is *not* outside the interval $-1.96 < z < 1.96$, although, again, it is close. (The number is negative because the calculations are doing the subtraction $\hat{p}_F - \hat{p}_M$, and the proportion for the males is bigger than the proportion for the females.) Both of these facts also say that we can*not reject* the null hypothesis of no difference between genders in proportion of well-travelled. We do not have enough evidence to say that in the college as a whole, the proportion of males who are well-travelled is higher than the proportion of females who are well-travelled.

For the second question, we can calculate a confidence interval. The Fathom output is shown below. Notice that one end of the confidence is negative and the other end is positive, although it is not very much larger than zero. Intervals that include zero are clumsy to put into elegant English (or perhaps other languages as well), so we have to say something like: "We are 95% confident that the likelihood of being well-travelled if female is anything from about 21% less than the likelihood for a male to just barely more likely (0.3%) of being well-travelled." Alternately, we could say: "We estimate with our data that there is a difference of 10.3% in the likelihood of being well-travelled between males and females for college students, but that comes with a margin of error of 10.8% so that it is possible that there is no difference at all, or that the difference is as much as 21%."

```
Estimate of Combined Class Data 09                     Difference of Proportions
Attribute (categorical): Travelled
Attribute (categorical or grouping): Gender

Interval estimate of the difference in proportions
56 out of 172, or 0.325581, Travelled where Gender is F are More than seven states.
59 out of 137, or 0.430657, Travelled where Gender is M are More than seven states.

Confidence level: 95.0 %
Estimate:   -0.105076 +/- 0.108532
Range:      -0.213608 to 0.00345651

¬(NumberStates = "missing")
```

The fact that the confidence interval includes *zero* is consistent with the conclusion that we did *not* reject the null hypothesis. That is, our sample data is plausible if in the population there is "no difference between males and females."

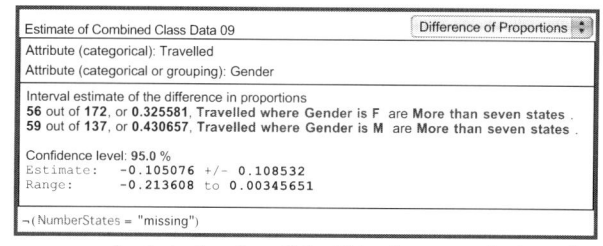

Summary and warning

The examples that we have looked at in this section were chosen because it was thought they would be fairly easy to understand. They illustrate the calculations nicely. However, they come with a big warning: we cannot be certain that the samples that were used are random samples of any population. The sample may *approach* being a random sample of students at a Northern California community college who are pursuing an academic program. This is because a great many students following an academic (rather than a vocational program) must take statistics. Some vocational programs (such as nursing) also require statistics. However, in the end, we do not know what population is being sampled.

All of the calculations that have been done presume that we have randomization.

The best research practice—such as the weight-loss experiment that was an example in §3.1—makes certain that randomization is used. However, many collections of data, including the data used in these sections, were not collected using randomization explicitly. The data may in effect be a random sample of some population, but we do not know. Proceed with caution.

Summary: Hypothesis Test for Comparing Proportions

Step 0: Decide which of the two categories of a categorical variable is to be called a "success" so as to define the proportions clearly and decide which group in the second categorical variable is "Group 1" and which is "Group 2."

Step 1: Set up the null and alternate hypotheses using an idea for the population proportion p_0.

H_0: $p_1 = p_2$ OR H_0: $p_1 - p_2 = 0$

H_a: $p_1 \neq p_2$ H_a: $p_1 - p_2 \neq 0$

Step 2: Check the conditions for a trustworthy hypothesis test.

1. The sample must be a simple random sample from a population ten times n.
2. All $n_1\hat{p}$, $n_1(1-\hat{p})$, $n_2\hat{p}$, and $n_2(1-\hat{p})$ must be five or greater.

Step 3: Calculate the test statistic $z = \dfrac{(\hat{p}_1 - \hat{p}_2) - 0}{\sqrt{\hat{p}(1-\hat{p})\left[\dfrac{1}{n_1} + \dfrac{1}{n_2}\right]}}$ where $\hat{p}$ is calculated from the sample by

$\hat{p} = \dfrac{\text{Total number of "successes" in the sample}}{\text{Total Number of cases in the sample}}$

Step 4: Calculate the p-value by getting $P(\hat{p} \text{ more extreme}) = 2P(Z \geq |z|)$ using the absolute value of the test statistic you calculated in Step 3.

Step 5: Evaluate the evidence that the p-value and the test statistic give you to determine whether your test successfully challenges the null hypothesized population proportion p_0 or not. Give an interpretation in the context of the data using the terminology of "statistical significance" and "rejecting the null hypothesis."

Summary: Confidence Interval for a Difference of Proportions

$$(\hat{p}_1 - \hat{p}_2) \pm z^* \sqrt{\dfrac{\hat{p}_1(1-\hat{p}_1)}{n_1} + \dfrac{\hat{p}_2(1-\hat{p}_2)}{n_2}}$$

where $z^* = 1.645$ for a 90% confidence interval
       $z^* = 1.96$ for a 95% confidence interval
       $z^* = 2.576$ for a 99% confidence interval

§4.3 Exercises on Two Proportions Inference

Special Exercise 1: Smoking and Premature Births

Here are some sample data for pregnancies (in England) showing the relationship between birth status (whether the birth was "premature" or "full-term") and the smoking status of the mother (smoker or not.) Our statistical question is:

Does mothers' smoking affect the likelihood of a premature birth?

Collection 1		Birth_Status		Row Summary
		Full Term	Premature	
Smoking_Status	Non-Smoker	4036	365	4401
	Smoker	465	49	514
Column Summary		4501	414	4915

S1 = count( )

Step 0: Decide what we are studying. We need to decide exactly the definition of the proportions we will study. The question "Does mothers' smoking affect the likelihood of a premature birth?" leads us to decide to think about:

p_S = The population proportion of premature births for mothers who are smokers (or we could say: "The probability of a premature birth for mothers who are smokers")

p_N = The population proportion of premature births for mothers who are *non*-smokers (or we could say: "The probability of a premature birth for mothers who are *non*-smokers")

1. a. What are the cases for these data? (There are two possible good answers and lots of wrong ones.)
   b. Which variable appears to be the explanatory variable and which variable is the response variable?
   c. In our statistical question (not in life!) is a "success" a *premature birth* or a *full-term* birth?
   d. Explain why it is that there are no "hats" in the notation used above.

Step 1: Write Null and Alternate Hypothesis.

2. Write the null hypothesis using the notation introduced above.
3. The form of the alternate hypothesis depends upon how you think smoking affects the proportion (i.e., the probability) of a premature birth.
   a. Do you think that mothers' smoking makes the proportion (probability) of a premature birth higher for all mothers in the population? Lower? Or have you no idea how smoking affects premature births?
   b. Does your answer in part a lead to a one-sided test or a two-sided test? Give a reason for your answer.
   c. If your answer to part b is one-sided (you may well have said $H_a : p_S > p_N$ where p_S is the probability of a premature birth for a smoking mother and p_N is the probability of a premature birth for a non-smoking mother), we will come back to that (perhaps as homework). Just for practice, we are going to test the two-sided alternative. Write the alternate hypothesis for a *two-sided* test.

Step 2: Check the Conditions.

4. Are we sure that we have independent random samples? Or are we not certain? (We will proceed with the test even if we are not certain, but then we will have some doubt about our conclusions.)
5. a. Calculate $\hat{p}_N$, $\hat{p}_S$ and $\hat{p}$, our sample estimate of the overall proportion of premature births.
   b. Determine whether all of $n_S\hat{p}$, $n_S(1-\hat{p})$, $n_N\hat{p}$, $n_N(1-\hat{p})$ are all at least 5.
6. One of the conditions that is usually not a problem is that the population from which the sample

was drawn should be at least ten times the size of the sample. Here we have a sample size of 4,915. Are there likely to be at least ten times this many births in England in a year?

Step 3: Compute the Test Statistic.

7. Look up the formula in the *Notes* for the test statistic. Write down the formula.
8. Using the numbers given, calculate the test statistic. You should have $\hat{p}_S = 0.0953$, $\hat{p}_N = 0.0829$, and $\hat{p} = 0.0842$, and you should come out with $z = 0.95744$ for the test statistic.
9. a. Judging from the value of the test statistic, does the test look like it is *statistically significant* or *not statistically significant*? Give a reason for your answer.
   b. Judging from the value of the test statistic, will we reject or *not* reject the null hypothesis?

Step 4: Compute a p-value.

10. We actually have enough information to say whether the *p* value will be bigger or smaller than a = 0.05, even before we find the *p*-value. Which will it be—bigger or smaller than a = 0.05? And why?
11. Make a sketch of a Normal distribution (the sampling distribution) showing the mean, the test statistic, and by shading the *p*-value. Consult the pictures in the **Notes**.
12. a. Calculate the *p*-value using the Normal Distribution Chart.
    b. Is the value you got consistent with your answer to question 9?
    c. Is the value you got consistent with your answers to question 10? (If it is not then you either made a mistake or your answer to question 9 is not correct.)

Step 5: Interpretation. Consult the guidelines in the **Notes for §4.2**. Remember that your conclusion should state the evidence for your conclusion and be written in the context of the problem. That is, you have to say something about smoking and premature births.

13. Write your interpretation for this hypothesis test in the context of the data.

Confidence Interval: Difference in Proportion of Premature Births between Smoking and Non-Smoking Mothers.

14. a. Using the formula for a confidence interval for two proportions,

    $$\hat{p}_1 - \hat{p}_2 \pm z^* \sqrt{\frac{\hat{p}_1(1-\hat{p}_1)}{n_1} + \frac{\hat{p}_2(1-\hat{p}_2)}{n_2}}$$, calculate a 95% confidence interval for the difference of the proportion (or probability) of premature births between children born to smoking mothers and children born to non-smoking mothers.
    b. Does your confidence interval include zero?
    c. Write a conclusion (in context) about your confidence interval that takes into account your answer to part b. (This may be difficult to put in elegant English; ask for help.)

One-Tailed Test.

15. You were asked to do a two-tailed test, but you may think that the proportion of premature births to smoking mothers should be higher than to non-smoking mothers.
    a. Write up the null and alternate hypotheses for this one-tailed test, using the correct notation.
    b. Which parts of the significance test will be different and which parts will be exactly the same for a one-tailed test as for a two-sided test? If they are the same, write "same"; otherwise, show what you would get for the one-tailed test.
       (i) Conditions? (ii) Test Statistic? (iii) Picture? (iv) *p*-value?

2. **Politics and Gender** [Fathom] It is generally thought that women are more liberal politically than men. Is this true for college students or is it that the proportion of students who have liberal political views is essentially the same, independent of gender.

- Open the Fathom file **CombinedClassDataY09.ftm**.
- Get a **Summary Table;** drag the variable *PoliticalView* to the right-pointing arrow and the variable *Gender* to the down-pointing arrow to get the table shown here.

Combined Class Data 09		PoliticalView			Row Summary
		Liberal	Moderate	Conservative	
Gender	F	95	69	11	175
	M	70	61	11	142
Column Summary		165	130	22	317

S1 = count ( )

We will consider these data to be from a *sample* of all the students at the college, recognizing that it is not quite the random sample that we would like to have.

a. Our interest is in the proportion of male students whose political view are "liberal" and the proportion of female students whose political views are "liberal." What symbols should be used to designate the *population proportions*? (What subscripts should the symbols have? Should the symbols have "hats"?)

b. Set up the null and alternate hypotheses for the test that we are doing, using the correct notation. You may either have a two-sided or a one-sided alternate hypothesis. That is up to you.

c. Calculate the sample proportion of males who have liberal political views and the sample proportion of females who have liberal political views and assign the correct symbols to these proportions.

d. Check whether the conditions for using the Normal distribution are met for this problem.

e. Calculate the test statistic.

f. From the value of the test statistic, does it appear that that the hypothesis is *statistically significant* or *not*? Give a reason for your answer.

g. Make a drawing of the Normal sampling distribution and show the mean and the location of your test statistic.

h. Considering the test statistic that you got, should the *p*-value be bigger or smaller than a = 0.05? Give a reason for your answer, perhaps aided by your drawing.

i. Use the Normal Distribution Chart to calculate the *p*-value. Show your work using correct notation.

j. Show the *p*-value on your drawing by shading (part g) in an area or areas. (If you have a one-sided alternate hypothesis, there should be just one shaded area; but if you have a two-sided then you will have two shaded areas.)

- Drag down a **Test** box from the shelf and change the "Empty Test" to **Compare proportions**.
- To the top "unassigned attribute" space, drag the variable *PoliticalView* and to the bottom "unassigned variable," the variable *Gender*.
- With the hypothesis test selected, go to the menu at the top under **Test** and deselect **Verbose**.
- Check to see that Fathom is doing the test you have done; that is, if you have decided on a two-sided test then Fathom should have the words "not equal to," but if you have decided on a one-sided test, Fathom should have "greater than."

- With the hypothesis test selected, go to the menu at the top under *Test>Show Test Statistic Distribution*.
- Collect your Fathom output and print it in a Word document to hand in with this exercise.
   k. Did your numbers agree with Fathom's calculations, and did your graphic agree with Fathom's graph?
   l. Put together all of your evidence to answer the question: "Can we say that in the population of *all* students at the college, female students are more likely to have liberal political view than male students?" Consider also the fact that the sample is not a random one.

3. **Tattoos and Politics** What is the relationship between having a tattoo and political views? Are students who have tattoos more likely to have political views that are liberal?

- Open the Fathom file **CombinedClassDataY09.ftm**.
- Get a **Summary Table**; drag the variable *PoliticalView* to the right-pointing arrow and the variable *Tattoo* to the down-pointing arrow to get the table shown here.

Combined Class Data 09

		PoliticalView			Row Summary
		Liberal	Moderate	Conservative	
Tattoo	N	114	108	17	239
	Y	51	22	5	78
Column Summary		165	130	22	317

S1 = count ( )

We will consider this to be a sample from all the students at the college, recognizing that it is not quite the random sample that we would like to have.

   a. Our interest is to compare and generalize about the proportion of students who have tattoos whose political view are "liberal" with the proportion of students who do *not* have tattoos whose political views are "liberal." If we were using the conditional notation of §1.2, should we be calculating $P(L \mid N), P(L \mid Y)$ or should we be calculating $P(N \mid L), P(Y \mid L)$? (Think: which variable are we taking as the explanatory variable?)
   b. Since we are not in §1.2 and are dealing with hypothesis tests, what symbols should be used to designate the *population proportions*? (What subscripts should the symbols have? Should the symbols have "hats"?)
   c. Set up the null and alternate hypotheses for the test that we are doing, using the correct notation. You may either have a two-sided or a one-sided alternate hypothesis. That is up to you.
   d. Calculate the sample proportion of students who have tattoos who have liberal political views and the sample proportion of students without tattoos who have liberal political views and assign the correct symbols to these proportions.
   e. Check whether the conditions for using the Normal distribution are met for this problem. Explain how you checked.
   f. Calculate the test statistic for the hypothesis test.
   g. From the value of the test statistic, does it appear that that the hypothesis is *statistically significant* or *not*? Give a reason for your answer.
   h. Make a drawing of the Normal sampling distribution and show the mean and the location of your test statistic.
   i. Considering the test statistic that you got, should the *p*-value be bigger or smaller than a = 0.05? Give a reason for your answer, perhaps aided by your drawing.

j. Use the Normal Distribution Chart to calculate the *p*-value. Use correct probability notation.
k. Show the *p*-value on your drawing by shading in an area or areas. (If you have a one-sided test, there should be just one shaded tail, but if you have a two-sided test, there should be two shaded tails.)

- Drag down a **Test** box from the shelf and change the "Empty Test" to **Compare Proportions**.
- To the top "unassigned attribute" space, drag the variable *PoliticalView* and to the bottom "unassigned variable," the variable *Tattoo*.
- With the hypothesis test selected, go to the menu at the top under **Test** and deselect **Verbose**.
- Check to see that Fathom is doing the test you have done; that is, if you have decided on a two-sided test then Fathom should have the words "not equal to," but if you have decided on a one-sided test, Fathom should have "greater than."
- With the hypothesis test selected, go to the menu: **Test>Show Test Statistic Distribution**.

l. Did your numbers and graphs agree with the Fathom output?
m. Put together all of your evidence to answer the question: "Can we from our data say that in the population of *all* students at the college, those with tattoos are also more likely to have liberal political views?" Consider also the fact that the sample is not a random sample.
n. Calculate a 95% confidence interval to show how much more likely the students with tattoos are to have liberal political views than the students without tattoos.
o. Give a good interpretation of your confidence interval.

- Get an **Estimate** box from the shelf and change the "Empty Estimate" to **Difference of Proportions**.
- To the top "unassigned attribute" space, drag the variable *PoliticalView* and to the bottom "unassigned variable," the variable *Tattoo*.
- With the hypothesis test selected, go to the menu at the top under *Estimate* and deselect *Verbose*.
- The confidence interval that Fathom calculates with take Tattoo = N first just because in the alphabet N comes before Y, so the confidence interval will have negative numbers. You can change this by changing "N" to "Y" and the confidence interval will be positive. Check your results.
- Collect your Fathom output and print it in a Word document to hand in with this exercise.

4. **Another Smoking and Childbirth Problem.**
From the same data as the example of smoking and premature births, they also recorded whether the child survived. Here are the data.

		Survival		Row Summary
		Died	Survived	
Smoking_Status	Non-Smoker	74	4327	4401
	Smoker	15	499	514
	Column Summary	89	4826	4915

S1 = count ( )

a. Test whether smoking affects the probability of child survival using a hypothesis test. Show all the five steps clearly: set up the hypotheses, check the conditions, calculate the test statistic and the *p*-value, and come to a conclusion. Write an interpretation of your conclusion in the context of the data. The test you carry out may be a one-sided or two-sided test, and you may choose the significance level (that is, you may choose the alpha).

•b. Check your calculations by opening the Fathom file **Smoking and Pregnancy A.ftm**. Get Fathom output for the hypothesis test.

5. **Better data for Smoking and Premature Births.** We have data that are actually better than the data in Exercise 1 because we are certain that the data are a simple random sample of all births in 2006. The source of the data used was not specified, except that they came from England.

- Use the file **QuattroMilleBirthsSample.ftm**. This is a simple random sample of from the entire population of births recorded in 2006.
- Get a **Summary Table** and make the table shown by dragging the variables *Premature* and *MotherSmokes*

QuatroMilleSample		Premature		Row Summary
		Full-Term	PreMature	
MotherSmokes	No	1673	174	1847
	Yes	195	30	225
Column Summary		1868	204	2072

S1 = count ( )

  a. What are the cases for these data?
  b. We will look at the proportion of premature (rather than full-term) births; calculate from the table the proportions $\hat{p}_N, \hat{p}_S$ and also $\hat{p}$. (You will be able to check your answers shortly.)
  c. What is the meaning of $\hat{p}$ compared with the meaning of $\hat{p}_N$ and $\hat{p}_S$?
  d. Checking the conditions. Here we are fortunate to have a simple random sample. The population of all births in 2006 must be bigger than 10 x 2072, so we just need to check whether our sample sizes are big enough. All of $n_N\hat{p}, n_N(1-\hat{p}), n_S\hat{p}, n_S(1-\hat{p})$ should be greater than 5. However, we need only check: the one that is likely to be smallest. Show the calculation.

- Get a **Test** from the shelf and change it to **Compare Proportions**.
- Drag the variables *Premature* and *MotherSmokes* to the top "Attribute" and "Grouping Attribute."
- Deselect **Verbose** from **Test**.
- Change the changeable (blue items) so that test looks like the one shown here.

```
Test of QuatroMilleSample                             Compare Proportions
Attribute (categorical): Premature
Attribute (categorical or grouping): MotherSmokes

Ho: Population proportion of PreMature in Premature where MotherSmokes is No equals that
of PreMature in Premature where MotherSmokes is Yes
Ha: Population proportion of PreMature in Premature where MotherSmokes is No is not
equal to that of PreMature in Premature where MotherSmokes is Yes

174 out of 1848, or 0.0941558, in Premature where MotherSmokes is No are PreMature
30 out of 225, or 0.133333, in Premature where MotherSmokes is Yes are PreMature
z:       -1.863
P-value:  0.062
```

  e. Write, in symbol form, the hypotheses for the test shown in the Fathom output. Is the test a one-sided or a two-sided test? How do you know?
  f. Using your calculations from part c (two of which you can check on the output), show how the test statistic was calculated. Show where the proportions go in the formula, without necessarily completing the calculations. (You may, of course, complete the calculations.)
  g. Based upon the test statistic and the *p*-value, give an interpretation of the hypothesis test. Discuss with great insight whether the results are consistent with what was found in Exercise 1.

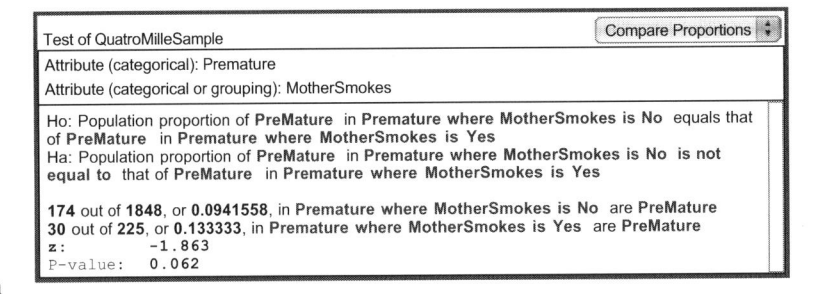

- h. Select the **Test** and from the menu get **Test>Show Test Statistic Distribution**. Make a sketch of what is shown, clearly labeling the test statistics and the *p*-value.
- Go to **Object>Duplicate Hypothesis Test** to get a duplicate test.
- Change the Fathom hypothesis test to reflect the alternate hypothesis $H_a : p_N - p_S < 0$. (Think!)
- Get **Test>Show Test Statistic Distribution** for this new one-sided test.
  i. Compared with the two-sided test, which numbers are different, and which the same?
  j. Is it possible for a two-sided test to be "not significant" but the one-sided test to be significant? Give a reason for your answer.

§4.4 Do We Have Independence? Chi-square

Introduction: One, two, three, more... One of the exercises in §4.3 looked at this table relating *PoliticalView* and *Gender*. We compared the proportions of males and females who have "Liberal" political views. We compared $P(L \mid M) = \frac{70}{142} \approx 0.493$ with $P(L \mid F) = \frac{95}{175} \approx 0.543$ (where L stands for "Liberal").

We then did a hypothesis test, so we used the notation $\hat{p}_M \approx 0.493$ and $\hat{p}_F \approx 0.543$. We tested the idea that females are more likely to hold liberal political views than males as against the null hypothesis that there is no difference by gender in political views. (As it happened, our test was not statistically significant.)

Combined Class Data 09

		PoliticalView			Row Summary
		Liberal	Moderate	Conservative	
Gender	F	95	69	11	175
	M	70	61	11	142
Column Summary		165	130	22	317

S1 = count( )

However, there are really three gender comparisons that could be made, not just one; we could compare $P(Moderate \mid M)$ with $P(Moderate \mid F)$ and also $P(Conservative \mid M)$ with $P(Conservative \mid F)$ as well as $P(L \mid M)$ with $P(L \mid F)$; it appears that statistical life got still more complicated. However, doing all of these comparisons would actually *not* be a good way to proceed. Instead, we ask a different question about the table *as a whole* and not about the various parts of the table. Our statistical question is:

Is there an association between the variables Gender and PoliticalView?

We will answer this question by building a model of "no association" between the variables and then seeing if the data that we have in the table successfully challenges the model of "no association." This sounds like a hypothesis test, and that is what it is. Where we have a table relating two variables (such as the one above) where the cells show the *counts* or *frequency* of cases in each intersection of categories (Female-Liberal, Male-Liberal, Female-Moderate, etc.), we call such a table a **r x c contingency table** because the table has **r rows** and **c columns**. Moreover, the numbers in the table may show how one variable is *contingent* or depends on the other.

> **Definition of an r x c Contingency Table**
>
> An $r \times c$ *contingency table* shows the counts (or frequencies) of cases in the intersections of the categories of two categorical variables.

The Opposite of Association: Independence

Our statistical question begins: "Is there an association ...?" To answer this, we start by looking at what "no association" would look like. If there is *no* association between variables, then we say that the variables are **independent**. No association between variables means independence between the variables; association means "not independence" or dependence between the variables.

In §1.2, we introduced the notion of **independence** for events. We will now expand this definition to apply it to variables. Here is the definition, once again (on the next page).

> **Independence of events**
>
> If $P(A \mid B) = P(A)$ then the events A and B are **independent.** In words, if the probability of A given B is the same as the probability of A then we say that the events A and B are **independent**.

So we could begin (tediously!) doing calculations to see if *events* are independent. Using the Gender and PoliticalView example shown above, we can calculate that $P(L) = \frac{165}{317} \approx 0.521$, but we see that $P(L \mid F) = \frac{95}{175} \approx 0.543$; so the events L and F—the events liberal political view and female gender—are *not* independent, since $P(L \mid F) \neq P(L)$. The events L and M (liberal and male) are also *not* independent because $P(L \mid M) \neq P(L)$. If we do the calculations for $P(Mod)$ and $P(Mod \mid M)$ and $P(Mod \mid F)$ we will see that the events Moderate and M are also *not* independent, and the events Moderate and F are *not* independent. Very seldom do we find exact independence when we are analyzing real data; we may come close, but, because of variation in samples, events that may seem to be *logically* independent will seldom meet the strict definition of independence. Moreover, things in the real world are simply too interconnected for strict independence; two variables may not be related in any simple fashion, but they may be in some extremely complicated way.

Combined Class Data 09					
		\multicolumn{3}{c	}{PoliticalView}	Row Summary
		Liberal	Moderate	Conservative	
Gender	F	95	69	11	175
	M	70	61	11	142
Column Summary		165	130	22	317
S1 = count( )					

But we want to transcend this tedious calculation and look at variables rather than events. The next section shows how we use the idea of independent events to build a "model of no association" — that is what the data would look like if there were *no* association between the variables.

Building the Model: Independence of Variables and Expected Counts

To build our model, we will do two things. First we will apply the idea of independence to *all* the categories of the categorical variables so that we will speak of variables (as well as the categories within the variables) as being independent. Then we pursue this idea and work backwards to answer the question: *What would the counts in the table be if the variables Gender and PoliticalView were independent?*

Here is how we reason. If the variables Gender and PoliticalView were independent then we must have $P(L \mid F) = P(L) \approx 0.521$ and also because $P(L) = \frac{165}{317} \approx 0.521$. What would the count for the "liberal females" have to be in order for this to be true? There are $n_F = 175$ female students. For $P(L \mid F) = P(L) \approx 0.521$, we need

$$P(L \mid F) = \frac{x}{175} = \frac{165}{317} = P(L),$$ where x is the number of "liberal females."

If we use algebra, and solve for x in this equation (by "cross multiplying"), we get

$x = 175 \times \frac{165}{317} \approx 91.088$ "liberal females" would be needed in the table to make $P(L \mid F) = P(L) \approx 0.521$.

Instead of 95 female students with liberal political views that we observe, we expect to have to be 91.1 for independence.

Likewise, for the male students, if we multiply $P(L) = \frac{165}{317} \approx 0.521$ by the number of male students $n_M = 142$, we get $x = 142 \times \frac{165}{317} \approx 73.912$. So, instead of 70 males with liberal views that we actually observe in our sample, we would expect to have 73.9 males if the variables *PoliticalView* and *Gender* were independent.

Then we do the same kind of thing for the "Moderate" and "Conservative" categories, but we use $P(Mod) = \frac{130}{317} \approx 0.410$ and $P(C) = \frac{22}{317} \approx 0.069$ and calculate $x = 175 \times \frac{130}{317} \approx 71.767$ for the moderate females and $x = 142 \times \frac{130}{317} \approx 58.233$ for the moderate males. The counts necessary for independence for the students with conservative political views are calculated in a similar fashion.

These counts are called **expected counts** (the symbol is **E**) and are shown in parentheses in the Fathom output here. Notice that even though we cannot have a fractional number of students (e.g. 58.233 students, for example) we do *not* round to the nearest whole number, and in the calculations, Fathom (and we) will use as many decimal places for the expected counts as we can manage. The expected counts are the *model* based upon the independence of the two variables. They are found by applying the overall proportion of one category of one variable (for example: $P(L) = \frac{165}{317} \approx 0.521$) to the overall numbers in the categories of the second variable.

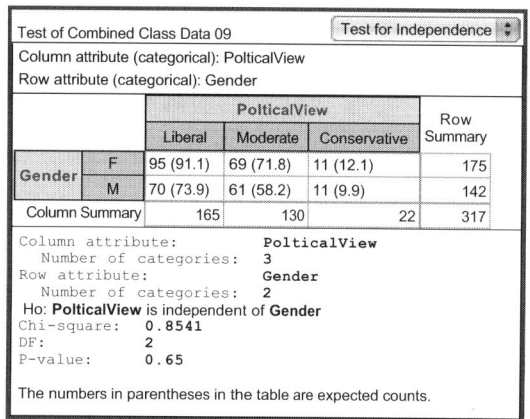

We can easily calculate that for any particular cell in a table, the expected count works out to be the row total for that cell multiplied by the column total for that cell and then divided by the total number of cases. Even more important is to realize that these counts *are the counts expected if the two variables are independent* of each other using the statistical definition of independence.

Expected Counts E for an r x c Contingency Table for the Model of Independence

The **expected counts** (or frequencies) for the model of independence are calculated by

$$E = \frac{(\text{Row total for that cell}) \times (\text{Column total for that cell})}{\text{Total Number of cases}}$$

and are the counts expected if the two variables in the contingency table are independent.

Observed Counts O for an r x c Contingency Table

The *observed counts* are the actual data observed in the contingency table.

323

Testing the Model with Data

We raised the question: *Is there an association between the variables Gender and PoliticalView?* If there is no relationship or association between the variables then it appears that the variables are *independent*. If the variables are independent then the probability that a student has (for example) moderate political views is the same whether the student is a male or a female. Our hypothesis test to answer this will have the same *five steps,* but what goes into these five steps will come from the model of independence — or *no association* — that we have been building.

Step 1: Hypotheses The structure of the hypotheses for our test is quite simple:

H_0 : *The two variables are independent*

H_a : *The two variables are not independent*

We could use symbols involving *p,* but the notation would be very complicated. For our example, we would write:

H_0 : *PoliticalView is independent of Gender*

H_a : *PoliticalView is not independent of Gender*

Step 2: Conditions The *expected counts* must not be too small. There are several rules of thumb that are used, but the simplest is that **all expected counts must be five or more.** Notice that it is the expected counts that must be five or more and not the **observed** counts for the condition. If we wish to infer from sample data to a population then the second and usual condition is that the data must be a **simple random sample** from the population of interest.

Step 3: Calculating the test statistic What we want to do is to compare the *observed counts* to the *expected counts.* If the observed counts differ much from the expected counts, the counts that we expect if the null hypothesis is true and the variables are independent, then we will have evidence *against* the null hypothesis of independence. But if the differences between the observed and expected counts are not great, we will have insufficient evidence against the null hypothesis of independence. We need a measure that will involve every single cell and count all the differences between observed and expected. That measure is the **chi-square goodness of fit statistic.** (Pronunciation of χ: **Chi** stands for the Greek letter χ and is pronounced as the first syllable in "Kaiser." It does not have the sound "chi" as in "chili.")

Chi-Square Goodness of Fit Statistic

$$\chi^2 = \sum_{\text{All cells}} \frac{(O-E)^2}{E}$$

where *O = Observed Counts* and *E = Expected Counts*

For our example, refer to the observed counts (95, 69, etc.) and the expected counts in parentheses (91.1, 71.8, etc.) in the Fathom output and see how they are used, and you should get an idea of how the calculation is done.

$$\chi^2 = \sum_{All\ cells} \frac{(O-E)^2}{E}$$

$$= \frac{(95-91.1)^2}{91.1} + \frac{(69-71.8)^2}{71.8} + \frac{(11-12.1)^2}{12.1} + \frac{(70-73.9)^2}{73.9} + \frac{(61-58.2)^2}{58.2} + \frac{(11-9.9)^2}{9.9}$$

$$\approx 0.1680 + 0.1067 + 0.1080 + 0.2070 + 0.1314 + 0.1331$$

$$\approx 0.8541$$

Think about how this statistic will behave; if all of the observed values were exactly the same as the expected, the value would be zero. The more the observed are *different* from the expected, the bigger will be the value of the χ^2. Notice also that the value cannot be negative, since all the terms in the sum are squared. Our calculation agrees with the value that Fathom calculated. (Had we used the rounded values, we might have encountered rounding error.)

		PoliticalView			Row Summary
		Liberal	Moderate	Conservative	
Gender	F	95 (91.1)	69 (71.8)	11 (12.1)	175
	M	70 (73.9)	61 (58.2)	11 (9.9)	142
Column Summary		165	130	22	317

Column attribute: PoliticalView
  Number of categories: 3
Row attribute: Gender
  Number of categories: 2
Ho: PoliticalView is independent of Gender
Chi-square: 0.8541
DF: 2
P-value: 0.65

The numbers in parentheses in the table are expected counts.

Sampling Distributions Again: The Chi-Square Distributions

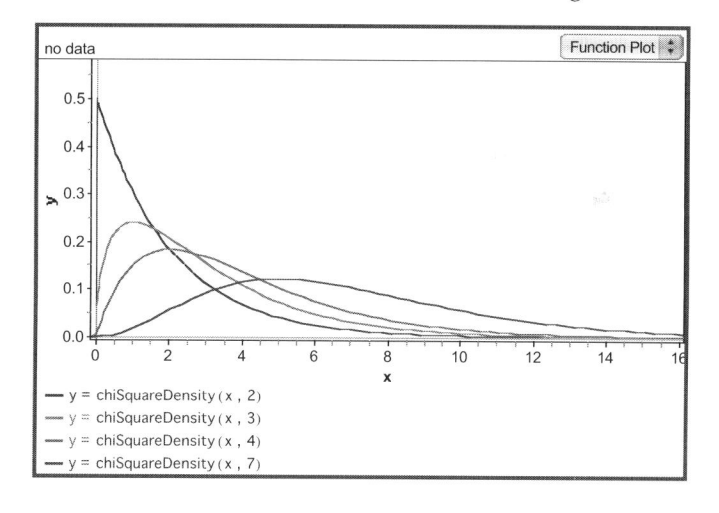

However, is what we have got a big value or a small value? To determine whether we have a big value of χ^2 or a small value of χ^2, we need a sampling distribution of the values of χ^2 that could come from all possible cell values with the same values for the row and column totals. The meaning of a sampling distribution is the same as we have seen before; we worked with the sampling distribution of $\hat{p}$s from all possible samples of size n and found that it was a Normal distribution, under certain conditions. When we worked with the differences of two sample proportions $\hat{p}_1 - \hat{p}_2$ we found that the sampling distribution was also Normal, again under certain conditions. We might expect that all sampling distributions are Normal; not so. The sampling distributions for the chi-square statistic are called the **chi-square distributions.** There is an entire *family* of distributions, and for that reason everything is in the plural (*chi-square distributions*). Here is the Fathom output showing the shapes of some of the *chi-square distributions*. The different distributions we see in the plot have different **degrees of freedom**, and ultimately the shapes go back to the mathematical definition of the distributions. Which degrees of freedom and therefore which distribution we use depends upon the use to which we are putting the distribution, but for our **Test of Independence for a Contingency Table** the degrees of freedom is defined by the formula $df = (r-1)(c-1)$, where r means the number of rows in the table, and c means the number of columns in the table. So, for our table relating *Gender* and *PoliticalView* we have

$$df = (r-1)(c-1) = (2-1)(3-1) = 1 \cdot 2 = 2 \text{ degrees of freedom}$$

> **Degrees of Freedom for an r x c Contingency Table**
> $$df = (r-1)(c-1)$$
> where *r* means the number of rows in the table, and *c* means the number of columns in the table.

We use this distribution in a similar way to the way we used the Normal table; we start by defining *reasonably likely* and *rare*. We have been using $\alpha = 0.05$ (that is 5%) as our definition of rare, which means that 95% or $1 - \alpha$ of the results are reasonably likely. For a chi-square distribution, the rare outcomes will be in the *right tail* because (in our application to contingency tables) these large values of χ^2 are the ones that challenge the null hypothesis of independence. Remember that a small value of χ^2 means that the observed counts are quite close to the expected counts and we have a sample result that could have come from two variables being independent. A large value of χ^2 means that the observed counts are far from the counts expected if the null hypothesis is true and the variables are independent. Work through this interaction to get some notion of the shapes of chi-square distributions and also what *rare* outcomes for the χ^2 look like.

- Open the Fathom file **ChiSquarePlot.ftm**.
- The slider indicates different degrees of freedom; move it so as to see how the shape of the Chi-Square Distributions change as the **df** changes. (Since we are dealing with tables, we will only have integer values of *df*, but the Fathom plot also accommodates values of df between the integers.)
- The vertical line in the plot shows the dividing line between *reasonably likely* and *rare*. To the right of this line are the rare values of a chi-square statistic χ^2 for a given *df*. Either by using the slider, or (better) by typing in 2.0 in the slider, get the chi-square distribution for *df* = 2. What you see should look like the plot shown here.

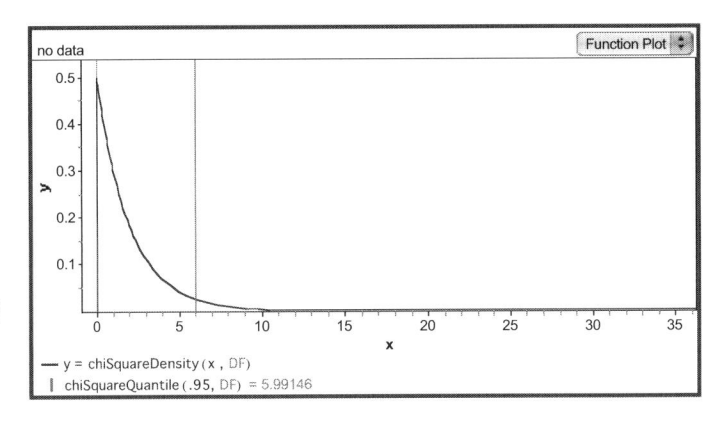

Read off the number at the foot of the table indicated by chiSquareQuantile(.95,DF)=5.991.

This number is the value that that is shown by the vertical line in the plot; to the right of this value are the rare values of a chi-square statistic χ^2 for a given *df* =2. The name that we give to this number is: **critical value**. For a Normal distribution (and for $\alpha = 0.05$) the *critical values* were the numbers $z = \pm 1.96$. Values of the chi-square statistic χ^2 that are greater than this critical value are rare (if we choose $\alpha = 0.05$), and values of χ^2 that are less than this 5.991 are reasonably likely. For the test whether the variables *PoliticalView* and *Gender* are independent or not, the test statistic is $\chi^2 = 0.8541$. Since this number is far less than the critical value 5.991, we certainly *do not* have evidence against our null hypothesis that the variables *Gender* and *PoliticalView* are independent. The test is *not* statistically significant; we do not have evidence that there is a difference in male and female college students' political views overall.

- Move the slider once again, but this time notice what happens to the vertical line and the value of chiSquareQuantile(.95,DF). You should see that as the DF gets bigger ("bigger tables"), the value of the *critical value* (the value that was 5.991 for df = 2) gets bigger.

There is a chart, the **Chi-Square Distribution Chart,** that records the *critical value*s for various *df* and for various choices of α— choices other than the α = 0.05 that we have been using.

Reading the Chi-Square Distribution Chart The chi-square chart is not like the Normal chart. In the Normal chart, the entries in the body of the table were probabilities, and the critical values for the test statistic z were on the margins (the left-hand side, with the second decimal place across the top.) The Chi-Square Distribution Chart is organized in exactly the opposite fashion, and the chart is much more "compressed" and gives less information because we have to cope with many different chi-square distributions, one for each value of degrees of freedom, *df*. Here is a picture of a part of the chart. The **degrees of freedom = df** are listed on the left-hand side, and the various choices of rare, the α, are listed on the top of the chart. For our example, we had *df* = 2, and we had chosen α = 0.05; if you look at the intersection of the row and the column for these *df* = 2 and α = 0.05, you will see the number 5.99, which is what we saw on the Fathom **ChiSquarePlot.ftm.** If we had *df* = 4 but still wanted α = 0.05 then our *critical value* would be 9.49. That is, in that situation, for us to say that our test was significant, we would have to have our test statistic to be bigger than 9.49, or, in other words, for $\chi^2 >$ 9.49.

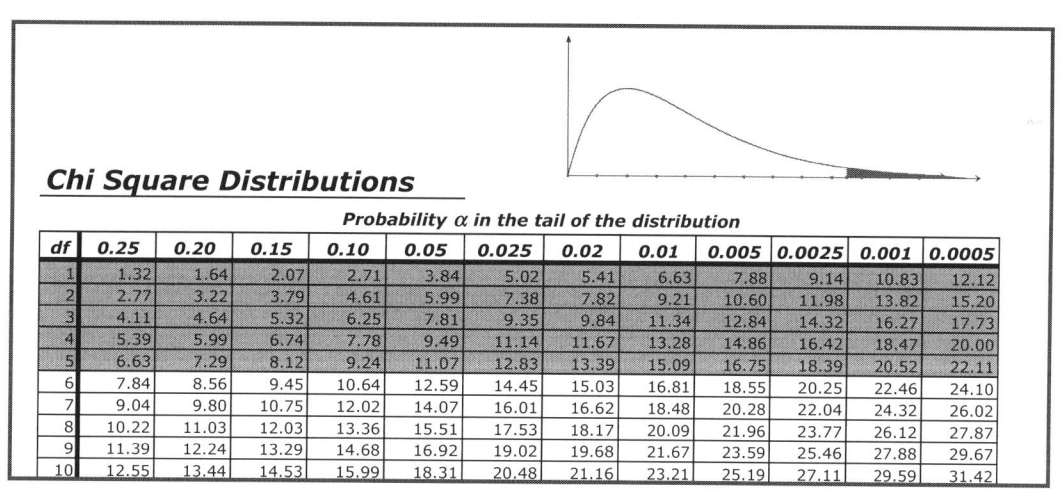

Chi Square Distributions

Probability α in the tail of the distribution

df	0.25	0.20	0.15	0.10	0.05	0.025	0.02	0.01	0.005	0.0025	0.001	0.0005
1	1.32	1.64	2.07	2.71	3.84	5.02	5.41	6.63	7.88	9.14	10.83	12.12
2	2.77	3.22	3.79	4.61	5.99	7.38	7.82	9.21	10.60	11.98	13.82	15.20
3	4.11	4.64	5.32	6.25	7.81	9.35	9.84	11.34	12.84	14.32	16.27	17.73
4	5.39	5.99	6.74	7.78	9.49	11.14	11.67	13.28	14.86	16.42	18.47	20.00
5	6.63	7.29	8.12	9.24	11.07	12.83	13.39	15.09	16.75	18.39	20.52	22.11
6	7.84	8.56	9.45	10.64	12.59	14.45	15.03	16.81	18.55	20.25	22.46	24.10
7	9.04	9.80	10.75	12.02	14.07	16.01	16.62	18.48	20.28	22.04	24.32	26.02
8	10.22	11.03	12.03	13.36	15.51	17.53	18.17	20.09	21.96	23.77	26.12	27.87
9	11.39	12.24	13.29	14.68	16.92	19.02	19.68	21.67	23.59	25.46	27.88	29.67
10	12.55	13.44	14.53	15.99	18.31	20.48	21.16	23.21	25.19	27.11	29.59	31.42

Getting an approximate p-value using the chart: "little boxes" A *p*-value is the probability of getting a result as extreme as or more extreme than the one we got from our sample *if* the null hypothesis is true. More extreme for a chi-square distribution means just one thing: area in the *tail on the right*, so it looks fairly simple. However, since the chi-square chart is compressed, we cannot get an exact *p*-value from the chart the way we were able to from the Normal chart. In fact, most of the time we will depend upon the calculation of a *p*-value from software. We can get an approximate *p* value from the chart by using a "little box." Here is an example. Suppose we had a contingency table with *df* = 4, and in our hypothesis test, our test statistic came out as χ^2 = 12.07. If we look at the chart, we do not see this number in the row for $df = 4$, but we see numbers near to it and their corresponding areas in the tail of the chi-square distribution, and we can put them into a small box:

Area (probability) in the tail:	**0.02**	**0.01**
Critical value:	11.67	13.28

We chose these values because our sample value $\chi^2 = 12.07$ falls between them. Then we can approximate our p value as $0.01 < p-value < 0.02$. We do not know exactly what the p-value is, but we know that it is somewhere in this interval $0.01 < p-value < 0.02$. This information about the p-value is all that we need; however, if we have chosen $\alpha = 0.05$ then this information tells us that the test will be *statistically significant*, since 0.02<0.05.

What can we say about the p-value for our example of the relationship between *Gender* and *PoliticalView*? Our test statistic is $\chi^2 = 0.8541$, which is far less than the critical value of 5.99, so we can say that the result is not rare, and is *not statistically significant*, and we would *not* be able to reject the null hypothesis. Can we approximate the p-value from the chart? Looking at the chart for the row $df = 2$, we see that the smallest value is 2.77 and that this value corresponds to a possible p-value of 0.25. Since we have $\chi^2 = 0.8541$, which is *smaller* than 2.77, we can say (from the chart) that the p-value is greater than 0.25. Fathom calculates that the p-value = 0.65, and that is certainly bigger than 0.25, so our approximation of p-value > 0.25 is true, if unimpressive as an approximation. It is helpful to get the picture that Fathom provides of the p-value shown on the chi-square distribution. Notice how much of the curve is shaded in; if the test had been statistically significant then the shaded portion would be very small and would show just a sliver of area in the tail of the distribution. Again, the p-value is shown as an area.

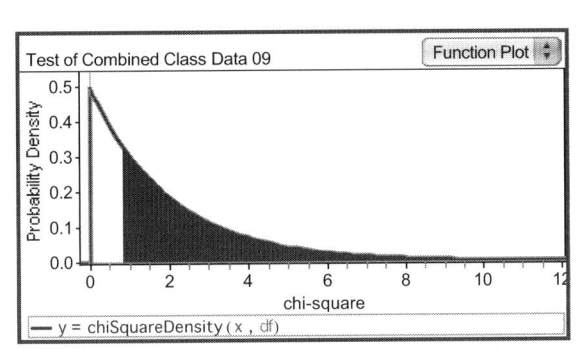

Step 5: Interpretation The ingredients are all here; looking at our example, we have:

➢ A test statistic $\chi^2 = 0.8541$ that shows a reasonably likely outcome (since $0.8541 < 5.99$) *if* the null hypothesis is true, meaning that the variables *Gender* and *PoliticalView* are independent.

➢ A p-value = 0.65 that also shows a reasonably likely outcome if the null hypothesis is true.

Therefore, we can say (as we have before) that our test is *not statistically significant* and we cannot reject the null hypothesis of independence between the variables *PoliticalView* and *Gender* for the population of students at the college. We do not sufficient evidence to say that there is any difference between male and female students in their political views (whether liberal, moderate, or conservative). As far as our data informs us, the political views of male and female students are similar. However, once again, our conclusions need to be taken cautiously because we know that our sample is *not* a random sample of the population, and so it is possible that the non-randomness of the sample has affected the results that we have seen in some way.

A Two-by-Two Table: Another (Familiar) Example

It is possible to apply the chi-square analysis to a table that has two rows and two columns. To illustrate this, we will look at the relationship between *Tattoo* and *Gender.* We have already done this with a comparison of proportions. We came to the conclusion that the test *is statistically significant* and that we could reject the null hypothesis of *no difference* in the proportions between male and female students, since the test statistic z = 3.13 > 1.96, and the *p*-value = 0.0017 < 0.05. Both the test statistic and the *p*-value indicate that what we have seen is very rare if the null hypothesis is correct. We conclude that we have evidence that there *is* a difference by gender in the probability of having a tattoo.

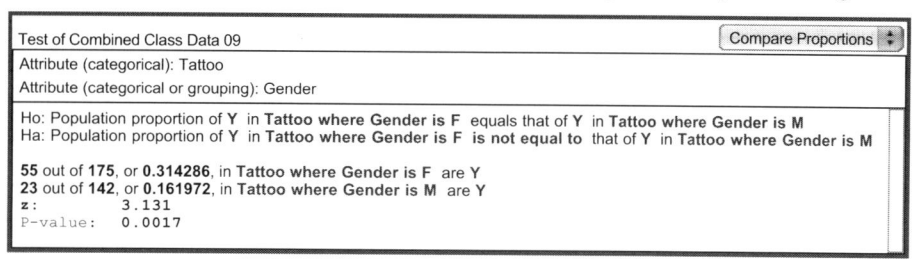

```
Test of Combined Class Data 09                                    Compare Proportions
Attribute (categorical): Tattoo
Attribute (categorical or grouping): Gender

Ho: Population proportion of Y in Tattoo where Gender is F equals that of Y in Tattoo where Gender is M
Ha: Population proportion of Y in Tattoo where Gender is F is not equal to that of Y in Tattoo where Gender is M

55 out of 175, or 0.314286, in Tattoo where Gender is F  are Y
23 out of 142, or 0.161972, in Tattoo where Gender is M  are Y
z:            3.131
P-value:      0.0017
```

We can analyze these same data using the chi-square techniques; here is the Fathom output we would get. We come to the same conclusion. Since the test statistic $\chi^2 = 9.803$ is bigger than the *critical value* for *df* = 1 of 3.84, and since the *p*-value is less than $\alpha = 0.05$, the hypothesis test is *statistically significant,* and we are led to *reject* the null hypothesis that *Tattoo* is independent of *Gender.* In fact, notice that the *p*-value is exactly the same; that is not a coincidence. There is another connection between the two tests. If we square z = 3.131, we will find that we get $\chi^2 = 9.803$. This is also not a coincidence.

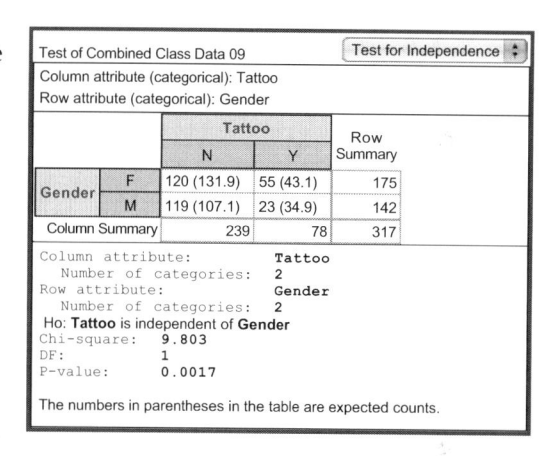

```
Test of Combined Class Data 09                  Test for Independence
Column attribute (categorical): Tattoo
Row attribute (categorical): Gender

                    Tattoo              Row
                 N          Y           Summary
         F   120 (131.9)  55 (43.1)     175
Gender
         M   119 (107.1)  23 (34.9)     142
Column Summary   239        78          317

Column attribute:        Tattoo
   Number of categories:    2
Row attribute:           Gender
   Number of categories:    2
Ho: Tattoo is independent of Gender
Chi-square:   9.803
DF:           1
P-value:      0.0017

The numbers in parentheses in the table are expected counts.
```

> **Test of Independence for a 2 x 2 Table**
> - The test of independence for a contingency table with two rows and two columns is equivalent to the *two-sided* hypothesis test where the null hypothesis is $H_0 : p_1 = p_2$ and alternate hypothesis is $H_a : p_1 \neq p_2$.
> - The *p*-values of the two tests are the same and $z^2 = \chi^2$.

Notice that the box above specifies that the tests are equivalent for a two-sided test. The chi-square test of independence does *not* test the same thing as a one-sided hypothesis test; with a one-sided test, we are essentially bringing in another preconceived idea.

Summary: Chi-Square Test for Independence

- The overall relationship between two categorical variables that have two or more categories can be analyzed using an *r x c contingency table* and a *model for independence of variables.*
  - An *r x c contingency table* shows the counts (or frequencies) of cases in the intersections of the categories of two categorical variables.
  - The *model for independence of variables* is based upon the definition for the independence of events, which says that events A and B are **independent** if $P(A \mid B) = P(A)$.
- The *model for independence of variables* builds an *r x c contingency table* whose counts in the cells of the table are the counts **expected** if the summary numbers for the two variables are the same but variables are independent.
  - The **expected counts** *(or frequencies)* for the model of independence are calculated by

  $$E = \frac{(\text{Row total for that cell}) \times (\text{Column total for that cell})}{\text{Total Number of cases}}$$

  and are the counts expected if the two variables in the contingency table are independent.
  - The **observed counts** are the actual data observed in the contingency table.
- The *observed* counts are compared with the *expected* counts of the *model for independence* by the **chi-square goodness of fit statistic** $\chi^2 = \sum_{\text{All cells}} \frac{(O-E)^2}{E}$ where O = observed and E = expected.
  - The values of the *chi-square goodness of fit statistic* χ^2 are always positive.
  - The larger the value of the *chi-square goodness of fit statistic* χ^2, the more difference there is between the observed and expected.
- The sampling distributions of the *chi-square goodness of fit statistic* χ^2 are the family of **chi-square distributions.**
  - There is a different **chi-square distribution** for different values of **degrees of freedom, df.**
  - For *r x c contingency tables*, the degrees of freedom is calculated by $df = (r-1)(c-1)$.
  - *Chi-square distributions* are typically right-skewed.
  - **Critical values** are values for a given *df*, and a given choice of level of significance α shows the smallest value for the *chi-square goodness of fit statistic* χ^2 that is considered rare.
  - **Critical values** can also be read off the **Chi-Square Distribution Chart.**
- **Test of Independence for a 2 x 2 Table**
  - The test of independence for a contingency table with two rows and two columns is equivalent to the *two-sided* hypothesis test where the null hypothesis is $H_0: p_1 = p_2$ and alternate hypothesis is $H_a: p_1 \neq p_2$.
  - The *p*-values of the two tests are the same, and $z^2 = \chi^2$.

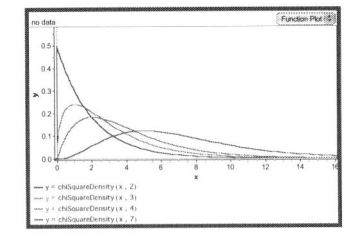

Summary: Test for Independence of Variables in an r x c Contingency Table

Step 1: Set up the null and alternate hypotheses using an idea for the population proportion p_0.

H_0: The two variables are independent
H_a: The two variables are not independent

Step 2: Check the conditions for a trustworthy hypothesis test.
  a. The sample must be a simple random sample.
  b. All of the expected counts (see step 3 below) must be five or greater.

Step 3: Find the test statistic
  a. Calculate the expected counts for the r x c table using:

  $$E = \frac{(\text{Row total for that cell})(\text{Column total for that cell})}{\text{Total Number of cases}}$$

  b. Calculate the chi-square goodness of fit statistic:

  $$\chi^2 = \sum_{\text{All cells}} \frac{(O-E)^2}{E}$$

Step 4: Calculate the p-value, either by using software or by approximating using the Chi-Square Distributions Chart and the "little box" method.

Step 5: Evaluate the evidence that the p-value and the test statistic give you to determine whether your test successfully challenges the null hypothesis of independence or not. Give an interpretation in the context of the data using the terminology of "statistical significance" and "rejecting the null hypothesis."

§4.4 Exercises for Testing Independence

Special Exercise 1: Smoking and Birth Status Again

Our statistical question can be expressed in two ways:

— Does the mother's smoking affect the likelihood of a premature birth?   OR

— Is birth status independent of smoking status?

We have seen these data before. We concluded that the probability of a premature birth is higher for births of smoking mothers than it is to births of non-smoking mothers; however, the difference of proportions was not big enough to reject the null hypothesis of *no* difference. Our hypothesis test was not *statistically significant*. Now we will analyze the same data but with chi-square.

		Birth_Status		Row Summary
		Full Term	Premature	
Smoking_Status	Non-Smoker	4036	365	4401
	Smoker	465	49	514
Column Summary		4501	414	4915

S1 = count ( )

1. Refer to the **Notes** for the definition of independence of two *events*. Call the event of a premature birth *Prem*, the event that the mother is a nonsmoker *N*, and the event that the mother is a smoker *S*.

   a. Use the correct conditional probability notation and show that the events *Prem* and *S* are *not* independent.

   b. Use the correct conditional probability notation and show that the events *Prem* and *N* are *not* independent.

Here is some background from the **Notes**.

> We call the numbers in the table "counts." The count for the Smoker-Premature cell is 49.

> We compare the *counts* (or frequencies) actually in the table (the observed counts) with the counts the table *would have* if *Smoking Status* and *Birth Status* were independent.

> The expected counts (or frequencies) are the frequencies (or counts) *expected* if the proportions premature are the same for the smoking mothers and the non-smoking mothers.

Step 1: Setting Up Hypotheses

2. Refer to the **Notes** about setting up hypotheses for a test for independence. Set up the hypotheses for our statistical question above.

Step 3a and Step 2: Getting Expected Counts and Checking the Conditions

How do we proceed? We begin by making the table of expected counts, which are the counts expected if there actually is *no* difference in proportion premature between smokers and non-smokers:

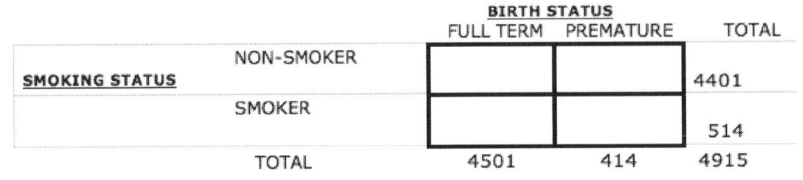

		BIRTH STATUS		
		FULL TERM	PREMATURE	TOTAL
SMOKING STATUS	NON-SMOKER			4401
	SMOKER			514
TOTAL		4501	414	4915

> The expected count for the "non-smoker premature cell" (the one with 365 in the actual table) is

$$\begin{pmatrix} \text{Overall Proportion} \\ \text{of Premature Births} \end{pmatrix} \times \begin{pmatrix} \text{Total Number} \\ \text{of NonSmokers} \end{pmatrix}$$ which is: $\frac{414}{4915} \times 4401$, which is

$0.084231943 \times 4401 = 370.70$

so that 370.70 is the expected count for that cell. Notice that we do not round to the nearest whole number, even though there cannot be .70 of a premature birth.

3. On the same pattern, get the expected count for the "smoker-premature cell." The only difference in the formula is that you will 514 in place of 4401.

4. Get the other two expected counts. (Again, show your work, although there is not much.) The expected counts that you should get are shown below in the Fathom output.

5. Add the expected counts vertically and horizontally. What do you notice compared to the original observed table? What you notice should always be true if you have calculated the expected counts correctly.

6. Are the conditions met for the hypothesis test to proceed?

Step 3b: Comparing Expected and Observed: The Chi-Square Goodness of Fit Test Statistic

> The formula is $\chi^2 = \sum_{\text{All cells}} \frac{(O-E)^2}{E}$ where O stands for "observed" and E stands for "expected," and the summation is for all the cells in the table.

7. If the expected and the observed happened to be the same numbers, what would the value of χ^2 be?

8. If chi-square is small, does that support the null hypothesis or the alternative hypothesis? Give a reason for your answer. Conversely, if chi-square is large, which hypothesis does that result support? Give a reason for your answer.

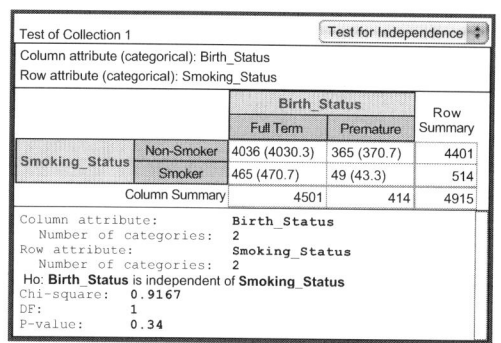

9. Here is part of the calculation of the test statistic. Copy the completed parts and fill in the blank parts.

$$\chi^2 = \frac{(4036 - 4030.30)^2}{4030.30} + \frac{(- 370.70)^2}{370.70} + \frac{()^2}{470.70} + \frac{()^2}{} = 0.917$$

Step 4: Getting the p-value

10. Calculate the degrees of freedom for our table for Smoking Status and Birth Status using the *df* formula.

11. a. Read the appropriate row on the **Chi-Square Distribution Chart** and determine the *Critical Value*–the value beyond which a c^2 that we have calculated would be considered *rare* if a = 0.05 and the df = 1.

    b. Judging by the test statistic and your answer to part a of this question, is our hypothesis test *statistically significant*? Give a reason for your answer.

12. a. Using the *Chi-Square Distribution Chart* and the procedure outlined in the **Notes**, get an approximation to the *p*-value for our hypothesis test.

    b. Does your approximation agree with the exact *p*-value that Fathom calculated?

    c. Does the *p*-value that Fathom calculated lead you to reject the null hypothesis or not reject the null hypothesis? Give a reason for your answer.

    d. Does your answer to 12c agree with your answer to 11b? Explain how the two answers agree.

Step 5: Interpretation
13. Write a good interpretation in the context of the data. "In the context of the data" means that you must say something about smoking and premature births.
14. We did this same hypothesis test using a Comparison of Proportions Test, where the null hypothesis was that there was no difference in proportion of premature births by the smoking status of the mother.

    a. Without doing that test again, can you predict what the test statistic would be using $\chi^2 = 0.916$? (*Hint:* See the box "Test of independence for a two-by-two table.")

    b. Can you predict the *p* value for the Comparison of Proportions Test? Give a reason for your answer.

2. **Tattoos and PoliticalView**

- Open the Fathom file **CombinedClassDataY09.ftm**.
- Get a **Summary Table** that relates the variables *Tattoo* and *PoliticalView*. The table should look like the one shown here.

    a. Show, using the definition of independence, that the *events L* (for liberal) and *Y* (for having a tattoo) are *not* independent. Use the notation correctly.

Combined Class Data 09		PolticalView			Row Summary
		Liberal	Moderate	Conservative	
Tattoo	N	114	108	17	239
	Y	51	22	5	78
Column Summary		165	130	22	317
S1 = count ( )					

- With the **Summary Table** selected, go to **Summary>Add Formula** and, in the dialogue box that comes up, choose **Special Measures>rowProportion**. (Double click on this and then click on **OK.**)

    b. Are the probabilities that are calculated $P(L \mid N), P(L \mid Y), P(Mod \mid N), P(Mod \mid Y)$ etc., or are they $P(N \mid L), P(Y \mid L), P(N \mid Mod), P(Y \mid Mod)$ etc.? (*Hint:* Which variable is being regarded as the "explanatory"?)

    c. Does it appear to you that the two variables are independent or not independent? Or (in different words) does it appear that there *is* a relationship (or *no* relationship) between having a *Tattoo* and *PoliticalView*?

    d. We will test whether the two variables *Tattoo* and *PoliticalView* are independent or not with a chi-square hypothesis test. We will let Fathom do most of the calculations, although you will confirm some of them. Set up the null and alternate hypotheses for this test.

- Drag down a **Test** icon from the shelf and change from an "Empty Test" to a **Test for Independence.**
- From the **Case Table,** drag the variable *PoliticalView* to be the "column attribute" and the variable *Tattoo* to be the "row attribute."
- With the test selected, *deselect* **Verbose.** You get should get the Fathom output shown here.

    e. Show how Fathom calculated the expected counts for the "Liberal-No" cell (to get 124.4) and the "Conservative-Yes" cell (to get 5.4).

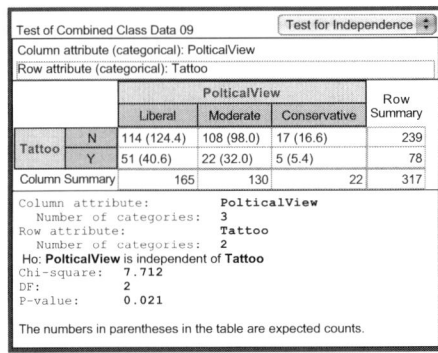

```
Test of Combined Class Data 09         Test for Independence
Column attribute (categorical): PolticalView
Row attribute (categorical): Tattoo
```

		PolticalView			Row Summary
		Liberal	Moderate	Conservative	
Tattoo	N	114 (124.4)	108 (98.0)	17 (16.6)	239
	Y	51 (40.6)	22 (32.0)	5 (5.4)	78
Column Summary		165	130	22	317

```
Column attribute:        PoliticalView
  Number of categories:  3
Row attribute:           Tattoo
  Number of categories:  2
Ho: PoliticalView is independent of Tattoo
Chi-square:   7.712
DF:           2
P-value:      0.021

The numbers in parentheses in the table are expected counts.
```

334

f. [Test-like question.] Complete this sentence in the context of this hypothesis test: "The expected counts are the counts expected if the _____ _____ were true and the variables were _____."

g. Are the conditions met for the test for independence? Consider both the numbers and the type of sampling. Explain fully.

h. Slightly Slack Silas answers question g by saying. "We are okay because all of the counts are five or above, even though one of the counts is exactly five." SSS is making a mistake; what is it?

i. [Test-like question.] Show how Fathom calculated the test statistic by putting the numbers in the formula. You can check yourself by doing the calculation, but the important thing is that you show where the numbers go. You have all the numbers you need in the Fathom output.

j. Fathom got the $df = 2$. Show how they calculated that.

k. For $df = 2$, consult the **Chi-Square Distributions Chart** and find the *Critical Value* for χ^2 for $\alpha = 0.05$. Is the χ^2 for this test above that *Critical Value*? Does that make out result "rare" or "reasonably likely"? **PTO**

l. Fathom calculated the *p*-value to be 0.021. Show that the approximation found by the "little box" method with the **Chi-Square Distributions Chart** incorporates the *p*-value Fathom found.

m. Is our test *statistically significant*? Give at least one reason for your answer. (You have multiple reasons.)

n. Do we have enough evidence to reject the null hypothesis of independence? Give at least one reason for your answer.

o. Give a good interpretation of the results of this test in the context of the data.

3. ***Pizza and Pasta across Australia*** Is there a difference in the probability that Australian high school students order take-away pizza depending upon where they live in Australia? Here is the analysis.

   a. Fathom has given you the null hypothesis. Express the meaning of the alternate hypothesis in the context of the data.

   b. Are the conditions for the test for independence met? If you think you have insufficient information, state what information you would like to have.

   c. Explain in the context of the data what the expected value of the "Queensland-Yes" means. Note: this question does not ask how it is calculated; it is asking what it means. You can choose your language, but it must be answered with language, not calculation.

   d. Is the test statistic calculated here bigger or smaller than the *critical value* for $df = 3$ for $\alpha = 0.05$ found in the ***Chi-Square Distributions Chart***?

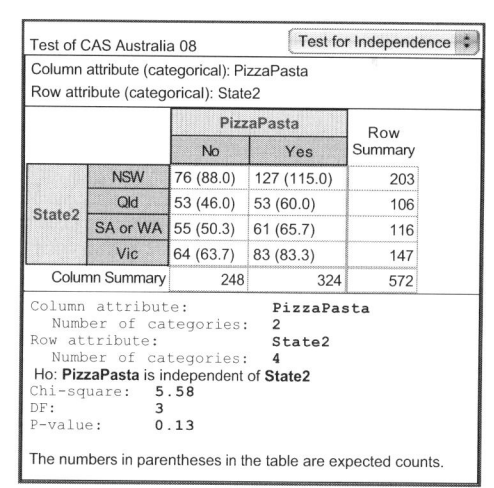

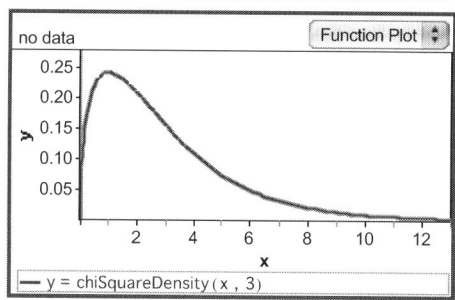

Test of CAS Australia 08 — Test for Independence

Column attribute (categorical): PizzaPasta
Row attribute (categorical): State2

	PizzaPasta No	PizzaPasta Yes	Row Summary
NSW	76 (88.0)	127 (115.0)	203
Qld	53 (46.0)	53 (60.0)	106
SA or WA	55 (50.3)	61 (65.7)	116
Vic	64 (63.7)	83 (83.3)	147
Column Summary	248	324	572

Column attribute: PizzaPasta
Number of categories: 2
Row attribute: State2
Number of categories: 4
Ho: PizzaPasta is independent of State2
Chi-square: 5.58
DF: 3
P-value: 0.13

The numbers in parentheses in the table are expected counts.

Function Plot: $y = \text{chiSquareDensity}(x, 3)$

e. [Test-like question] Sketch a copy of this graphic of the chi-square sampling distribution for $df = 3$ and show the test statistic and the p-value on it.
f. Is this test *statistically significant*? Give at least one reason for your answer.
g. Should we reject the null hypothesis of independence? Give at least one reason for your answer.
h. Give a good interpretation of the results of this test in the context of the burning question posed in the introduction to this exercise—that is, the question: is there really a difference in different parts of Australian for high school students to order pizza?
i. The data for the Census @ School were collected for schools where the school administration decided to participate. A very high proportion of the schools did participate in the five states listed. Is this sample:
   (I)   A simple random sample of all high school students in Australia
   (II) A good approximation to a census of high school students in the states covered
   (III) A convenience sample
   Choose one and give a reason for your answer.
j. How does your answer relate to the conditions for the test for independence?
k. What does Vic stand for?

4. **Which weight-loss program is best? Another weight-loss experiment** In the UK, a large experiment was conducted to evaluate the effectiveness of weight-loss programs as against not following a weight-loss program. [Helen Truby, et al. *Randomised controlled trial of four commercial weight loss programmes in the UK: initial findings from the BBC "diet trials"* BMJ, doi:10.1136/bmj.38833.411204.80 (published 23 May 2006)—this is a different experiment than the one examined in §3.1.] Here is how the study in the UK was conducted:
   – Overweight and obese volunteers from across Great Britain were *randomly allocated* to four different weight-loss programs or to no weight-loss program.
   – The cost of the weight-loss programs was paid for by the researchers.
   – Those who were allocated to the "no-program" treatment were promised that, after the study, the researchers would then pay for a weight-loss program for them.
   – Records of weight, as well as other medical variables, were kept over a period of time.
   Here we are looking at the participants who lost 10% or more of their weight compared with those who lost less than 10%. The results are shown in the Fathom output below. (Rosemary Conley is a UK-based weight-loss program.)
   a. What makes this study an experimental study and not an observational study?
   b. What is the factor?
   c. What is the response variable?
   d. What are the null and alternate hypotheses for this test of independence? Write them in the context of the data.
   e. Are the conditions met for this test of independence? (See the Fathom output on the next page.)

f. What is the meaning (in words) of the expected counts in the context of the experiment?

g. Copy this graphic of the chi-square sampling distribution for *df* = 3 and show the test statistic and the *p*-value on it.

h. If α = 0.05, is this hypothesis test statistically significant? Give at least one reason for your answer.

```
From Summary Statistics                                    Test for Independence
Column attribute (categorical): unassigned
Row attribute (categorical): unassigned
                                  Programme
                        Atkins  Weight Watchers  Slim-Fast  Rosemary Conley  Row Summary
WeightLoss  Lost less than 10%  22 (25.2)  30 (29.6)  33 (26.4)  22 (25.8)  107
            Lost 10% or more    18 (14.8)  17 (17.4)   9 (15.6)  19 (15.2)   63
            Column Summary         40         47         42         41      170

Column attribute:     Programme
  Number of categories:   4
Row attribute:        WeightLoss
  Number of categories:   2
Ho: Programme is independent of WeightLoss
Chi-square: 7.011
DF:         3
P-value:    0.072

The numbers in parentheses in the table are expected counts.
```

i. If we decided to be less strict and set α = 0.10, so that a rare result would be one that would come about by sampling variation 10% instead of 5% of the time, would the test be statistically significant?

j. Consider the *p*-value and the test statistic and use them to give a good interpretation of the results of this test in the context of the experiment.

k. If someone from the UK asked you: "Does it really make a difference which program I choose?" how would you answer this person?

l. If the *p*-value had been 0.023, would your answer to the question in k differ or be the same?

m. If the *p*-value had been 0.023, would you be able to tell from that which program is best or worst? Explain why or why not. (*Hint: Would you need to do more analysis?*)

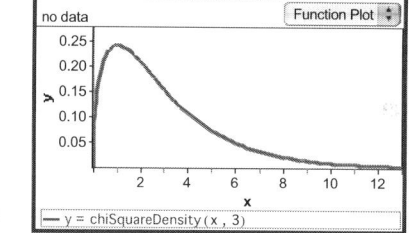

5. **The Titanic** [Fathom Exercise] Who survived on the *Titanic*?
   - Use the Fathom file **TitanicPassengers.ftm.**
   - Get a **Summary Table** that relates the variables *Survival* and *PClass*. (= "Passenger Class," whether first-, second-, or third-class).

	Survival No	Survival Yes	Row Summary
PClass 1st	129	193	322
PClass 2nd	161	119	280
PClass 3rd	573	138	711
Column Summary	863	450	1313

   S1 = count ( )

   a. Do you think that survival was independent of class? Do some calculations of proportions to investigate your ideas.

   b. Formulate your ideas into a statistical question appropriate for a test for independence.

   - With the **Summary Table** selected, go to **Summary>Add Formula** and in the dialogue box that comes up, choose **Special Measures>rowProportion.** (Double click on this and then click on **OK.**)

   c. Does it appear to you that the two variables are independent or not independent? Or (in different words) does it appear that there is a relationship between the variables *Survival* and *PClass*?

   d. We will test whether the two variables *Survival* and *PClass* are independent or not. We will let Fathom do most of the calculations, although you will confirm some of them. Set up the null and alternate hypotheses for this test.

- Drag down a **Test** icon from the shelf and change the test from an "Empty Test" to a **Test for Independence.**
- From the **Case Table,** drag the variable *Survival* to "column attribute" and the variable *PClass* "to row attribute."
- With the test selected, go to the menu for **Test** and *deselect* **Verbose.** What you get should look like the Fathom output shown here.

```
Test of Titanic passengers    Test for Independence
Column attribute (categorical): Survival
Row attribute (categorical): PClass
```

		Survival		Row Summary
		No	Yes	
PClass	1st	129 (211.6)	193 (110.4)	322
	2nd	161 (184.0)	119 (96.0)	280
	3rd	573 (467.3)	138 (243.7)	711
Column Summary		863	450	1313

```
Column attribute:       Survival
  Number of categories: 2
Row attribute:          PClass
  Number of categories: 3
Ho: Survival is independent of PClass
Chi-square:  172.3
DF:          2
P-value:     < 0.0001

The numbers in parentheses in the table are expected counts.
```

e. Show how the expected counts were calculated for the "Yes-1st Class" cell and the "No-Third Class" cell.

f. [Test-like question.] Complete this sentence in the context of this hypothesis test: "The expected counts are the counts expected if the _____ _____ were true and the variables were _____."

g. Are the numbers in the cells big enough for the conditions to be met (for the Chi-Square Test for Independence)? Give a reason for your answer. (Will you look at the observed or the expected counts?)

h. Show how Fathom calculated the test statistic by putting the numbers in the formula. You can check your self by doing the calculation, but the important thing is that you show where the numbers go. You have all the numbers you need in the Fathom output.

i. For $df = 2$, consult the **Chi-Square Distributions Chart** and find the *Critical Value* for χ^2 for $\alpha=0.05$. Is the χ^2 for this test above that *Critical Value?* Does that make out result "rare" or "reasonably likely"?

j. The *p*-value fathom gives is < 0.0001. Is our test *statistically significant?* Give a reason for your answer.

k. Give a good interpretation of the results of this test in the context of the data—surviving on the *Titanic*. What do the data say? Who survived on the *Titanic?* (Note: We have an entire population and not a sample, but the test for independence still applies. We are in effect saying: "If the survivors were chosen at random without regard to their class on the ship, what would the distribution of survivors look like?")

§4.5 Hypothesis Test, Confidence Interval, Both, or Neither?

Introduction: The Assignment

This section starts with an assignment to work with and make sense of data using most of what we have been looking at. The sections you have studied so far have presented various *descriptive* techniques (Units 1 and 2) and *inferential* techniques (Units 3 and 4), and this section's goal is to pull all of that together. When confronted with a statistical question and some data, what technique should be used? How do you decide? And what can go wrong? What mistakes can be made, and how can you spot them?

Perhaps more importantly (but less enjoyably!), suppose it is not you that had the assignment but someone else. If you are reading the analysis and interpretation of some data that someone else has made, what helps you to know if that person has done the job well or has blundered in some way? One way you can spot others' possible mistakes is to be exposed to the possibility of making them yourself—even actually making the mistakes! So we will analyze some data to answer some statistical questions.

The Assignment   You have been working with the Census at School data for Australia. The same kind of data (see http://www.censusatschool.com/en/about) have been collected in Canada, New Zealand, South Africa, and the UK, and some of the same data—the same variables—were collected in all of these places. So our *assignment* is:

What are the similarities and differences in high school students and their lives in these different places?
You can easily imagine that this could be an assignment for a sociology course or an education course or it could be a journalist's assignment: "Write us an article that shows how the lives of high school students are basically the same in different countries" or "write us an article about how the lives of high school students differ in different countries." The idea is to use the data to say something.

We begin with three essential questions.
- What are the cases?
- What are the variables?
- How were the data collected?

The data. The cases for the data are students in schools. All students at participating schools answer the questions online rather than using paper questionnaires. It is called the Census at School because all the students in a school participate. The schools include all grade levels, although the variables measured for the secondary-level students differ from those measured for the elementary-level students. It is important to know that school participation is voluntary and the schools that participate are *not* a random sample of schools in the country. Here is what the **case table** looks like for the data we will analyze.

	Gender	Age...	NumLan...	Hand	Reflex	Height	ArmSpan	RightFoot	NumHH	TimeConc	Transport	NoBreak...	Cereal	Sport	StatePr...	Country
199	Female	14	(a) One	Right han...	0.4	164	2	6	5	33	(a) Walk o...	No	Yes	Running,...	NSW	Australia
200	Male	14	(a) One	Right han...	0.37	173.5	190.5	26.5	5	41	(c) Public ...	No	Yes	Football/...	NSW	Australia
201	Female	16	(a) One	Right han...	0.41	158	165	22	4	25	(b) Car	No	No	Netball	Vic	Australia
202	Female	13	(a) One	Right han...	0.43	160	159	23	3	46	(b) Car	No	Yes	Basketball	Vic	Australia
203	Male	14	(b) Two	Right han...	0.3	167.5	169	25.5	4	63	(b) Car	No	No	Other	Vic	Australia
204	Male	17	(b) Two	Right han...	0.31	175	183	27	4	32	(b) Car	No	Yes	Basketball	Vic	Australia
205	Male	15	(a) One	Left han...	0.45	177	178	27	5	33	(c) Public ...	No	No	Football/...	Vic	Australia
206	Female	15	(b) Two	Right han...	0.34	160	156	24	4	55	(c) Public ...	Yes	No	Dancing	Vic	Australia
207	Female	16	(a) One	Right han...	0.33	155	155	24	2	32	(a) Walk o...	No	Yes	Swimming	Vic	Australia

You will find a complete list of the variables and their definitions at the end of this section of the **Notes**; you should be familiar with some of the variables in any case, as you have previously worked with them. Some of the variables are **quantitative** and some of the variables are **categorical.** Recall that there are different techniques for analyzing quantitative and categorical variables.

Statistical questions with no inference: descriptive analyses The basic question for this assignment, *"What are the similarities and differences in high school students and their lives in these different places?"* is far too broad and must be broken down into specific questions about variables. Statistical questions generally involve comparisons between groups or relationships between variables, or a comparison of a variable to some standard. Here are some examples for the first two or three variables listed in the case table.

Example 1: *Are there differences in the ages of the students in our Australian and Canadian samples?* How do we answer this question? First, we notice that *Age* is a quantitative variable, and, secondly, we notice that our question is only about our samples and not the population from which the samples were drawn. The analysis appropriate for comparing groups on a quantitative variable is to compare **distributions.** The graphics we use are dot plots or box plots or histograms. For numerical summaries, we could compare the means, medians, standard deviations, and IQRs for the two samples. The means and the medians tell us that, on average, the Canadian students sampled are slightly older than the Australian students sampled. For our measures of spread, the standard deviation indicates a slightly greater spread for the ages of the Canadian students, but recall that the *s* can be sensitive to skewness and outliers. The IQR for the two samples shows us the spread for the middle 50%; this spread is the same, and the appearance of the box plots confirms this. Does this indicate that secondary-school students are younger generally (not just in our sample)? That would be an **inferential** question because it is a question about Australian students and Canadian students in general and not only about our samples. We cannot answer that question yet; inference for quantitative variables is the topic of Unit 5.

Example 2: *Is there a relationship between age and the scores on the concentration game (TimeConc)?*

The students were asked to play a "concentration game"; the time to complete the task was measured. Shorter times indicate a deeper level of concentration on the game. Shown below is the "prompt" that was given to the Australian students. As a square is selected, an image appears, which is then hidden until that image's double is located under one of the other squares. Success at this game requires that the player remember (concentrate on) the location of the various images.

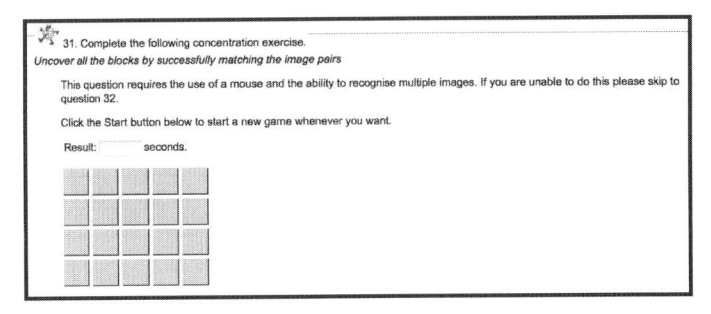

The variable *TimeConc* measures how well students do on this game. It is natural to ask (since we are thinking of age) whether there is a relationship between *Age* and *TimeConc*; do older students do better than younger students? Is their schooling having an effect at all? Or are the scores on this game relatively unrelated to age and dependent on other variables that we have not measured? These questions lead to our general question about the relationship between *TimeConc* and *Age*.

We consider whether the variables are categorical or quantitative. Here we have *two quantitative* variables, and since we are interested in the relationship between two quantitative variables, the appropriate techniques involve linear models, and specifically the *correlation coefficient r* and the fitting of a least squares linear model, evaluated by the *coefficient of determination* R^2, which were studied in Unit 2. Here are the graphics and the results of the calculations.

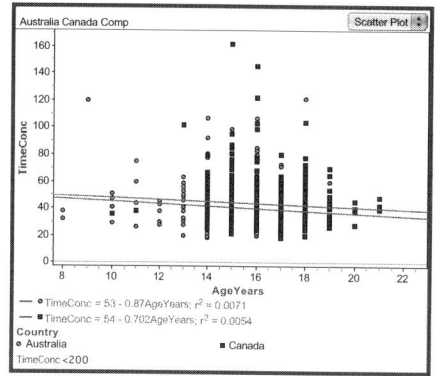

Age, apparently, does *not* affect the scores on the concentration game. The most important numbers here are the *coefficient of determination R2*. The R^2 are nearly zero for the students in both countries, and those small numbers indicate that the variation in the *TimeConc* scores cannot be explained by age at all. We have to look elsewhere for explanations of differences in the scores. We have treated this analysis as a *descriptive* analysis and not an *inferential* analysis concerned about generalization to a population beyond the samples we have. Inference for linear models uses some of the same tools as in Unit 5, but we leave it to the next course in statistics.

Both *Example 1* and *Example 2* require descriptive analysis and interpretation based on that analysis. These questions involve no formal inference to a larger population. The focus is on the samples and not a population beyond the samples. In the next sub-section, we ask questions where formal inferential analysis is implied in the question or the background to the question. Descriptive analysis is still worthwhile; indeed, descriptive analysis should always precede formal inferential analysis.

Statistical Questions with Inference Here is an interesting summary table comparing students in Australia and in Canada on the number of languages they speak. Both Australia and Canada are countries peopled by immigrants, and so you would expect a sizable proportion of high school students to speak more than one language. Moreover, Canada has two official languages, English and French. In Quebec, French is the predominant as well as the official language. French speakers are found in all parts of Canada, and it is common for students to be bilingual if they are a minority (e.g., English-speaking in a majority French area or French in a majority English area). For that reason alone we would expect that a sizable proportion of students will speak more than one language.

In these **Notes** we have also looked at the proportion of students in a California community college who are at least bilingual and found that proportion to be over 50%. With all of these things in mind, we can write down a number of statistical questions that require formal *inferential* procedures.

Example 3: Determine whether the proportion of monolingual students (those who speak just one language) in Australia differs significantly from the proportion of monolingual students in Canada.

Example 4: If there is a difference, how big is the difference between Australia and Canada in the proportion of monolingual students?

Example 5: Determine if, in the population of all students in the two countries, there are differences between the countries in the proportions of students who speak one, two, or three or more languages? (Or is NumLanguages independent of Country?)

Example 6: Does the proportion of students in Canada who speak two or more languages differ from the standard of 50%, which we think describes the typical percentage among college students in Northern California?

Example 7: Can we estimate the proportion of all high school students in Australia who speak two or more languages?

We have broken down our overall question—*What are the similarities and differences in high school students and their lives in these different places?*—into specific questions about specific variables. For all of these examples we will use *inference*. We will consider each of these examples below.

Conditions and a potential problem One of the first things that we do for any inferential technique is to check the **conditions.** One condition that is common to all of the inferential procedures is that the data come from random samples. For these data, we know that they are *not* random samples. First, they are certainly not simple random samples of the student populations; we would not expect them to be, as random cluster samples are much more practical to collect. However, we know that the schools participating in both countries were not randomly selected. The school authorities had to volunteer to participate. This is a big enough problem that we have devoted a special section to it. (See below: **Problems with non-randomness.**)

The sample size **condition** differs among the various procedures, so we will address that as we go. But how do we choose which procedure to use? We have done *hypothesis tests* and *confidence intervals* and we have done these for one proportion and for the difference of proportions. And when we were faced with more than two proportions, we conducted a hypothesis test called a *test of independence.* How do we decide which of these to use?

Choosing between estimating and testing We have done two kinds of procedures—*hypothesis tests* and *confidence intervals*—and although there are connections between these procedures, the way these questions are worded will direct us to either a hypothesis test or calculating a confidence interval. Notice that some of the questions ask to **estimate** a population proportion or to find "how big" a difference is. A *confidence interval* is an estimate, and so for these questions we want a *confidence interval.* Other questions have the form: "Are there differences...?" or "Is the proportion different or the same as...?" or "Do the proportions differ...?" *All* of these kinds of questions indicate hypothesis tests. They all ask whether we have evidence that a statement is true or not true. So the first thing to do will be to see what kind of procedure is implied by the language.

> **Guideline to choosing a procedure**
> Decide, by reading carefully, whether the question implies a *confidence interval* or a *hypothesis test* by looking at the structure of the question and whether it appears to be asking for an estimate or a test.

Examples: choosing a procedure. Example 3 : *Determine whether the proportion of monolingual students (those who speak just one language)in Australia differs significantly from the proportion of monolingual students in Canada.* We will of course calculate proportions of monolingual high school students in the two countries. However, the words *whether* and "differs significantly" suggest a hypothesis test. Since we have *two* proportions, we can use a **compare proportions test.** "Monolingual" means speaking just one language, and we see from the Fathom output that the test statistic is large and that the *p*-value is small. This shows that the test is *statistically significant* and that we do have evidence that the proportions of monolingual students are different in the two countries.

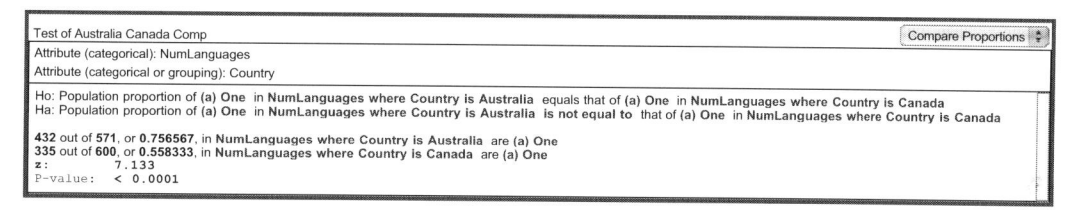

```
Test of Australia Canada Comp                                           Compare Proportions
Attribute (categorical): NumLanguages
Attribute (categorical or grouping): Country
Ho: Population proportion of (a) One in NumLanguages where Country is Australia equals that of (a) One in NumLanguages where Country is Canada
Ha: Population proportion of (a) One in NumLanguages where Country is Australia is not equal to that of (a) One in NumLanguages where Country is Canada
432 out of 571, or 0.756567, in NumLanguages where Country is Australia are (a) One
335 out of 600, or 0.558333, in NumLanguages where Country is Canada are (a) One
z:        7.133
P-value:  < 0.0001
```

Example 4: *If there is a difference, how big is the difference between Australian and Canada in the proportion of monolingual students?* To infer about "how big" a difference is, the question is asking for an estimate rather than asking *whether* there is a difference. We need a confidence interval for the difference of the proportion of monolingual students in Australia as compared with Canada. Here is the Fathom output for the difference in proportions.

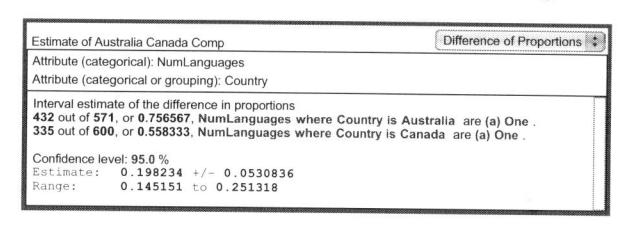

```
Estimate of Australia Canada Comp                                       Difference of Proportions
Attribute (categorical): NumLanguages
Attribute (categorical or grouping): Country
Interval estimate of the difference in proportions
432 out of 571, or 0.756567, NumLanguages where Country is Australia are (a) One .
335 out of 600, or 0.558333, NumLanguages where Country is Canada are (a) One .
Confidence level: 95.0 %
Estimate:   0.198234 +/- 0.0530836
Range:      0.145151 to 0.251318
```

We can interpret this confidence interval by saying: "With 95% confidence, we can say that the proportion of monolingual high school students in Australia is 19.8% higher than the proportion of monolingual high school students in Canada, with a margin of error of 5.31%." Or we could say, "We are 95% confident that the proportion of monolingual high school students in Australia is from 14.5% to 25.1% higher than the proportion of monolingual high school students in Canada."

How a confidence interval can tell you about a hypothesis test Notice that both ends of the confidence interval in the calculation of Example 5 are positive. If one end of the confidence interval were negative and the other end were positive then the interval would include zero percent, or no difference, which would mean that, in the population, it is plausible that there is no difference in the proportions. In that scenario—where the confidence interval of two proportions *includes zero*—the hypothesis test would be *not statistically significant*. In our scenario—where the confidence *excludes zero*—a zero difference in proportions in the population is not plausible, and the hypothesis test will be *statistically significant*. So here is a situation in which getting the confidence interval also informs us about a hypothesis test implied by the confidence interval. One has to be careful, as the formulas for the estimated sampling distribution standard deviation are slightly different; however, in most cases, a confidence interval for the difference of two proportions that includes zero indicates that the hypothesis test result will be *not significant*, and a confidence interval that *excludes zero* indicates a *significant* result to a hypothesis test. If you draw the sampling distribution for two proportions you will see why this will be so; a significant difference will come when the $\hat{p}_1 - \hat{p}_2$ is rare for the null hypothesized $p_1 - p_2 = 0$. However, a confidence interval based upon a rare $\hat{p}_1 - \hat{p}_2$ should *miss* (*exclude*) zero, whereas a confidence interval based upon a reasonably likely $\hat{p}_1 - \hat{p}_2$ will "hit" or include the null hypothesized $p_1 - p_2 = 0$.

The box summarizing this phenomenon is below. There are some situations where the confidence interval is more useful than the hypothesis test, since the hypothesis test can be read from the confidence interval.

Confidence Intervals and Hypothesis Tests for Differences of One Proportion:

If a confidence interval for one proportion
- *Excludes* zero then the associated hypothesis test will be *statistically significant*.
- *Includes* zero then the associated hypothesis test will be *not statistically significant*.

This principle works only if
- The level of confidence used is $1 - \alpha$ (for example, 95% with $\alpha = 0.05$), and
- The hypothesis test is a **two-sided test.**

More Examples: choosing a procedure. Example 5: *Determine if, in the population of all students, there are differences between the countries in the proportions of students who speak one, two, or three or more languages? (Or is NumLanguages independent of Country?)*

This particular question asks about all the proportions for the variable *NumLanguages*, and we see that there are three such proportions. The words "determine if" suggest a hypothesis test, but we cannot use a comparison of proportions test because we have three proportions. Actually, the alternate form of the question shows what is needed: *Is NumLanguages independent of Country?* indicates a *test of independence.*

Here is the Fathom output for the *test of independence*. Notice that the sample size conditions are met since all the *expected counts* are bigger than five. Since the *p*-value is very small, we can say that the differences of proportions that we see in the samples are unlikely to have arisen just from sampling variation from random sampling (if we had random sampling!). There would appear to be a *statistically significant* difference in the proportions of students in Canada and Australia in the proportions of student who speak one, two, or three or more languages. We are led to reject the null hypothesis that Australian and Canadian high school student populations have essentially the same proportions speaking one, two, and three or more languages. The students' language backgrounds differ.

Test of Australia Canada Comp — Test for Independence
Column attribute (categorical): NumLanguages
Row attribute (categorical): Country

		NumLanguages			Row Summary
		(a) One	(b) Two	(c) Three or More	
Country	Australia	432 (374.0)	110 (146.3)	29 (50.7)	571
	Canada	335 (393.0)	190 (153.7)	75 (53.3)	600
Column Summary		767	300	104	1171

```
Column attribute:      NumLanguages
  Number of categories: 3
Row attribute:         Country
  Number of categories: 2
Ho: NumLanguages is independent of Country
Chi-square:   53.26
DF:           2
P-value:      < 0.0001
```
The numbers in parentheses in the table are expected counts.

Example 6: *Does the proportion of students in Canada who speak two or more languages differ from the standard of 50%, which we think describes the typical percentage among college students in Northern California?* Despite the absence of the words "determine whether," the wording "*Does the proportion...*" implies "does" or "does not," and that is the language of a hypothesis test. Moreover, a specific standard (50%), and just one population (Canada), is mentioned. All of this leads us to a two-sided hypothesis test for one proportion. The sample proportion can be calculated from

$P(\text{Two or More} \mid C) = 1 - P(\text{One} \mid C) = 1 - \frac{335}{600} \approx 1 - 0.558 = 0.442$. Since we are doing a hypothesis test

for one proportion, we will call this $\hat{p} = 0.442$ and compare this to our idea that $p_0 = 0.50$. Here is the Fathom output for this test. Again the *p*-value is extremely small, indicating statistical significance, so that we have evidence that the proportion of bilingual (or more) high school students in Canada is different from what we are accustomed to seeing in our California community colleges. Note: if the word "differ" in the question had been "greater than," we would be led to a one-sided and not a two-sided hypothesis test.

Example 7: *Can we estimate the proportion of all high school students in Australia who speak two or more languages?* The word "estimate" indicates that we want a confidence interval, and the other specifications indicate that we want a confidence interval for just one proportion. Here is the Fathom output, whose interpretation is that we can be 95% confident that between 40.2% and 48.1% of all Australian students speak two or more languages, if all the conditions are met. We can apply the principle that connects confidence intervals with hypothesis tests for two proportions. The principle is: does the confidence *include* or *exclude* the null hypothesized value? Here the null hypothesized value is $p = 0.50$ (and not zero), so we see if the confidence interval *includes* or *excludes* 0.50. The confidence interval *excludes* $p = 0.50$, so the associated two-sided hypothesis test will be statistically significant.

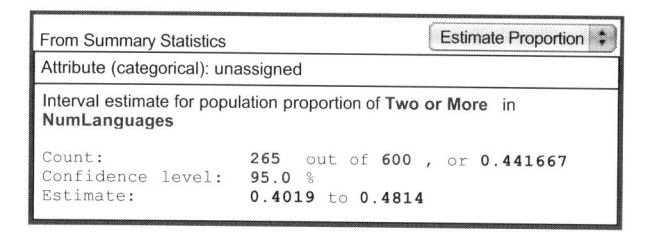

```
From Summary Statistics                    Estimate Proportion
Attribute (categorical): unassigned
Interval estimate for population proportion of Two or More in
NumLanguages

Count:              265     out of 600 , or 0.441667
Confidence level:   95.0 %
Estimate:           0.4019 to 0.4814
```

Confidence Intervals and Hypothesis Tests for One Proportion:

If a confidence interval for one proportion

– *Excludes* the hypothesized p_0 then the associated hypothesis test will be *statistically significant.*

– *Includes* the hypothesized p_0 then the associated hypothesis test will be *not statistically significant.*

This principle works only if

– The level of confidence used is $1 - \alpha$ (for example, 95% with $\alpha = 0.05$), and

– The hypothesis test is a **two-sided test.**

Problems with non-randomness

The most common thing that can go wrong is to use *non-random samples*. Strictly random samples are hard to collect, and it is quite common for all of these formal inferential procedures to be applied to samples that are not random samples, as we have done here—partly so that we can make this point. (Note, the software cannot tell if our samples are random or not; it is up to the researcher.) We said at the outset that our samples were not strictly random samples of students and not even random cluster samples of schools. Had the schools been randomly selected then the analyses here would be approximately correct, although an organization such as Gallup would probably use slightly different formulas. Our samples are *voluntary response samples* in the sense that the school authorities volunteered to participate. For the Canadian Census at School, we are able to get some idea of how the participation of the schools under- and overrepresents various parts of Canada. (The Census at School data in the first column is from http://www19.statcan.gc.ca/04/04_0809/prov/04_0809_prov_001-eng.htm, and the Enrollment by Province is from Statistics Canada, *Education Indicators in Canada: Report of the pan-Canadian Educators Indicators Program*, catalogue 81-582-XIE, 2007.)

We can see by comparing the percentages of enrollments by province to the percentages by province of secondary students who participated in the Census at School program that students in the provinces of Ontario, Nova Scotia, and Manitoba were overrepresented, whereas those from Saskatchewan and British Columbia and to some extent Quebec are underrepresented. How does this affect our conclusions about number of languages spoken by high school students? The answer is that we cannot say. It may be that if students in British Columbia and Quebec were proportionally represented (through random sampling), the percentage of bilingual students would be higher, but we are just guessing here. If that were so, it would make the contrast between Canada and Australia more striking. However, the contrast between Canada and California would not be as striking.

What does one do in the face of no-random samples? There are several choices.

One choice is to do no inferential analyses at all. Since the conditions are not met, the calculations may be meaningless. Or, if you discover that another researcher has done inferential analyses where the conditions (especially the randomness condition) are not met, you can ignore the inferential part.

A second choice is to try to assess the direction of non-randomness. That is what we have done here (but only in a rudimentary way). Sometimes when this is done, one can argue that the direction of under- or

Province	C@S participation (% by Province)	Enrollments (% by Province)
Newfoundland	0.00	1.56
Prince Edward Island	0.00	0.45
Nova Scotia	6.50	2.95
New Brunswick	2.16	2.38
Quebec	18.90	21.66
Ontario	50.56	40.85
Manitoba	5.95	3.62
Saskatchewan	1.34	3.46
Alberta	9.49	10.80
British Columbia	4.74	11.79
Yukon Terr.	0.35	0.11
North West Territory, Nunavut	0.00	0.37
	100.00	100.00

overrepresentation would just make the contrasts sharper, as we have done here with the comparison with Australia. Other times, making a judgment about the direction of under- or overrepresentation will not be possible. In all situations, in reporting results, one has to be forthcoming and transparent about the quality of the data.

A third (and wise) choice is to consult an expert.

However, know that the application of inferential techniques to non-random data is widespread.

On the next two pages there is a summary of formulas for inference for categorical variables.

Summary of Inference Procedures for Categorical Variables

One Proportion

Type	Confidence Interval	Test of Significance	Comments (Mostly Conditions)
One Proportion	$\hat{p} \pm z^* \sqrt{\dfrac{\hat{p}(1-\hat{p})}{n}}$ where $z^* = 1.645$ for a 90% Confidence Interval, $z^* = 1.96$ for a 95% Confidence Interval, $z^* = 2.576$ for a 99% Confidence Interval, and $\hat{p}$ is the sample proportion, and n is the sample size.	1. Hypotheses: $H_0: p = p_0$ $H_a: p \begin{cases} < p_0 \\ > p_0 \\ \neq p_0 \end{cases}$ 2. Test Statistic $z = \dfrac{\hat{p} - p_0}{\sqrt{\dfrac{p_0(1-p_0)}{n}}}$ 3. Use the Standard Normal Distribution Chart to find p-values	The sample from which $\hat{p}$ is calculated is a simple random sample where the population is at least 10 times the sample size. For Confidence Intervals: Both $n\hat{p} \geq 10$ and $n(1-\hat{p}) \geq 10$ For Hypothesis Tests: Both $np_0 \geq 10$ and $n(1-p_0) \geq 10$

Comparing Two Proportions

Type	Confidence Interval	Test of Significance	Comments (Mostly Conditions)
Comparing Two Proportions	$(\hat{p}_1 - \hat{p}_2) \pm z^* \sqrt{\dfrac{\hat{p}_1(1-\hat{p}_1)}{n_1} + \dfrac{\hat{p}_2(1-\hat{p}_2)}{n_2}}$ where z^* as for one proportion, $\hat{p}_1, \hat{p}_2$ are the sample proportions, and n_1, n_2 the sample sizes for the two samples.	1. Hypotheses: $H_0: p_1 - p_2 = 0$ $H_a: p_1 - p_2 \begin{cases} < 0 \\ > 0 \\ \neq 0 \end{cases}$ 2. Test Statistic $z = \dfrac{(\hat{p}_1 - \hat{p}_2) - 0}{\sqrt{\hat{p}(1-\hat{p})\left[\dfrac{1}{n_1} + \dfrac{1}{n_2}\right]}}$ where $\hat{p} = \dfrac{\text{Total count of successes over both samples}}{\text{Total number of cases in both samples}}$ 3. Use the Standard Normal Distribution Chart to find p-values	The samples from which $\hat{p}_1$ and $\hat{p}_2$ are calculated are simple random samples where the populations are at least 10 times the sample size. For Confidence Intervals: $n_1\hat{p}_1 \geq 5$, $n_1(1-\hat{p}_1) \geq 5$, $n_2\hat{p}_2 \geq 5$, $n_2(1-\hat{p}_2) \geq 5$ For Hypothesis Test: $n_1\hat{p} \geq 5$, $n_1(1-\hat{p}) \geq 5$, $n_2\hat{p} \geq 5$, $n_2(1-\hat{p}) \geq 5$

Type	Test of Significance	Comments (Mostly Conditions)
Test for Independence	1. Hypotheses: H_0: The two variables are independent H_a: The two variables are not independent 2. Test Statistic: $\chi^2 = \sum_{All\ cells} \frac{(O-E)^2}{E}$ where O = observed frequencies, E = expected frequencies And the expected frequencies are the frequencies expected if the null hypothesis is true. They can be calculated for a specific cell by: $$E = \frac{(\text{Row Total})(\text{Column Total})}{\text{Grand Total}}$$ 3. Use the Chi-Square Distribution Chart to approximate *p*-values	If inferring to a population, the sample containing the data must be a random sample. The expected frequencies must be ≥ 5.

Summary: Hypothesis Tests and Confidence Intervals

- **Guidelines to using inferential or descriptive procedures**
  - Descriptive analyses are always appropriate (even if inferential procedures are also used) and serve to illuminate the data at hand.
  - Inferential procedures should be chosen if the conditions are met, and there is an interest in generalizing *beyond* the data at hand.
  - If the conditions for inferential are not met, abandon inferential procedures or proceed knowing that the results may not be able to be generalized.
- **Guidelines to choosing an inferential procedure** Decide, by reading carefully, whether the question implies a *confidence interval* or a *hypothesis test* by looking at the structure of the question and whether it appears to be asking for an estimate or a test.
  - Language that implies making a decision ("whether, " "if," "Are the differences or not") are indicators that a hypothesis test is appropriate.
  - Language that implies an estimate (i.e., "estimate," "how big...") are indicators that a confidence interval is required.
- **Confidence Intervals and Hypothesis Tests for the Difference of Two Proportions:**
  If a confidence interval for one proportion
  - *Excludes* zero then the associated hypothesis test will be *statistically significant*.
  - *Includes* zero then the associated hypothesis test will be *not statistically significant*.

  This principle works only if
  - The level of confidence used is $1 - \alpha$ (for example, 95% with $\alpha = 0.05$), and
  - The hypothesis test is a **two-sided test.**
- **Confidence Intervals and Hypothesis Tests for One Proportion:**
  If a confidence interval for one proportion
  - *Excludes* the hypothesized p_0 then the associated hypothesis test will be *statistically significant*.
  - *Includes* the hypothesized p_0 then the associated hypothesis test will be *not statistically significant*.

  This principle works only if
  - The level of confidence used is $1 - \alpha$ (for example, 95% with $\alpha = 0.05$), and
  - The hypothesis test is a **two-sided test.**

§4.5 Exercises: Confidence Interval, Hypothesis Test...?

1. **Who Skips Breakfast?** In both the Australian and Canadian Census at School questionnaires, there were questions concerning what students had for breakfast. We will be interested in those brave souls among the high school students who reported that they skipped breakfast. Our general question is "Who is more likely to skip breakfast: an Australian or a Canadian? Or are they equally likely to skip breakfast?" And if there is a difference in the proportions then we want to estimate how big the difference is.

   - Open the Fathom file **Australia Canada Comparison.ftm** and get a *case table*.
   - Make a **Summary Table** showing the frequencies of the students who said yes to having skipped breakfast and the students who said no to *SkipsBreakfast* on one side and the variable *Country* on the other.

   Australia Canada Comparison

		SkipsBreakfast		Row Summary
		No	Yes	
Country	Australia	480	91	571
	Canada	475	125	600
Column Summary		955	216	1171

   S1 = count ( )

   - [Optional] Get a **Graph** and drag the variable *SkipsBreakfast* to the vertical axis. Then drag the variable *Country* to the horizontal axis. (You can also get a ribbon chart by dragging the variable *Country* to the horizontal axis and then *Country* to the *body* of the graph and then changing the type in the upper right-hand corner of the graph to "Ribbon Chart.")

   a. Calculate the proportion who skipped breakfast (that is, "yes" to *SkipsBreakfast*) among the Australian students and the proportion who skipped breakfast among the Canadian students. Use the correct *conditional probability* notation (the notation from §1.2) to express these probabilities.

   b. Our statistical question is: "Is there a difference in the proportion who skips breakfast between the population of high school students in Australia and the population of high school students in Canada?" For this question, choose the best procedure and explain your choice.
      - A confidence interval for one proportion? (§4.1)
      - A hypothesis test for one proportion? (§4.2)
      - A confidence interval for the difference of proportions? (§4.3)
      - A hypothesis test to compare proportions? (§4.3)

   c. For your choice, use notation (not the conditional probability notation) that will distinguish between the *population proportions* of students who skip breakfast in Australia and students who skip breakfast in Canada. (*Hint:* Will you have "hats" or will you have "no hats"?)

   d. Express the sample proportions with the correct notation. (*Hint:* "hats" or no "hats"?)

   e. We want a hypothesis test (so that helps you with question b). Set up the null and alternate hypotheses.

   f. Set up the test statistic for your test, showing the symbol for it and the formula. You do not actually have to complete the calculation (although you may) because we will have Fathom do the calculation. But to show the calculation, you will have to calculate a third proportion and give it the correct symbol.

   - Drag down a **Test** box from the shelf in Fathom; it will show "Empty Test."
   - From the small scroll on the **Test,** choose the correct test.

- If you have chosen the correct test, there should be two "unassigned" categorical attributes to fill. Drag *SkipsBreakfast* to the top one and *Country* to the lower one.
- With the **Test** selected, go to the menu for **Test** and deselect **Verbose.**
- Change all the blue instances of *No* to *Yes* (that is, "yes" to *SkipsBreakfast*; actually it would not matter to the outcome of the test if we left them as "no," but we are testing who skipped breakfast).

  g. What is the value of the test statistic? Using $\alpha = 0.05$ and the appropriate critical value, is the value of the test statistic "reasonably likely" or "rare"? Give a reason for your answer.

  h. What is the *p*-value? Is it bigger or smaller than $\alpha = 0.05$?

  i. Does the evidence of parts g and h lead you to declare the test *statistically significant* or not? (The answers to parts g and h should point in the same direction; either both should give you reason to say that the test is significant or both should say that the test is not significant.)

  j. Do you have evidence against the null hypothesis? Explain.

  k. From the results, can you say that there *is* a difference or that there is *no* difference in the likelihood of skipping breakfast between Canadian students and Australian students?

  l. Now we want to have an *estimate* of the difference in the proportions of Australian and Canadian students who skip breakfast. Choose the best procedure and give a reason for your choice.
   - A confidence interval for one proportion? (§4.1)
   - A hypothesis test for one proportion? (§4.2)
   - A confidence interval for the difference of proportions? (§4.3)
   - A hypothesis test to compare proportions? (§4.3)

  m. Your estimate will be in the form of an interval. You should be able to predict from the results of your hypothesis test whether the interval will *exclude* or *include* zero. Choose the correct answer (exclude or include) and give a reason for your answer.

- Drag down an **Estimate** box from the shelf in Fathom; it will show "Empty estimate."
- From the small scroll on the **Estimate,** choose the estimate that you want to use.
- If you have chosen the correct test, there should be two "unassigned" categorical attributes to fill. Drag *SkipsBreakfast* to the top one and *Country* to the lower one.
- With the **Estimate** selected, deselect **Verbose.**
- Change all the blue instances of *No* to *Yes*. If you hate negative numbers, you can also get Fathom to put Canada before Australia: change the blue *SkipsBreakfast where Country is Australia* to *SkipsBreakfast where Country is Canada.*

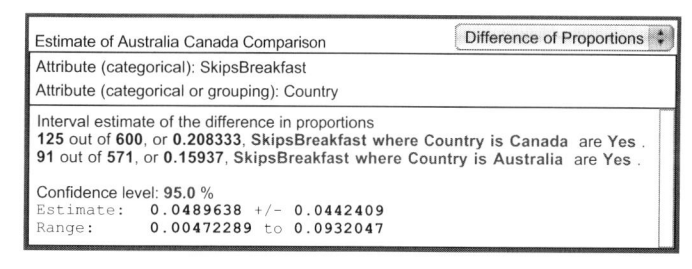

```
Estimate of Australia Canada Comparison          Difference of Proportions
Attribute (categorical): SkipsBreakfast
Attribute (categorical or grouping): Country
Interval estimate of the difference in proportions
125 out of 600, or 0.208333, SkipsBreakfast where Country is Canada are Yes.
91 out of 571, or 0.15937, SkipsBreakfast where Country is Australia are Yes.

Confidence level: 95.0 %
Estimate:   0.0489638 +/- 0.0442409
Range:      0.00472289 to 0.0932047
```

  n. Here is an interpretation of the confidence interval that is wrong. "We can be 95% confident that in the populations of all high school students in Australia and Canada, between 0.5% and 9.3% of students skip breakfast." What is wrong? (*Hint:* the proportion of students who skip breakfast is about 15% or 20%, not less than 1% to 9%. What do the numbers 0.5% to 9.3% mean?)

  o. Give a correct interpretation of your estimate in the context of the data.

p. We want an estimate of the proportion of all *Australian* high school students who skip breakfast. Choose the correct procedure and carry out the procedure by hand (it is chosen to be simple).
- A confidence interval for one proportion? (§4.1)
- A hypothesis test for one proportion? (§4.2)
- A confidence interval for the difference of proportions? (§4.3)
- A hypothesis test to compare proportions? (§4.3)

2. **Who Skips Breakfast: Males or Females?** Our statistical question is:

Are male or female students more likely to skip breakfast? Or is there no gender difference?

We intend this statistical question to be an inferential question, using the Census @ School data.

a. Do we want a hypothesis test or a confidence interval for our question? Give a reason for your answer.

b. Are we looking at the difference of proportions or are we looking at a single proportion against some standard? Give a reason for your answer.

We are actually going to do the analysis twice, once for the Canadian sample and once for the Australian sample. So for the Canadian students we will compare the proportions of male and female students who skip breakfast and then, for the Australian students, we will again compare the proportions of male and female students who skip breakfast. To do this with Fathom, follow the directions.

- Open the Fathom file **Australia Canada Comparison.ftm**.
- Select the **Collection icon** and go to **Object>Add Filter** (shown here) and in the dialogue box, type: *Country = "Canada"*. This will confine everything we do to only the Canadian students.
- Make a **Summary Table** showing the frequencies of the students who said yes to having skipped breakfast and the students who said no to *SkipsBreakfast* by the variable *Gender* on the other side. Your summary table should resemble the one here.

Australia Canada Comparison

		SkipsBreakfast		Row Summary
		No	Yes	
Gender	Female	227	59	286
	Male	248	66	314
Column Summary		475	125	600

S1 = count ( )

c. Using appropriate notation, set up the null and alternate hypotheses for a hypothesis test to test whether males or females are more likely to skip breakfast. Use the correct notation.

d. Calculate the sample proportions from your summary table and give them the correct symbols.

e. Also calculate $\hat{p}$ and say what this $\hat{p}$ means in the context of the data.

- Drag down a **Test** box from the shelf in Fathom and choose the correct test.
- If you have chosen the correct test, there should be two "unassigned" categorical attributes to fill. Drag *SkipsBreakfast* to the top one and *Gender* to the lower one.
- With the **Test** selected, go to the menu for **Test** and deselect **Verbose**.
- Change all the blue instances of *No* to *Yes* (that is, "yes" to *SkipsBreakfast*).
- With **Test** selected, in the menu choose **Test>Show Test Statistic Distribution**. (It should look like the graphic shown here.)

f. With the proportions that you have calculated, show how the test statistic was calculated.

g. Is the test *statistically significant*? Give reasons for your answer in terms of the *p*-value shown in the Fathom output and also the test statistic that Fathom has calculated.
h. Interpret the results of the hypothesis test for Canadian students in the context of the question.
i. Explain why the **Test Statistic Distribution** shows so much shading.

- Select the **Collection icon** again and change (by double clicking on it) the **Add Filter** to Country = "Australia." The Fathom output should show the test for the Australian students instead of the Canadian students.

j. Is the test *statistically significant*? Give reasons for your answer in terms of the *p*-value shown in the Fathom output and also the test statistic that Fathom has calculated.
k. Is there a gender difference in skipping breakfast for Australian high school students? Explain.
l. How and why is the **Test Statistic Distribution** different?

3. **San Mateo Real Estate I** Unlike the samples from the Census at School, the sample that you will analyze here is a simple random sample, so we can have confidence in our results. The sample size is $n = 400$. The cases are houses that were sold in San Mateo County in 2007–2008.

- Open the Fathom file **San Mateo RE Sample Y0708.ftm**.
- Make a **Summary Table** showing the relationship between the variable Style_2 and Region. It should look this.

Our statistical question is:

Is there an association between the style of houses sold and the region? Another way of putting this is whether there is a particular region in which a particular style of house predominates or whether most areas have a similar mix of styles.

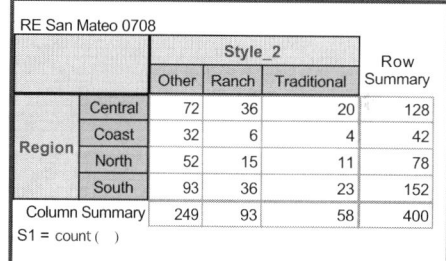

RE San Mateo 0708		Style_2			Row Summary
		Other	Ranch	Traditional	
Region	Central	72	36	20	128
	Coast	32	6	4	42
	North	52	15	11	78
	South	93	36	23	152
Column Summary		249	93	58	400
S1 = count ( )					

a. Choose which analysis is appropriate for this question, and for each of the other proposed procedures state why each of them would not be correct.
   – Difference of proportions hypothesis test
   – Test of independence
   – Least squares regression line
   – Single proportion hypothesis test
b. Set up the null and alternate hypotheses for the procedure you have chosen.

- Drag down a **Test** box from the shelf in Fathom and choose the correct test.
- If you have chosen the correct test, there should be two "unassigned" categorical attributes to fill. Drag Region to be the Row attribute and Style_2 to be the column attribute.
- With the **Test** selected, go to the menu for **Test** and deselect **Verbose**.

c. Are the conditions met for the test you have chosen? Be complete and include both types of sample and whether the numbers in the cells are sufficiently big.
d. Show how the **df** was calculated.
e. Show how the test statistic was calculated; you do not have to complete the calculation but show enough of it so that a reader can see that you know where the test statistic comes from.

353

f. Here is the sampling distribution for the test statistic. Copy the plot and show the test statistic and the *p*-value on it. (Be certain you can do this for a quiz or test.)

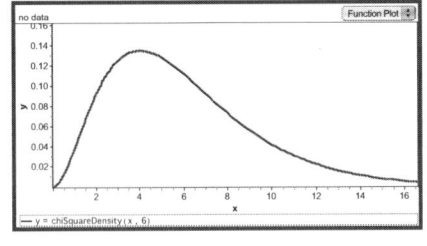

g. From the evidence that you have, is the test statistically significant? Give a reason for your answer.

h. Give an interpretation of this test in terms of the statistical question stated above.

- With **Test** selected, in the menu choose **Test>Show Test Statistic Distribution**. Compare what you got to your answer to part f. The shading should be the shading that you showed, although you should also have an arrow pointing at the shading and saying: "*p*-value."

4. **San Mateo Real Estate II** Here is another exercise with the same sample of data. The sample size is $n = 400$. The cases are houses that were sold in San Mateo County in 2007–2008.

- Open the Fathom file **San Mateo RE Sample Y0708.ftm**.
- Make a **Summary Table** showing the relationship between the variable *ListSale* and *Region*. It should look like the one shown here.

The variable *ListSale* is a categorical variable that records whether the house sold for "over the listed price," "the same as the listed price," or "under the listed price." Our statistical question is:

Are there differences by region in the proportions of houses sold over, the same as, or under the list price?

a. Choose which analysis is appropriate for this question and give a reason for your answer.

b. Set up the null and alternate hypotheses for the procedure you have chosen.

- Drag down a **Test** box from the shelf in Fathom and choose the correct test.
- If you have chosen the correct test, there should be two "unassigned" categorical attributes to fill. Drag *Region* to be the row attribute and *ListSale* to be the column attribute.
- With the **Test** selected, go to the menu for **Test** and deselect **Verbose.**

c. For the correct procedure, the numbers in parentheses are expected counts. Explain what these counts *mean* in terms of the null hypothesis. Make certain that your answer speaks of the variables specifically.

d. Are the conditions met for the test you have chosen? Be complete and include both types of sample and whether the numbers in the cells are sufficiently big.

e. In this case you should have said that the conditions are *not* met because one of the cells has an expected count smaller than five. Notice that fathom merely reports that one cell is less than five. The rule of thumb that we have given in the **Notes** is conservative. Another rule of thumb says that for a table with twelve cells, as many as 0.2 x 12 = 2.4 or 20% of the cells can be smaller than five. If we use this rule of thumb, we can proceed. If we do, is the test statistically significant? Give a reason for your answer.

f. Here is the sampling distribution for the test statistic. Copy the plot to show the test statistic and *p*-value on it.

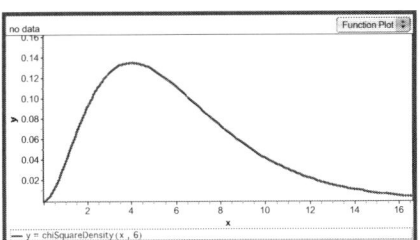

354

Check your answer by following the Fathom instructions below.
- With **Test** selected, in the menu choose **Test>Show Test Statistic Distribution**.
    g. Give an interpretation of this test in terms of the statistical question stated above.
    h. Your cousin, a Realtor in Seattle, is curious about the housing market in San Mateo County. She would like to have an estimate of the proportion of houses for the entire county that sold for "over the list price" for 2007–2008. You have your sample of $n = 400$, and you should be able to give her that estimate. What procedure will you use? Give a reason for your answer.
    i. Carry out your procedure and give your cousin the results with an explanation. You can assume that your cousin is well-educated and has had a course in statistics. She may need a bit of reminding, however, about what a margin of error means.

5. **Roller Coasters** Here is a table showing the numbers of steel and wooden roller coasters in two regions. The data are from the roller coaster database. The roller coaster database collects data on as many of the roller coasters in the world that the database builders can have access to.
    a. Calculate proportions to indicate whether the proportion of wooden roller coasters differs between the two regions. Use good notation.
    b. Is there evidence from your calculations that there may be a difference in the proportion of wooden roller coasters by region? Explain.

    Roller Coasters Summer 09

	Construction Steel	Wood	Row Summary
Region Great Lakes	50	18	68
West Coast	84	9	93
Column Summary	134	27	161

    S1 = count ( )
    ¬(Region = "West")

    c. If the conditions are met, there would be two ways to use formal inference procedures to generalize to the population of all roller coasters in these two regions. What are these two procedures? Explain.
    d. Check out the conditions for at least one of the two procedures referred to. Do you have any doubts about whether the conditions are met? What questions would you want answered before proceeding?
    e. For both of these procedures (assuming that it is wise to carry them out—which it may not be), the p-value comes out to be 0.0048. What does that tell you?

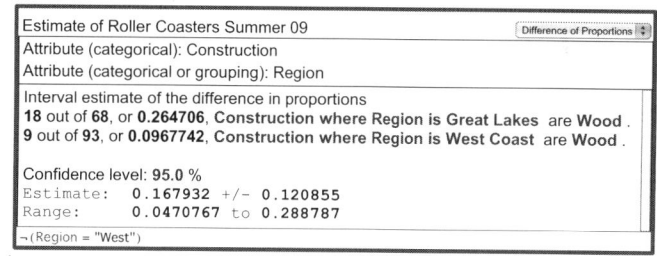

    Estimate of Roller Coasters Summer 09 — Difference of Proportions
    Attribute (categorical): Construction
    Attribute (categorical or grouping): Region
    Interval estimate of the difference in proportions
    18 out of 68, or 0.264706, Construction where Region is Great Lakes are Wood.
    9 out of 93, or 0.0967742, Construction where Region is West Coast are Wood.
    Confidence level: 95.0 %
    Estimate: 0.167932 +/- 0.120855
    Range: 0.0470767 to 0.288787
    ¬(Region = "West")

    f. Here is the Fathom output for the confidence interval for the difference of proportions of wooden roller coasters in the two regions. State how this confidence interval agrees with the p-value of 0.0048. **PTO**

To answer the next two questions, you may wish to open the Fathom file **Roller Coasters Summer 09A.ftm.**

g. Suppose you wanted to see if wooden roller coasters in the database were *longer* on average than the steel roller coasters. (We are not worried about the population of all roller coasters; we are only interested in the ones that happen to be in the database in our sample.) Which procedure would you use and why?
   - Chi-square test of independence on *Construction* (Wooden and Steel) and *Length* of roller coaster
   - Compare means and medians of the variable *Length* by the variable *Construction*
   - Get a least squares regression line between the variable *Length* by the variable *Construction*
   - Do a difference of proportions hypothesis test on *Length* by the variable *Construction*

h. Suppose you wanted to see if there is a difference in the relationship between the *Speed* and the *Length* of a roller coaster according to its *Construction*. Which procedure would you use and why?
   - Chi-square test of independence on *Construction* (wooden and steel) and *Speed* of roller coasters
   - Compare means and medians of the variable *Length* and *Speed* by the variable *Construction*
   - Get a least squares regression lines relating *Speed to Length* for the two types of *Construction* of roller coasters
   - Do a difference of proportions hypothesis test on *Speed and Length* by the variable *Construction*

§5.1 Inference for Quantitative Variables

Moving from categorical variables to quantitative, from proportions to means

All of Unit 4 was about confidence intervals and hypothesis tests for categorical variables. We looked in detail at sample proportions $\hat{p}$ and how we can infer beyond sample proportions to a population proportion p. Now, we move on to quantitative variables, and our focus will be on sample means $\bar{x}$ and how we can infer to population means μ. (For these sections, we leave behind proportions—temporarily.)

We looked at inferring from sample means $\bar{x}$ to population means μ in §3.2, when we studied the sampling distribution for a sample mean $\bar{x}$. Here is some useful notation to remember from that section. Again, we are moving from the world of

Notation for means and standard deviations

	Population Distribution	Sample Distribution	Sampling Distribution
Mean	μ	$\bar{x}$	$\mu_{\bar{x}}$
Standard Deviation	σ	s	$\sigma_{\bar{x}}$

proportions, $\hat{p}$ and p, to the world of quantitative variables and $\bar{x}$, s, and the population mean μ.

The summary box about the sampling distribution for sample means $\bar{x}$ is also relevant to this section.

Facts about Sampling Distributions of Sample Means Calculated from Random Samples

Shape: The shape of the sampling distribution of $\bar{x}$ will be approximately Normal if the sample size n is sufficiently large, even if the population distribution from which the samples were drawn is not Normal.

Center: The mean of the sampling distribution of $\bar{x}$ is equal to the mean of the population distribution. That is, $\mu_{\bar{x}} = \mu$.

Spread: The standard deviation of the sampling distribution of $\bar{x}$ is $\sigma_{\bar{x}} = \dfrac{\sigma}{\sqrt{n}}$.

Although the graphic is not big enough to read here, the graphic showing the difference between a population distribution, a sample distribution, and a sampling distribution is a good picture to have in mind. You can refer to the large version in §3.2.

To see how we *infer* from $\bar{x}$ to μ, how the ideas of confidence intervals and hypothesis tests are worked out with quantitative variables, we consider an example.

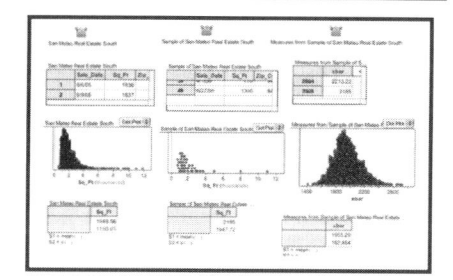

Example: How long does it take to get to Seattle?

If you log on to https://www.alaskaair.com/shopping/ssl/shoppingstart.aspx for Alaska Airlines and specify Oakland to Seattle (as an example), you will see the screen something like the one displayed here. Besides the times that the flights depart and arrive, they also show the duration of the flight. For Oakland to Seattle, the duration of flight is just over two hours; the flights appear to last about 125 minutes. How could we check that these numbers for flight durations are correct? If we had a random sample of flights between Oakland and Seattle then, using hypothesis tests and confidence intervals, we could answer the questions:

> Is the average duration of the flights between Oakland and Seattle really 125 minutes?
> Can I get an estimate of how long, on average, the flights really are?

The first looks like a question for a hypothesis test, and the second question (since it asks for an estimate) looks like it could be answered by a confidence interval.

The Bureau of Transportation Statistics (BTS) collects detailed information on every commercial flight (see http://www.bts.gov/xml/ontimesummarystatistics/src/dstat/OntimeSummaryDepaturesData.xml), and we do have a random sample of flights from Oakland to Seattle. One of the variables measured by the BTS is the actual duration of each flight, which in our data has the name *ActualDuration*. The variable *ActualDuration* is a **quantitative** variable, and therefore the population mean for *ActualDuration* should be denoted with the Greek letter μ. Once again, we do not know the value of the population mean μ, but we *do* need a symbol to refer to what we do not know.

Using the Facts about Sampling Distributions to Get a Confidence Interval

To answer the second question above ("Can I get an estimate...?"), we want a confidence interval, and the formula for the confidence interval should be based on the sampling distribution for sample means $\bar{x}$. (Refer to the box about sampling distributions for sample means $\bar{x}$ on the previous page.) For a confidence interval, the basic idea was to calculate the interval

(Sample Estimate)±(Margin of Error) or, in terms of the sampling distribution:

(Sample Estimate)±z*(Standard deviation of the sampling distribution). For proportions, we had

$\hat{p} \pm (\text{Margin or Error}) = \hat{p} \pm z * \sqrt{\frac{\hat{p}(1-\hat{p})}{n}}$ because the standard deviation of the sampling distribution

was $\sqrt{\frac{p(1-p)}{n}}$. So, it makes sense that the formula for the confidence interval to estimate a population

mean μ from a sample mean $\bar{x}$ to be: $\bar{x} \pm z * \left(\frac{\sigma}{\sqrt{n}}\right)$. This formula makes sense because the **Facts about Sampling Distributions of Sample Means...** tells us that the sampling distribution will be approximately Normal (so that is why the z* is there), and the standard deviation of the sampling distribution is $\frac{\sigma}{\sqrt{n}}$.

A sample of flights: To make things more concrete, we have flight data from Oakland to Seattle for March 2009, and from these data we have a simple random sample of just $n = 20$. The first thing we must do is a descriptive analysis of the sample data. Here is a dot plot of the duration of our twenty flights (the variable is *ActualDuration*). Looking at the plot, it appears that most of the flights actually took less time than the hypothesized two hours and five minutes (125 minutes), although there were some flights that were longer. The next thing is to calculate the sample mean $\bar{x}$ and the sample standard deviation s. We find that $\bar{x} = 117.75$ minutes, and that $s \approx 10.151$ minutes. You may begin to

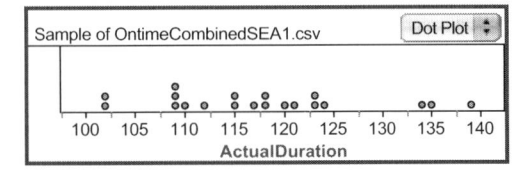

Sample of OntimeCombinedSEA1.csv	
	ActualDuration
	117.75
	10.1508
S1 = mean ( )	
S2 = s ( )	

358

think that perhaps the mean duration of 125 minutes is wrong, if we go by this small sample.

Calculating the Confidence Interval: Trouble! When we go to calculate a 95% confidence interval, we can calculate $\bar{x} \pm z * \frac{\sigma}{\sqrt{n}} = 117.75 \pm 1.96 \frac{\sigma}{\sqrt{20}}$. However, we realize that we do not know the population standard deviation σ. What we have is the sample standard deviation s. If we use the sample standard deviation $s \approx 10.151$ in place of the unknown population standard deviation σ, we would then calculate:

$$\bar{x} \pm z * \frac{\sigma}{\sqrt{n}} = 117.75 \pm 1.96 \frac{10.151}{\sqrt{20}} \approx 117.75 \pm 1.96(2.2698) \approx 117.75 \pm 4.45.$$

This calculation gives us an estimated population mean of $\mu = 117.75$ minutes with a margin of error of 4.45 minutes.

However, can we do this? Are we allowed to simply substitute the sample standard deviation s for the population standard deviation σ? The trouble we have is that there is another source of variability—another source of randomness—in our calculations, and we do not know whether the sampling distribution will still have the shape of a Normal distribution or something different. If we knew the population standard deviation σ then we would also know that the sampling distribution of the sample mean will be approximately Normal. But if we do *not* know the population standard deviation σ then it is possible that the sampling distribution is not Normal but some other shape. Our next job is to understand what happens to the sampling distribution when we do not know σ and want to use the sample standard deviation in its place.

Finding the Sampling Distribution of $\bar{X}$ When Sigma Is Not Known

Once again, you may find it helpful to follow along with the Fathom simulation; we will do a similar "simulation" of the sampling distribution to what we have done before. Recall that we started with a population and had Fathom take many random samples from the population, and for each sample that was taken, Fathom calculated the sample mean. We then collected together these sample means into a collection. This collection was our approximation of the sampling distribution; it was an approximation because we knew that the complete sampling distribution would require something like 10^{80} samples, which would take too much of our time.

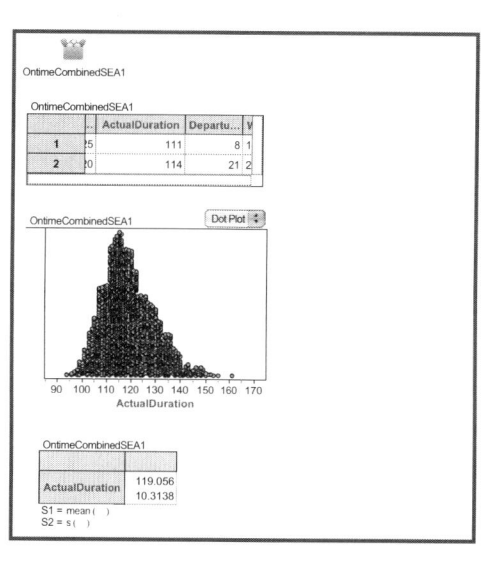

In this simulation, we will have Fathom calculate what is called a **t-score** $t = \frac{\bar{x} - \mu}{\frac{s}{\sqrt{n}}}$ instead of calculating just $\bar{x}$ each time a sample is taken. (You may notice that a **t-score** has a form like a **z-score** using the mean and standard deviation of the sampling distribution.) We have Fathom calculate the t-score so that we can see the influence that the sample standard deviation s has on the sampling distribution. For any sample that we have, we want to include the variability that comes from the sample standard deviation s as well as the variability that comes from the sample mean $\bar{x}$. Follow the bullets; what you see on the screen should be similar to the graphics shown here. We start with *all* of the data for a month; we are treating all these data as the population.

- Open the Fathom file **OntimeCombined4SEA.ftm**.
- Get a **Dot Plot** of the variable *ActualDuration* and a **Summary Table** showing the mean and the standard deviation of the variable *ActualDuration*. Your screen should look like the picture here.

What we see is the population distribution of the variable *ActualDuration*. We see that the shape of the distribution is nearly symmetrical, with perhaps just a hint of right skewness.

- With the collection icon selected, go to the menu **Collection>Sample Cases.** You will get a sample of $n = 10$ flights from the population, and the **Inspector** for the sample should appear.
- In the **Inspector,** <u>de</u>select the *Animation* and the *With Replacement* boxes and change the number of cases from $n = 10$ to $n = 20$ then press *Sample More Cases*.
- Select the **Sample of Online...** and get a **Case Table** and then a **Dot Plot** of the variable *ActualDuration*.
- Get a **Summary Table** showing the mean and the standard deviation of the variable *ActualDuration*. Your screen should look like the picture here but will differ slightly since you have a different random sample.

This second set of results is the sample distribution for one random sample of size $n = 20$. Now we are ready to get Fathom to start building a sampling distribution. We will use the actual population mean $\mu = 119.056$, but because we want to see the effect of using *sample standard deviation s*, we will not use the population standard deviation.

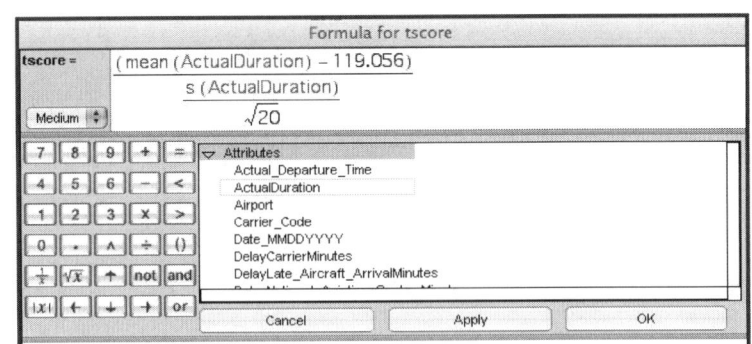

- In the **Inspector** of the **Sample of Online...,** select **Measures** (it may look like *Meas...*) and in the box below *Measure* (where it says <*New*>), type in **tscore** and press the '*tab*' or the '*return*' key. A box should appear.
- Double click on the space below **Formula** and, inside the parentheses, type in (mean(ActualDuration) -119.056).
- Then on the outside of the parentheses, press the "division" key, and a space will appear for you to type in the denominator.
- In the denominator, type: s(ActualDuration) and then the "division" key and then the $\sqrt{}$ key and inside that 20. Your formula box should look like the one shown above. Press **OK**.

360

These instructions have made Fathom calculate a ***t-score***, which is the formula $t = \dfrac{\bar{x} - \mu}{\dfrac{s}{\sqrt{n}}}$. In the next steps, Fathom will repeatedly sample randomly, and each time it gets a sample of $n = 20$, Fathom will calculate the *t*-score. Then Fathom will collect together all of the *t*-scores that it has collected. Follow the bullets.

- Select the **Sample of Online...** collection icon and go to the menu **Collection>Collect Measures.** You should get the **Inspector** of the **Measures of Sample of Online...** as well.
- Select the **Measures of Sample of Online...** icon and get a **Case Table.** You should have just five measures that Fathom has collected.
- Get a **Dot Plot** of the variable *tscore* and also a **Summary Table** showing the mean and the standard deviation of the variable *tscore*.
- Return to the **Inspector** of the **Measures of Sample of Online...** and *de*select the animation; change the number of measures from 5 to 1000 and then click on *Collect More Measures.* At this point you will wait.

Your screen should resemble (but be different from) what you see here.

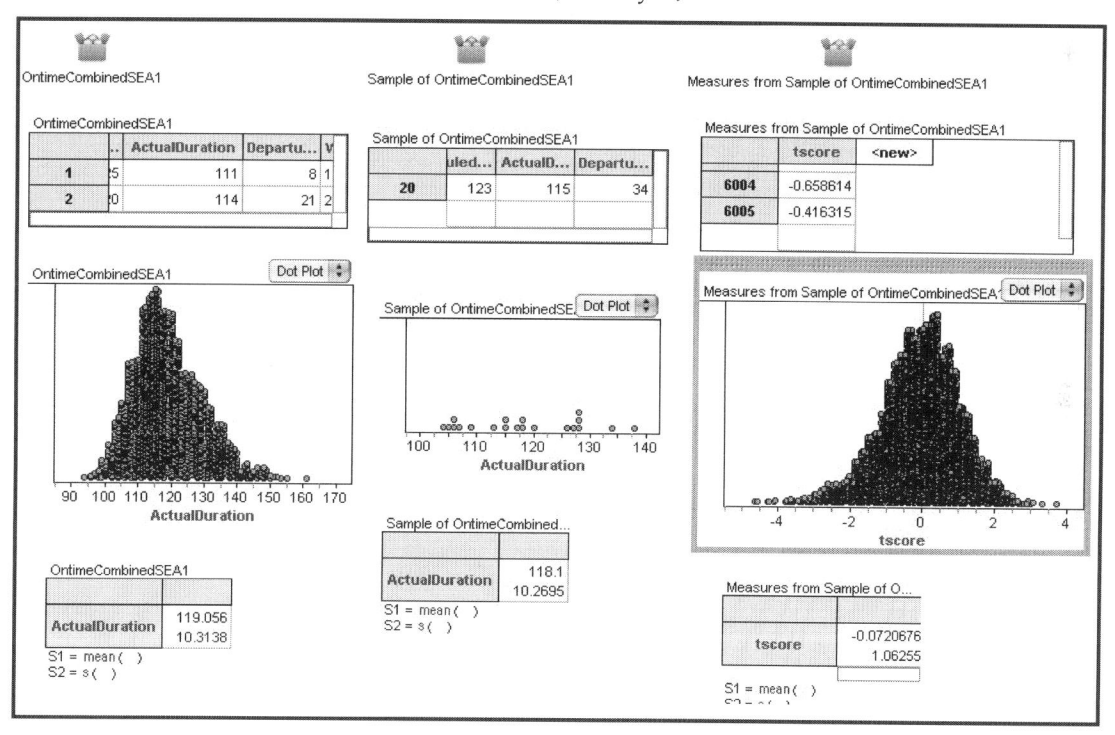

What we see from our simulation. Look at the plot on the right, which is our simulation of the sampling distribution. Notice that it is nearly symmetrical, although there is a hint of left skewness. If this were a Normal distribution then the empirical rule tells us that we expect 99.7% of the distribution (that is nearly 100%) to be within three standard deviations. But here we see that (especially on the left-hand side) the distribution stretches beyond three standard deviations to four standard deviations.

The effect of using the sample standard deviation *s* instead of the population standard deviation σ gives a distribution that still looks like a Normal distribution but has **thicker tails** than a Normal distribution.

This distribution is called a **t-distribution** or a **"Student's t-distribution."** More importantly for our calculations: the **t-distribution** is the sampling distribution that arises when we substitute the standard deviation s instead of the population standard deviation σ in our calculations for a hypothesis test or a confidence interval.

Introduction to the t distributions

Like the chi-square distributions, there are many different *t* distributions. Below are graphics of some.

Notice that all of the distributions "look Normal" in that all of the *t* distributions are symmetrical. However, some of the *t* distributions have markedly thicker tails than the Normal distribution does. They differ by their **degrees of freedom,** an idea we encountered when studying the **chi-square distributions.**

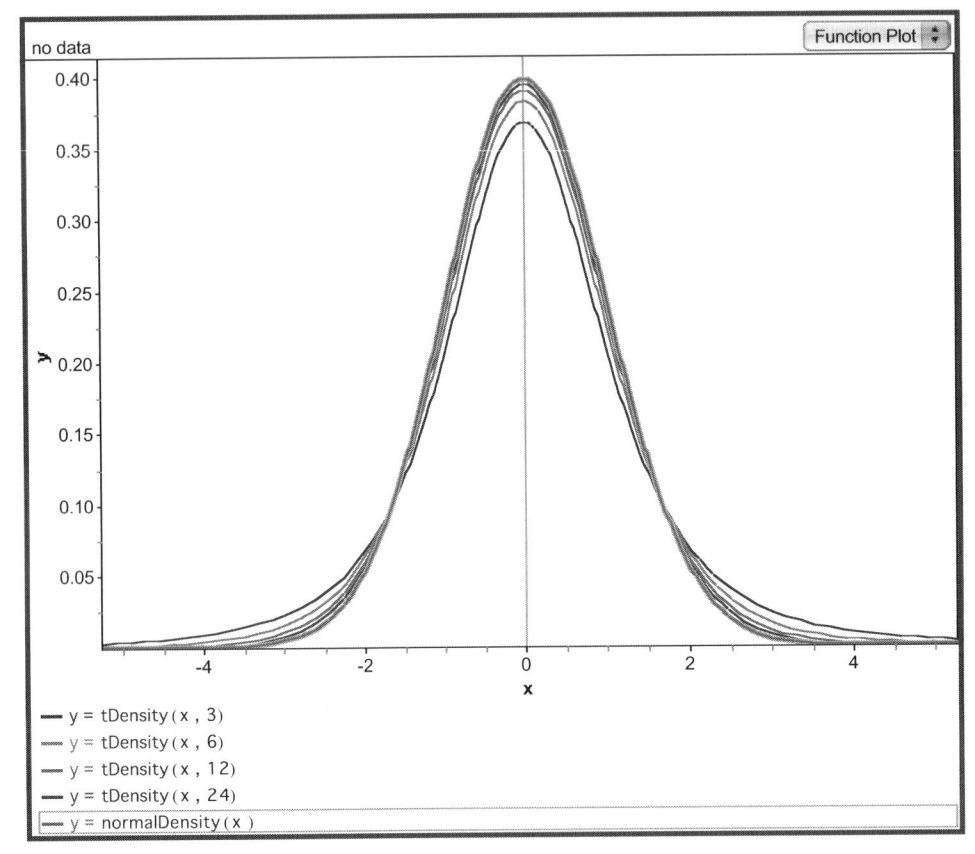

Here are more facts about *t* distributions that we will use.

Facts about t-distributions

Compared with the standard Normal distribution, Normal with $\mu = 0$ and $\sigma = 1$, the **t distributions**:

1. Are perfectly symmetrical with $\mu = 0$ (which is also true for the standard Normal distribution)
2. Have "thicker tails" than the standard Normal distribution
3. The "thickness of the tails" is greater the *smaller* the **degrees of freedom** where **df = n – 1**.
4. The *bigger* the **degrees of freedom,** the more a **t distribution** approximates the Normal distribution.

With the **t distributions**, the degrees of freedom depend upon the sample size, $df = n - 1$, and so for the example that we have been looking at, we should have $df = n - 1 = 20 - 1 = 19$. Our simulation calculated a *t* score that acts something like a *z* score. What a *z* score does is to translate to the standard

Normal distribution, in the Normal Distribution Chart, the Normal distribution with $\mu = 0$ and $\sigma = 1$; the t score does a similar translation when we are using the sample standard deviation s.

Using the t distributions: the plot. First we need a definition. Recall that for the Normal distribution we could say that between $z^* = -1.96$ and $z^* = 1.96$ there would be 95% of the distribution and that in each of the tails beyond the $z^* = -1.96$ and $z^* = 1.96$ there will be 2.5%, adding up to 5%. Because the t distributions have thicker tails than the standard Normal distribution, the corresponding number that includes 95% of the t distribution will always have an absolute value bigger than 1.96. To extend this idea to the t distributions, we use the term **critical value** and the symbol **t*** as defined in the box just below.

Definition of a Critical Value, t*

A **critical value t*** indicates a value of t that gives a specified percentage (or probability) of a t distribution in the tails—to the right of **t*** or to the left of **–t***—and therefore a specified percentage between **–t*** and **t***.

Notice that the definition of a critical value is very general. The picture illustrates the area between $-t^*$ and t^* for 95%, and the Fathom file **Tplot.ftm** will allow you to see the idea extended. For our example, with

$df = n - 1 = 20 - 1 = 19$ instead of

$z^* = -1.96$ and $z^* = 1.96$, the values of the t distribution that enclose 95% are the critical values $t^* = -2.093$ and $t^* = 2.093$.

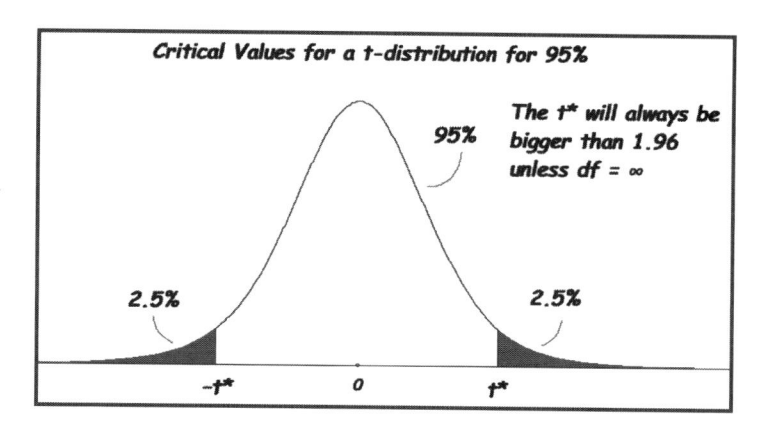

How do we know this number $t^* = 2.093$? As with the chi-square distributions, we have a table that shows these critical values. Part of the table is shown below, where we discuss how to use the table.

What do we do with this number? It is the number that we must use in a confidence interval *instead* of $z^* = 1.96$, so that the confidence interval for the actual duration of flights to Seattle (for *all* flights) will be $\bar{x} \pm t^* \frac{s}{\sqrt{n}} = 117.75 \pm 2.093 \frac{10.151}{\sqrt{20}} \approx 117.75 \pm 2.093(2.2698) \approx 117.75 \pm 4.75$ or the interval 113.0 to 122.50 minutes.

Using the t distributions: the chart. For the infinitely many t distributions, we have a very condensed chart that shows the relationship between the probabilities (or percentages) in the tails of the distributions and the values of the critical values and therefore the probability between **–t*** and **t*** for many different degrees of freedom. Here (on the next page) is a piece of the chart, but to follow the discussion here, you should have the entire chart in your hand or on the screen.

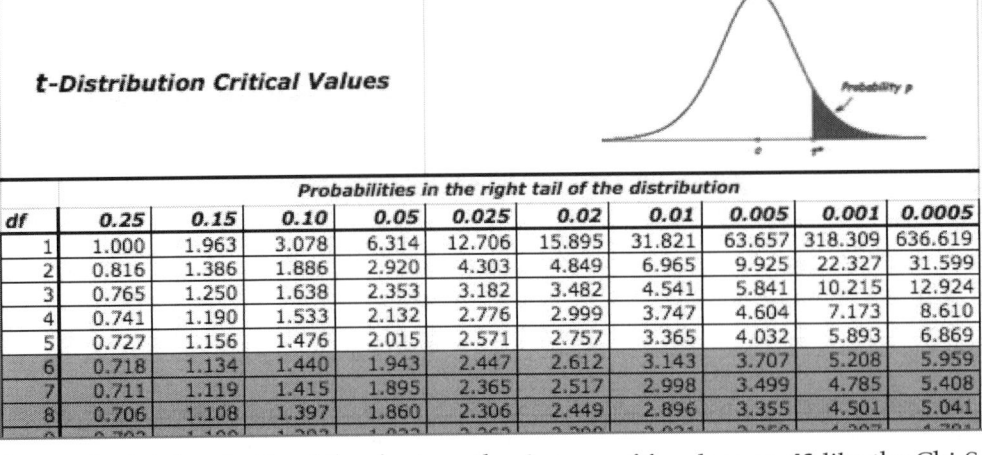

t-Distribution Critical Values

Probabilities in the right tail of the distribution

df	0.25	0.15	0.10	0.05	0.025	0.02	0.01	0.005	0.001	0.0005
1	1.000	1.963	3.078	6.314	12.706	15.895	31.821	63.657	318.309	636.619
2	0.816	1.386	1.886	2.920	4.303	4.849	6.965	9.925	22.327	31.599
3	0.765	1.250	1.638	2.353	3.182	3.482	4.541	5.841	10.215	12.924
4	0.741	1.190	1.533	2.132	2.776	2.999	3.747	4.604	7.173	8.610
5	0.727	1.156	1.476	2.015	2.571	2.757	3.365	4.032	5.893	6.869
6	0.718	1.134	1.440	1.943	2.447	2.612	3.143	3.707	5.208	5.959
7	0.711	1.119	1.415	1.895	2.365	2.517	2.998	3.499	4.785	5.408
8	0.706	1.108	1.397	1.860	2.306	2.449	2.896	3.355	4.501	5.041

Notice that on the left-hand side of the chart are the degrees of freedom or **df**, like the Chi-Square Distribution Chart. (The construction of this chart is basically the same as the Chi-Square Chart, but the distributions are quite different.)

The numbers across the top of the chart give the probabilities (or proportions, or areas) in the right tail of the *t* distribution. Hence, for the probability of 0.025 in the tail of the distribution, we see that for $df = 5$ (which means that the sample size must have been $n = 6$), we have a $t^* = 2.571$. Since the *t* distributions are perfectly symmetrical, we know that the probability to the left of $t^* = -2.571$ —that is, in the left tail—will also be 0.025, and the probability in the two tails together is 0.05 or 5%. Hence the probability between $t^* = -2.571$ and $t^* = 2.571$ is 0.95 or 95%. So if we really did have only $n = 6$ then in the confidence interval formula $\bar{x} \pm t^* \frac{s}{\sqrt{n}}$ we would use $t^* = 2.571$ and *not* $z^* = 1.96$. It is worthwhile to look at the foot of the *t* Distribution Chart and see that the connection between the probability on the right-hand tail and the confidence level is made explicit. The last row shows the values of z^* showing that "at the limit" the *t* distributions get closer and closer (as close as we want for sufficiently large *n*) to the Normal distribution.

250	0.675	1.039	1.285	1.651	1.969	2.065	2.341	2.596	3.123	3.330
400	0.675	1.038	1.284	1.649	1.966	2.060	2.336	2.588	3.111	3.315
1000	0.675	1.037	1.282	1.646	1.962	2.056	2.330	2.581	3.098	3.300
∞	0.674	1.036	1.282	1.645	1.960	2.054	2.326	2.576	3.091	3.291
	50%	70%	80%	90%	95%	96%	98%	99%	99.5%	99.9%
					Confidence Level C					

Things to notice and rounding down. Notice that not all the possible degrees of freedom are shown on the chart and also notice that the values for the t^* are not very different for large degrees of freedom. It is good practice in using the *t* Distribution Chart to "round down," giving a slightly larger t^*. So, if we had a sample size of $n = 140$, so that the $df = 139$, we would use $t^* = 1.657$, which is the value for $df = 125$. Most often we will be using software, so the *t* Distribution Chart and the Fathom file **Tplot.ftm** are here primarily to get the idea of the relationships between the *critical values t^**, the tail probabilities, and the degrees of freedom.

Applying the t distributions, including conditions for use

Confidence Intervals Here is the Fathom output for the 95% confidence interval for the sample of $n = 20$ of the flight times from the San Francisco Bay Area to Seattle. One of the big reasons for studying the t distributions in some detail is that Fathom and other statistical software use the t distributions in their calculations of confidence intervals and the test statistics for hypothesis tests involving means of quantitative variables. The software does this because *in practice* researchers do not know the population standard deviation σ.

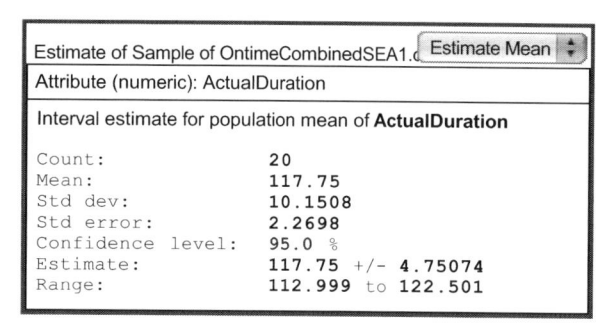

```
Estimate of Sample of OntimeCombinedSEA1.c  Estimate Mean
Attribute (numeric): ActualDuration
Interval estimate for population mean of ActualDuration
Count:               20
Mean:                117.75
Std dev:             10.1508
Std error:           2.2698
Confidence level:    95.0 %
Estimate:            117.75 +/- 4.75074
Range:               112.999 to 122.501
```

Researchers have samples (or experimental data), so they can always calculate the sample standard deviation *s*. (Even though, with large sample sizes, the *t* distributions are nearly the same as the standard Normal distribution so that it would be feasible to work with the Normal distributions if we were working by hand, statistical software developers tend to use the *t* distributions for all the calculations involving sample means $\bar{x}$.)

Formula for Confidence Intervals for One Population Mean

$$\bar{x} \pm t^* \frac{s}{\sqrt{n}}$$

where t^* depends upon the $df = n - 1$ and the confidence level chosen.

This formula can be written $\bar{x} \pm t^*(SE)$, where $SE =$ Standard Error $= \frac{s}{\sqrt{n}}$, so

$ME =$ Margin of Error $= t^*(SE)$.

Be careful not to confuse **standard error** with **margin of error**. *Standard error* refers in this context to the sample estimate of the standard deviation of the sampling distribution, and so $SE = \frac{s}{\sqrt{n}}$, whereas **margin of error** refers to $t^* \frac{s}{\sqrt{n}}$. Fathom output gives both the standard error and also the margin of error, but the margin of error needs to be read as the ± portion of the "estimate" line; in our example shown above, $ME = 4.75074$.

Interpretation of Confidence Intervals for Means The interpretation of a confidence interval for a population mean follows the same pattern as for proportions, so for our estimate of the population mean duration of flight from the San Francisco Bay Area to Seattle, we would say either:

➢ "With 95% confidence, we can say that the mean duration of all flights from the SF Bay Area to Seattle is 117.75 minutes with a margin of error of 4.75 minutes," or:

➢ "We are 95% confident that the mean duration of all flights from the SF Bay Area to Seattle is between 113.0 minutes and 122.5 minutes."

Conditions with t distributions

There are two basic conditions for using the *t* distributions for inference.

Conditions for Using the t Distributions for Inference
1. The sample or samples used must be random.
2. The population distribution or distributions must be Normal.

These conditions look rather restrictive; we have encountered many variables whose sample distributions are quite skewed, and we know that random samples are hard to get. It turns out that one—but only one—of these conditions can be "relaxed" or "violated" and the *t* distributions will still yield trustworthy results. The **random sampling condition** *cannot* be relaxed; to have trustworthy confidence intervals or results from hypothesis tests, we *must have random sampling.*

Robustness and the 15/40 Rule of Thumb Statisticians call a procedure (such as a formula for a confidence interval) **robust** if the procedure gives trustworthy results even when one or more of the conditions for use are relaxed or violated. There are limits to the violation of the conditions, and therefore **rules of thumb** have been developed to guide the researcher in using the procedure. Sometimes there are different rules of thumb followed by different statisticians, and sometimes there are arguments about the robustness of procedures. For the *t* procedures, the Normality condition can be relaxed according to the **15/40 rule of thumb.** This rule of thumb relates sample size to the Normality condition of the *t* procedures. The basic idea is that the bigger the sample size, the less important is the Normality condition.

The 15/40 Rule of Thumb
- *Small Samples, n < 15* For samples where $n < 15$, the data *must* come from a Normal or nearly Normal distribution. The *t* procedures will not be trustworthy otherwise.
- *Moderate-sized Samples, 15 ≤ n ≤ 40* The *t* procedures may cope with some skewness but will not be trustworthy in the face of strong skewness or outliers.
- *Large Samples, n > 40* The *t* procedures are *robust* in the face of skewness or outliers where the sample is large and the more so the larger the sample size.

Assessing whether a population is Normal How does one know if the population distribution is Normal or not? One way is by examining the sample distribution to see if there is skewness in the sample distribution. This is a good guide if the sample size *n* is large enough so that the shape of a distribution can be assessed from a dot plot or stem plot. However, with small samples (such as the $n = 20$ we were working with), it may not be possible to detect shape. If that is so then what you are left with is thinking carefully about what the population distribution should be; this is actually what statisticians regularly do. For example, for the duration of flight data, we had reason to believe that there would be some flights whose durations would be longer than the scheduled durations because of bad weather and other circumstances, but because the flights were between California and Seattle, we thought that there would be very few such flights.

Meaningful data and nonsensical data It is always wise to look at data carefully and think about the data. We expect the variable *Height* to be Normally distributed, but when looking at the Australian Census at School data, we find some high school students who are apparently one thousand centimeters tall. Not likely! More likely some Australian high-schoolers were not taking the exercise seriously. In the flight data, flights that were cancelled obviously had an ActualDuration of zero minutes—completely true, but this is not meaningful for the analysis.

Why we have the t distributions and why they are called "Student's t"

The answers to these questions are related. The *t* test and other procedures were developed in the early part of the twentieth century by William Sealy Gosset. Gosset had degrees in both chemistry and mathematics and worked for the Guinness Brewery in Dublin, Ireland. At the brewery he developed the *t* procedures to aid quality control. His problem was that he only had small samples to work with, and, as we have seen, if we have small samples, we dare not simply use the sample standard deviation *s* in place of the population standard deviation σ. Gosset's work was original, but when he went to publish his results in scientific journals, the brewery insisted that he not publish under his own name; Gosset chose to publish under the name "Student," and so the *t* procedures have (since that time) sometimes been referred to as "Student's t." For more information, see http://www.gap-system.org/~history/Biographies/Gosset.html.

Summary: Introduction to Inference for Quantitative Variables

- With quantitative variables, we infer from sample means $\bar{x}$ to population means μ. The notation used is shown here.

  Notation for means and standard deviations

	Population Distribution	Sample Distribution	Sampling Distribution
Mean	μ	$\bar{x}$	$\mu_{\bar{x}}$
Standard Deviation	σ	s	$\sigma_{\bar{x}}$

- **Facts about Sampling Distributions of Sample Means Calculated from Random Samples**
    - **Shape:** The shape of the sampling distribution of $\bar{x}$ will be approximately Normal if the sample size *n* is sufficiently large, even if the population distribution from which the samples were drawn is not Normal.
    - **Center:** The mean of the sampling distribution of $\bar{x}$ is equal to the mean of the population distribution. That is, $\mu_{\bar{x}} = \mu$.
    - **Spread:** The standard deviation of the sampling distribution of $\bar{x}$ is $\sigma_{\bar{x}} = \dfrac{\sigma}{\sqrt{n}}$.

- The sampling distribution of sample means $\bar{x}$ presents the researcher with a practical problem in that the population standard deviation σ is not known, so the formulas implied by the **Facts** above cannot be applied directly, with the sample standard deviation *s* substituted for σ. Instead, if *s* is used then the family of ***t* distributions** must be used.

- **Facts about t-distributions** Compared with the standard Normal distribution, Normal with μ =0 and σ =1, the ***t* distributions**:
    - Are perfectly symmetrical with $\mu = 0$ (which is also true for the standard Normal distribution)
    - Have "thicker tails" than the standard Normal distribution
    - The "thickness of the tails" is greater the *smaller* the **degrees of freedom** where ***df = n − 1***.
    - The *bigger* the *degrees of freedom*, the more a ***t* distribution** approximates the Normal distribution.

- **A critical value t*** indicates a value of *t* that gives a specified percentage (or probability) of a *t* distribution in the tails—to the right of **t*** or to the left of **−t***—and therefore a specified percentage between **−t*** and **t***.

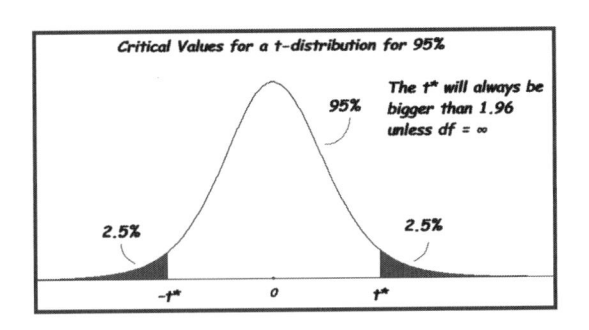

  Critical Values for a t-distribution for 95%

  The t* will always be bigger than 1.96 unless df = ∞

- **Formula for Confidence Intervals for One Population Mean**

$$\bar{x} \pm t^* \frac{s}{\sqrt{n}}$$

  where t^* depends upon the $df = n-1$ and the confidence level chosen. This formula can be written

  $\bar{x} \pm t^*(SE)$, where $SE =$ Standard Error $= \frac{s}{\sqrt{n}}$, so $ME =$ Margin of Error $= t^*(SE)$ **PTO.**

- **Conditions for Using the t Distributions for Inference**
  - The sample or samples used must be random.
  - The population distribution or distributions must be Normal.
- A **robust** procedure in statistics is one that gives trustworthy results even if the conditions for its use are not met. The procedures for inference based upon *t distributions* are robust if the population distributions are not Normal but not robust if randomization is not used. In the situation that the population distributions are not Normal, the **15/40 rule of thumb** (below) may be used as a guide.
- **The 15/40 Rule of Thumb**
  - *Small Samples, n < 15*   For samples where $n < 15$, the data *must* come from a Normal or nearly Normal distribution. The *t* procedures will not be trustworthy otherwise.
  - *Moderate-sized Samples, 15 ≤ n ≤ 40*   The *t* procedures may cope with some skewness but will not be trustworthy in the face of strong skewness or outliers.
  - *Large Samples, n > 40*   The *t* procedures are *robust* in the face of skewness or outliers where the sample is large and the more so the larger the sample size.

§5.1 Exercises: Inference for Quantitative Variables

1. t Distribution Plot and Chart Exercise

- Open the Fathom file **Tplot.ftm** and have the **t distribution Critical Values Chart** ready at hand. Here is a picture of what you should see. The standard Normal distribution is shown and one of the infinitely many t distributions. There are two "sliders." One of these sliders controls the degrees of freedom **df** and the other controls the **Critical Value t***. The **Summary Table** near the foot of the screen shows:

    S1: The probability in the right tail of the t distribution with the degrees of freedom that you choose, and

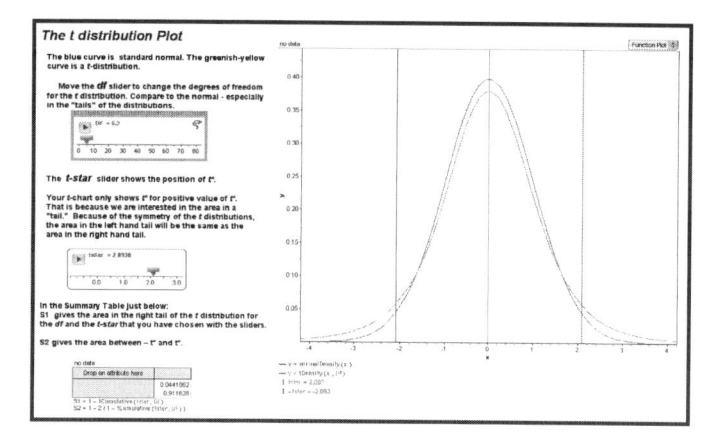

    S2: The probability between $-t^*$ and t^* for the t distribution that you choose

- Move the **df** slider to the right so that the t distribution and the Normal distribution are almost the same.

    a. Find the value of the df where you perceive no difference between the Normal and the t distribution. (Look for the df so that the t curve and the Normal curve are indistinguishable.)

    b. Look at the facts about t distributions in the **Notes**. Which of the four facts is shown by part a?

- Move the **df** slider to the left so that you get $df = 4$. You can actually type in 4.

    c. If $df = 4$, what must have been the sample size n?

    d. Describe the differences and similarities between the t distribution and the standard Normal distribution that you see. Consult the facts about t distributions.

    e. For the standard Normal distribution, what is the proportion in the tail to the right of $z^* = 1.96$?

    f. For the standard Normal distribution, what is the proportion between $z^* = -1.96$ and $z^* = 1.96$?

- With the **df** slider set to $df = 4$ set the **tstar** slider to 1.96. You have set $t^* = 1.96$.

    g. From the Summary Table at the foot of the screen, what is the proportion (or area) in the tail to the right of $t^* = 1.96$ for the t distribution with $df = 4$? (Look at the value for s_1.)

    h. From the Summary Table at the foot of the screen, what is the proportion (or area) between $t^* = -1.96$ and $t^* = 1.96$? (Look at the value for S2; this area in the middle gives the "percent confidence in a CI.")

    i. Forgetful Fiona, with her pitiful sample size of $n = 5$, uses 1.96 in her confidence interval out of habit, thinking (or not thinking) that she is calculating a 95% confidence level. What "percent confidence" is she actually calculating for her confidence interval? Give a reason for your answer.

- Now we will find the t^* Fiona should have been using for a 95% confidence interval, using **Tplot.ftm**. Keep the $df = 4$ but move the **tstar** slider so the area in the tail is as near to 0.025 as you can and the area between $-t^*$ and t^* is as near to 0.95 as you can make it. It will help also to watch what is happening to the t^* vertical lines.

   j. Record the value of t^* you found.

   k. The **t distribution Critical Values Chart** will allow us (and Fiona) to get the value exactly for the row $df = 4$ and the column that indicates the proportion (or area) to the right of the t^* we are seeking. (Which is it? Look at the foot of the table). What t^* should Fiona be using?

   l. Change the **tstar** value (by typing in the correct value) and read off the values of the area to the right and also the area between $-t^*$ and t^* by looking at S1 and S2 in the Summary Table.

   m. What will happen to the margin of error of Fiona's confidence interval when she uses the correct t^* instead of 1.96? Give a reason for your answer.

   n. Suppose that Fiona manages to increase her sample size to $n = 50$. She still wants a 95% confidence interval. What is the df and what should be her t^* now? (Remember the advice that is given about rounding in the **Notes.**)

   o. Either by using **Tplot.ftm** or by using the **t distribution Critical Values Chart,** complete this sentence choosing the correct word and give examples of how what you have written is true: "As we increase sample size n but keep the confidence level the same, the value of t^* [increases/decreases]."

   p. Fiona still has her sample size of $n = 50$ but decides that she is satisfied having a 90% confidence interval. Find what the t^* should be.

   q. Will Fiona's confidence interval for 90% be wider or narrower than the one for 95%? Give a reason for your answer.

2. **Confidence Intervals for Duration of Flight: A Bigger Sample Size**

   In the **Notes** we calculated the confidence interval for the population mean *ActualDuration* of flights from the San Francisco Bay Area to Seattle using our sample of $n = 20$ flights. (See the **Notes.**) Now suppose you manage to have a sample size of flights to Seattle of $n = 36$ times instead of just $n = 20$ times.

   a. What are the cases for our data: Passengers? Airlines? Flights? Durations of flights? (Remember that we could collect data on more than one variable.)

- Open the Fathom file **OntimeCombined4SEA.ftm.**
- With the collection icon selected, go to the menu **Collection>Sample Cases.**
- In the **Inspector,** deselect the *Animation* and the *With Replacement* boxes and change the number of cases from $n = 10$ to $n = 36$ then press *Sample More Cases*.
- Get a **Summary Table** and get the sample mean and sample standard deviation for the variable *ActualDuration*. These are the $\bar{x}$ and s for your SRS of $n = 36$ of *ActualDuration* of flights to Seattle from the SF Bay Area.

   b. Record the sample mean and sample standard deviation using the correct notation.

   c. Consult the **t distribution chart** to get the correct t^* to calculate a 95% confidence interval.

   d. Calculate, by hand using a calculator, a 95% confidence interval.

- From the shelf, drag down an **Estimate** box and change the box from "Empty" to "Estimate Mean."

- Drag the variable *ActualDuration* to the "Unassigned Attribute"; deselect **Verbose** in the **Estimate** menu. Check that your calculations are very close to what Fathom calculated.
    e. State what the margin of error is and also what the standard error is.
    f. Explain from the formula for a confidence interval why the margin of error for your CI using $n = 36$ should be smaller than the margin of error for the CI from a sample where $n = 20$.
    g. Give a good interpretation of your confidence interval in the context of flights to Seattle.

3. **Area of Houses for Sale in San Mateo County (SqFt Again)** [Fathom Exercise] We will look at a *simple random sample* of houses in San Mateo County that were sold in the years 2007–2008. The variable *SqFt* measures the area inside a house. Our goal here is to estimate the population mean area (measured in square feet) of the all of the houses that were sold in the *Central Region* of San Mateo County in 2007–2008. A part of a random sample is still a random sample.
- Open the Fathom file **San Mateo RE Sample Y0708.ftm** and get a **Case Table.**
- Select the **Collection icon** for the collection and go to the menu **Object>Add Filter.**
- In the **Add Filter** dialogue box, type in: Region = "Central". Doing this will guarantee that all of the Fathom work that we do will be for just the Central Region. You should see a sample of $n = 128$ houses.
- Get a **dot plot** for the variable *SqFt* for the Central Region.
- Get a **Summary Table** with the sample mean, the sample standard deviation, and the count (the *n*) for the variable *SqFt*.
    a. What are the cases for these data?
    b. Using the correct notation, write down the sample mean and sample standard deviation.
    c. Is the sample distribution left-skewed, symmetric, or right-skewed?
- Read the **Notes** on the **Conditions** for using the *t* distributions and also read the paragraphs and the box on **Robustness** and the **15/40 Rule.**
    d. According to the information on **Conditions,** determine whether the two conditions for use of the *t* distributions are met. Refer to the way the sample was drawn and the shape of the sample distribution.
    e. Can we safely proceed with the calculations? Why or why not? (*Hint:* Consider the 15/40 rule.)
    f. Consult the **t distribution chart** to get the correct *t** to calculate a 95% confidence interval.
    g. Perplexed Percy does not understand why we can't continue to use 1.96 instead of your answer to part f. "That was so simple—just one number to remember," says PP. Explain to PP why using 1.96 would be wrong. (See the **Notes** or arrange to see Forgetful Fiona of Exercise 1.)
    h. Calculate, by hand using a calculator, a 95% confidence interval for the population mean *SqFt* for the Central Region for the population of all houses in San Mateo County sold in 2007–2008.
- From the shelf, drag down an **Estimate** box and change the box from "Empty" to "Estimate Mean."
- Drag the variable *SqFt* to the "Unassigned Attribute"; deselect **Verbose** in the **Estimate** menu. Check that your calculations are very close to what Fathom calculated.
    i. What is the margin of error for your confidence interval?
    j. What is the standard error for your confidence interval?

k. Confused Conrad's answers to questions i and j are the same number. Should his answers be the same or is Confused Conrad confused? Explain.

l. Give a good interpretation of your confidence interval in the context of the data.

- Go to the Fathom estimate and change the 95% to 99%.

m. What has happened to the margin of error? Explain (in terms of the *t distribution chart*) why the change that you have seen is in the direction that you have seen.

n. Does a bigger confidence level lead to a wider or a narrower confidence interval? Explain (mostly to convince yourself that the relationship makes sense).

4. **Average BMI level for college-age people.** We have a random sample from the NHANES (National Health and Nutrition Examination Survey) with various health measures for the sample of people. We will look at the people surveyed between that are of "college age," which we will take to be the interval $17 \le Age \le 24$. We will examine the variable Body Mass Index, or **BMI**, which is defined as $BMI = \dfrac{Weight}{(Height)^2}$, where the measures are in metric units. BMI is a common measure to assess obesity.

- Open the Fathom file **NHANES Data A.ftm** and get a **Case Table**.
- Select the **Collection icon** for and go to the menu **Object>Add Filter.**
- In the **Add Filter** dialogue box, type in: *(Age >16) and (Age < 25)*. To get the "and," use the "and" on the Add Filter dialogue box keypad. If you have done this correctly, you should have $n = 158$.
- Get a **dot plot** for the variable *BMI* for the age interval we have selected.
- Get a **Summary Table** with the sample mean, the sample standard deviation, and the count (the n) for the variable *BMI*. You will find that $n = 157$ for those having a BMI measurement.

a. What are the cases for these data?

b. Using the correct notation, write down the sample mean and sample standard deviation.

c. Is the sample distribution left-skewed, symmetric, or right-skewed?

- Read the **Notes** on the **Conditions** for using the *t* distributions and read the paragraphs and the box on **Robustness** and the **15/40 Rule.**

d. Determine whether the two conditions for use of the *t* distributions are met. Explain with reference to the way the sample was drawn and the shape of the sample distribution.

e. Can we safely proceed with the calculations? Why or why not?

f. Consult the **t distribution chart** to get the correct t^* to calculate a 95% confidence interval.

g. Calculate a 95% confidence interval for the population mean *BMI* for people in the age range of seventeen to twenty-four. Do the calculations "by hand" with the aid of a calculator. You will check with Fathom below.

- Drag down an **Estimate** box and change the box from "Empty" to "Estimate Mean."
- Drag the variable *BMI* to the "Unassigned Attribute"; deselect **Verbose** in the **Estimate** menu. Check that your calculations are very close to what Fathom calculated.

h. What is the margin of error and what is the standard error for your confidence interval?

i. Give a good interpretation of your confidence interval in the context of the data.

Obesity is defined as a BMI ≥ 30. We will make a variable that distinguishes the obese from those not.

- Scroll to the end of the **Case Table** and, in the <new> attribute, create a new variable called *Obese*.
- Select the variable so that it turns blue and go to **Edit>Edit Formula.** A dialogue box will appear.
- Type in: *If (…)* and in the parentheses type what you see here.
- Get a **Summary Table** for the variable *Obese*. 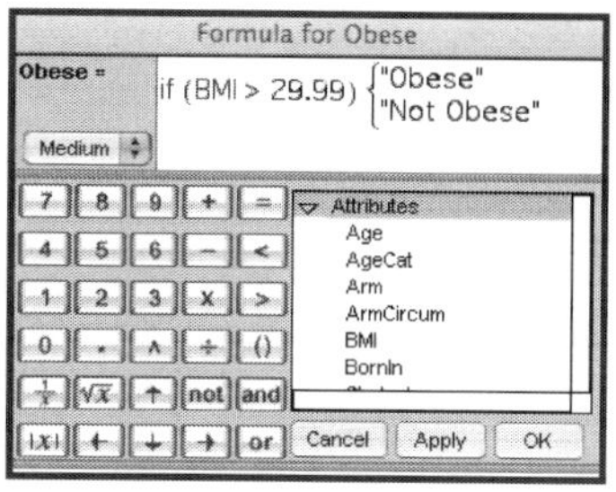
    j. What kind of variable is *Obese*?
    k. Use everything you know from your course in statistics to calculate an estimate for the population *proportion* of seventeen- to twenty-four –year-olds who are obese.
    l. Give an interpretation of your estimate in the context of the data.

5. **Gosset and Roller Coaster G-Force** G-Force is one of the variables for which the Roller Coaster Database (www.rcdb.com/) has some (but not much) data. In our sample from the database, there are just $n = 31$ roller coasters where this information was provided. Remember that www.rcdb.com depends on others—either the operators or the builders of the roller coasters—to provide information, so the information they get is not a random sample. Here is some Fathom output showing the distribution of G-Force, including the sample mean and sample standard deviation.

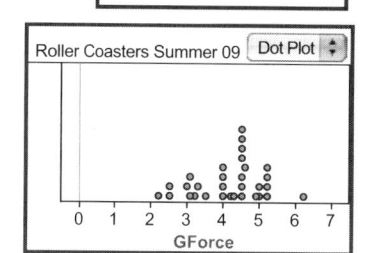

    a. What are the cases for these data?
    b. Is *GForce* a quantitative or categorical variable?
    c. Use the correct notation for the sample mean and sample standard deviation.
    d. Show exactly how Fathom calculated the 90% confidence interval, showing the correct t^* and where the numbers go. Work out and show the calculations so that you can be certain that you have the right numbers. Use many decimal places in your calculations.
    e. Confused Conrad expresses the confidence interval as $3.387094 \leq \bar{x} \leq 4.44519$, seeing that $\bar{x} = 4.15805$ is in that interval. Even so, CC is wrong. What is his mistake? Correct his mistake.
    f. Will a 95% confidence interval be wider or narrower? Give a reason for your answer.
    g. One of the conditions for a trustworthy confidence interval is that the sample be random. In this instance, do we have a random sample? If you think so, support your case. If not, say why you think the sample is not a random sample.
    h. Does the **15/40 rule of thumb** allow you to do the calculation of the confidence interval even if you have doubts about whether the conditions are met? Explain completely why or why not.
    i. Who is William Sealy Gosset and why is he important for this problem?

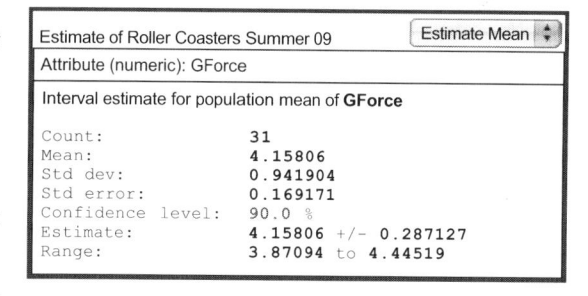

6. **Age of Mother at First Birth and Length of Pregnancy.**
   The data are a simple random sample from the population of all births registered in 2006.
   - Open the Fathom file **MilleBirthsSample.ftm.**
   - With the **Collection Icon** selected, go to **Object>Add Filter** and in the dialogue box type: *FirstBirth ' "First Birth"*. All of the analyses will be for the first births only.
   - Get a Graph showing the distribution of the variable *AgeMother*.
   - Get a **Summary Table** with the sample mean, sample standard deviation, and the count. (You should have a sample size of $n = 428$.)
     a. One of the conditions for doing a confidence interval is that the population distribution be Normal, which we judge by looking at the sample distribution. Is this sample distribution Normal? If not, can we proceed to get a trustworthy confidence interval? Give a reason.
     b. The other condition is that the sample be a random sample. Is that condition met?
   - Get an **Estimate;** choose the kind of estimate that you want.
   - Drag the variable *AgeMother* to the Attribute.
   - With **Estimate** selected, from the **Estimate** menu deselect **Verbose.**
     c. Fathom expresses the confidence interval as a range. If you write this as $24.52 < \quad < 25.73$, what symbol should be placed between the "less than" signs?
     d. Is the margin of error the number that is approximately 0.308 or approximately 0.606?
     e. What in the formula for the confidence interval is the "standard error" and what in the formula is the "margin of error"?
     f. Give a good interpretation of the confidence interval in the context of the data.
   - Go to: www.cdc.gov/nchs/data/databriefs/db21.pdf .
     g. Do the data here agree with what this download is saying? Give a reason for your answer.

The next part looks at the length of pregnancy, which is expressed in weeks in the variable Gestation3.

   - Select the Collection Icon, and go to **Object>Remove Filter.** The graphs and estimates are now for all births.
   - Select the **Graph** and go to **Object>Duplicate Graph.**
   - Select the **Estimate** and go to **Object>Duplicate Interval Estimate.**
   - Drag the variable Gestation3 to the duplicated graph and estimate.
     h. Write, using the correct notation, the confidence interval for the mean gestation time in weeks for all births.
     i. The distribution of length of pregnancy (Gestation3) is quite left-skewed. Does the left-skewness make our estimate not trustworthy or not?
   - Get a **Summary Table** and get the mean and standard deviation of *Gestation3*.
   - For the second variable, drag the variable *FirstBirth*. You should have this Summary Table.
   - Drag the variable *FirstBirth* to the left-hand axis of your graph.
     j. Judging from the graphs and the numbers, does there appear to be difference in length of pregnancies between first and later births overall? Give a reason for your answer.

§5.2 Hypothesis Testing for One Mean

How long it takes to get to Seattle, continued

In the last section we examined the duration of flights from the San Francisco Bay Area to Seattle. Looking at the advertised duration of flights, we posed two inferential statistical questions:

> Is the average time of all flights 125 minutes or not?
> Can I get an estimate of how long, on average, the flights really are?

Is there? Are there? Or estimate... We found that to do inference for quantitative variables (inferring from sample means $\bar{x}$ to population means μ, we had to use procedures based on the *t distributions* and not the Normal distribution). We were able to calculate an estimate of the population mean duration of flights μ:

$$\bar{x} \pm t^* \frac{s}{\sqrt{n}} = 117.75 \pm 2.093 \frac{10.151}{\sqrt{20}} \approx 117.75 \pm 2.093(2.2698) \approx 117.75 \pm 4.75$$

This gave us an interval estimate for the population mean of $113.00 < \mu_{\text{Actual Duration}} < 122.50$ minutes, 95% confidence interval with a margin of error of 4.75 minutes. To do this calculation, we found that we used the *t* distributions, with the sample standard deviation *s* in the calculation.

However, the first question posed above implies a hypothesis test where $H_0 : \mu = 125$ minutes, and $H_a : \mu \neq 125$ since it is a version of an "Is there..." or "Are there..." type of statistical question.

For proportions, we had **five steps** for a hypothesis test. We will follow these five steps, with an additional preliminary step to look at the data.

Step 0: Looking at our sample: First, we do a descriptive analysis of the sample data. We have already seen that one of the **conditions** for using the *t* distributions is that the population from which we sample is Normal. One of the ways that we can sometimes judge the Normality of the population is shape of the sample distribution. We also know that the shape of the population distribution is important if the sample size is small—the **15/40 rule**. However, with a small sample size, it is sometimes hard to detect the shape of the distribution; all we can say from the sample about the shape of the distribution is that there does not appear to be marked skewness; what we see suggests symmetry.

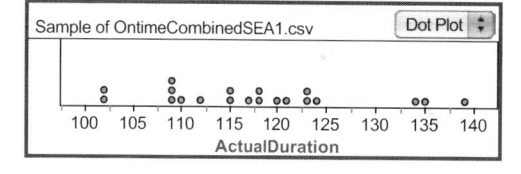

Next, we calculate the sample mean $\bar{x}$ and the sample standard deviation *s*. Again, we need to have notation to help distinguish the sample means and standard deviations from the population means and standard deviations. For a hypothesis test, we must have notation to identity our **hypothesized population mean.** The notation is shown below.

Step 1: Setting up the hypotheses: The first step in doing a hypothesis test is to set up the hypotheses, and with that we will need notation to keep track of the hypothesized population mean. On the next page, the notation is introduced.

> **Notation for Hypothesis Testing for Means**
>
> | | Population | Sample |
> |---|---|---|
> | Mean: | μ | $\bar{x}$ |
> | Standard Deviation: | σ | s |
>
> ➤ The *null hypothesis* that the population mean μ is equal to μ_0 is written: **H_0: $\mu = \mu_0$**
>
> ➤ The *alternate hypothesis* that the population mean μ is not equal to μ_0, written: **H_a: $\mu \neq \mu_0$**
> where μ_0 is the value of the hypothesized mean.

The μ_0 here functions in the same way that the p_0 did in hypothesis tests for proportions; it represents the value of the population mean μ that we are testing, but now we are dealing with *means* and *not* proportions. For our example of the duration of flights to Seattle, we would write:

$$H_0 : \mu = 125 \text{ minutes}$$
$$H_a : \mu \neq 125 \text{ minutes}$$

Step 2: Checking the conditions: The first condition says that the sample must be drawn randomly from the population. These data are a random sample of the information on the Bureau of Transportation Statistics website, so we have satisfied the randomness condition. For the Normality condition, we would be a bit uncertain if we had only the sample, although (see **Step 0** above) the sample does not reveal extreme skewness or outliers. The **15/40 rule** says that with $n = 20$ we should be wary of skewness in the population distribution.

Step 3: Calculating the test statistic: When we did hypothesis tests for proportions, we used the z score (to get the p-value) because the sampling distribution for $\hat{p}$ is a Normal distribution. If we use the sample standard deviation s in our calculations then the sampling distribution involves the t distributions, and so the test statistic will be the **t score** (introduced in §5.1), which the equivalent of the z score. The t score has the same form as the z score but uses the sample standard deviation instead of the population standard deviation σ.

$$t = \frac{(\text{Sample result}) - (\text{Mean of Sampling Distribution})}{\text{Standard Deviation of Sampling Distribution}} = \frac{\bar{x} - \mu_0}{\frac{s}{\sqrt{n}}}$$

In our example, the sample mean is $\bar{x} = 117.75$ minutes, the sample standard deviation is $s = 10.151$, the sample size is $n = 20$, and our hypothesized mean is $\mu_0 = 125$ minutes, so that we calculate:

$$t = \frac{\bar{x} - \mu_0}{\frac{s}{\sqrt{n}}} = \frac{117.75 - 125}{\frac{10.151}{\sqrt{20}}} \approx -3.194.$$

> **Test Statistic for Testing a Single Population Mean μ**
>
> $$t = \frac{\bar{x} - \mu_0}{\frac{s}{\sqrt{n}}}$$ where μ_0 is the value used in the null hypothesis H_0: $\mu = \mu_0$.

Step 4: Finding the p-value: A reminder of what a p-value is (see §4.2): a p-value is the probability of getting a *test statistic as extreme as or more extreme* than the one observed when compared with the null hypothesized value for the population. It measures the "rarity" of our observed value if the null hypothesis is true and just random variation is operating.

Our test statistic in our example comes out to be $t = -3.194$; if this were a z score for a Normal distribution, we would know that the *p*-value would be very small. But we are dealing with the *t* distributions, and we know that the *t* distributions have thicker tails than the Normal, so it *may* be that the *p*-value is not as small as we think. So how do we determine the *p*-value when we are using a *t* distribution? There are two answers; one is to approximate using the **t Distribution Chart** and the second way is to use software.

We use the **t Distribution Chart** in the same way we used the **Chi-Square Distribution Chart**. We first locate our degrees of freedom in the left-hand column of the chart, so that we look at $df = 19$. Then, for our example, ignoring the negative sign (we are using the absolute value of

df	0.25	0.15	0.10	0.05	0.025	0.02	0.01	0.005	0.001	0.0005
1	1.000	1.963	3.078	6.314	12.706	15.895	31.821	63.657	318.309	636.619
2	0.816	1.386	1.886	2.920	4.303	4.849	6.965	9.925	22.327	31.599
3	0.765	1.250	1.638	2.353	3.182	3.482	4.541	5.841	10.215	12.924
4	0.741	1.190	1.533	2.132	2.776	2.999	3.747	4.604	7.173	8.610
5	0.727	1.156	1.476	2.015	2.571	2.757	3.365	4.032	5.893	6.869
6	0.718	1.134	1.440	1.943	2.447	2.612	3.143	3.707	5.208	5.959
⋮	⋮	⋮	⋮	⋮	⋮	⋮	⋮	⋮	⋮	⋮
18	0.688	1.067	1.330	1.734	2.101	2.214	2.552	2.878	3.610	3.922
19	0.688	1.066	1.328	1.729	2.093	2.205	2.539	2.861	3.579	3.883
20	0.687	1.064	1.325	1.725	2.086	2.197	2.528	2.845	3.552	3.850
21	0.686	1.063	1.323	1.721	2.080	2.189	2.518	2.831	3.527	3.819

Probabilities in the right tail of the distribution

$|t| = |-3.194| = 3.194$), we find the two numbers that surround 3.194 are 2.861 and 3.579. Then, reading the probabilities in the right tail from the top of the chart, we see that our 3.194 will have a probability between 0.001 and 0.005. However, this probability is the probability only in the right tail, and because our alternate hypothesis is two-sided (we hypothesized that the population mean could be either larger or smaller than $\mu_0 = 125$), we need to consider both sides of the *t* distribution. Hence, we need to multiply this interval by two. When we do this, we get $0.002 < p - value < 0.010$, since $2(0.001) = 0.002$, and $2(0.005) = 0.010$. From the **t Distribution Chart** we can only approximate that the *p*-value is between 0.002 and 0.01.

Actually, the information that the *p* value is in the interval $0.002 < p - value < 0.010$ is sufficient for us to interpret the hypothesis test. If our level of significance is α=0.05, the interval for the *p* value tells us that we have a rare test statistic. The *p*=value indicates that *if* $\mu_0 = 125$, we have observed a *rare* rather than *reasonably likely* event in our random sample.

Why can we ignore the negative sign in $t = -3.194$ and use the absolute value of *t*? We can ignore the negative sign because we know that the *t* distributions are perfectly symmetrical, so that the area (or probability) in the right tail, to the right of 3.194, is the same as the area (or probability) to the left of $t = -3.194$. The **t Distribution Chart** has been constructed to take advantage of the symmetry of the *t* distributions; the chart only needs to show one side.

This way of approximating the *p*-value can be referred to as the "little box" method because we can depict the values of the *t* score test statistic and the probabilities in the tail, as a little box.

	Probability in tail	
	0.005	0.001
t-score	2.861	3.579

Alternately, we can let software do a more accurate calculation, as in the Fathom output shown. Fathom will also show a picture of the sampling distribution of the test statistic (the command in the menu for the test, is **Test>Show Test Statistic Distribution**).

```
From Summary Statistics                    Test Mean
Attribute (numeric): unassigned
Ho: population mean of ActualDuration equals 125
Ha: population mean of ActualDuration is not equal to 125

Count:        20
Mean:         117.75
Std dev:      10.1508
Std error:    2.26979
Student's t:  -3.194
DF:           19
P-value:      0.0048
```

We see that the *p*-value = 0.0048 and that agrees with our approximation from the **t Distribution Chart**, since 0.0048 is in the interval $0.002 < p-value < 0.010$. The shaded-in portions of the *t* distribution with $df = 19$ show (together) the small *p*-value. The shaded areas are barely visible because the *p*-value = 0.0048 is so small, indicating that our sample ActualDuration mean $\bar{x} = 117.75$ would be very rare *if* it were true that $\mu_0 = 125$.

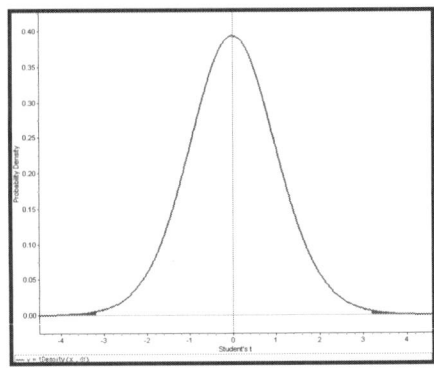

Step 5: Interpret the results in the context of the data: Our hypotheses were that for all flights starting at SF Bay Area and ending at Seattle, the population mean ActualDuration $\mu = 125$ minutes; the alternate hypothesis claimed that the mean duration $\mu \neq 125$ minutes. Our random sample of $n = 20$ flights gave us a sample mean of $\bar{x} = 117.75$ minutes for the variable ActualDuration and we found that the probability of seeing this sample mean $\bar{x} = 117.75$ or one more extreme than this was only 0.0048—very rare, *if* it were true that $\mu = 125$ minutes. Hence, since the *p*-value is so small, we can say that the test was *statistically significant* and that we have sufficient evidence contrary to the null hypothesis that the mean actual duration of flights to Seattle is $\mu = 125$ minutes.

In our interpretation, we related the essentials of the evidence (the *p*-value) and we did this in the context of our test—that is, in the context of the amount of time that flights take to get to Seattle from the SF Bay Area.

How we can read a two-sided hypothesis test from a confidence interval

We met the idea of the connection between a confidence interval and hypothesis test before in the context of tests for the difference of two proportions (§4.3). Here is the reasoning: if a hypothesis test is significant then we know that the $\bar{x}$ is in the *rare* region (as opposed to the region of *reasonably likely* $\bar{x}$) for the μ_0 we are testing. Then we also know that the confidence interval that we calculate based on that $\bar{x}$ will *miss* or *not capture* the μ_0. If the $\bar{x}$ is in the "rare" region, the distance between the $\bar{x}$ and the μ_0 is bigger than the margin of error that gets added and subtracted from the $\bar{x}$.

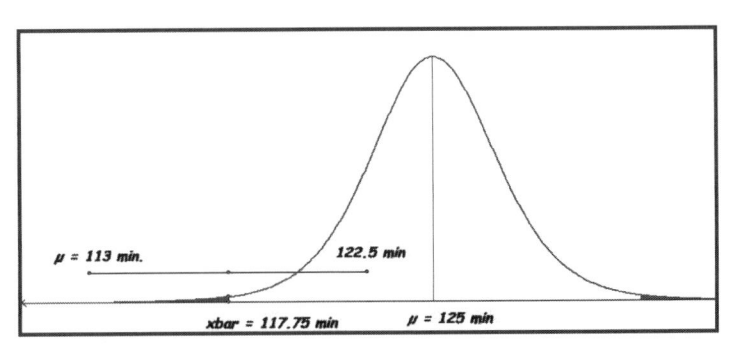

Our analysis of the *ActualDuration* of flights to Seattle illustrates this connection. The confidence interval for the population mean *ActualDuration* came out to be $113.00 < \mu_{ActualDuration} < 122.5$ minutes; this interval does *not* include the hypothesized mean time of $\mu_0 = 125$ minutes. And we found that our hypothesis test of $H_0 : \mu = 125$ versus $H_a : \mu \neq 125$ was *statistically significant* (the p-value was 0.0048). Now, think of what would have happened if we had got a reasonably likely $\bar{x}$; in that case, the confidence interval would have captured (or crossed over or hit) the $\mu_0 = 125$ minutes, since the distance between the $\bar{x}$ and μ_0 is smaller than the margin of error.

> **Confidence Intervals and Hypothesis Tests** When a confidence interval
> - **excludes** an hypothesized value then the related **two-sided** hypothesis test will be *statistically significant*;
> - **includes** an hypothesized value then the related **two-sided** hypothesis test will *not* be *statistically significant*.

However, the principle only works for **two-sided** hypothesis tests.

This principle connecting confidence intervals and hypothesis tests works in reverse as well; if we have a *statistically significant* two-sided hypothesis test then we know that if we used the same sample data to calculate a confidence interval, that confidence interval would *exclude* the hypothesized value. This principle means that in some ways confidence intervals are more useful inferential techniques than hypothesis tests, since confidence intervals give all the information we would get from a two-sided hypothesis test and an interval of plausible values for the population value.

Another Example: A One-Sided Test

Despite the nice connection between confidence intervals and hypothesis tests for two-sided tests, one-sided hypothesis tests are also important. Here is an example. Let us suppose someone (call her Cindy) has another sample of *ActualDuration* data and wants to use these data to do her own hypothesis test. There will be one difference, however; Cindy has concluded that the proper way to challenge the null hypothesis $H_0 : \mu = 125$ should be a one-sided alternate hypothesis $H_a : \mu < 125$. She thinks that $H_a : \mu < 125$ in her alternate hypothesis because she thinks that airlines give themselves more time on average than they actually need to make the flight. Therefore, the actual flight durations should, on average, be shorter than the announced flight durations. What Cindy is doing is bringing more information to the test, and that is common in the use of hypothesis tests.

Step 0: Descriptive analysis: Here are the results of the descriptive side of Cindy's analysis. The sample mean $\bar{x} = 120.95$ minutes for Cindy's sample is higher than $\bar{x} = 117.75$ minutes for the sample that we had, but it is still less than the hypothesized $\mu_0 = 125$ minutes. The standard deviation $s = 10.4553$ is similar, and the plot does not show obvious skewness, although it could be there.

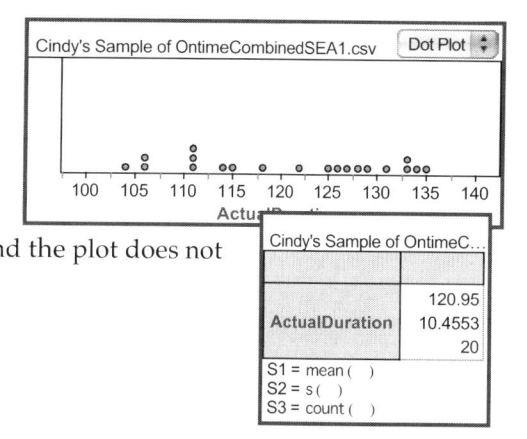

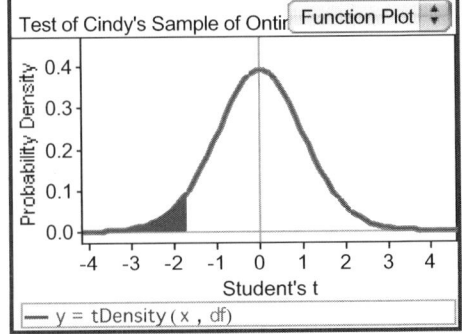

Step 1: Setting up the hypotheses: Cindy wants a one-sided test, since she has reason to believe that the population mean μ will be less than the $\mu_0 = 125$ minutes, so her hypotheses will be:

$$H_0 : \mu = 125 \text{ minutes}$$
$$H_a : \mu < 125 \text{ minutes}$$

It is in the alternate hypothesis that the one-sided test differs from the two-sided test; the two-sided test would have $\mu \neq 125$ minutes.

Step 2: Checking the conditions: The conditions and their assessment will be the same as with the example we went through. Once again, Cindy is using data that come from a random sample of the population of flight records. Once again, Cindy has a small sample of $n = 20$, so we need to be cautious about outliers or evidence of extreme skewness in the population distribution of the variable *ActualDuration*.

Step 3: Calculating the test statistic: For a one-sided test, the calculation of the test statistic is the same as for a two-sided test. For Cindy's sample where $\bar{x} = 120.95$, $s = 10.4553$, and $n = 20$, she will calculate

$$t = \frac{\bar{x} - \mu_0}{\frac{s}{\sqrt{n}}} = \frac{120.95 - 125}{\frac{10.4553}{\sqrt{20}}} \approx -1.732$$

This does not look nearly as extreme as the test statistic we got earlier. To see whether it is extreme enough to challenge the null hypothesis, we need to calculate the *p*-value.

Step 4: Finding the p value: Again, we can approximate the *p*-value using the **t distribution Chart**, with the "little box" method. We will ignore the negative sign of our test statistic since the right hand tail (which the chart shows is the same as the left-hand tail). Here is the "little box" for $t = -1.732$, and this shows that our approximate *p*-value is between 0.025 and 0.05. We would express this as $0.025 < p-value < 0.050$. However, here we did *not double* the probabilities shown in the chart; we did not double because we are conducting a one-sided test instead of a two-sided test. As before, we could let software do the work. Here is the Fathom output.

The plot showing the sampling distribution of the test statistic *t*-score shows the *p*-value shading only on the left side. This is because the hypothesis test we are doing is *one-sided* and not *two-sided*.

Step 5: Interpret the results in the context of the data: For the Fathom output, the *p*-value appears to be "exactly" 0.05, which is the most common value of α, the "line" that divides "significant" from "not significant." However, the *p*-value shown is rounded, and Fathom can show the *p*-value to more decimal places. If we do "round" to seven decimal places, then we see that the *p*-value is 0.0497388.

(To get this in Fathom, get a **Summary Table** and go to:
Add Formula>Functions>Distributions>Student's t>tCumulative ; type in the value of the *t*-score and *df*.)

So what conclusion do we come to "on the boundary"? Is the test statistically significant or not? One lesson we learn from getting an example "on the boundary" is that there is nothing magical or "cut in stone" about the number $\alpha = 0.05$ (or the number $\alpha = 0.01$ or any other value of α.) What we can say is that we *do* have some evidence for the alternate hypothesis that the mean flight time is less than 125 minutes. The evidence is that *if* the mean flight time were really $\mu_0 = 125$ *then* the probability that we would get sample mean as small as $\bar{x} = 120.95$ or smaller is about 5%.

In the conditional probability notation of §1.2, we could write the *p*-value as

$P(\bar{x} \leq 120.96 \mid \mu_0 = 125) \approx 0.05$. This rather elegant way of writing the *p*-value says: "Given the null hypothesized value $\mu_0 = 125$ minutes, the probability that we will see a sample mean $\bar{x}$ of 120.96 minutes or less is 5%." Another way of putting our conclusion is that the sample mean of $\bar{x} = 120.95$ is unlikely (probability about 5%) to have come about by random sampling variation *and* a population mean of $\mu_0 = 125$.

Lessons from Cindy's sample results   One lesson is (as mentioned above) not to look for hard and fast rules that tell you: "When you see this number, say this..." or "When you see that number, say that..." Real data analysis is often not that simple and must take into account many pieces of evidence. The kinds of statistical analysis we are doing here primarily answer the question: "Is what I am seeing consistent or inconsistent with random variability with the null hypotheses I have set up?"

Second lesson: In research that uses these inferential techniques, you may well find that *p*-values rather than declarations of statistical significance or non-significance are reported. Reporting *p*-values rather than stating that a result is statistically significant or not allows the reader (often another researcher) to make up his or her own mind on the basis of what that reader considers *rare*.

Can anything else go wrong? Data analysis in the face of random variation

The answer is yes and not because of bad or wrong calculations. We can do all of the calculations correctly but because of random variation still come to the wrong conclusions. We can do the calculations for confidence intervals correctly and still "miss" the population mean *µ* in our interval, and we can make the same kinds of "errors stemming from randomness" in hypothesis tests.

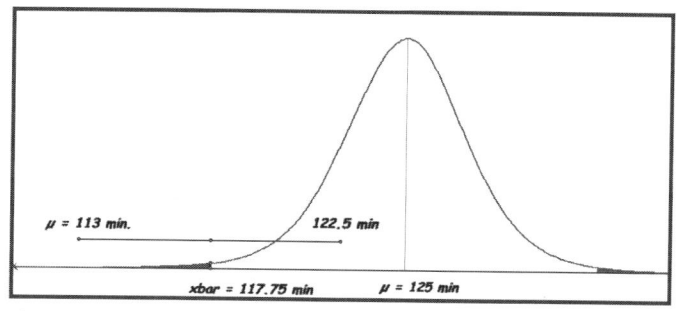

Confidence Intervals.   The issues can be understood by considering confidence intervals. The confidence interval for our first sample was $113.00 < \mu_{ActualDuration} < 122.5$. We would say that we are "95% confident that the mean flight duration from the SF Bay Area is between 113 minutes and 122.5 minutes." Can we be wrong? Yes. We could have, just by random chance, chosen a *rare* sample. If we did, our confidence interval would not capture the true population mean *µ* duration of flight. The graphic shown here illustrates this. We are 95% confident that the mean duration of flight to Seattle is *not* as long as (for example) 124 minutes, but we are not 100% confident.

Hypothesis Testing Can we "miss the mark" with hypothesis testing in a similar fashion? Yes, it is possible, and it is such a possibility that statisticians have a special terminology for the ways in which we can miss the mark. There are essentially two kinds of "errors" (and they are *not* errors in calculation) that can be made. They are called **Type I** and **Type II errors.**

Type I and Type II Errors

Type I Error: If the null hypothesis is *actually true,* but our test *rejects* the null hypothesis

Type II Error: If the alternate hypothesis is *actually true,* but our test *does not reject* the null hypothesis

---

We can illustrate these two types of errors using Cindy's hypothesis test.

$\mu = 125$ true, then what? Suppose it *actually* is true that the mean duration of flight is $\mu = 125$ minutes, but Cindy decides (based on her sample and calculations) to reject the null hypothesis and go with her alternate hypothesis, which is that $\mu < 125$ minutes. In this situation, Cindy would be making a **Type I error.** She has rejected the null hypothesis when it is actually true.

$\mu < 125$ true, then what? Suppose it is *actually* true that that the mean duration of flight is less than 125 minutes, that is $\mu < 125$, but Cindy decides on the basis of her test to *not to reject* the null hypothesis. She did not reject the null hypothesis when she should have. Cindy would be making a **Type II error.** She has failed to reject the null hypothesis when she should have.

For any hypothesis test, you should be able in the *context* of any hypothesis test to identify what a **Type I error** and what a **Type II error** will be. To keep these two types of errors straight, some people find this diagram useful. Since Cindy is on the boundary, she could be in danger of making either one of these two errors, but we are always in danger of making one or the other of these errors when doing hypothesis testing.

Power and how it works Statisticians are concerned about both types of errors, but in general a **Type II error** is the more problematic of the two. For one thing, it is very easy to see the probability of making a Type I error. That probability is just α; if we have chosen $\alpha = 0.05$ then the probability that we will get a rare result by random sampling variability is just 5%. However, Type II errors are harder to control. The smaller one tries to make the probability of making a Type I error, the bigger will be the probability of making a Type II error. Because of this, statisticians have a more useful term to get at what they are trying to achieve. The technical term is called **power,** and its definition is given below.

Power

The **power** of the hypothesis test is the *probability* that a hypothesis test opts for the alternate hypothesis when the alternate hypothesis is *actually* true.

In general, other things being equal, the bigger the sample size *n*, the greater the power of a test.

Calculations of the power of a hypothesis test are very, very complicated, partly because they depend upon the unknown reality behind the alternate hypothesis. The calculations are complicated, but they are also important. Expect to see references to the power of a test in technical reports. The calculations are beyond the scope of an introductory course. (There is specialized software to do power calculations!)

However, one result of these calculations is that by increasing sample size n, the *power* is also increased. Bigger samples are better and give more confidence in our results. Bigger samples often cost more as well.

Summary: Hypothesis tests for one mean

- The logic and terminology for hypothesis tests for means are the same as for proportions, but the formulas are different, being based upon the sampling distribution that arises when the sample mean $\bar{x}$ and sample standard deviation s are used, namely the t distributions. Specifically,
  - The form of the null and alternate hypotheses is the same but refer to population means μ rather than population proportions p.
  - The meaning of the test statistic is the same, but the formula is different (see below).
  - The meaning of a p-value is the same.
  - The interpretation of the results of a hypothesis test follows the same pattern; the term *statistically significant* has the same meaning.

Five steps for hypothesis tests for one mean

Step 1: Set up the null and alternate hypotheses using an idea for the population mean μ_0.

$$H_0 : \mu = \mu_0 \qquad H_a : \begin{cases} \mu \neq \mu_0 \text{ or} \\ \mu < \mu_0 \text{ or} \\ \mu > \mu_0 \end{cases}$$

Step 2: Check the conditions for a trustworthy hypothesis test.
3. The sample must be a simple random sample.
4. The population must be Normal; however, since the t procedures are robust for large sample sizes, judge whether to proceed using the 15/40 rule.

Step 3: Calculate the test statistic $t = \dfrac{\bar{x} - \mu_0}{\dfrac{s}{\sqrt{n}}}$ using the sample mean $\bar{x}$ and sample standard deviation and the hypothesized μ_0.

Step 4: Calculate the p-value by getting $P(\bar{x} \text{ more extreme} | \mu = \mu_0)$ using the absolute value of the test statistic t you calculated in Step 3.

Step 5: Evaluate the evidence that the p-value and the test statistic give you to determine whether your test successfully challenges the null hypothesized population mean μ_0 or not. Give an interpretation in the context of the data using the terminology of "statistical significance" and "rejecting the null hypothesis."

- **The 15/40 rule of thumb** for the use of the *t* distributions introduced in connection with confidence intervals also applies to the use of *t* distributions for hypothesis tests.
    - <u>Small Samples, n < 15</u> For samples where n < 15, the data must come from a Normal or nearly Normal distribution. The t procedures will not be trustworthy otherwise.
    - <u>Moderate-sized Samples, 15 ≤ n ≤ 40</u> The t procedures may cope with some skewness but will not be trustworthy in the face of strong skewness or outliers.
    - <u>Large Samples, n > 40</u> The t procedures are robust in the face of skewness or outliers where the sample is large and the more so the larger the sample size.
- **Confidence Intervals and Hypothesis Tests** When a confidence interval
    - *includes* an hypothesized value then the related hypothesis test will <u>not</u> be *statistically significant*
    - *excludes* an hypothesized value then the related hypothesis test <u>will</u> be *statistically significant*
    - This principle works only for *two-sided* hypothesis tests.
- **Type I and Type II Errors**
    - *Type I Error:* If the null hypothesis is *actually true,* but our test *rejects* the null hypothesis
    - *Type II Error:* If the alternate hypothesis is *actually true* but our test *does not reject* the null hypothesis
    - The statements above are general and should be translated into the context of the data to be useful.
- **Power**
    - The *power* of the hypothesis test is the *probability* that a hypothesis test opts for the alternate hypothesis when the alternate hypothesis is *actually* true.
    - Generally, the larger the sample size used, the larger the power, other things being equal.

§5.2 Exercises on Hypothesis Testing for One Mean

1. **Flying to LAX** [Fathom]   This exercise is similar to the example in the *Notes* about the duration of flights but to Los Angeles (whose airport code is LAX). Here are the non-stop flights from SFO to LAX for December 18, 2009, for Southwest Airways according to http://www.southwest.com/. Enjoyable as it may be, rather than actually flying to LAX $n = 36$ times, we will randomly sample from the records kept by the Bureau of Transportation Statistics, which are found online at:

   http://www.bts.gov/xml/ontimesummarystatistics/src/dstat/OntimeSummaryDepaturesData.xml.

   - Open the Fathom file **OntimeCombined3LAX.ftm**.
   - Select the collection icon for **OntimeCombined3LAX.ftm** and go to **Collection>Sample Cases.** Fathom will randomly sample from the all flights in our collection from the BTS, but its sample will be $n = 10$.
   - In the **Inspector** for the sample, change the sample size to $n = 36$ and make certain that "With Replacement" is *de*selected. Click on **Sample More Cases**.
   - Get a **case table** for your sample and check that you do have $n = 36$.
   - Pull down a **Graph** from the shelf and get a dot plot for the variable *ActualDuration* for your sample.
   - Pull down a **Summary Table** from the shelf and get a sample mean and sample standard deviation for the variable *ActualDuration* for your sample of $n = 36$.

   a. What are the cases for these data? (Airlines, airports, flights, travelers, tickets?)

   b. Record the sample mean and sample standard deviation for the variable *ActualDuration* for your sample, using the correct notation.

   c. Inspect the dot plot for whether there is evidence of extreme skewness or outliers. State why, in terms of the **conditions** for using the *t* distributions and the **15/40 rule,** it is important to check for skewness and extreme outliers.

   Inspecting the sample distribution puts you in the position of almost all researchers who cannot see the shape of the population distribution. For this exercise, the "population" shows that while there is some skewness in the variable *ActualDuration,* it is not extreme. So, because of the *robustness* of the *t* procedures, we can proceed with a hypothesis test. It appears that at least Southwest flights take about eighty minutes to get from SFO to LAX. We will take this as our hypothesized mean and test whether the sample data suggest that the mean duration of flight is different than eighty minutes.

   d. Which symbol should be used for the eighty minutes: $\bar{x}$, $\hat{p}$, μ_0, σ, μ, s? Read the last sentence in the paragraph just above. Give a reason for your answer.

   e. Confused Conrad can't decide between $\bar{x}$ and $\hat{p}$ to answer question d. You say: "Neither of those!" Convince CC why these two symbols cannot be the answer to part d. (Be gentle but firm.)

   f. **Step 1:** Set up the null and alternate hypotheses for the test announced in the paragraph before part d.

   g. **Step 2:** Conditions. How are the *two* conditions for using the *t* conditions met for this test?

h. **Step 3:** Calculate the test statistic for the test. See the example in the **Notes.**

i. **Step 4:** Find an approximation to the *p* value using the **t distribution Chart** and the "little box" method. See the example in the **Notes.**

- From the shelf, pull down a **Test** box and change it from an "Empty Test" to **Test Mean.**
- Drag the variable *ActualDuration* to the "Attribute (numerical) unassigned" space.
- The **Test** box will say that the population mean of "*ActualDuration is not equal to 0*" is the alternate hypothesis. Zero here implies taking no time to get to LAX. Change zero to the number we are testing. Deselect **verbose** in the **Test** menu.

j. Check your calculation of the test statistic against what Fathom has given. Correct any errors or self-congratulate.

k. What is the *p*-value according to Fathom? Does it agree with the approximation you determined for part i? How does it agree?

l. Make a sketch of the sampling distribution and show the *p* value and test statistic.

m. Is the test *statistically significant*? Give a reason for your answer.

n. **Step 5:** Interpret the results of the hypothesis test *in the context of the data* on duration of flights to Los Angeles from the SF Bay Area.

o. According to the section **How we can read an hypothesis test from a confidence interval,** we may be able to predict the outcome of a confidence interval from an hypothesis test. For this problem, will a 95% confidence interval *include* or *exclude* the $\mu_0 = 80$ minutes? Explain why.

p. Using the sample mean $\bar{x}$ and sample standard deviation *s* from your sample of $n = 36$, calculate ("by hand" using a calculator, not software) a 95% confidence interval for the mean duration of flight μ from the SF bay Area to LAX. Show your work.

q. Forgetful Fiona still forgets and uses "1.96" in her calculation of a confidence interval. Gently tell FF what the t^* should be for the correct degrees of freedom for your sample.

- From the shelf, pull down an **Estimate** box and change it from an "Empty Test" to **Estimate Mean.**
- Drag the variable *ActualDuration* to the "Attribute (numerical) unassigned" space.
- If you wish, deselect **verbose** in the **Estimate** menu.

r. What is the margin of error for your confidence interval?

s. Check your confidence interval calculation with the Fathom output. Is $\mu_0 = 80$ minutes included in or excluded from your confidence interval? Does what happened agree with your answer to part o?

t. Give a good interpretation of your confidence interval in the context of the data.

u. Suppose we did a one-sided hypothesis test so that $H_a : \mu < 80$ instead of the two-sided test. Which of the following would be the same and which would be different. If different, how would it be different? You may use your Fathom **Test** to help your answer.
  - the test statistic?
  - the *p*-value?

v. With the alternative hypothesis for the one-sided test, $H_a : \mu < 80$, state in the *context of the hypothesis test* (which means that something has to be said about durations of flights) what a **Type I error** for the one-sided test would be.

2. Pulse Rate for Children According to http://www.medindia.net/Patients/Calculators/pulse_chartresult.asp, the average (or mean) pulse rate for children aged eight to fourteen should be eighty-four beats per minute. We have a data set from the Centers for Disease Control that is part of a national random sample of health indicators. The cases in the data set are children ages four to fourteen, but pulse rate was measured only for the children aged eight through fourteen. A different measure was used for the younger children. Our statistical question is:

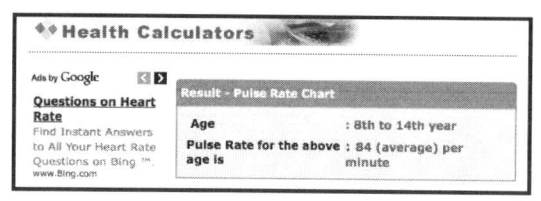

For children eight to fourteen, is the mean pulse rate equal to or different from eighty-four beats per minute?

a. Set up the null and alternate hypotheses for the hypothesis test implied by statistical question above.

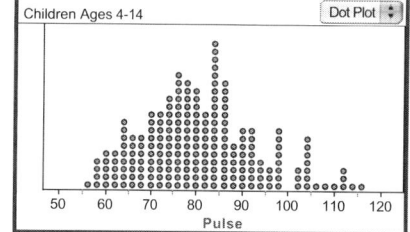

- Open the Fathom file **NHANES Children 4 – 14.ftm.**
- Pull down a **Graph** from the shelf and get a dot plot for the variable *Pulse* for our sample.
- Pull down a **Summary Table** from the shelf and get the sample mean, the sample standard deviation, and the count for the variable *Pulse* for our sample.

b. The *two* conditions for our hypothesis test are met with the NHANES sample that we have. Explain why this is. Use the graph and the information in the introduction.

- From the shelf, pull down a **Test** box and change to **Test Mean.**
- Drag *Pulse* to the "Attribute (numerical) unassigned" space.
- The **Test** box will say that the alternate hypothesis is that the population mean of *Pulse is not equal to 0.* This will be in blue, so it can be changed. Change it to what we are testing. If you wish, deselect **verbose** in the **Test** menu. You should see the box shown.

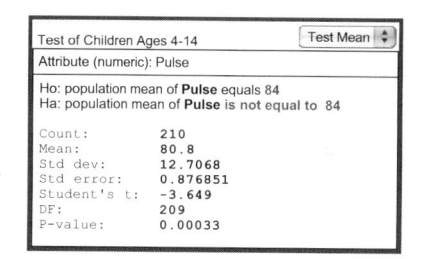

c. Show how Fathom calculated the test statistic. Make certain that you get the same answer.

d. Stat Sam locates the test statistic on the **t-distribution chart** and sees that the test statistic is "off the chart." SS says, "Ah, yes, significant!" She is right. Explain how she knows the test is significant just from the test statistic being "off the chart." (You may wish to draw a picture.)

e. Interpret the *p*-value in the context of the hypothesis test.

f. What should you do to get an *estimate* of the population mean pulse rate for children aged eight to fourteen based upon our data? After explaining what you should do, either do the calculation or get Fathom to do the calculation.

g. Interpret the results of your calculation in part f in the context of the data.

h. "Since the sample size is large, the **power** of this hypothesis test is big." Explain the meaning of the word **power** in this sentence and in the context of the data.

3. **Flying to Las Vegas** Here is another take on the duration of flights from the SF Bay Area. This time we will look at flights to Las Vegas. We have a random sample of $n = 41$ flights from the SF Bay Area to Las Vegas drawn from the Bureau of Transportation Statistics. Here is some Fathom output for this sample.

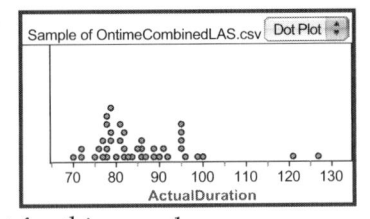

   a. We will start with getting an *estimate* of the mean population duration of flight for the flights from the SF Bay Area to Las Vegas. Do we want (I) a hypothesis test OR (II) a confidence interval? Choose one and give a reason for your choice.
   b. With the information that you have (the dot plot and what you are told about the sample), determine whether it is possible to have trustworthy results using the *t* procedures. (*Hint:* Consider **conditions** and the **15/40 rule.**)
   c. Do the calculations for what you decided in part a.. You can check your calculations by opening the Fathom file **OntimeSampleLasVegas.ftm** and getting Fathom to calculate.
   d. Give a valid interpretation of the results of what you have done.

We have been basing our hypothesis tests on the advertised durations that we happened to see online on the airline websites. The Bureau of Transportation actually records the *scheduled* duration of each flight, and the *scheduled* mean duration for all of the flights in our "population" of flights is eighty-eight minutes (rounded to the nearest minute). We will use this eighty-eight minutes as our hypothesized mean for our hypothesis test. Just because the scheduled mean is eighty-eight minutes does not mean that the *actual* duration of flight is the same. Our statistical question is:

Is the mean of ActualDuration of flights equal to or different from eighty-eight minutes?

This, incidentally, tells you that you should have calculated a confidence interval in the first part of this exercise. Your estimate should be: $82.37 < \mu_{ActualDuration} < 89.68$ minutes for the population mean duration.

   e. Set up the null and alternate hypothesis test implied by the italicized sentence above.
   f. You should be able to predict whether this hypothesis test will be significant or not from your confidence interval. (See the section entitled **How we can read an hypothesis test from a confidence interval.**) Explain your conclusion.
   g. Here is part of the Fathom output for the hypothesis test. (You can see the entire output by opening the Fathom file **OntimeSampleLasVegas.ftm** and getting Fathom to do the calculations.) Show how Fathom calculated the test statistic.
   h. Interpret the *p*-value shown here for this hypothesis test in the context of the data.
   i. Make a sketch of the sampling distribution and show on that distribution the test statistic and the *p*-value by shading. (Even before you start: will the amount of shading be big or small? Once again, you can see the answer to this question by getting **Test>Show Test Statistic Distribution.**)
   j. Does your interpretation agree with your answer to question f? Explain very briefly.
   k. Explain what a **Type II error** would be in the context of this hypothesis test. You need to say something about the mean duration of flights!

Special Exercise 4: Real Estate Data Work on Means

The scenario: You have a cousin who is a Realtor in Seattle, but she is getting tired of the long, rainy winters and she is considering moving her business to San Mateo County. She has some questions about the real estate market in San Mateo County, and she has learnt that you have a random sample of data on recent house sales in the county (you *do* have such a sample). She is asking you to use this random sample to give her some idea of the *population* of recent house sales in San Mateo County.

- Open the Fathom file **San Mateo RE Sample Y0506.ftm.** This is a simple random sample of $n = 300$.

Your cousin from Seattle says that there has been a trend in the last twenty years for houses to be bigger in square feet. Her idea is that the mean square feet for houses built in the last twenty years is greater than 2,200 square feet.

- Get a summary table for the variable *Sq_ft* and get the mean and standard deviation and the count. Also get a dot plot for the variable *Sq_ft*. This gives you the data for all the houses in your sample of $n = 300$.

1. What is the shape of the distribution? Record the mean and the standard deviation for the *entire sample* of $n = 300$, including

- With the **Collection icon** selected, use the **Object>Add Filter** to get the statistics and the dot plot for the houses built in the last twenty years. In the dialogue box for the filter, use *Age* < 21. Find the mean and standard deviation and the count for the variable *Sq_ft*. for this subsample.

2. Record the mean and standard deviation and the count (the n) of this subsample. Does this distribution have a mean higher than 2,200 square feet?

3. a. Are the analyses you have done to answer parts questions 1 and 2 *descriptive* or *inferential* analyses? Give a reason for your answer.

   b. To answer the question: "Is the mean area (in square feet) of houses built in the last twenty years in all of San Mateo County greater than 2,200 ft² or not?" should we use (I) descriptive analyses only? (II) A hypothesis test for one population proportion? (III) a chi-square test of independence? (IV) a hypothesis test for one mean? In answering this question, state why each of the techniques you have "rejected" should not be used.

4. Your cousin wants you to do a hypothesis test. Set up the null and alternate hypotheses for your cousin's question. (*Hint:* Consider what your cousin says to be a challenge to the accepted idea that houses have 2,200 square feet or less.) Will the test be a one-sided or two-sided test?

5. Check whether the conditions are met. Relate the appearance of the sample distribution to the condition that the population distribution must be Normal and to the **15/40 rule**. (Note: the sample is a SRS from the population, and taking a filter just means that your SRS is a SRS of the houses built in the last twenty years.)

6. Using the information recoded in answer to question 2, you should be able to calculate the test statistic. Show your work.

7. Use the *t* table to get an estimate for the *p* value for this test, using the "little box" procedure. Show your work.

Checking your calculations: Check whether you got the correct test statistic and *p* value.

- Drag down a **Test icon** (the scale) from the shelf, change "Empty test" to **Test mean**, and drag the attribute *Sq_ft* to the "unassigned attribute." This will give you the hypothesis test that tests the sample mean against an H_0 that the population mean square feet is *zero* unless you change the μ_0.
- In the dialogue box for **Test Mean**, anything in blue can be changed: change what is in blue to reflect your alternate hypothesis of your hypothesis test.
- If you wish, in the **Test** menu, deselect **Verbose.**
- Check the test statistic and check that the *p*-value calculated by Fathom is inside of your approximation.

8. Forgetful Fiona looks at the Fathom output and sees that the test statistic is much, much bigger than the $t = 3.128$ she calculated by hand. What do you think she forgot to do in her Fathom work?

9. Write a conclusion to the test for your cousin. Interpret the *p*-value in the context of her ideas. If you have any doubts about the validity of the test, express those as well and state why you have these doubts.

10. a. In the context of the scenario, explain what a **Type I error** would be.
    b. In the context of the scenario, explain what a **Type II error** would be.

11. Your cousin wants an *interval estimate* for the population mean size of houses built in the last twenty years.
    a. Which procedure should be used? (I) a hypothesis test for one mean? (II) a confidence interval for one proportion? (III) a confidence interval for one mean?
    b. Carry out the procedure you have selected in part a, showing your calculations. Check your answer with Fathom by following the bullets below.

- Drag down an **Estimate** (the scale) from the shelf, change "Empty test" to **Estimate mean**, and drag the attribute *Sq-Ft* to the "unassigned attribute." This will give you a confidence interval for the population mean *Sq_ft* based upon your sample mean.
- If you wish, in the **Test** menu, deselect **Verbose.**
- Check that the estimate is close to what you have calculated.

    c. What is the t^* that you used in your calculation?
    d. Confused Conrad answers 11c with 3.128. This is wrong. What is CC's confusion? (CC's confusion is a very common confusion: t and t^* are not the same!)
    e. What is the margin of error for your estimate?

12. Interpret the confidence interval in a way that makes sense to your cousin in Seattle (your answer must say something about the size of houses).

13. You think of another way to analyze your cousin's idea that the newer houses are larger. You recognize that both variables *Age* and *Sq_Ft* are quantitative. To analyze the relationship between two quantitative variables, which techniques will you use? (I) compare means, medians, standard deviations between two or more groups? (II) chi-square test of independence? (III) a scatterplot and least squares regression line, r^2? (IV) a confidence interval for the variable *Age*?

- Go back to the **Collection icon** and go to **Object>Remove Filter.** This will give you the entire sample of $n = 300$ houses and you can begin your new adventure.

[14. ***Extra Credit Adventure.*** Do a <u>complete</u> job of analyzing the *relationship* between the variables *Age* and *Sq_Ft*. Write a paragraph explaining to your cousin what you have found. (*Hint:* From your analysis, can you say that newer houses are larger? Can you say that there must be other variables besides when the house was built that affect the size of the house? What statistic informs you? What do the graphics tell you?)]

5. Flights to Hawaii: One-sided and two-sided hypothesis tests

This exercise is designed especially to show the differences between one- and two-sided hypothesis tests. It again looks at the variable *ActualDuration* for a random sample of flights. This time the random sample of flights is from San Francisco to Hawaii, and all of them are United Airlines flights.

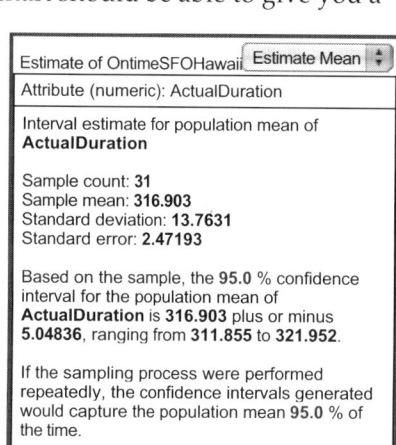

- Open **OntimeSFOHawaiiUnited.ftm** and get a **case table.**
- Get either a dot or box plot of *ActualDuration* for the $n = 31$ cases.
- Get a **summary table** showing the mean, median, standard deviation, and IQR for the sample. What you get should look like this.

  a. We will use t procedures for confidence intervals and hypothesis tests. Judging from what is said in the introduction, and considering the **15/40 rule,** can we proceed? Explain exactly why you know we can proceed.

  b. By hand, calculate a 95% confidence interval for the population mean *ActualDuration* using the information you have so far. Since $n = 31$, the t distribution chart should be able to give you a number to use for the t^*.

- From the shelf, drag down an **Estimate** icon and get the 95% confidence interval for population mean of the variable *ActualDuration*. Here is the **verbose** version of the output.

  c. Give a good interpretation of the confidence interval in the context of the variable being measured.

  d. There is a principle that the result of a *two-sided* hypothesis test can be read from whether the confidence interval *excludes* or *includes* the hypothesized μ_0; determine whether the hypothesis test whose hypotheses are:

  $H_0 : \mu = 312$ min
  $H_a : \mu \neq 312$ min   would be *statistically significant* or *not statistically significant*. Explain your reasoning in the context of the test and the confidence interval.

[e. Calculate the *test statistic* and use the result to get an approximation of the *p*–value using the t Chart.]

- From the shelf, drag down a **Test** icon and set up the hypothesis test whose null and alternate hypotheses for the population mean of the variable *ActualDuration* are $\begin{array}{l} H_0 : \mu = 312 \text{ min} \\ H_a : \mu \neq 312 \text{ min} \end{array}$. (In the Fathom **Test** box, you will have to change "is not equal to 0" to "is not equal to 312.") (Here is the *verbose* version.)

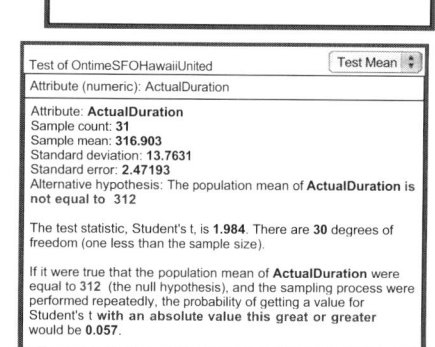

- Select the Fathom hypothesis test, go to the menu, and get **Test: Show Test Statistic Distribution.** You should have a shaded-in distribution like the one shown here.

    f. Check whether your calculations agree with the Fathom output and whether the *p*-value calculated by Fathom agrees with your answer to part d above. Do the shaded areas of the graph that Fathom gave you represent (i) *"rare"* test statistics or (ii) the *p*-value for this test?

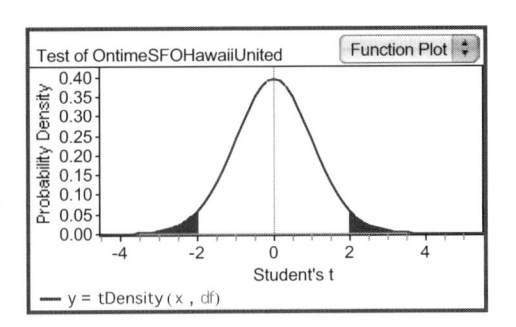

    g. Someone has the idea that the hypothesis test should be one-sided so that the null and alternate hypotheses are $H_0: \mu = 312 \text{ min}$, $H_a: \mu > 312 \text{ min}$. For this hypothesis test, will the test statistics be the same or different than the one that you calculated? Give a calculation or a reason for your answer.

    h. For the one-sided hypothesis test of part g, what should the *p*-value be? (You should be able to determine the answer by using the *p*-value that Fathom calculated for the two-sided test.)

- To get Fathom to carry out the one-sided test, go back to the **Test** box and change "is not equal to 312" to "is greater than 312." The non-verbose version of the output is shown below. Fathom should also automatically update the graph showing the *p*-value to show the *p*-value for the one-sided test.

    i. Does the fathom output agree with your calculations in parts g and h? Is the one-sided test statistically significant? Was the two-sided test statistically significant?

    j. Does the principle connecting whether the null hypothesized value is included in or excluded from confidence interval work with the one-sided test? The answer to this question should be slightly more than a single word; you should give some reason for your answer. However, the answer is given in the note below.

    k. For your two-sided test (the one that was not statistically significant), explain in the *context* of the data what a **Type II error** would be.

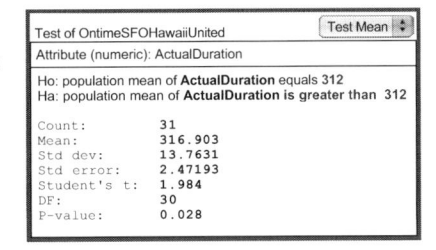

Note:

The reason that the principle connecting confidence intervals and hypothesis tests does not work for one-sided hypothesis tests is that confidence intervals are essentially "two-sided." When we calculate a confidence interval, we are saying that the random variation that is inherent in our data can occur on either side of our mean (or proportion). A one-sided test brings another preconceived idea to the test; a one-sided test says: "No, the random variation can affect only one side of the mean or proportion." A one-sided test restricts the operation of random variation to one side. Of course, the preconceived idea that provokes the one-sided test may be wrong; if so, the researcher can only be embarrassed.

§5.3 Comparing Two Measures in One Collection

Blood pressure measurements: once or twice?

When you visit your physician, you may have your blood pressure measured more than once. For some people, visiting the doctor is a scary experience, and so the first time the measurement is made, blood pressure may be elevated just because they are nervous. It makes sense to take the measurement a second time when the patient may be somewhat more relaxed. Our statistical questions are:

> *If we measure the blood pressure of a person twice, do we find a difference between the measurements that is bigger than what we would expect by sampling variation?*

> *If there is a difference (beyond just sampling variation) between the first and second measurements of a person's blood pressure, can we estimate the size of this difference?*

The first question looks as though it should be answered by a hypothesis test. If we do *not* find a significant difference between the measurements then we may conclude that, for many people, one measurement should be sufficient. However, finding a significant difference would suggest that there is something beyond sampling variation (being scared? the "white coat" effect?) that is causing the difference in the two measurements, and we would like to have an estimate of the size and direction of the difference. The second question looks like a job for a confidence interval.

A short introduction to blood pressure measurement. There are two numbers that come from the measurement of blood pressure. One number is called **systolic blood pressure**, and the second one is called **diastolic blood pressure**, and both measurements are in millimeters of mercury, or mm Hg. Normal blood pressure is expressed with both numbers in a "fraction" **systolic/diastolic** and the numbers that are ideal for adults are 120/80. For more information (especially on the medical difference between the two) see http://www.new-fitness.com/Blood_Pressure/numbers.html. When we speak of *two* measurements of blood pressure, what we mean is that we will have two values for the *systolic* and two measures for the *diastolic*. In the case table above, case 502 is a twenty-nine-year-old married female, and for her first measurement she had 130/74; for her second measurement she had 128/76. In all, there are four numbers, but the important thing for us is that there are two measures being made. This section is how we analyze such data. First we need a definition for these kinds of data.

NHANES Blood Pressure Data

	Sex	Age	Marital	Systolic1	Diastolic1	Systolic2	Diastolic2	Pulse	DiffSys2Sys1
499	Male	26	Living wi...	106	54	106	58	84	0
500	Male	28	Married	118	80	112	78	80	-6
501	Male	39	Never m...	140	74	132	68	50	-8
502	Female	29	Married	130	74	128	76	86	-2
503	Male	53	Married	142	86	144	82	54	2
504	Male	21	Never m...	114	62	118	70	60	4

> **Paired comparison data** *consist of two comparable measures within essentially one collection.*

When it makes sense to get the difference between two measures (to subtract one measure from another) then we say we have **two comparable measures**. We will usually express these two comparable measures as **two comparable variables** in a Fathom case table so they will appear as two columns. In a Fathom case table, we will end up analyzing another variable that is the *difference* of the two variables' values. As an example, it makes sense to determine whether there is a difference between the two measurements of systolic blood pressure. For case 502, the difference between the *second* measurement of systolic blood pressure and the *first* measurement is $128 - 130 = -2$. It would *not* make sense to subtract the person's *Age* from that person's *Pulse*; these two variables are *not* comparable variables.

Examples of Paired Comparison Data:
- The cases are students, and each student takes a mathematics placement test twice; it makes sense to see if the score on the second attempt differs from the first.
- The cases are places, and every month a measure of air pollution is taken; for each place, it makes sense to see if there is a change in pollution (a difference) between a later time and an earlier time.
- The cases are gas stations selling both gasoline and diesel fuel; it makes sense to find the difference between the price of regular gasoline and diesel fuel.

All of these examples involve **comparable measures** (usually expressed as variables), and in each situation, there is *one* collection of cases: *one* collection of students, *one* collection of places, *one* collection of gas stations. The definition says "essentially **one collection**"; in §5.4, we will look at comparisons between **two independent collections**. We can change the three examples above so that we would have one measure (or one variable) but from *two independent collections*.

Examples of Two Independent Collections: <u>Not</u> Paired Comparison Data:
- The cases are students, and each student takes a mathematics placement test twice; we compare male and female students for one of the tests.
- The cases are places, and every month a measure of air pollution is taken; we compare places in the northern and southern parts of a state or province for the same month.
- The cases are gas stations selling both gasoline and diesel fuel; we compare the prices of gasoline for stations in two different places.

For the first example of *two independent collections*, comparing males and females is considered two independent collections even if they are all in one data set. If random sampling has been used, the fact that a particular male is in the sample is independent of whether a particular female is there. Moreover, it would not make sense to subtract a female score on the placement test from a male score on a placement test. In the second example, the same principle operates: even though the data may have been collected together, the places chosen can be considered independent.

Answering our statistical questions

Descriptive analysis. The first thing that we should always do is to look at our sample data. As an example, we will examine the blood pressure data for the females in our sample and leave the analysis of the data for males as one of the exercises. We will look at three variables initially: *Systolic1*, which is the first measure of systolic blood pressure, *Systolic2*, which is the second measure of systolic blood pressure, and the difference between the second and first measures, calculated as *DiffSys2Sys1 = Systolic2 − Systolic1*. Shown below are the dot plots of each of these for the entire sample. (The one for the difference is on the next page.)

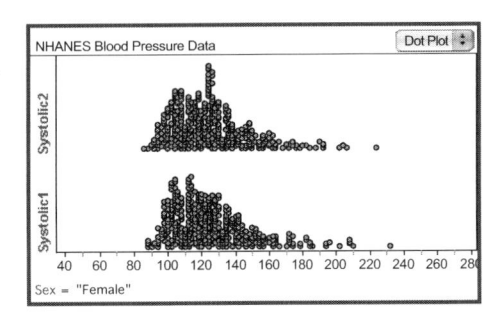

Recall that for case 502, the twenty-nine-year-old married female, the difference between the *second* measurement of systolic blood pressure and the *first* measurement is $128 - 130 = -2$, so that she is one of the dots in the stack of dots in the plot for the variable *DiffSys2Sys1*. To compare two measures in one collection, we will work entirely with the differences, so, in this instance, we will look entirely at the variable *DiffSys2Sys1*. We continue our descriptive analysis by getting the mean and standard deviation of the variable *DiffSys2Sys1*. It appears that, on average, the mean difference between the second measurement and the first measurement is $\bar{x}_{Diff} = \bar{x}_{DiffSys2Sys1} \approx -1.807$ mm Hg, and that the standard deviation is $s_{Diff} = s_{DiffSys2Sys1} \approx 6.484$ mm Hg. On average, the second measure is slightly lower than the first (by –1.81 mm Hg), but there is some variability in the differences, as measured by the standard deviation of the differences of 6.484 mm Hg. The shape of the distribution of differences between the measures shows some left skewness but has a single peak, like a Normal distribution.

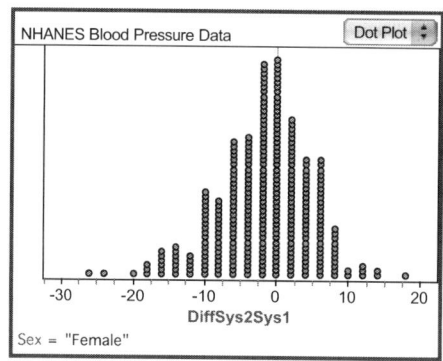

Inferential analysis. Our first statistical question was whether we find a difference between the measurements that is bigger than what we would expect by sampling variation. We shall see that the hypothesis test follows the form of the hypothesis test for one mean using the *t* procedures, although the mean that we are examining is a mean of differences; that is because we are dealing with essentially one collection. We need some notation to keep things straight for our hypothesis test.

Notation for Paired Comparison Data.		
	Sample	Population
Mean	$\bar{x}_{Diff}$	μ_{Diff}
Standard deviation	s_{Diff}	σ_{Diff}

Step 1: Setting up hypotheses: Our basic idea is that there should be *no* difference between the measurements apart from sampling variation. A challenge to that idea is that perhaps the first measure is different—maybe higher but maybe lower—than the second measure. We would therefore set up:

$$H_0 : \mu_{Diff} = 0$$
$$H_a : \mu_{Diff} \neq 0$$

Here we are using a preconceived idea that the mean difference in measurements is *zero*. We would normally indicate the hypothesized value for a hypothesis test with the symbol μ_0, so here we are saying that our preconceived idea is that $\mu_0 = 0$. It is possible that we would have a preconceived idea for the difference that is not zero (no difference), but the most usual test tests the idea of no difference against the idea that the difference is not equal to, or greater than, or less than zero. In symbols, for the alternate hypothesis we will have either $H_a : \mu_{Diff} \neq 0$ or $H_a : \mu_{Diff} > 0$ or $H_a : \mu_{Diff} < 0$ and for the null hypothesis: $H_0 : \mu_{Diff} = 0$.

Step 2: Checking the conditions: Just as the hypotheses followed the pattern of the one mean test, so do the conditions but with respect to the *differences*. Recall that we have two conditions; one is that data be a random sample, and the second is that the population (of differences) be Normal. However, the Normality condition can be relaxed if the sample size is sufficiently large, according to the 15/40 rule. Our data are a random sample, and the sample size ($n = 322$) is large enough that the *t* procedures are *robust* with the amount of skewness we see in the sample data. We can proceed.

Steps 3 and 4: Calculating the test statistic, getting the p-value: Since we are using the *t* procedures, the test statistic is a *t* score calculated with the data on the differences between the two measurements. Hence, using $\bar{x}_{Diff} = -1.80745$ and $s_{Diff} = 6.48412$,

$$t = \frac{\bar{x}_{Diff} - 0}{\frac{s_{Diff}}{\sqrt{n}}} = \frac{-1.80745 - 0}{\frac{6.48412}{\sqrt{322}}} \approx \frac{-1.80745}{0.361346} \approx -5.002$$

```
Test of NHANES Blood Pressure Data        Test Mean
Attribute (numeric): DiffSys2Sys1

Ho: population mean of DiffSys2Sys1 equals 0
Ha: population mean of DiffSys2Sys1 is not equal to  0

Count:              322
Mean:               -1.80745
Std dev:            6.48412
Std error:          0.361346
Student's t:        -5.002
DF:                 321
P-value:            < 0.0001

Sex = "Female"
```

The value of $t = -5.002$ looks extreme (the *t* distribution for $n = 322$ is not too different from the Normal distribution). Checking the **t Distribution Chart** for $df = 250$ shows that $t = -5.002$ is off the chart, indicating a very small *p*-value.

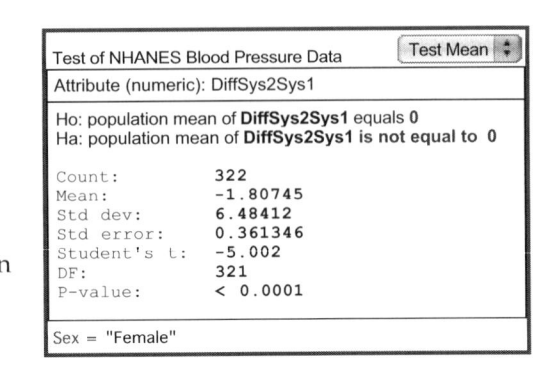

The Fathom output is shown here; the Fathom command for this test is **Test Mean**, which is the same command as for one mean. That the command is the same as for one mean makes sense, since the computations are the same.

Step 5: Interpreting the Result It does appear that there is a tendency for the first reading of blood pressure to be higher—making the difference between the second and first measures negative. With our sample size, we are confident that this result is not just sampling variability. On the other hand, the difference is not large, being less than 2 mm Hg on average. We can go on and get an estimate of the mean size of the difference. For that we need a confidence interval.

Calculating and Interpreting a Confidence Interval Once again, the *t* procedures will follow the pattern of the *t* procedures for one mean. Using $\bar{x}_{Diff} = -1.80745$ and $s_{Diff} = 6.48412$, and $t^* = 1.969$ for $df = 250$, we have

$$\bar{x}_{Diff} \pm t^* \frac{s_{Diff}}{\sqrt{n}} = -1.80745 \pm 1.969 \frac{6.48412}{\sqrt{322}} \approx -1.80745 \pm 0.71149$$

and this leads to a confidence interval of $-2.52 < \mu_{Diff} < -1.10$ mm Hg. Here is the Fathom output for the confidence interval, which, once again, uses the procedures for a single mean.

As expected, the two measures *exclude* the null hypothesized $\mu_0 = 0$ since the associated two-sided hypothesis test was statistically significant. The numbers are negative because the *first* measure *Systolic1* is higher than *Systolic2* for many people.

```
Estimate of NHANES Blood Pressure Data     Estimate Mean
Attribute (numeric): DiffSys2Sys1

Interval estimate for population mean of DiffSys2Sys1

Count:              322
Mean:               -1.80745
Std dev:            6.48412
Std error:          0.361346
Confidence level:   95.0 %
Estimate:           -1.80745 +/- 0.710905
Range:              -2.51836 to -1.09655

Sex = "Female"
```

We can certainly say with 95% confidence that we expect (for females) the second measure of systolic blood pressure to be about 1.8 mm Hg lower than the first measure, with a margin of error of about 0.72 mm Hg. However, the numbers are quite small; our hypothesis test shows that the difference does not stem from random sampling variation—from chance variation—but the difference is small nonetheless. The lesson is that statistically significant differences are not necessarily practical differences.

Statistical significance and practical significance Hypothesis testing tests whether what we actually see in our data could have been "merely" the product of the random variation. Hypothesis testing does not indicate the importance of what we have seen in the context of our data. Is a difference of 2 mm Hg an important difference? It may not be an important difference in the context of measuring blood pressure. The result may lead us to look for the kinds of people or the kinds of situations that produce bigger differences. For example, it may be that the "routine" testing of the NHANES survey is different from the situation that people face when they "go to see their doctor." Or it may be that we could examine the NHANES data to find the characteristics of people for whom there is a big difference between the second and first measurements.

Creating Pairs: Matched Pair Designs

Often, data come with comparable measures "built in," as in the examples shown at the beginning of this section. But not always. When it is impossible to have a paired comparison, it is sometimes possible to approximate a paired comparison by making what are almost pairs. Matched pair designs are quite common in medical studies, where the idea can be called a **case control study.** Here are two examples of a matched pair design, one a medical study and the other involving houses for sale.

Does using a cell phone cause cancer? If you think that there is a possibility that using cell phones leads to certain kinds of cancer, it is obviously impractical (and unethical) to create an experiment where people are randomly assigned to a group in which each person is given a cell phone and told to use the cell phone as much possible or to a second group that would be prohibited from ever using a cell phone. Then wait (years, perhaps?) to see whether the incidence of cancer is higher in the cell phone group. Instead, what is more likely to be done is to observe the cell phone usage of those who already have cancer and compare it to the cell phone usage to those who do not have cancer. However, in medical studies of this kind, the cancer cases are *matched up* by sex, age, and other variables to people who have cancer. Here is how the matched pair or case controls is described in one such study[7] that was carried out in two parts of the UK on the relationship between risk of glioma (a type of brain cancer) and cell phone usage:

> ...one control per case was individually matched on age, sex and general practice after the patient with glioma had been interviewed. Nonparticipating controls were replaced.

This means that a patient without glioma (a control) was chosen so that the control patient was the same age and sex as the glioma patient and had the same physician—and that if the "control" decided not to participate in the study, another was found. That the cases and controls had the same general practitioner meant that they probably lived in the same area. The idea was to control at least three

---

[7] Hepworth, Sarah J, *et al*. Mobile phone use and risk of glioma in adults: case control study *British Medical Journal,* 2006, 332:883–887

variables that might have some effect on the development of glioma. The matching process can be quite complicated or it can be quite simple.

Example of Matched Pairs: Location, Location, Location The average sale price of houses for sale in two different places may be different for any number of reasons. One way of assessing the effect of location on the price of a house is to compare houses that are alike (or as alike as we can get them) in all respects *except* location. Here is a very crude attempt to do just that. Houses in Menlo Park and houses in Palo Alto that were part of a sample of the houses for sale in 2007–2008 were arranged in rank order by the measure of the size of the house, *SqFt*. Then the biggest house—the one with biggest *SqFt*—in Palo Alto was matched (or paired) with the biggest house in Menlo Park, the second-largest house in Palo Alto paired with the second-largest house in Menlo Park, and so on, until the smallest house in Palo Alto was matched with the smallest house in Menlo Park. Our matching procedure is extremely crude; it would be better to consider other variables, such as the number of bedrooms, the number of bathrooms, the age of the house, in making the matched pairs. However, even with this crude matching, what we have done is to calculate, for each matched pair of houses—one in Palo Alto, one in Menlo Park—the difference in sale prices. The plot here shows the distributions of the areas of the houses (measured by the variable *SqFt*) for Palo Alto and Menlo Park. (The top distribution is for Palo Alto, and the second distribution is for Menlo Park.)

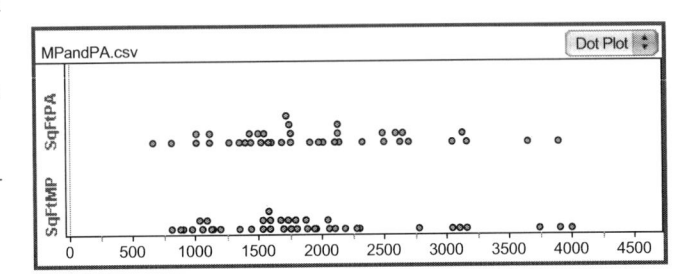

The second plot compares the sale prices of the houses in the two places; again, the first distribution is for Palo Alto, and the lower one is for Menlo Park.

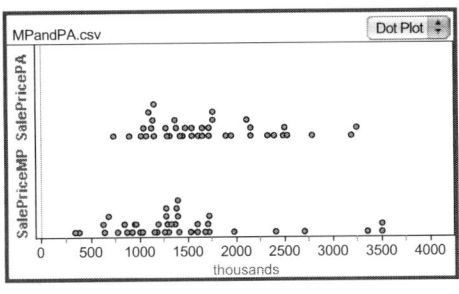

Now that we have the data matched, the variable that we will calculate for each matched pair is *DiffSalePrice = SalePricePA − SalePriceMP*. That is, we will calculate the difference in price between two houses with very similar sizes. It is for this variable (the difference between two measures) that we will calculate a confidence interval. The confidence interval for the variable: *DiffSalePrice = SalePricePA − SalePriceMP* with the calculations done by Fathom are shown here. Using the formula for paired comparisons, and t^* for $df = 39$

$$\bar{x}_{Diff} \pm t^* \frac{s_{Diff}}{\sqrt{n}} = 271507 \pm 2.023 \frac{610256}{\sqrt{40}}$$

$$\approx 271507 \pm 195169$$

The interval does not include zero; if we were doing an hypothesis test where the null hypothesis was that there was no difference in prices for the same size houses then our $\mu_0 = 0$ would be rejected. Hence, we see that there *is* a significant difference in price between Palo Alto and Menlo Park.

Summary: Paired Comparison Data

- The formulas for inference for paired comparison data are essentially the same as the formulas for one mean, but the variable that is being analyzed is the difference of *two comparable measures* for *one collection* of data.
- The **notation for paired comparison data** reflects the fact that the variable being measured is the difference between two measures (or variables).

	Sample	**Population**
Mean	$\bar{x}_{Diff}$	μ_{Diff}
Standard deviation	s_{Diff}	σ_{Diff}

- ***Paired comparison*** data may either come from:
  - One collection in which there are comparable measures whose difference may be calculated, or
  - Two collections in which the cases have been matched, usually because the matched cases have similar or the same values for one or more variables. *Paired comparisons* that have been created in this fashion are called ***matched pairs***.

Summary: Hypothesis Test for Paired Comparison Data

Step 0: Do a descriptive analysis of the two quantitative variables in the pair and also the variable that represents the difference between the variables for each pair.

Step 1: Set up the null and alternate hypotheses for the mean difference μ_{Diff} using an idea for the population mean difference. It is very common to have $\mu_0 = 0$, indicating no mean difference between the measures, but it is possible to test the hypothesis using a null hypothesis with a mean that is something other than zero. (Here, on the left, $\mu_0 = 0$ is shown—on the right, something other than zero; μ_{Diff_0} can be some other number.)

$$H_0 : \mu_{Diff} = 0 \qquad H_a : \begin{cases} \mu_{Diff} \neq 0 \\ \mu_{Diff} < 0 \\ \mu_{Diff} > 0 \end{cases} \qquad H_0 : \mu_{Diff} = \mu_{Diff_0} \qquad H_a : \begin{cases} \mu_{Diff} \neq \mu_{Diff_0} \\ \mu_{Diff} < \mu_{Diff_0} \\ \mu_{Diff} > \mu_{Diff_0} \end{cases}$$

Step 2: Check the conditions for a trustworthy hypothesis test.
  5. The sample must be a simple random sample.
  6. The population must be Normal; however, since the t procedures are robust for large sample sizes, judge whether to proceed using the 15/40 rule.

Step 3: Calculate the test statistic $t = \dfrac{\bar{x}_{Diff} - \mu_{Diff_0}}{s_{Diff}/\sqrt{n}}$ using the sample mean difference $\bar{x}_{Diff}$, sample standard deviation for the differences s_{Diff}, and the hypothesized μ_{Diff_0}.

Step 4: Calculate the p-value by getting $P\left(\bar{x}_{Diff} \text{ more extreme} \mid \mu_0 = \mu_{Diff_0}\right)$ using the absolute value of the test statistic t calculated in Step 3.

Step 5: Evaluate the evidence that the p-value and the test statistic give you to determine whether your test successfully challenges the null hypothesized population mean μ_{Diff_0} or not. Give an interpretation in the context of the data using the terminology of "statistical significance" and "rejecting the null hypothesis."

Summary: Confidence Interval for Paired Difference Data

$$\bar{x}_{Diff} \pm t^* \frac{s_{Diff}}{\sqrt{n}}$$

where t* depends on the df = n − 1 and the level of confidence. The conditions for trustworthy confidence interval are the same as for a hypothesis test.

§5.3 Exercises on Paired Comparisons

1. **More Blood Pressure Data and Other Analyses** This exercise will investigate the second component of blood pressure, **diastolic blood pressure,** but it also reviews what we have done in the past weeks and even before that. Expect to reach back into what we have done before. Since in the **Notes** we only looked at the data for females, this exercise will also look only at females. The sample is a random sample of the NHANES data, which itself can be regarded as a random sample of the American population.

   a. What are the cases for this collection?

   - Open the Fathom file **NHANES Blood Pressure.ftm** and get a **case table.**
   - With the **Collection icon** for **NHANES Blood Pressure.ftm,** go to **Object>Add Filter.** In the dialogue box, type: Sex = "Female". This filter will mean that all of the analyses will be for females.
   - From the shelf, bring down a **Graph** and get a dot plot of the variable Diastolic1.
   - Drag the variable Diastolic2 to a small ✚ at the base (the horizontal axis) of the graph, which allows you to put both dot plots on the same graph.
   - Get a new **Graph** and get a dot plot of the variable DiffDist2Dist1. This variable is the difference between the second and first measurements: DiffDist2Dist1=Diastolic2 − Diastolic1.
   - In the dot plot of DiffDist2Dist1=Diastolic2 − Diastolic1, find the column of dots where DiffDist2Dist1 = − 6.
   - Highlight the top-most dot on the plot of DiffDist2Dist1 for the value of DiffDist2Dist1 = − 6 so that the dot turns red and notice that two red dots appear on the dot plots of Diastolic1 and Diastolic2. In the lower left-hand corner of your screen, the values of the red dots should appear when your cursor is on them.

   b. In the context of the data, what does DiffDist2Dist1 = − 6 mean? Back up your answer by getting the values of Diastolic2 and Diastolic1 for this case that you have selected.

   c. Read the definition of **paired comparison data** in the **Notes.** Explain how the variables Diastolic2 and Diastolic1 fit into that definition.

   - Get a **summary table** with the sample mean, the sample standard deviation, and the count for DiffDist2Dist1. The summary table should look like this.

NHANES Blood Pressure Data	
	DiffDis2Dist1
	−0.36646
	4.98745
	322
S1 = mean ( )	
S2 = s ( )	
S3 = count ( )	

   d. Write the numbers in the **summary table** using the correct notation for paired comparisons. (You should have "Diff" subscripts.)

   e. We want to have an estimate of how *much* the second measure of diastolic blood pressure differs from the first for the population of people from which our sample is drawn. What inferential procedure should we use? Give a reason for your answer.

   f. Are the conditions met for the inferential procedure you have chosen? Or do you have doubts about one or more of the conditions? Be complete.

   g. Confused Conrad, when he answers the question about conditions, writes: "Since $0.36646 * 322 = 118 > 10$, we know that the sample size is large enough and the conditions are met." CC has made at least two errors. First, CC is confused about what procedure he is using,

400

and, secondly, CC has also forgotten to consider one of the conditions that need to be considered. Explain what his two mistakes are. (The two mistakes are common ones.)

h. Use the information in the summary table to calculate the kind of inferential procedure that you decided upon in part e.

- Check your calculation with Fathom by dragging down an **Estimate** box from the shelf. (This should give you an idea of the answer that was expected to part e.) It should be an "Empty estimate." Change the empty estimate to the appropriate estimate that we want.

i. What is the margin of error for the estimate?

j. What is the standard error for the estimate? The answer to this question should not be the same as the answer to part i.

k. Does your confidence interval include zero? What does this say about the outcome of the hypothesis test where $H_0 : \mu_{Diff} = 0$ and $H_a : \mu_{Diff} \neq 0$? Specifically, will the hypothesis test be statistically significant or not? Give a reason for your answer.

l. What does the null hypothesis $H_0 : \mu_{Diff} = 0$ mean in the context of the measuring blood pressure?

m. For the hypothesis test for the variable where *DiffDist2Dist1* and $H_0 : \mu_{Diff} = 0$ and $H_a : \mu_{Diff} \neq 0$, calculate the test statistic.

n. Use the **t Distribution Chart** to get an estimate of the *p*-value.

- Check your calculation of the test statistic with Fathom by dragging down a **Test** box from the shelf. It should be an "Empty test." Change the empty test to the appropriate test that we want.

o. Does the *p*-value agree with your calculation from the *t* chart done in part n? Is the test statistically significant? Do we reject the null hypothesis?

p. Interpret the outcome of your hypothesis test and the information from the confidence interval in the context of measuring blood pressure. (Your answer to part l should help you.)

The website http://www.new-fitness.com/Blood_Pressure/numbers.html says that the optimal blood pressure should be 120/80 or less, and "normal blood pressure" is 130/85, where the convention is that the "fraction" refers to **systolic/diastolic.** We will use these as the standard and ask the question:

Do American women (or men) have optimal (or normal) blood pressure?

In our data we have the average of all the systolic measurements in the variable *Systolic* and the average of all the diastolic measurements in the variable *Diastolic*. You have two choices to make: whether to look at systolic or diastolic and whether to look at "Optimal" or "Normal" blood pressure. (If you choose to analyze the data on males, you will go to the **Add Filter** in the **collection icon** and changing Sex = "Female" to Sex = "Male".)

q. Decide whether you will use a confidence interval or a hypothesis test to answer the statistical question. Give a reason for your answer. Your answer should look forward to the possible results that you can get. (It is possible to answer our statistical question with either a confidence interval or a hypothesis test.) **PTO**

r. If you decide on an hypothesis test, clearly state your null and alternate hypotheses, do some descriptive work to look at the distribution of *Systolic* or *Diastolic*, get the sample mean and sample standard deviation, and calculate the test statistic (using the correct notation). You may use Fathom to aid you, including getting the test. **OR**

r'. If you decide on a confidence interval, do some descriptive work to look at the distribution of *Systolic* or *Diastolic*, get the sample mean and sample standard deviation, and calculate the confidence interval (using the correct notation). You may use Fathom to aid you, including getting the test.

s. Clearly, in a short paragraph, state how your analysis answers the statistical question.

2. **City and Highway Miles per Gallon (MPG)** This exercise uses a random sample of the Federal Government's Guide to Fuel Economy for 2008 cars. For most cars (but not all), highway MPG is bigger than city MPG, but how much bigger? In this exercise you will answer the statistical question:

    How much bigger on average is highway MPG than city MPG?

- Open the Fathom file **FuelEconomySample2008.ftm** and get a **case table.**
- From the shelf, bring down a **Graph** and get a dot plot of the variable *CityMPG*.
- Drag the variable *HwyMPG* to a small **+** at the base (the horizontal axis) of the graph. This will allow you to put both dot plots on the same graph.

    a. What are the cases for these data?

    b. Read the definition of **paired comparison data** in the **Notes**. Explain how the variables *CityMPG* and *HwyMPG* fit into that definition.

- In the **case table,** go to the right end of the variable list and create a new variable called *DiffHwyCity*.
- Select the new variable, go to **Edit>Edit Formula,** and in the dialogue box type in *HwyMPG − CityMPG*.
- Get a new **Graph** and get a dot plot of the variable *DiffHwyCity*.
- In the dot plot of *DiffHwyCity* find the column of dots where *DiffHwyCity* = 6.
- Highlight the top-most dot on the plot of *DiffHwyCity* for the value of *DiffHwyCity* = 6 so that the dot turns red and notice that two red dots appear on the dot plots of *HwyMPG* and *CityMPG*. In the lower left-hand corner of your screen, the values of the red dots should appear when your cursor is on them.

    c. In the context of the data, what does *DiffHwyCity* = 6 mean? Back up your answer by getting the values of *HwyMPG* and *CityMPG* for this case that you have selected.

- Get a **summary table** with the sample mean, the sample standard deviation, and the count for *DiffHwyCity*. The summary table should look like this.

Sample of 2008FuelEconomyGuide
DIFFHWYCITY
6.35313
2.26261
320
S1 = mean ( )
S2 = s ( )
S3 = count ( )

    d. Choose a good inferential procedure to answer the question in italics in the introductory paragraph to this exercise. Give a reason for your choice.

e. Are the conditions met for the inferential procedure you have chosen? Or do you have doubts about one or more of the conditions? Be complete about your judgment.

f. Use the correct notation for the sample mean and sample standard deviation for a paired comparison.

g. Do the calculation for what you decided in part d.

- Use Fathom to check your calculations; your numbers should be very close but not exactly Fathom's because your t^* from the **t Distribution Chart** will be slightly different.

h. Interpret the results in the context of the data. (You need to say something about fuel economy in the city and on the highway.)

[Parts i and j are review.]

i. Suppose the government decreed that the average HwyMPG for pickup trucks should be 19 MPG. That is the standard. You suspect that the average is less than that, and you have these sample data. Is it true that pickup trucks average 19 MPG or is it less, as you think? Carry out an appropriate inferential procedure, showing all steps.

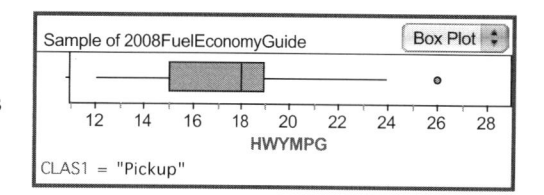

j. Is what you have just done a "paired comparison" problem? Give a reason for your answer.

3. **Cost of Fuel** [Fathom] This exercise uses data collected on prices of gasoline and diesel fuel by the Energy Information Administration. The data are collected each week from gas stations (and other sellers), and the data we have start from 1994 and run to 2008, although some of the data collection started later in 1994. We can regard the data as randomly sampled from the sellers of fuel. Generally, there are three grades of gasoline: regular, mid-grade, and premium, and we expect that each higher grade will cost more than the grade below. The relationship between the price of diesel fuel and gasoline is more difficult to predict. At times it appears to be higher, at times lower. (The Energy Information Administration is a US government agency; see: http://www.eia.gov/dnav/pet/pet_pri_gnd_dcus_nus_w.htm.)

- Open the Fathom file **FuelPrices.ftm** and get a **case table.** (It should resemble the table below, but the data on mid-grade and premium did not start until November 1994; scroll down in the Fathom file.)

Each row in the case table is a date for a *week* and what is shown are the prices for the different types of fuel for that week. The weeks actually range from March 28, 1994, to November 24, 2008.

The first four variables give the price for that week of regular, mid-grade, premium, and diesel. The prices are in cents. The last four variables represent the *differences* in cost on a specific date between different types of fuel for a specific week: "diesel and premium," etc. Here are the definitions:

DieselPrem = Diesel − Premium   DieselReg = Diesel − Regular
PremReg = Premium − Regular   MidReg = MidGrade − Regular

PTO

a. What are the cases for these data? (Do not confuse the variables with the cases. What are the rows?)
b. Give a reason that these data are paired comparison data
c. From your experience (and before looking at the data), what is your *guess* about the average difference between mid-grade gasoline and regular gasoline: ten cents, nine cents, eight cents, twelve cents? If we do a hypothesis test to test your idea with these data, what symbol should be used for your idea?
d. Set up the null and alternate hypotheses to test your idea.
- e. Use Fathom to do a descriptive analysis of the data to determine whether the conditions are met for the hypothesis test. The conditions are met, but you need to say exactly *why* they are met.
- f. Use Fathom to get the sample mean and sample standard deviation for the relevant variable and assign to these the correct symbols.
- g. Calculate the test statistic by hand and use it to approximate a *p*-value. Then use Fathom to check your calculations.
h. Interpret the hypothesis test in light of your hypotheses. Do the data shoot down your idea about the difference between regular and mid-grade, or are the data consistent with your idea?
i. Explain what a **Type II error** would be in the context of your guess for the population mean difference.
j. We want to get an estimate of the mean difference between diesel and regular. Which inferential procedure will we use (CI or HT), and which variable will we use?
k. Check the dot plot of the distribution of the variable you chose in part j. This is always a good practice. However, why can we proceed with the analysis whatever shape the dot plot has?
l. Either by hand or using Fathom, carry out the procedure you chose in part j and interpret your results in the context of the data.

§5.4 Comparing Means: One Measure, Two Collections

Blood pressures for two independent collections

To provide some continuity with the last section, but also to *contrast* this section with the last one, our examples will continue to be about blood pressure. Our initial statistical question is:

Is there a difference between men and women in average (or mean) systolic blood pressure?

We have **one quantitative variable**—the variable *systolic* blood pressure— but now we are comparing two groups or **two collections:** males and females. In the last section (**Paired Comparison**), we analyzed *two measures* but in *one collection*. The two collections we are comparing must be **independent samples.** If the groups that we compare are *parts* of a random sample, such as the males and females in our sample here (where we are analyzing blood pressure), then the groups *are* independent. Parts of random samples are independent. This situation of independent groups—or collections—is quite different from what we analyzed in §5.3. In §5.3 we had two measures or variables for just one group. Here we have *two* independent groups but just *one* measure; we have *one* variable that is measured in each of the groups. The situation here is the same that we encountered for the difference of proportions for two independent groups or collections.

Our statistical question is the kind (an "Is there...?" type of question) that should be handled by a hypothesis test, and so we have a situation comparable to the situation when we compared two proportions. As in that situation, we need notation for all of the numerical quantities we will encounter.

Notation for Comparing a Quantitative Variable between Two Collections:

	Sample Size	Population Mean	Sample Mean	Population Standard Deviation	Sample Standard Deviation
Collection 1:	n_1	μ_1	$\bar{x}_1$	σ_1	s_1
Collection 2:	n_2	μ_2	$\bar{x}_2$	σ_2	s_2

As before, we will replace the numbered subscripts with more informative letters that relate to the context, but we need the numbered subscripts to write formulas in general. So we will write null and alternate hypotheses using μ_F and μ_M rather than μ_1 and μ_2; the important thing is that we need to have distinct names for all the quantities we use.

Step 0: Some Descriptive Analyses We will restrict our statistical question to men and women between the ages of twenty and thirty. Our question now becomes:

Is there a difference between men and women in mean systolic blood pressure (for people between the ages of twenty and thirty)?

From the box plot of the distributions of *Systolic* by *Sex*, the distributions look nearly symmetrical and from the summary table showing means and standard deviations, it does appear that women in this age group have lower levels of blood pressure than men. We see that, using our notation: $\bar{x}_F = 108.544$, $s_F = 10.9601$, $n_F = 79$ $\qquad \bar{x}_M = 119.873$, $s_M = 10.6609$, $n_M = 71$

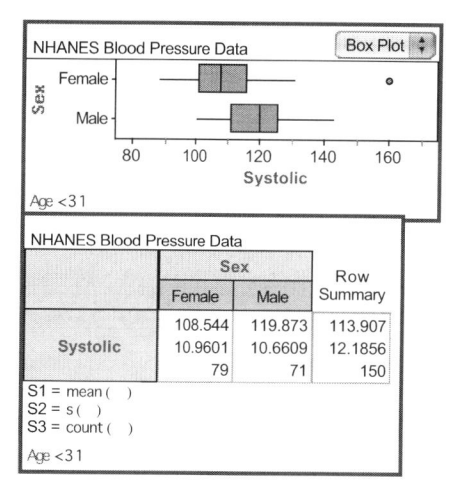

Step 1: Null and Alternate Hypotheses The question we originally asked implies a two-sided alternate hypothesis, since we made no prediction about the direction of the difference between men and women. (It would be bad practice to base the test on what we happen to see in the data at hand.) So we have:

$$H_0 : \mu_F - \mu_M = 0$$
$$H_a : \mu_F - \mu_M \neq 0$$

These hypotheses can also be expressed as $H_0 : \mu_F = \mu_M$, $H_a : \mu_F \neq \mu_M$, but it is a good idea to get accustomed to thinking of the difference as being zero (or not zero) since that is how we will calculate the test statistic. In words, our test is based upon the null hypothesis that the mean *systolic* blood pressure does *not* differ between men and women in the age group twenty to thirty years.

Step 2: Checking the Conditions We will again be using the *t* procedures, and the conditions for the *t* procedures for *two* collections are similar to the conditions we have seen before:

> - The samples must be random samples independently drawn from two populations.
> - The population distributions must both be Normal.

The words "independently drawn" mean that we cannot have data that are paired up in some way. The samples of men and women cannot be from a matched pair design, for example. The men and women cannot be (for example) husbands and wives or brothers and sisters. Our NHANES data are a random sample, and so within that sample, the men and the women can be regarded as independently drawn random samples.

The second condition we have seen before, and we have seen that the *t* procedures are *robust* if the sample sizes are large enough; to decide what is "large enough," we have the **15/40 rule**. For comparing means, we can use the *15/40 rule* with the sum of sample sizes $n_1 + n_2$. So here, although both $n_F = 79$ and $n_M = 71$ are each over 40, what is important is $n_F + n_M = 79 + 71 = 150$. In any case, in our example, the sample distributions of the variable *Systolic* are both nearly symmetric, with one outlier. We can safely proceed.

> **Conditions for one measure, two independent collections comparisons**
> - The samples must be simple random samples independently drawn from two populations.
> - The populations must have Normal distributions, however, since the t procedures are robust for large sample sizes. Use the 15/40 rule but with the sum of sample sizes $n_1 + n_2$.

Step 3: Calculating the test statistic The *form* of our test statistic should be:

$$\frac{(\text{Difference of Sample Means}) - (\text{Hypothesized Difference of Population Means})}{\text{Standard deviation of Sampling Distribution for Difference of means}}$$

If we knew the population standard deviations, this general formula would be a z score:

$$z = \frac{(\bar{x}_1 - \bar{x}_2) - (\mu_1 - \mu_2)}{\sqrt{\frac{\sigma_1^2}{n_1} + \frac{\sigma_2^2}{n_2}}}$$

However, researchers generally do not know the values of the population standard deviations σ_1 and σ_2 while researchers *do* know the values of the sample standard deviations s_1 and s_2. Hence, substituting the sample values for the population values, and therefore using the *t* distributions, the **test statistic** will be:

$$t = \frac{(\bar{x}_1 - \bar{x}_2) - (\mu_1 - \mu_2)}{\sqrt{\frac{s_1^2}{n_1} + \frac{s_2^2}{n_2}}}$$

So for our example we will have:

$$t = \frac{(\bar{x}_F - \bar{x}_M) - 0}{\sqrt{\frac{s_F^2}{n_F} + \frac{s_M^2}{n_M}}}$$

$$= \frac{(108.544 - 119.873) - 0}{\sqrt{\frac{(10.9601)^2}{79} + \frac{(10.6609)^2}{71}}}$$

$$\approx \frac{-11.329}{\sqrt{1.52055 + 1.60077}}$$

$$\approx \frac{-11.329}{1.76673}$$

$$\approx -6.412$$

In the formula, the zero after the $(\bar{x}_F - \bar{x}_M)$ indicates that our null hypothesis of *no* difference in the population means between men and women; that is $H_0 : \mu_F - \mu_M = 0$.

A value of *t* = -6.412 looks fairly extreme, and we think that the value of the test statistic shows that *if* the null hypothesis were true and there were really no difference between the mean *Systolic* blood pressure for men and women then what we have seen in our sample is very rare just by sampling variation—so rare as to prompt us to doubt the null hypothesis of no difference.

Step 4: Getting a p-value. However, it turns out that the sampling distribution of the *test statistic* shown above does *not* follow a *t* distribution. Fortunately, the *t* distributions can be used with this test statistic with the appropriate choice of **degrees of freedom.** We have two choices of degrees of freedom. One of these choices is appropriate for working by hand. This is **Choice 1:** the degrees of freedom will be calculated by taking the smaller of the two $df = n - 1$ for our two samples. For our example we will have $df = n - 1 = 71 - 1 = 70$. **Choice 2**, used by software such as Fathom, calculates the

degrees of freedom by the formula: $df = \dfrac{\left(\frac{s_1^2}{n_1} + \frac{s_2^2}{n_2}\right)^2}{\frac{1}{n_1-1}\left(\frac{s_1^2}{n_1}\right)^2 + \frac{1}{n_2-1}\left(\frac{s_2^2}{n_2}\right)^2}$. When this formula is used, the

result is often not an integer. This formula looks formidable to calculate, but it is not at all a problem for software and is actually the better choice. The disadvantage of *Choice 1* is that for small sample sizes, the potential power of the hypothesis test can be low.

Here is the Fathom output for this hypothesis test. Notice that the degrees of freedom is $df = 147.062$ and

```
Test of NHANES Blood Pressure Data                Compare Means
First attribute (numeric): Systolic
Second attribute (numeric or categorical): Sex

Ho: Population mean of Systolic for Female equals that for Male
Ha: Population mean of Systolic for Female is not equal to that for Male

                Female      Male
Count:          79          71
Mean:           108.544     119.873
Std dev:        10.9601     10.6609
Std error:      1.2331      1.26522

Using unpooled variances
Student's t:    -6.412
DF:             147.062
P-value:        < 0.0001

Age < 31
```

407

notice that the *p* value is very small, as we predicted it would be.

Step 5: Interpretation Even if we had gone with Choice 1 for the calculation of the degrees of freedom, the test statistic $t = -6.412$ is "off the chart," and so the *p* value is clearly less than 0.0005. Hence, our test is *statistically significant* and we have evidence that there *is* a difference in the mean *Systolic* blood pressure for men and women. From the evidence that we have, it appears that women between the ages of twenty and thirty have, on average, lower *systolic* blood pressure. So, then, the question is: how much lower is the mean for the women than for the men? We need an estimate, and so we will calculate a confidence interval.

Confidence Intervals: One measure with two collections

The *form* of a confidence interval should be predictable. In general, we have been calculating

$$(\text{Sample Estimate}) \pm t^*(\text{Standard Deviation of the sampling Distribution})$$

and so we should be able to predict, given what we have seen for the hypothesis test, that the formula will be

$$(\bar{x}_1 - \bar{x}_2) \pm t^* \sqrt{\frac{s_1^2}{n_1} + \frac{s_2^2}{n_2}}$$

For our example, if we have Choice 1, where the $df = n - 1 = 71 - 1 = 70$, we see that for a 95% confidence interval, we should have $t^* = 1.994$. So we will calculate the confidence interval for the population mean difference in *systolic* blood pressure between men and women—that is, our estimate of $\mu_F - \mu_M$ as shown below.

$$(\bar{x}_F - \bar{x}_M) \pm t^* \sqrt{\frac{s_F^2}{n_F} + \frac{s_M^2}{n_M}} = (108.544 - 119.873) \pm 1.994 \sqrt{\frac{(10.9601)^2}{79} + \frac{(10.6609)^2}{71}}$$

$$\approx -11.329 \pm 1.994 \left(\sqrt{1.52055 + 1.60077} \right)$$

$$\approx -11.329 \pm 1.994 (1.76673)$$

$$\approx -11.329 \pm 3.523$$

We have a **margin of error** of 3.523 mm Hg, and this can be written as an interval as $-14.852 < \mu_F - \mu_M < -7.806$. We can say, with 95% confidence, that women in the age group twenty to thirty have mean *systolic* blood pressure that is between 7.8 mm Hg and 14.9 mm Hg lower than the mean *systolic* blood pressure for men in that age group. The signs are negative because we calculated $\bar{x}_F - \bar{x}_M$, and the sample mean for men was higher than the sample mean for women. Here is the Fathom output for the confidence interval.

Fathom's confidence interval is just slightly narrower than the one we calculated "by hand.". The reason for that is that Fathom has used the t^* based upon the degrees of freedom of $df = 147.062$, rather than $df = 70$.

Comparing means from two *independent* collections, either with confidence intervals or with hypothesis tests, is an extremely common form of analysis.

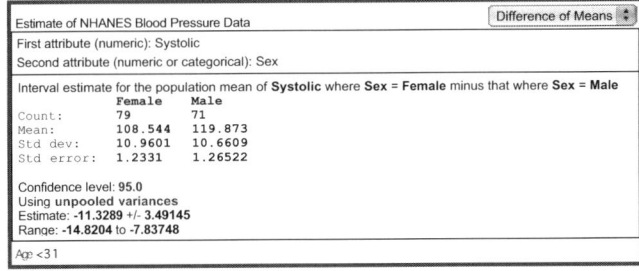

```
Estimate of NHANES Blood Pressure Data                    Difference of Means
First attribute (numeric): Systolic
Second attribute (numeric or categorical): Sex
Interval estimate for the population mean of Systolic where Sex = Female minus that where Sex = Male
              Female      Male
Count:        79          71
Mean:         108.544     119.873
Std dev:      10.9601     10.6609
Std error:    1.2331      1.26522

Confidence level: 95.0
Using unpooled variances
Estimate: -11.3289 +/- 3.49145
Range: -14.8204 to -7.83748

Age <31
```

408

Summary: One measure, two independent collections

- It is quite important to distinguish **one measure, two independent collections** (the subject of this section) from **two measures, essentially one collection** (the subject of the preceding section, §5.3).
    - Another name for **one measure, two independent collections** is **a two independent sample comparison** (this section: §5.4).
    - Another name for **two measures, essentially one collection** is **paired comparisons** (§5.3).

Summary: Hypothesis Test for Comparing One Measure for Two Collections

Step 0: Do a descriptive analysis of the quantitative variable in the two samples. Get graphics and summary statistics.

Step 1: Set up the null and alternate Hypotheses for the population mean difference $\mu_1 - \mu_2$.

$$H_0 : \mu_1 - \mu_2 = 0 \qquad H_a : \begin{cases} \mu_1 - \mu_2 \neq 0 \\ \mu_1 - \mu_2 < 0 \\ \mu_1 - \mu_2 > 0 \end{cases}$$

Step 2: Check the conditions for a trustworthy hypothesis test.
  7. The samples must be simple random samples independently drawn from two populations.
  8. The populations must be Normal; however, since the t procedures are robust for large sample sizes, judge whether to proceed using the 15/40 rule, but with the sum of sample sizes $n_1 + n_2$.

Step 3: Calculate the test statistic $t = \dfrac{(\bar{x}_1 - \bar{x}_2) - 0}{\sqrt{\dfrac{s_1^2}{n_1} + \dfrac{s_2^2}{n_2}}}$ using the sample means and sample standard deviations.

Step 4: Calculate the p-value by getting $P(\bar{x}_1 - \bar{x}_2 \text{ or more extreme} \mid \mu_1 - \mu_2 = 0)$ using the absolute value of the test statistic t calculated in Step 3.

Step 5: Evaluate the evidence that the p-value and the test statistic gives in order to determine whether the test successfully challenges the null hypothesized population mean $\mu_1 - \mu_2 = 0$. Give an interpretation in the context of the data using the terminology of "statistical significance" and "rejecting the null hypothesis."

Summary: Confidence Interval for Comparing One Measure for Two Collections

$$(\bar{x}_1 - \bar{x}_2) \pm t^* \sqrt{\dfrac{s_1^2}{n_1} + \dfrac{s_2^2}{n_2}}$$

where t^* depends on the level of confidence and degrees of freedom calculated either from:
- the smaller of $df = n_1 - 1$ and $df = n_2 - 1$ or

- $df = \dfrac{\left(\dfrac{s_1^2}{n_1} + \dfrac{s_2^2}{n_2}\right)^2}{\dfrac{1}{n_1 - 1}\left(\dfrac{s_1^2}{n_1}\right)^2 + \dfrac{1}{n_2 - 1}\left(\dfrac{s_2^2}{n_2}\right)^2}$

The conditions for trustworthy confidence interval are the same as for a hypothesis test.

§5.4 Exercises: Comparing One Measure for Two Collections

1. **Blood Pressure Data Again** In the *Notes* for this section we found that there was a significant difference between males and females in mean *systolic* blood pressure. Will we find the same pattern for the second component of blood pressure, that is, *diastolic* blood pressure? Our statistical question is:

   Is there a difference between men and women in average (or mean) diastolic blood pressure?

   - Open the Fathom file **NHANES Blood Pressure.ftm** and get a *case table.*
   - With the **Collection icon** for **NHANES Blood Pressure.ftm,** go to **Object>Add Filter.** In the dialogue box, type *Age < 31*. This filter will mean that all of the analyses are for the age group twenty to thirty.
   - From the shelf, bring down a **Graph** and get a dot plot of the variable *Diastolic*. (This variable is actually the average of two measures for each case.)
   - Drag the variable *Sex* to the vertical axis of the dot plot. When you do this, you will be able to compare the distributions of *Diastolic* for the males and the females.
     a. How will you characterize the shapes of the two distributions?
     b. Just from the distributions, does it appear that, on average, men have higher or lower *Diastolic* blood pressure than females?
   - Get a **summary table** with the sample mean, the sample standard deviation, and the count for *Diastolic* for the two groups male and female. The summary table should look like this.

     NHANES Blood Pressure Data

	Female	Male	Row Summary
Diastolic	63.6203	67.9014	65.6467
	11.7553	12.0442	12.0453
	79	71	150

     S1 = mean ( )
     S2 = s ( )
     S3 = count ( )
     Age < 31

     c. Write the numbers in the **summary table** using the correct notation for comparisons of one measure for two groups.
     d. We want to know *whether* (or *if*) there is a difference between males and females in mean *Diastolic* blood pressure. What inferential procedure should we use? Give a reason for your answer.
     e. Are the conditions met for the inferential procedure you have chosen? Or do you have doubts about one or more of the conditions? Be complete.
     f. The intended answer to part d is that you should do a hypothesis test. Set up the null and alternate hypotheses for the hypothesis test, using good notation.
     g. For your alternate hypothesis, do you want a two-sided test or a one-sided test? Choosing either one is defensible, but you must choose one, and you must give a reason for the one you chose.
     h. Both this exercise and the exercise in the last section were on blood pressure. If these were test questions, write how you would know that you should use the methods of this section rather than the methods of paired comparison to do the hypothesis test. What about the context tells you this is a §5.4 exercise and not a §5.3 exercise?
     i. Use the information in the summary table to calculate the test statistic for the hypothesis test.

- Check your calculation with Fathom by dragging down a **Test** box from the shelf. It should be an "Empty test." Change the empty test to the appropriate test that we want, which should be "Compare Means." If you are doing a one-sided test, you will have to change the Fathom box to reflect this.

  j. How did Fathom get the degrees of freedom that are shown? You may just give the formula.

  k. Use the **t Distribution Chart** to get an approximation of the *p*-value using the test statistic and the degrees of freedom that Fathom has calculated. Make certain that the actual *p*-value that Fathom has calculated fits in with your approximation.

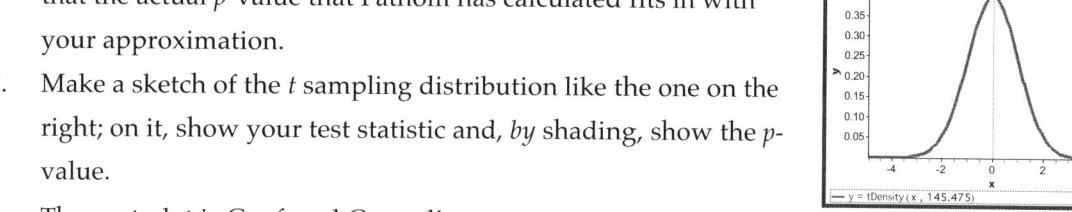

  l. Make a sketch of the *t* sampling distribution like the one on the right; on it, show your test statistic and, *by* shading, show the *p*-value.

  m. The next plot is Confused Conrad's answer to question l, showing his *p* value of 0.03 with a vertical line just next to the mean of the sampling distribution. What is CC's confusion? How should he show the *p*-value?

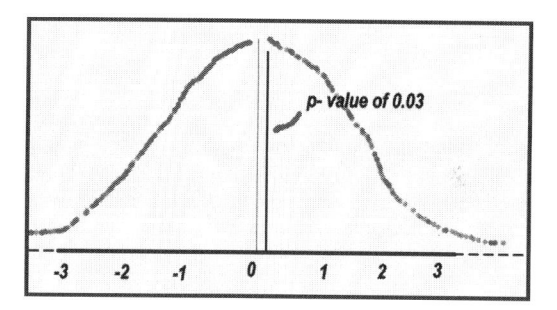

- Confirm your picture by selecting your Fathom test and going to the menu **Test>Show test statistic distribution.**

  n. Interpret your hypothesis test in the context of the data.

  o. The next step will be to calculate a 95% confidence interval. Why is the *t** that you will use *not* 2.198? What should the *t** be and how will you find it? (This could be a Confused Conrad type question, because confusing *t* calculated from an hypothesis test and *t** for a confidence interval is a very common confusion.)

  p. Calculate the 95% confidence interval.

- Check your calculation with Fathom by dragging down an **Estimate** box from the shelf. It should be an "Empty estimate." Change the empty estimate to the appropriate estimate that we want, which should be "Difference of Means."

  q. Give a good interpretation of the confidence interval in the context of the data.

- Make certain that you are finished with what you need for this analysis. Select to the collection icon and go to **Object>Remove Filter**. Everything on your screen should be for all ages rather than for just the age group 20–30.

  r. There should be a hypothesis test for the difference between men and women on the mean level of diastolic blood pressure, but now it is for all ages. Interpret this hypothesis test in the context of the data.

  s. Describe a Type II error in the context of the hypothesis test in part q. (Context means that the answer must say something about diastolic blood pressure and gender.)

Special Exercise 2: More Work on the Real Estate Data

- This work will also be done using a Fathom file that is a sample of the Real Estate Data for the population of houses in San Mateo County that were sold between June 2005 and June 2006.
- <u>Here is the scenario</u>: You have a cousin who is a Realtor in Seattle, but she is getting tired of the long, rainy winters, and she is considering moving her business to San Mateo County. She has some questions about the real estate market in San Mateo County, and she has learned that you have a random sample of data on recent house sales in the county (you *do* have such a sample). She is asking you to use this random sample to give her some idea of the *population* of recent house sales in San Mateo County.

  <u>Background</u>: In Seattle, your cousin specializes in selling "big, old" houses. Her idea is to see if she can concentrate her business on big, old houses here in San Mateo County. So here are your Realtor cousin's questions:

<u>First Question</u>: I want to know whether the average (that is the mean) sale price of big, old houses is significantly greater than the houses for sale that are not big and old. By "big, old," I mean houses whose area is more than 1,800 square feet in the house and also whose age is more than fifty years old.

<u>Second Question</u>: If you find that the big, old houses are selling for more on average, I would like to have an estimate of how much more they sell for, on average.

1. Let us get some direction as to where we are going.
   a. To answer her first question, does your cousin want a significance test or a confidence interval? Give a reason for your answer based on what she says.
   b. To answer her *second* question, does your cousin want a significance test or a confidence interval? Give a reason for your answer based on what she says.
   c. Will we use the formulas for means or for proportions? Give a reason for your answer.
   d. Will we use the formulas for *one* sample/population or the formulas for *two* samples/populations? Give a reason for your answer.

   To answer her questions, you must make a variable in the Fathom file that distinguishes "Big Old" houses from "Not Big Old houses" on the Fathom file. Here are some instructions.

- Open the Fathom file **San Mateo RE Sample Y0506.ftm.** This is a sample with $n = 300$.
- Create a new variable in Fathom and call it **BigOld.**
- Select this variable, go to **Edit>Edit Formula:** and use an **If ( )** statement: It should look like this:

$$If\big((Age > 50) \text{ and } (Sq\_ft > 1800)\big) \begin{cases} \text{"BigOld"} \\ \text{"NotBigOld"} \end{cases}$$

- Drag down a **graph** from the shelf and get either comparative dot plots or comparative box plots to show the distributions of the variable *Sale_Price.*
- Now, get a **summary table** and get the sample mean and sample standard deviation of the variable *Sale_Price* for the BigOld and NotBigOld houses.
- Select the **summary table**, go to the menu for "Summary," choose "Format Numbers" and change to a fixed number of decimal places, and specify two decimal places. This will rid you of scientific notation.

2. Record the mean and standard deviation and the count (the n) of the BigOld and NotBigOld houses, using the correct notation.

3. What shapes do the distributions have? Can we safely proceed with our analysis? In other words, are the conditions met for using the *t* procedures? Consult the **Notes** about conditions and apply the **15/40 rule.**

4. To answer one of your cousin's questions, you will use a significance test. Set up the null and alternate hypotheses for that test using good notation.

5. From the information in the **summary table**, calculate the test statistic.

Check your answer with Fathom. You should know how to do this, but here are some reminders.

- Now, go to Fathom, bring down a **Test icon** (the scale) from the shelf, and select the appropriate test.
- You have two variables to drag: one is *Sale_Price* and the other is *BigOld*. Drag them to the correct places; if you get it wrong, Fathom will tell you.
- With the **Test** selected, go to the **Test menu** and deselect **verbose**.
- Is Fathom using the alternate hypothesis you have chosen? If not, you can change it by changing the blue part.

6. **Reading the Fathom output.** First you should know that numbers like 1.63691e+06 mean $1.63691 \times 10^6 = 1636910$, or, in our context, $1,636,910.
   a. Which number is the test statistic?
   b. What does DF stand for?
   c. Why is this number not a whole number? (*Hint:* How was the number calculated?)
   d. Is the test statistically significant? Give a reason for your answer.

7. Write a conclusion for your cousin (her name is Emily) stating what you have found from your hypothesis test. Make certain that it makes sense to a Realtor.

8. Your cousin's second question was to get an estimate of *how much more* the "BigOld" houses sell for (on average) than the "NotBigOld" houses. For this you will need a confidence interval.
   a. Why do we need a confidence interval? Why can't we just subtract the sample means: $1636910 - 1048706 = 588204$ and say that the BigOld houses sell for $588,000 more on average?
   b. Write the formula that will be used for this confidence interval and calculate the confidence interval using the correct value for t^*. (The correct number is not 5.281 and it is not 1.96.)
   c. Calculate the confidence interval.

- Get Fathom to get an **Estimate** to check your work.

9. a. You should see Range: 365462 to 810944. Write this in the form: 365462 < ____ < 810944. What should be in the blank space?
   b. Which number on the Fathom output is the margin of error?
   c. Change the 95% to 90%. Does the interval get wider or narrower when you do this?
   d. Why does the change to 90% change the width of the interval the way it does?

10. Write a conclusion for the confidence interval output for your cousin, stating what you have found from your confidence interval calculations. Make certain that it makes sense to a Realtor.

11. One step further: there is a variable called *OnMarket* that records the number of days on the market. Was there a significant difference between BigOld and NotBigOld houses in the time on the market in 2005–2006? Answer with the appropriate procedure.

Special Exercise 3: Hypothesis Testing and Confidence Interval Review

We have a small (*n* = 120) random sample of all the types of cars and small trucks sold in the USA in 2008. Here is what the data look like:

	MFR	CAR_LINE	Class2	CityMileage	HwyMile...	DiffHwyCityMi...	Combined_...	ANL_FL...
1	AUDI	RS4 CAB...	Small Car	14.8	26	11.2	18.3588	3213
2	LINCOLN...	MKZ FWD	Large Car	22.6	39.6	17	28.0113	1911
3	VOLKSW...	RABBIT	Small Car	24.394	39.6511	15.2571	29.5025	1751
4	JEEP	GRAND ...	SUV	15.8	24.6	8.8	18.8314	2801
5	JEEP	GRAND ...	SUV	16.3	25.6	9.3	19.4854	2801
6	PONTIAC	G8	Large Car	18.8235	32.8199	13.9964	23.2937	2502

1. Here is a hypothesis test from this sample.
    a. Is this a one-mean test, an independent two-collection test, or a paired-comparison test? Give a reason for your answer.
    b. Write the null and alternate hypotheses using appropriate symbols.
    c. Show the formula that was used to get the test statistic.
    d. Just below the Fathom output are box plots showing the two distributions. From this evidence, are the conditions for the test met?
    e. Use the *p*-value to interpret the results of the test in the context of the data. (In the context of the data means that you need to say something about small cars and SUVs.)

   Test of Collection 1 — Compare Means
   First attribute (numeric): Combined_MPG
   Second attribute (numeric or categorical): Class2
   Ho: Population mean of **Combined_MPG** for **Small Car** equals that for **SUV**
   Ha: Population mean of **Combined_MPG** for **Small Car** is not equal to that for **SUV**

   ```
                  Small Car   SUV
   Count:         38          35
   Mean:          25.3891     22.4428
   Std dev:       5.44011     5.47095
   Std error:     0.882502    0.92476

   Using unpooled variances
   Student's t:   2.305
   DF:            70.4411
   P-value:       0.024
   ```
   (Class2 = "Small Car") or (Class2 = "SUV")

2. From the sample of cars, we would like to estimate the mean amount that highway MPG is more than mean city MPG for the population of all cars.
    a. For this question, do we want a hypothesis test or do we want a confidence interval? Give a reason for your answer.
    b. Here are some summary data, and here are some graphs, some of which you need to use to answer the question. Do we have a one-mean procedure, a two-collection procedure, or a paired-comparison procedure? Give a reason for your answer.
    c. Decide on what you will do and carry out the procedure, using the numbers given to you.
    d. Check the conditions for your procedure.
    e. Give a good interpretation of the results of your calculation.

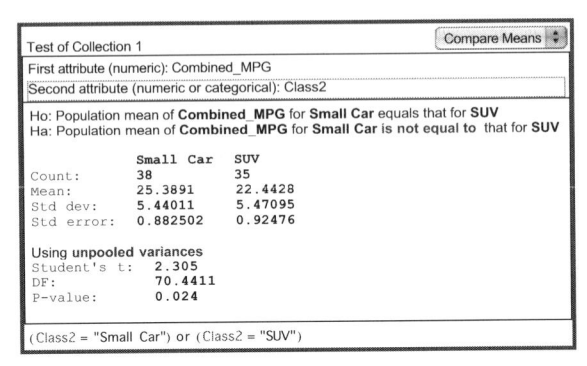

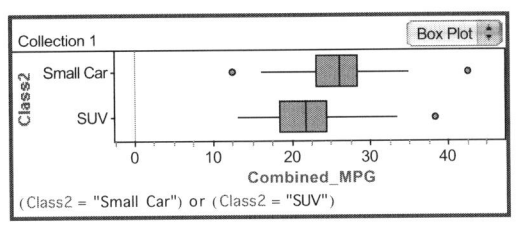

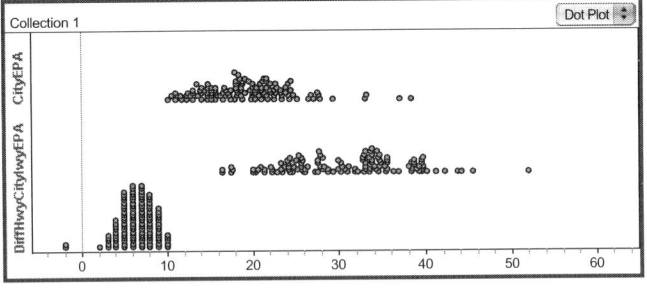

	HwyEPA	CityEPA	DiffHwyCityEPA
	30.569	19.5	11.069
	6.93245	5.18991	3.16758
	120	120	120

   S1 = mean( )
   S2 = s( )
   S3 = count( )

414

3. You want to know whether there is a difference in the mean annual cost of running an SUV versus the mean annual cost of running a pick-up truck. And here you have some sample data.

Collection 1	Class2		Row Summary
	Pick-up Truck or Van	SUV	
ANL_FL_CST	2757.67	2569.97	2646.32
	569.015	534.053	551.576
	24	35	59

S1 = mean( )
S2 = s( )
S3 = count( )

(Class2 = "Pick-up Truck or Van") or (Class2 = "SUV")

   a. Do you want a hypothesis test or a confidence interval, or could you go either way? Give a reason for your answer.
   b. Is this problem a one-mean problem, an independent two-collection problem, or a paired comparison? Give a reason for your answer.
   c. Carry out your procedure and answer the question, using excellent and elegant statistical language.
   d. Is there more information that you would like to have? If so, what is that information and why do you need that information? (*Hint:* Think what you need to see to check conditions.)

- The data are in the Fathom file **Sample of Fuel Economy Data.ftm.** Open this file and get the information or descriptive analyses that you need. You can also use it to check your calculations.

§5.5 Analysis of Variance

Introduction: which types of cars are more economical?

Cars and trucks come in various shapes and sizes, and they differ in their fuel economy. Every year the US Department of Energy and the Environmental Protection Agency publish their Fuel Economy Guide for all of the cars sold in the United States (see www.fueleconomy.gov).

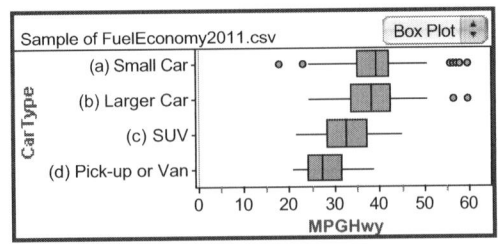

The Fuel Economy Guide reports fuel consumption as city miles per gallon (MPG), highway MPG, and a combined measure. Here are box plots showing the distributions of various types of vehicles for highway MPG and a summary table showing the means, standard deviations, and counts for each of the type of vehicles for highway MPG.

It appears (as we would expect) that small cars are more economical (greater miles per each gallon) than SUVs, pick-up trucks, or vans. Are the differences statistically significant? We would be tempted to compare the types of vehicles pair-wise—small cars compared with larger cars, small cars compared with SUVs—but there is a hypothesis test, called *analysis of variance*, or **ANOVA,** that looks at all of the categories at once.

ANOVA is a hypothesis test that answers the question: "Is there a significant difference in three or more means?" The null and alternate hypotheses for this test are always the same, no matter how many groups we have, as shown below.

Null and alternate hypotheses for ANOVA

$H_0 : \mu_1 = \mu_2 = \cdots = \mu_k$

H_a : At least one of the k means differs from the others

where k indicates the number of different groups in the test.

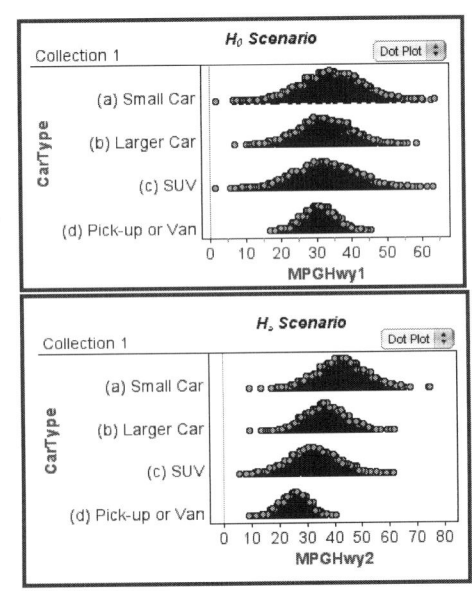

It is useful to think graphically what the populations would look like if the null hypothesis is true (the "H_0 scenario") and what the populations would look like if the alternate hypothesis is true (the "H_a scenario"). In the "H_0 scenario" the distributions all have the same center, but in the "H_a scenario" (shown below) at least one of the distributions has different center. It is the job of ANOVA to detect which of the scenarios our data most resembles, given that there will be sampling variation that may make data from populations that are the "H_0 scenario" look like an "H_a scenario" and vice versa.

Ideas behind the ANOVA test and the test statistic. Instead of comparing means as we did with the "comparing independent means test," we will compare two different *variances* (hence, "analysis of variance"). The two variances are **mean square groups (MSG)** and **mean square error (MSE)**. **MSG** measures how much the group means vary around the mean of all the groups taken together (the grand mean of all of the data). It is sometimes known as variance *between* the groups. **MSE** measures the average variance *within* the groups.

Then, the idea is if the MSG (variance *between* the group means) is large compared with the MSE (variance *within* the groups), we have evidence that the means are more varied—more different—than the data as a whole and, hence, evidence in accord with the H_a, the alternate hypothesis, that there are differences between the means. This is what is pictured in the left-hand graphic below. On the other hand, if the MSG and the MSE come out to be about the same then we have evidence that the means are no more varied than the data as a whole and evidence consistent with the null hypothesis that there is no difference between the group means. This is picture in the right-hand graphic below.

The way we compare the MSG to the MSE is by means of a ratio, called the **F statistic**, as defined in the box below. This becomes the test statistic for the ANOVA test (instead of a *t* or *z*).

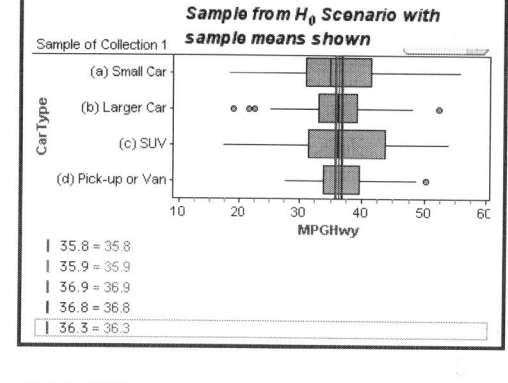

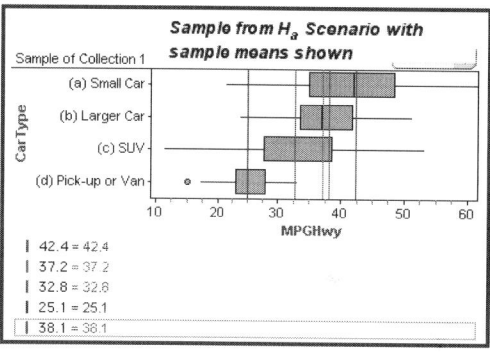

Definition of the F test statistic

$$F = \frac{MSG}{MSE} = \frac{\text{Variation betweeen the means of the groups}}{\text{Variation within the groups}}$$

From what was said above, it follows that if **F** is large (so that the MSG is much bigger than the MSE), we have evidence for the alternate hypothesis, and if **F** is near to one (or even less than one) then we have evidence consistent with the null hypothesis. The two examples above are deliberately extreme; the calculations show that $F = \frac{MSG}{MSE} = \frac{26.84}{56.64} \approx 0.474$ for the left-hand graphic and $F = \frac{MSG}{MSE} = \frac{3436.61}{61.87} \approx 55.546$. The formulas for the MSG and the MSE are given in the box just below.

Formulas for MSG and MSE

$$MSG = \frac{n_1(\bar{x}_1 - \bar{x})^2 + n_2(\bar{x}_2 - \bar{x})^2 + \cdots + n_k(\bar{x}_k - \bar{x})^2}{k-1} \qquad MSE = \frac{(n_1 - 1)s_1^2 + (n_2 - 1)s_2^2 + \cdots + (n_k - 1)s_k^2}{N-k}$$

where: $\bar{x}$ is the mean for all of the data taken together;

$\bar{x}_1, \bar{x}_2, \cdots, \bar{x}_k$ are the means of the 1, 2,...k groups;

$n_1, n_2, \cdots, n_k$ are the sample sizes in each of the groups, and

$N = n_1 + n_2 + \cdots + n_k$ is the total sample size for all the groups together.

Of course, we need a sampling distribution to tell us how big an *F* statistic must be to detect a difference in the means, but let us work on some real data to see how the formulas work.

An example of the calculations: a sample from the Fuel Economy Guide

Since ANOVA is a hypothesis test, we will follow the five steps of the hypothesis test. For ANOVA, there is a standard way of displaying the calculations and the *p*-value that we must also discuss. Here are the box plots for the variable *MPGHwy* (highway MPG). The plots suggest that there are differences in the fuel economy of the different types of vehicles. SUVs and truck-based vehicles do poorly (have lower MPG) compared with cars, especially small cars, although the variability with the small car group appears to be great.

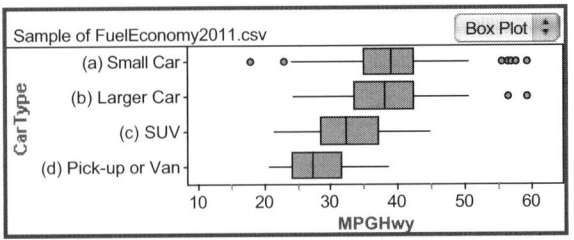

Step 1: Setting up the hypotheses. As mentioned above, the null and alternate hypotheses are standard, although it is a good idea to label the groups for reference. Hence:

$$H_0 : \mu_{Small} = \mu_{larger} = \mu_{SUV} = \mu_{Pick-up}$$
$$H_a : \text{At least one of the four means differs from the others}$$

Step 2: Checking the Conditions. We will return to this step.

Step 3: Calculating the F test statistic. To calculate *F*, we obviously must first calculate the MSG and the MSE, and for that we need the means and the standard deviations for each of the groups. The Fathom Summary Table below gives these. Below are the calculations:

$$MSG = \frac{n_{Small}(\bar{x}_{Small} - \bar{x})^2 + n_{larger}(\bar{x}_{Larger} - \bar{x})^2 + n_{SUV}(\bar{x}_{SUV} - \bar{x})^2 + n_{Pick-up}(\bar{x}_{Pick-up} - \bar{x})^2}{k-1}$$

$$= \frac{88*(39.117 - 35.096)^2 + 86*(38.308 - 35.096)^2 + 87*(32.903 - 35.096)^2 + 63*(28.122 - 35.096)^2}{4-1}$$

$$= \frac{88*(4.021)^2 + 86*(3.212)^2 + 87*(-2.193)^2 + 63*(-6.974)^2}{3}$$

$$= \frac{1422.823 + 887.257 + 418.405 + 3064.111}{3} = \frac{5792.596}{3} = 1930.87$$

$$MSE = \frac{(n_{Small}-1)s^2_{Small} + (n_{Larger}-1)s^2_{larger} + (n_{SUV}-1)s^2_{lSUV} + (n_{Pick-up}-1)s^2_{Pickup}}{N-k}$$

$$= \frac{(88-1)*(8.385)^2 + (86-1)*(6.747)^2 + (87-1)*(5.729)^2 + (63-1)*(4.899)^2}{324-4}$$

$$= \frac{6116.816 + 3869.371 + 2822.644 + 1488.012}{320} = \frac{14296.843}{320} = 44.678$$

From these calculations, we calculate the test statistic: $F = \frac{MSG}{MSE} = \frac{1930.865}{44.678} \approx 43.22$.

The Display Table for ANOVA. For the ANOVA test there is a (fairly) standard way of laying out all of these numbers, once they have been calculated. For our example above, here is the Fathom output **ANOVA Table.** To become familiar with this ANOVA table, it is good to relate the numbers to the calculations.

You should be able to recognize the value of the ***F statistic*** by comparing the numbers in the table with the calculations laid out above. In the column labeled ***Mean Square***, you should see the values of the MSG and MSE. The value of MSG is at the intersection of the ***Mean Square*** column and ***Groups*** row. Likewise, the value of the MSE is at the intersection of the ***Mean Square*** column and ***Error*** row. The column labeled ***Degrees of Freedom*** shows the denominators of the calculations of MSG and MSE. For our example, the MSG denominator is $k - 1 = 4 - 1 = 3$ and the MSE denominator is $N - k = 324 - 4 = 320$. You see both these numbers in the ***Degrees of Freedom*** column. The numbers in the column labeled ***Sum of Squares*** are the numerators of the calculations of MSG and MSE. The last row (***Total***) represents the calculation for the overall variance in the collection. Below there is a graphic showing how the entries in the table relate to the formulas. The graphic is actually part of Fathom output and also shows an interpretation of the results.

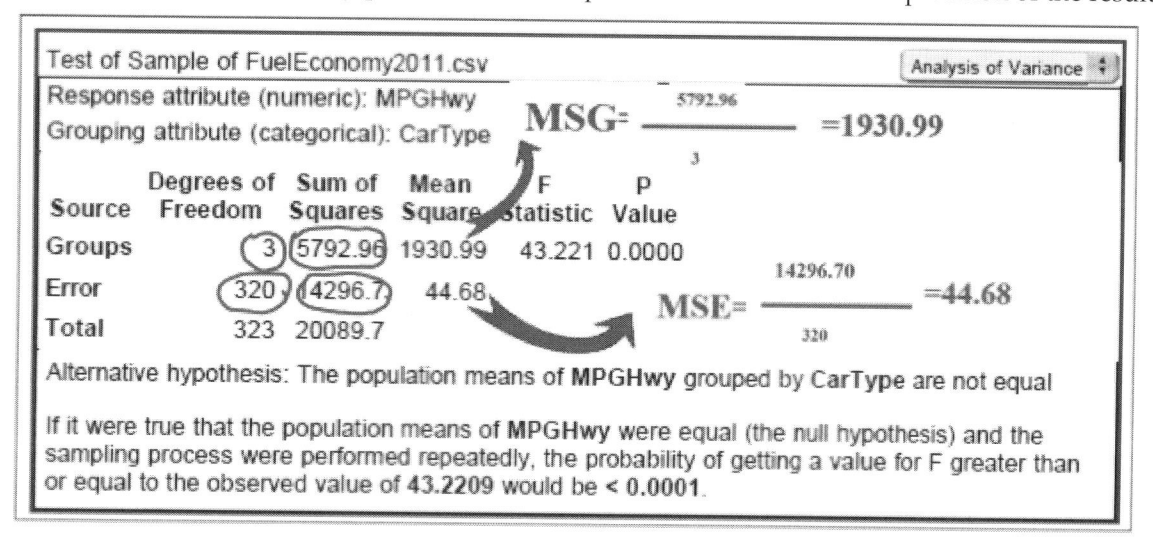

Step 4: Getting the p-value. The ***p-value*** is also shown in the ANOVA table. In this instance, the *p*-value is shown as 0.0000, which indicates $p < 0.0001$, and this is the information we need. The test is significant, and the data show evidence that there are differences among the means of MPG for the different types of cars.

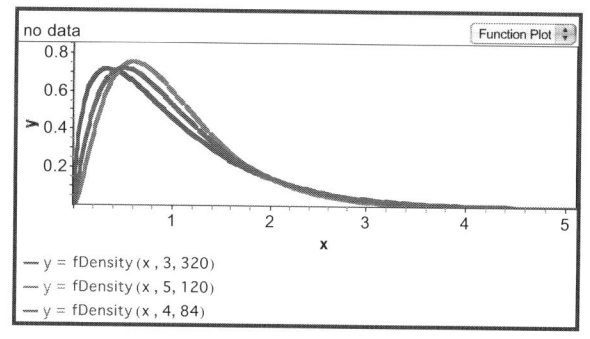

From what sampling distribution was this *p*-value calculated? ANOVA uses a family of distributions called the **F distributions** whose shape depends upon the degrees of freedom. Here are some examples from Fathom. Instead of relying upon a chart, we will depend upon software to give the *p*-value for a given **F** test statistic value. There is an exercise showing how to do this with Fathom. Our graphic includes the F distribution for numerator $DF = 3$ and denominator $DF = 320$, and we can see quite clearly that the F test statistic of $F = 43.22$ is "way off the chart" to the right; the reported $p < 0.0001$ is consistent with what we see. The shaded portion showing the *p*-value would be so small as to be invisible. However, if we had a test with a big *p*-value, we would be able to see the *p*-value as a shaded-in area.

Another Example: automatic or manual?

It is an old idea that cars with manual transmissions rather than automatic transmission get better gas mileage—their MPH should be higher. We can test this with the category of small cars. The Summary Table (above) shows the means and standard deviations for MPGComb (the combination of city and highway MPG). There appears not to be much difference between the MPG for the transmission types, and in fact the automatic cars do slightly better. We should expect that the ANOVA test will be consistent with the null hypothesis, and that the *p*-value will be relatively large. Here is the Fathom output.

The test statistic *F* is near to the value "one," which is what we should see for the equal means situation, and the *p*-value reported by Fathom is large: $p = 0.35$. Using Fathom, we can get a picture of the sampling distribution with the *p*-value shown shaded in by selecting the ANOVA test output and going to the menu: **Test>Show Test Statistic Distribution**. Here is what we will see for this example. The shaded-in portion shows the size of the *p*-value.

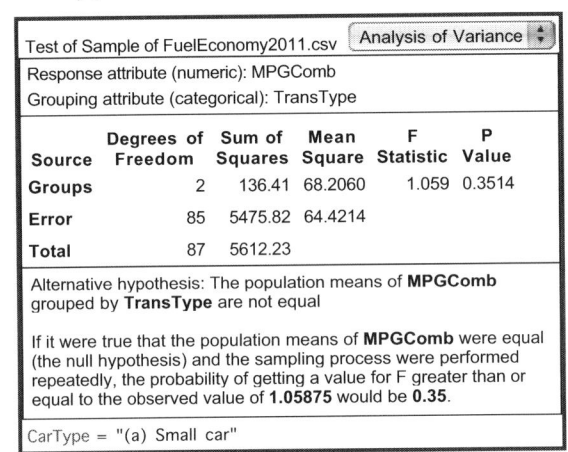

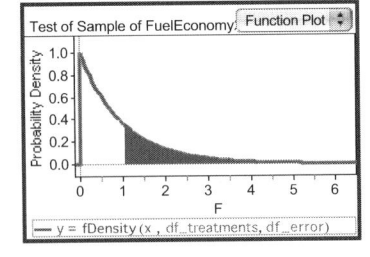

Getting the p-value using Fathom

The Fathom ANOVA output always gives the *p*-value, but there is a way to get the exact *p*-value for any value of a test statistic *F*. Here are the instructions.

- From the shelf, drag down a **Summary Table** and with it selected go to **Summary>Add Formula**.
- Scroll down to **Functions>Distributions>F** and in the dialogue box type: *1 –fCumulative*. Your dialogue box should look like the graphic shown. What fCumulative gives is the proportion of the distribution to the *left* of a given point *x*. The dialogue box works in exactly the same way as the Normal Distribution Chart: it gives the proportion from the left up to a given value.

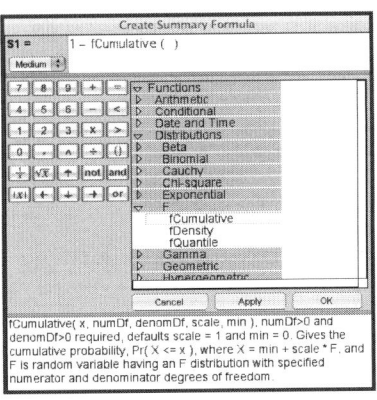

- The directions at the foot of the box show what to type in the parentheses: in general, it says *(x, NumDF, DenomDF, scale, min)*. The *x* is the F test statistic, and the two DFs are the degrees of freedom for the MSG and the MSE. We will ignore scale and min and leave them at the default values. So in our situation, you should type in the parentheses: *1.059, 2, 85*. Here is what you should see.

```
no data
Drop an attribute here
                                     0.351333
S1 = 1 − fCumulative (1.059, 2, 85)
```

Conditions for ANOVA: Robustness again. Here are the conditions that must be met for the ANOVA procedure to work well. You will see that two of the conditions can be "relaxed" under certain circumstances. In other words, ANOVA procedures are to some extent robust.

Conditions for ANOVA
- The samples must be k *independent simple random samples*.
- The populations from which each of the *k* samples is drawn must be **normally distributed**.
- The populations from which each of the *k* samples is drawn must have the **same standard deviation**.

The first condition regarding random sampling is "not negotiable"; samples that are convenience samples or other types of non-random collections make the ANOVA procedures suspect. There are ANOVA procedures for other types of random sampling that can be used, but those are beyond the scope of a first course. The second condition should be investigated in the same way it was for the *t* procedures; outliers and extreme skewness can cause problems if the sample sizes are very small. There is a rule of thumb regarding the third condition, which is expressed in the following box.

Rule of thumb for equal standard deviations
ANOVA has been found to give reliable results if the largest of the *k* standard deviations is no greater than twice the smallest of the *k* standard deviations.

Step 2: Checking the conditions. We see that since the sample that includes the four types of vehicles was randomly taken from the Fuel Economy Guide, the first condition is met. The box plots show outliers on both sides of some of the vehicle groups' distributions, but the distributions appear symmetrical, and the sample sizes are large. The largest standard deviation is for the small cars $s_{Small} = 8.385$ but it is less than *twice* the size of the smallest standard deviation for the pick-ups: $2 * s_{Pick-up} = 2 * 4.899 = 9.798$. For the transmission types example, since the largest standard deviation $9.235 < 2*(6.849) = 13.698$, the rule of thumb shows that we can proceed safely.

Step 5: Interpretation: Car Type Example We have chosen a rather uncontroversial example; of course, small cars are more fuel-efficient than larger cars, and these in turn more economical than SUVs or pick-up trucks and vans. Our data bear out by the small *p*-value what we knew. By the appearance of the box plots, it seems that there are differences among all the types of cars; however, ANOVA (like the chi-square test) can only tell us that there is some significant difference somewhere and cannot pinpoint where that difference is. That must be done with further analysis.

Step 5: Interpretation: Transmission Type Example. For the transmission types, the ANOVA test shows that our original idea that manual transmission cars get better mileage was not correct. Now it may be that we are looking at different kinds of cars. A better way to analyze this question is to choose just those cars that are offered with either a manual or an automatic transmission and compare them.

Summary: Analysis of Variance

- **ANOVA** is a hypothesis test to detect whether data show a difference among group means for more than two independent means in a population. If there are just two means, use a *t* test.

- **Null and Alternate Hypotheses for ANOVA**

$H_0 : \mu_1 = \mu_2 = \cdots = \mu_k$

H_a : At least one of the k means differs from the others

- **Test Statistic for ANOVA**

$$F = \frac{MSG}{MSE} = \frac{\text{Variation betweeen the means of the groups}}{\text{Variation within the groups}}$$

where $MSG = \dfrac{n_1(\bar{x}_1 - \bar{x})^2 + n_2(\bar{x}_2 - \bar{x})^2 + \cdots + n_k(\bar{x}_k - \bar{x})^2}{k-1}$ and

$$MSE = \frac{(n_1 - 1)s_1^2 + (n_2 - 1)s_2^2 + \cdots + (n_k - 1)s_k^2}{N - k}$$

- **Sampling distributions for the ANOVA test are the family of F distributions.** There is a different distribution for each combination of numerator and denominator degrees of freedom.

- **ANOVA Display Table**

Source	Degrees of Freedom	Sum of Squares	Mean Square	F Statistic	P Value
Groups	Denominator of MSG	Numerator of MSG	MSG	MSG/MSE	
Error	Denominator of MSE	Numerator of MSE	MSE		
Total	Denominator of Variance	Numerator of Variance			

- **Conditions for ANOVA**
  - The samples must be k *independent simple random samples*.
  - The populations from which each of the k samples is drawn must be **Normally distributed**.
  - The populations from which each of the k samples is drawn must have the **same standard deviation**.

- **Rule of thumb for equal standard deviations**

ANOVA has been found to give reliable results if the largest of the k standard deviations is no greater than twice the smallest of the k standard deviations.

§5.5 Exercises for Analysis of Variance

1. **Formulas and the ANOVA Table.** The purpose of this exercise is just to get accustomed to the ANOVA table and how its entries reflect the calculations of the test statistic. Here are the formulas for **MSG** and **MSE** and a copy of the ANOVA table in the **Notes**.

$$MSG = \frac{n_1(\bar{x}_1 - \bar{x})^2 + n_2(\bar{x}_2 - \bar{x})^2 + \cdots + n_k(\bar{x}_k - \bar{x})^2}{k-1} \qquad MSE = \frac{(n_1-1)s_1^2 + (n_2-1)s_2^2 + \cdots + (n_k-1)s_k^2}{N-k}$$

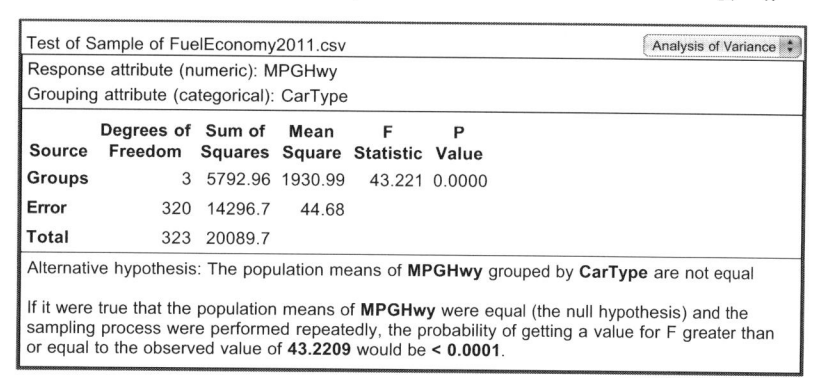

Test of Sample of FuelEconomy2011.csv — Analysis of Variance
Response attribute (numeric): MPGHwy
Grouping attribute (categorical): CarType

Source	Degrees of Freedom	Sum of Squares	Mean Square	F Statistic	P Value
Groups	3	5792.96	1930.99	43.221	0.0000
Error	320	14296.7	44.68		
Total	323	20089.7			

Alternative hypothesis: The population means of **MPGHwy** grouped by **CarType** are not equal

If it were true that the population means of **MPGHwy** were equal (the null hypothesis) and the sampling process were performed repeatedly, the probability of getting a value for F greater than or equal to the observed value of **43.2209** would be **< 0.0001**.

a. For the **Groups** row, divide the number for the **Sum of Squares** (or **SS**) by the value for the **Degrees of Freedom** (or **DF**). What number did you get? What is the name of that number?

b. For the **Error** row, divide the number for the **Sum of Squares** (or **SS**) by the value for the **Degrees of Freedom** (or **DF**). What number did you get? What is the name of that number?

c. Divide the value for **Mean Square** for **Groups** by the value for the **Mean Square** for **Error**. What is the name of the number that you got?

d. Add the **Groups** and **Error** values for **Degrees of Freedom**. Add the **Groups** and **Error** values for **Sum of Squares**. What do you notice about the results? (Are they shown in the ANOVA table?)

e. Here are box plots showing the *MPGCity* by whether a car has "front-wheel drive," "rear-wheel drive," or "all-wheel drive" but just for small cars. If we use an ANOVA test, do the box plots look like the "H_0 scenario" or the "H_a scenario." Put into English a reason for your answer.

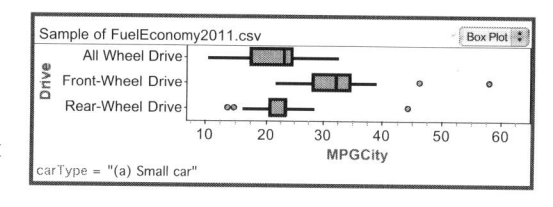

f. Using appropriate notation, write the null and alternate hypotheses for this test.

g. For this test for just the small cars, $N = 88$. What is the value of k?

h. Copy the table and fill in the missing spaces, using the information given and what you have found about the structure of the ANOVA table above and the formulas above.
(*Hint:* Start with $k - 1$ for **DF**.)

Source	Degrees of Freedom	Sum of Squares	Mean Square	F Statistic	P Value
Groups		2070.38			0.0000
Error		3187.24			
Total	87	5257.61			

To check your answers, follow the bulleted instructions.

- Open the file **FuelEconomyGuideSample.ftm** and get a **Case Table.**
- From the shelf, pull down a **Test** box and scroll to get **Analysis of Variance**.
- Drag the *MPGCity* to the **Response Attribute** space and *Drive* to the **Grouping Attribute** space.
- Filter by going to **Object>Add Filter** and typing *CarType="(a) Small car"*.

  Your answers should be the same as the Fathom answers. The *p*-value was given as less than 0.0001, but the next instruction shows how you can get to Fathom to calculate the *p*-value.

- From the shelf, drag down a **Summary Table** and with it selected go to **Summary>Add Formula**.
- Scroll down to **Functions>Distributions>F** and in the dialogue box type: *1 –fCumulative*. Your dialogue box should look like the graphic shown. What fCumulative gives is the proportion of the distribution to the *left* of a given point *x*. The dialogue box works like the Normal Distribution Chart: it gives the proportion from the left up to a given value.
- The directions at the foot of the box show what to type in the parentheses: in general, it says *(x, NumDF, DenomDF, scale, min)*. The *x* is the F test statistic, and the two DFs are the degrees of freedom for the MSG and the MSE. We will ignore scale and min and leave them at the default values. So in our situation, you should type in the parentheses: *27.607, 2, 85*. Here is what you should see.

  The number 5.77658e–10 is $5.77658 \times 10^{-10} = 0.000000000577658$.

  i. The number 0.000000000577658 is the *p*-value. Interpret the results of the hypothesis test in the context of the data. Does the test confirm what you guessed looking at the graphic?

- From the shelf, pull down a **Summary Table** and drag the variables *MPGCity* and *Drive* to the right-hand and down-pointing arrows of the Summary Table.
- Filter by going to **Object>Add Filter** and typing *CarType="(a) Small car"*. You should see this Summary Table.

  j. **Step 2: Checking the conditions**. Because we wanted to look at the ANOVA table, we did not check the conditions. The third condition says that the standard deviations in the population should be equal. However, there is a **rule of thumb** about this condition. Read the rule of thumb and how it is applied and apply it to the standard deviations shown. Are the standard deviations here close enough?

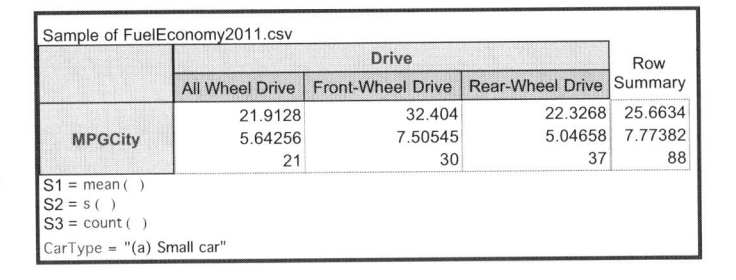

  k. What are the other two conditions for ANOVA? Are they met for our test?

  l. If you do not know much about cars, consult someone who does to answer this question. The question is: is it appropriate to say that a car being a front-wheel drive *causes* the car's fuel economy to be better? Or is reality more complicated?

m. Here is the calculation of the **MSG**. Using the Summary Table on the previous page, give the values that should be placed at A, B, C, D, E, and G.

$$MSG = \frac{n_{AW}(\bar{x}_{AW} - \bar{x})^2 + n_{FW}(\bar{x}_{FW} - \bar{x})^2 + n_{RW}(\bar{x}_{RW} - \bar{x})^2}{k-1}$$

$$= \frac{(21)*([A] - 25.663)^2 + ([B])*([C] - 25.663)^2 + ([D])*(22.327 - [E])^2}{G-1}$$

n. Use the numbers in the Summary Table and the formula for MSE to calculate the MSE. The ANOVA table gives you the answer.

2. **Full-term births**. This exercise looks a simple random sample of full-term births (that is, births that were not premature). For the full-term births: $n = 983$.

- Open the Fathom file **TwelveHundredBirthsSample.ftm**.
- Get a **Summary Table** showing the mean and standard deviation of the variable *AgeMother* broken down by *EducationMother*.
- Filter using **Object>Add Filter** and typing: *First Birth = "First Birth"*. The table should look like this.
- Also, get a **graph** showing the four *AgeMother* distributions for the variable *EducationMother*, filtered by *First Birth = "First Birth"* either as dot plots or box plots. Here is the box plot version.

TwelveHundredBirthsSample		AgeMother
EducationMother	(a) Less than High School	17.7778
		3.382969
	(b) High School	21.6846
		4.87144
		149
	(c) College or University	26.8144
		5.1198
		167
	(d) Post-Graduate	29.8077
		4.49417
		52
	Column Summary	24.9841
		5.8446
		377

S1 = mean ( )
S2 = s ( )
S3 = count ( )
FirstBirth = "First Birth"

a. What are the cases for this collection of data? (Be careful; do not confuse the variables with the cases.)

b. We will analyze these data using ANOVA. **Step 1:** Set up the null and alternate hypotheses for this hypothesis test using appropriate notation for the four groups.

c. From the numbers and the appearance of the graph, do the data seem to be consistent with the "H_0 scenario" or the "H_a scenario"? Give a reason for your answer.

d. **Step 2:** For the ANOVA test to be trustworthy, three conditions must be met. Check that these conditions are met. For the third condition, use the **rule of thumb** given in the **Notes**.

e. **Step 3:** The calculation of the test statistic means calculating the **Mean Square Groups (MSG)**, the **Mean Square Error (MSE)**, and then their ratio, the **F** test statistic: $F = \frac{MSG}{MSE}$. Calculate the MSG using the formula $MSG = \frac{n_1(\bar{x}_1 - \bar{x})^2 + n_2(\bar{x}_2 - \bar{x})^2 + \cdots + n_k(\bar{x}_k - \bar{x})^2}{k-1}$ (*Hint:* Identify the overall mean $\bar{x}$ and the k. For more accuracy, use more "decimal places" in the calculations; three decimal places will give good accuracy. You will be able to check your answers by seeing the Fathom output shortly. Keep a record of the sum in the numerator.)

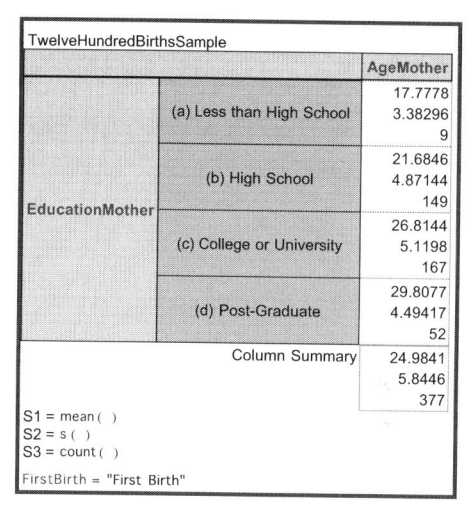

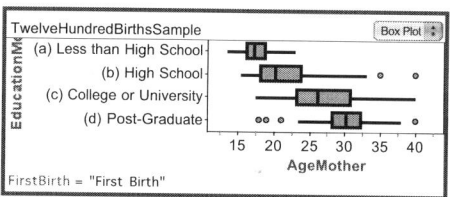

f. **Step 3, Cont**. Calculate MSE using $MSE = \dfrac{(n_1-1)s_1^2 + (n_2-1)s_2^2 + \cdots + (n_k-1)s_k^2}{N-k}$. (Identify N, and n_{LessHS}, n_{HS} etc. Once again, you will be able to check your answers in this exercise. Keep a record of the sum in the numerator.)

g. **Step 3, cont**. Calculate the test statistic $F = \dfrac{MSG}{MSE}$. From the **Notes**, you should be able to say whether the value is "big" and consistent with the H_a, or whether the value is "small" and consistent with H_0.

h. Check your calculations by getting Fathom to do the calculations and lay out the results in the ANOVA table. You may see "rounding errors," but your answers should be close to the Fathom answers. You can check the parts of your calculations by checking Sums of Squares, especially.

- From the shelf, pull down a Test, and change the Empty Test to Analysis of Variance.
- From your **Case Table**, drag the variable *AgeMother* to "Response Attribute" and *EducationMother* to the "Grouping Attribute."
- Filter using **Object>Add Filter** and typing: *First Birth = "First Birth"*.

   i. **Step 5: Interpretation**. Do the data support the idea that the mean age at first birth differs by the education of the child's mother? Give a reason.

   j. Below the ANOVA output are box plots showing the relationship between length of pregnancy (the variable *Gestation3*) and the education of the mother. State why ANOVA is the appropriate inferential technique for analyzing this relationship and not: (I) a *t* test for comparing means OR (II) a chi-square test for independence.

- Get the box plots shown. An easy way to do this is to select the graph showing the box plots for *AgeMother* and go to **Object>Duplicate Graph**. Then from the **Case Table** drag the variable *Gestation3* to the foot of the graph to replace *AgeMother*.
- Get a **Summary Table** showing the means and standard deviations for *Gestation3* by the categories of *EducationMother*. Go to **Object>Dupicate Summary Table** and replace *AgeMother* by *Gestation3*.

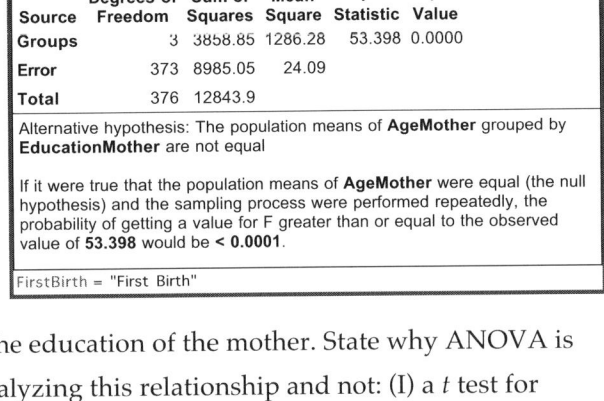

   k. Set up the null and alternate hypotheses for the ANOVA test for the relationship between *Gestation3* and *EducationMother*. From what you see in the graph and Summary Table, do you think that the data are consistent with H_0 or H_a? Give a reason for your answer.

   l. Use the information about standard deviations from the **Summary Table** and the **rule of thumb** to check that the third condition for ANOVA is met.

- Get the Fathom output for the ANOVA test for the relationship between *Gestation3* and *EducationMother*. An easy way to do this is to go to **Object▶Duplicate Hypothesis Test** and replace *AgeMother* by *Gestation3*.

    m. From the ANOVA display table in the Fathom (and not by calculating), give the values of MSG, MSE, and F.

    n. Use the *p*-value given in the ANOVA display table to give an interpretation of the ANOVA hypothesis test in the context of the data. You should say something about whether the mean length of pregnancies differs according to the education of the mother.

- Select the Fathom ANOVA hypothesis test for the relationship between *Gestation3* and *EducationMother* and go to **Test>Show Test Statistic Distribution.** You should see the graph here.

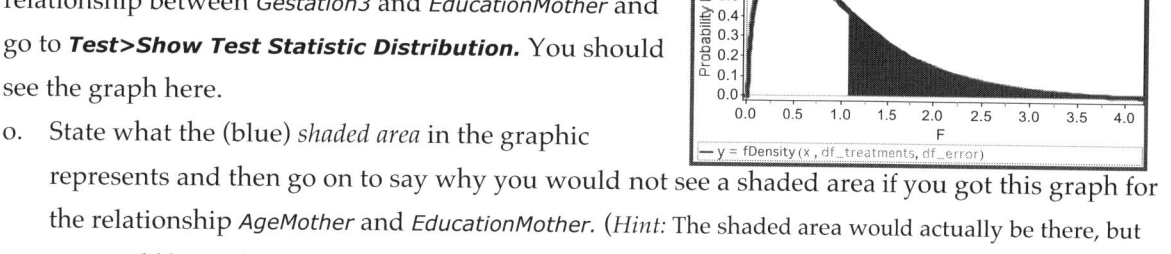

    o. State what the (blue) *shaded area* in the graphic represents and then go on to say why you would not see a shaded area if you got this graph for the relationship *AgeMother* and *EducationMother*. (*Hint:* The shaded area would actually be there, but you would be unable to see it.)

    p. **Choosing the correct procedure.** If you wish to investigate whether mean *AgeMother* for first births differs between the mothers who are married and the mothers who are not married, which inferential procedure would you use: (I) ANOVA (II) a *t* test for comparing two means (III) chi-square (IV) some other inferential procedure? Give a reason for your answer. (The variable ParentsMarried has two categories.)

3. **Weight Loss in Israel**   Dr. Iris Shai and her colleagues carried out a study in Israel that compared three types of diets in a randomized controlled experiment (www.nejm.org/doi/full/10.1056/NEJMoa0708681). The subjects ($n = 322$; 277 men and 45 women) were randomly assigned to one of three diets: either a "low-fat" diet, a "low-carbohydrate" diet, or a "Mediterranean diet." All of the subjects were employed at the same workplace and ate in a cafeteria at work, so the diets for the main meal of the day (the mid-day meal in Israel) could be closely monitored. They also had relatively high adherence rates for the dieters; they kept to their diets, perhaps because they worked with each other and saw each other every day.

    For studies of change, it is important for the research report to include the "baseline" characteristics (the measures at the beginning of the study) of the subjects (or experimental units). Here are the baseline values for the weight-loss study. The numbers are in the form "mean ± sd."

PTO

Table 1. Baseline Characteristics of the Study Population.*

Characteristic	Low-Fat Diet (N=104)	Mediterranean Diet (N=109)	Low-Carbohydrate Diet (N=109)	All (N=322)
Age — yr	51±7	53±6	52±7	52±7
Male sex — no. (%)	89 (86)	89 (82)	99 (91)	277 (86)
Current smoker — no. (%)	19 (18)	16 (15)	16 (15)	51 (16)
Weight — kg	91.3±12.3	91.1±13.6	91.8±14.3	91.4±13.4
BMI	30.6±3.2	31.2±4.1	30.8±3.5	30.9±3.6
Blood pressure — mm Hg†				
Systolic	129.6±13.2	133.1±14.1	130.8±15.1	131.3±14.5
Diastolic	79.1±9.1	80.6±9.2	79.4±9.1	79.7±9.2
Waist circumference — cm‡	105.3±9.2	106.2±9.1	106.3±9.1	105.9±9.1

a. The means of weight (in kg) are almost the same for the three groups at the beginning of the study. Is this a good thing or not a good thing? Give a reason for your answer.

b. If the mean weights at the start are almost the same then should the result of an ANOVA hypothesis test on the *baseline weights* be consistent with the ANOVA null hypothesis or the ANOVA alternate hypothesis? Give a reason for your answer.

c. Here are the results from Shai, et. al., for the weight loss for the three groups after twenty-four months. Our question is: "Is there a significant difference in weight loss among the three diets?" Set up the null and alternate hypotheses for an ANOVA test using appropriate notation.

IsraelieightLossStudy

		WeightLoss
Diet	Low-Carbohydrate	-4.7
		6.5
		109.0
	Low-Fat	-2.9
		4.2
		104.0
	Mediterranean	-4.4
		6.0
		109.0
	Column Summary	-4.0
		5.7
		322.0

S1 = mean ( )
S2 = s ( )
S3 = count ( )

d. Check the conditions for the ANOVA to be trustworthy. Consider:

- The participants in the experiment were randomly assigned to the three diets.
- The box plots to assess normality
- The standard deviations to assess equal standard deviations in the population

e. Find the *MSG, MSE,* and *F* either by using the formulas (and the means and standard deviations given in the **Summary Table**) or by filling in the ANOVA display table shown below. (You may wish to copy the ANOVA table for your submission, unless you are provided with a worksheet.)

Source	Degrees of Freedom	Sum of Squares	Mean Square	F Statistic	P Value
Groups		196.60			0.049
Error		10267.92			
Total	321	10464.52			

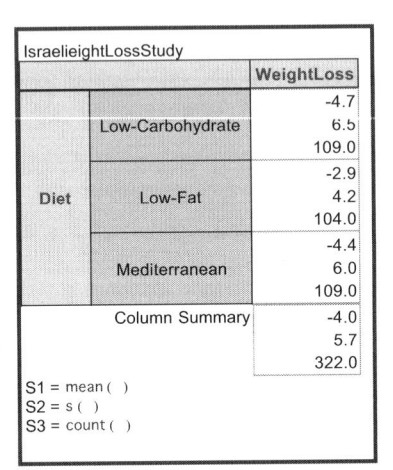

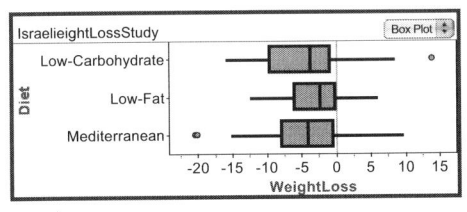

f. Here is the *F* distribution for Numerator DF = 2, and Denominator DF = 319 (which you should have used in answering part e). Copy this graph (or make one using Fathom) and on it show the *F* and the *p*-value.

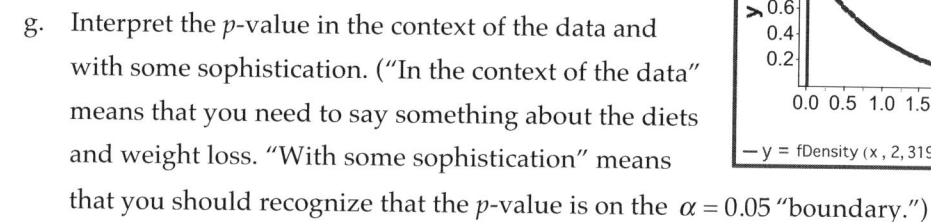

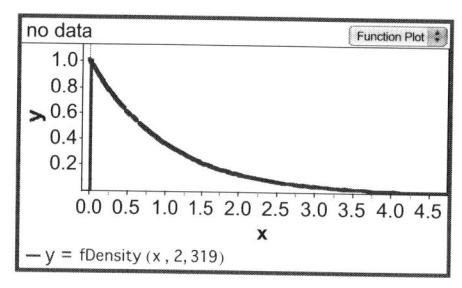

g. Interpret the *p*-value in the context of the data and with some sophistication. ("In the context of the data" means that you need to say something about the diets and weight loss. "With some sophistication" means that you should recognize that the *p*-value is on the $\alpha = 0.05$ "boundary.")

h. **Reading the research report**. On page 235 of Shai, *et. al.* (2008), there is the following sentence. (It may be useful to answer this part and the next part i together.)

    All groups had significant decreases in waist circumference and blood pressure, but the differences among the groups were not significant.

    Does the part of the sentence that says "All groups had significant decreases in waist circumference…" refer to (I) (perhaps multiple) *t* tests on one mean; (II) (perhaps multiple) comparisons of two means; or (III) an ANOVA test? Read carefully, think carefully, and give reasons for your answer.

i. **Reading the research report**. Does the part of the sentence quoted in h that says: "…but the differences among the groups were not significant" refer to: (I) (perhaps multiple) *t* tests on one mean; (II) comparison of two means; or (III) an ANOVA test? Give reasons for your answer.

j. **Reading the research report**. The sentence following the one above says:

    "The waist circumference decreased by a mean [±sd.] of 2.8±4.3 cm. in the low-fat group, 3.5±5.1 cm. in the Mediterranean-diet group, and 3.8±5.2 cm in the low-carbohydrate group (P = 0.33 for the comparison among groups.)"

    Is "P = 0.33" a (I) *p*-value or (II) a test statistic, and is it for: (III) (perhaps multiple) *t* tests on one mean; (IV) comparison of two means; or (V) an ANOVA test? Based upon your decision, give an interpretation of the "P = 0.33."

k. **Review.** The correct answer to part h is that the sentence is referring to three different "one mean tests" and is saying that in all (and in each) of the diet groups there was a significant reduction in waist size. Choose one of the groups and carry out the one mean *t* test; find the *test statistic* to confirm that the *t* test is significant. (Use $n_{LF} = 104$, $n_{Med} = 109$, or $n_{LC} = 109$; what should be the logical value of μ_0 for weight loss?)

l. **Reading the research report**. Compared with most dieting studies, this Israeli study had relatively high retention or "adherence" rates (that is, the proportion who continued with their diets). Here is what the authors say (p. 232):

    "…the 24-month adherence rates were 90.4% in the low-fat group, 85.3% in the Mediterranean-diet group and 78% in the low-carbohydrate group (P = 0.04 for the comparison among diet groups.)"

    (The actual numbers were 94 out of 104 for *LF*, 93 out of 109 for *Med*, and 85 out or 109 for *LC*.)
    What inferential procedure gave the *p*-value of 0.04: (I) ANOVA (II) Comparison of two means (III) Chi-square or (IV) Comparing two proportions? Give a reason for your answer.

m. Interpret the *p*-value of 0.04 referred to in part l in the context of the study.

4. **Weight-Loss Director** For this exercise, you will get to determine how many people lose weight—well, only to some extent, as there is randomness involved.

The Fathom file **WeightLossExperimentSimulated.ftm** simulates the Israeli comparison of diets experiment (Shai et. al. 2008) for the male participants ($n = 277$). The file contains data for 277 men aged forty to sixty-five and with BMI > 27 randomly sampled from the NHANES data. Then these 277 men were randomly allocated to three diet groups, as in the experiment carried out in Israel. For this exercise, you will do an ANOVA analysis several times, and you should have a Word document ready to copy the ANOVA test and graph showing the test statistic sampling distribution each time. In the first analysis, the response variable will be the initial *Weight* of the $n = 277$ men.

Source	Degrees of Freedom	Sum of Squares	Mean Square	F Statistic	P Value
Groups	2	351.796	175.898	0.777	0.4607
Error	274	62019.9	226.350		
Total	276	62371.7			

Alternative hypothesis: The population means of **Weight** grouped by **Diet** are not equal

If it were true that the population means of **Weight** were equal (the null hypothesis) and the sampling process were performed repeatedly, the probability of getting a value for F greater than or equal to the observed value of **0.777107** would be **0.46**.

a. If the random allocation of the men into the three diets has been "successful," we should expect that there should be very little difference in the weight distributions of the men in each of the three groups. Here is the ANOVA test for the weight distributions for the three diet groups. If the random allocation has been successful, should the ANOVA be "statistically significant" or "not statistically significant"? What do the results tell you?

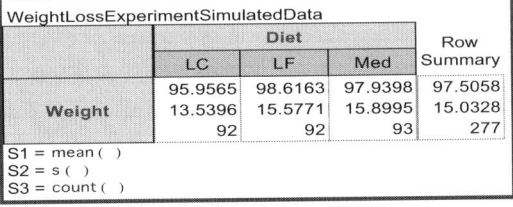

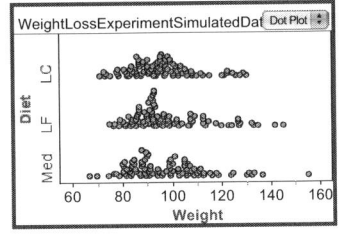

- Open the Fathom file **WeightLossExperimentSimulated.ftm**. You will see the Case Table and sliders labeled **MeanLF, MeanMed,** and **MeanLC.** These sliders will allow you to control the *mean weight change* for the three groups. At opening, all three are set at the value – 5.00, which means that Fathom is set up to randomly simulate a five-kilogram weight loss for all three groups. That is, there is no difference between the groups.

  b. With the sliders set up as they are, do we expect an ANOVA test to be consistent with the H_0 or the H_a? Give a reason for your answer. Get Fathom to do the ANOVA test with the variables *WeightLoss* and *Diet* by following the directions below.

- Get a **graph** of the **response variable** *WeightLoss* by the categories of the variable *Diet*.
- Get a **Summary Table** showing the means and standard deviations of the **response variable** *WeightLoss* by the categories of the variable *Diet*.
- From the shelf, pull down a **Test** and change it from an **Empty Test** to Analysis of Variance.
- Drag the specified variable to the **Response Attribute** and the variable *Diet* to the **Grouping Attribute.**
- With the **Test** selected, go to the menu and get **Test>Show Test Statistic Distribution.**

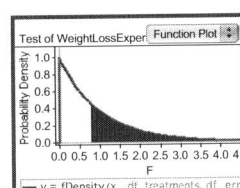

c. You should have the **Summary Table**, the **graph**, the **ANOVA test**, and the **test statistic sampling distribution graph**. Copy the ANOVA test and the sampling distribution graph, with the *p*-value indicated. Report the *F* and the *p*-value and whether the test shows evidence of no differences between the diets or some differences between the diet groups. Keep your output; you will need it below.

d. Now the simulation is doing things randomly, and so the results will *not* be exactly the way we specified. If the test showed a difference between the groups even though you had set it up to show no difference, would that be a Type I error or a Type II error? Give a reason.

e. **Controlling Weight I** Now, you get to control the weight. Move the sliders so that two of the diet groups are nearly the same mean weight change, but one is much different in its mean weight change. You can make two of the diets ineffective and one of the diets effective or you can make two diets that work and one that does not work; that is up to you. However, **report what you have done.** With the sliders set up as you have set them, do you expect an ANOVA test to be consistent with the H_0 or the H_a? Give a reason for your answer. You may have to re-randomize—in the menu: **Collection>Rerandomize.**

- **Follow the Fathom instructions above to do the ANOVA test.**

f. Copy the ANOVA test and the sampling distribution graph, with the *p*-value indicated. Report the *F* and the *p*-value and whether the test shows evidence of no differences between the diets or some differences between the diet groups.

g. If the test showed no difference in the three diet groups even though you intended one group to be much different, will that be a Type I error or a Type II error? Give a reason for your answer.

h. **Controlling Weight II** Now move the sliders so that there appears to be a difference between each of the diet groups in the mean weight change are nearly the same but one is much different. The diets you choose to be best and worst are up to you. (If you want, you can even make the people gain weight.) However, **report what you have done.** With the sliders set up as you have set them, do you expect an ANOVA test to be consistent with the H_0 or the H_a? Give a reason for your answer. Your graph may look something like this one.

i. The answer to the last question about Type I and Type II errors should be "a Type II error" since the test did not detect the difference that it should have. In the last set-up (part h), was your test correct or did you get a Type II error?

j. Now make a difference between the diets that is not zero but is small and keep the ANOVA test open so that you can watch it as you re-randomize (in the menu: **Collection>Rerandomize**). Record how many of ten re-randomizations you get a big p-value (so no difference) instead of a small p-value, as it should be. What proportion of times was the ANOVA correct?

k. Now make a big difference between the diet groups and repeat the re-randomizing process. What proportion of the time is ANOVA correct now? Put into words the lesson that comes out of the exercises in j and k. (Discuss with your instructor.)

5. **Playing with Cars** The data in the Fathom file ***Summer 09 2006 Cars.ftm*** has all of the 2006 model year cars that were being sold through the Internet site www.cars.com for these makes and these places.

   Place: Boston, Chicago, Dallas, SF Bay Area
   Make: Audi, BMW, Infiniti, Lexus, Mercedes-Benz

   Our statistical question is whether there are differences in the mean miles driven (the variable *Miles*) for these 2006 used cars in 2009 (when the cars were about three years old) among the places: Boston, Chicago, Dallas, and the SF Bay Area (the variable *Place*). We will be using ANOVA.

   - Open the Fathom file ***Summer 09 2006 UsedCars.ftm*** and get a ***case table***.
   - Get the ***graph*** and the ***Summary Table*** shown below.

   a. State why ANOVA is the appropriate inferential procedure to answer our statistical questions and not a (I) a t test to compare means or (II) chi-square test for independence.

   b. **Step 1:** Set up the null and alternative hypotheses for the first statistical question, using notation that shows the context (that is, *Place*).

   c. **Step 2:** There are three conditions for doing ANOVA. Show that two of them are met with no difficulty but that we have reservations about one of the conditions.

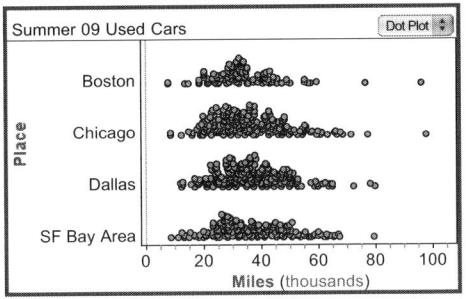

   d. **Step 3:** Without actually doing the calculations (we will get Fathom do that), show where the numbers in the Summary Table should go in the formula for MSG. It will be sufficient to show just two terms. (What is k?)

   e. **Step3, cont.** Without actually doing the calculations (we will get Fathom do that), show where the numbers in the Summary Table should go in the formula for MSE. It will be sufficient to show just two terms. (What is N?)

   - From the shelf, pull down a ***Test*** and change it from an ***Empty Test*** to Analysis of Variance.
   - Drag the variable *Miles* to the ***Response Attribute*** and the variable *Place* to the ***Grouping Attribute.***
   - With the ***Test*** selected, go to the menu and get ***Test>Show Test Statistic Distribution.***

   f. From the ANOVA display table, give the value of the MSG, the MSE, and the test Statistic F.

   g. **Step 4** The ANOVA display gives the p-value, and it is shown in the graph of the sampling distribution. Copy the sampling distribution graph and indicate what shows the p-value.

   h. **Step 5: Interpretation**. Interpret the p-value in the context of the data. Is there evidence of a difference in the average miles driven in three-year-old cars between Boston, Chicago, Dallas, and the SF Bay Area?

   i. Do the complete ANOVA analysis to answer the statistical question: "Is there a difference in the average miles driven among the makes Audi, BMW, Infiniti, Lexus, and Mercedes Benz?"

Summer 09 Used Cars

	Place				Row Summary
	Boston	Chicago	Dallas	SF Bay Area	
Miles	33900.7	36336.5	37247.6	37287.8	36423.1
	12354	13361.2	12509.4	13661.1	13065.8
	109	213	188	160	670

S1 = mean ( )
S2 = s ( )
S3 = count ( )

§5.6 Reading the Mental Map

Getting the Connections Right

For a complex subject like statistics, it is helpful to have a mental map that will organize the material so that it "hangs together." The map can be schematic, like a transit map. Even without having ridden the train, you know that the tracks are not likely to be the straight lines shown in the diagram; however, the diagram still works to show the relationships of the various lines and where they go. As a review, we explore a "transit map" for the introductory statistics course. The map is shown in reduced form below.

Stat Land

We can think of the map as a map of a place called "Stat Land," whose geography is the content of our introductory statistics course. The course has two related "regions"— **Descriptive Statistics** and **Inferential Statistics**—separated by a "river," which is the theory of sampling distributions.

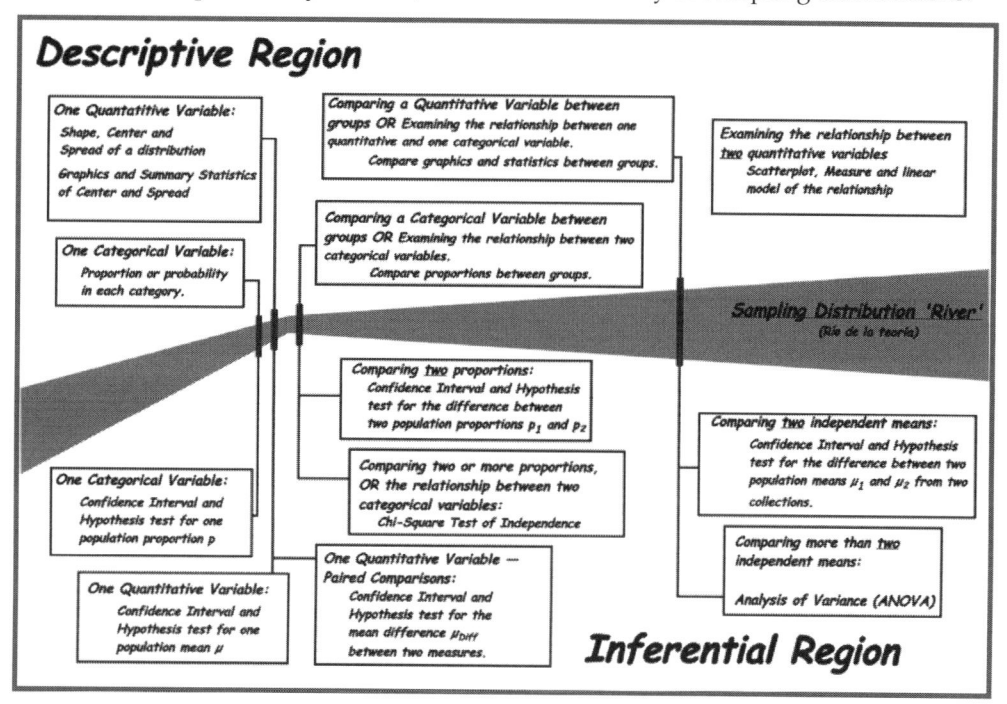

Descriptive Region

- One Quantitative Variable: Shape, Center and Spread of a distribution. Graphics and Summary Statistics of Center and Spread
- Comparing a Quantitative Variable between groups OR Examining the relationship between one quantitative and one categorical variable. Compare graphics and statistics between groups.
- Examining the relationship between *two* quantitative variables. Scatterplot, Measure and linear model of the relationship
- One Categorical Variable: Proportion or probability in each category.
- Comparing a Categorical Variable between groups OR Examining the relationship between two categorical variables. Compare proportions between groups.

Sampling Distribution 'River' (Río de la teoría)

- One Categorical Variable: Confidence Interval and Hypothesis test for one population proportion p
- Comparing *two* proportions: Confidence Interval and Hypothesis test for the difference between two population proportions p_1 and p_2
- Comparing *two* independent means: Confidence Interval and Hypothesis test for the difference between two population means μ_1 and μ_2 from two collections.
- Comparing two or more proportions, OR the relationship between two categorical variables: Chi-Square Test of Independence
- One Quantitative Variable: Confidence Interval and Hypothesis test for one population mean μ
- One Quantitative Variable — Paired Comparisons: Confidence Interval and Hypothesis test for the mean difference μ_{Diff} between two measures.
- Comparing more than *two* independent means: Analysis of Variance (ANOVA)

Inferential Region

Descriptive and Inferential Statistics The goal of **descriptive statistics** is to understand data and be able to communicate its form by using graphics and summary statistics. The goal of formal **inferential statistics** is to generalize beyond the data at hand; however, this can only be done with the theory of sampling distributions, with the procedures of **hypothesis testing** and **confidence intervals**, but only under specified **conditions**. You can think of the "bridges" in the diagram as representing the conditions, but the bridges are there also because proper inferential statistical analysis *starts* with descriptive statistical analysis. After description, then, if the samples are random and sufficiently large, we can "cross the bridge" to inferential analysis.

Categorical and Quantitative Variables The other important distinction seen in the map is the difference between **categorical** and **quantitative** variables. The summary statistic that we use with categorical variables is a proportion, whereas, with quantitative variables, it makes sense to depict their distributions with graphics such as dot plots, box plots, or histograms and then to calculate measures of the location (center) such as **mean** and **median** and of spread, such as the **IQR** and **standard deviation**.

Comparing Groups or Examining Relationships between Variables When we are comparing groups or collections, we compare the distributions and the summary statistics for a quantitative variable in the two (or three or four or more) groups; or if we have a categorical variable, we compare the proportions between the two (or three or four or more) groups. However, there are two ways of looking at what we are doing; both ways are legitimate and often end up with the same calculations, but they do involve a slightly different stance.

- We can regard the comparisons of either quantitative or categorical variables as being between groups. If we compare the hours that male students work with the hours that female students work then we are regarding our comparison as between groups.
- We can regard the comparison as indicating the relationship between variables; in our example, we would speak of the relationship between the variable *Students Work Hours* and the variable *Gender*.

Four Kinds of Statistical Questions If you look at the map carefully, you will see that there are four kinds of comparisons that we typically ask, and for the first three of these we have developed formal inferential techniques.

- Comparing data from either a *categorical* or *quantitative* variable with a standard. Are 10% of people left-handed or is the mean flight time to Seattle 125 minutes?
- Comparing a *quantitative* variable by the categories (or groups) of a *categorical* variable. Is there a difference in the mean age of day community college students and evening community college students?
- Comparing a *categorical* variable by the categories (or groups) of a *categorical* variable. Are female college students more likely or less likely to a have a tattoo?
- Examining the relationship between two *quantitative* variables. At what rate does the price of a house increase with the size of the house?

To analyze the relationship between two quantitative variables (the last question above), we use graphic called a **scatterplot** and then (if appropriate) applied a *linear model* to the data. Using a linear model, we were able to find a best-fitting line—the *least squares regression* line—and also get both an assessment of the direction and strength of the linear model (using the *correlation coefficient r*) and also get a measure of how well the linear model fits the data using the *coefficient of determination* R^2. There are important formal inferential techniques for linear models not covered in this course.

Mathematical models   We have used mathematical models several times. One use was mentioned just above; the simplest model for the relationship between two variables is a linear—simply a straight line—model, which we expressed as $\hat{y} = a + bx$. The other model that we used extensively is the *Normal distribution* model. We use the Normal model to characterize some distributions but more in the "inferential region" part of the course. We used the Normal model there because sampling distributions often (under certain conditions) follow a Normal distribution. The *t distributions* and the *chi-square distributions* are also formal mathematical models.

The Language of Stat Land: "Probability spoken here." Throughout Stat Land, the language and symbols of probability are used. (If Stat Land really had roads, the road signs would be in "probability.") The symbols that are used may differ; for example, if we are describing the proportion of female students who speak two or more languages, we would express this as $P(X \geq 2 \mid F)$, but if we were conducting a hypothesis test we would express the sample proportion as $\hat{p}_F$ where it is understood that the symbol refers to the proportion who speak two or more languages. However, wherever we use the language of probability, it can always refer to the *likelihood* or the *chance* of an event occurring. Wherever we use the language, it must be true that a probability is between zero and one since "*p*-values" are probabilities, and it makes no sense to write, "The *p*-value is 2.13."

Reading the Stat Land Map in the Face of a Statistical Question

From www.cars.com we have data on used cars that are advertised there. Data were collected from Boston, Chicago, Dallas, and the San Francisco Bay Area. Data were collected on Audi A4, BMW 3 Series, Mercedes C-class, Infiniti G-35, and Lexus IS. The data were collected in July 2009. Before we answer any statistical question, we need to see what kind of sample we have.

Summer 09 Used Cars

	Make1	Place1	Price	Miles	Age	Convert...	Body	Seller	Distance
1131	BMW 3 S...	Chicago	34995	5654	0.416667	Not Conv...	Sedan	Dealer	21
1132	BMW 3 S...	Chicago	34991	10637	0.416667	Not Conv...	Sedan	Dealer	29
1133	BMW 3 S...	Chicago	34991		0.416667	Not Conv...	Sedan	Dealer	29
1134	BMW 3 S...	Chicago	33995	8252	0.416667	Not Conv...	Sedan	Dealer	30
1135	BMW 3 S...	Chicago	33995	8590	0.416667	Not Conv...	Sedan	Dealer	30
1136	BMW 3 S...	Chicago	33991	10882	1.33333	Not Conv...	Sedan	Dealer	18
1137	BMW 3 S...	Chicago	33900	14000	1.33333	Not Conv...	Coupe	Private S...	15

What kind of sample do we have? Do we have simple random sample? What can we regard as our population? The answer to the first question is no because the cars were not chosen by a random process from a complete list of used cars for sale—a list that does not exist in any case. Since *all* of the cars (that is, all Audi A4, all BMW 3 Series, etc.) advertised on www.cars.com were selected on the dates and places that were chosen, we have something like a cluster sample. Is it random? We need to think about what the population could be. The population is not *all* cars (that is, all Audi A4, all BMW 3 Series, etc.) since we have selected only cars that were offered for sale. And the population is not all cars that were offered for sale because cars are sold in other ways than through www.cars.com—there are other Internet sites that have used cars, and there are many other ways of selling used cars. Cars that are sold by dealers that are not put on Internet sites may be different; they may be older, or less expensive, or more expensive. We do not know since we have no data on those cars. We appear to have a cluster sample (over four places) of cars that are sold through one specific Internet site, www.cars.com. It may even be that the cars that are offered in the summer differ in some systematic way (age, body style, etc.) from those that are offered in the winter. We may be safe in saying that what we have can be regarded as a random cluster sample of cars (again, that is all Audi A4, all BMW 3 Series, etc.) offered for sale in the summer— perhaps just in the summer of 2009. When we interpret our results, we have to take into account what our sample is.

A statistical question and how we answer it

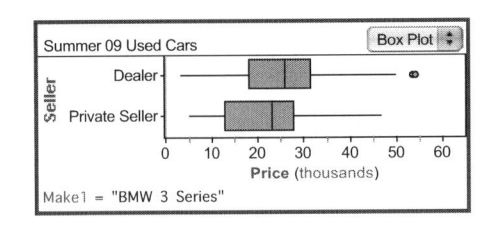

One of the variables measured is whether the seller is a dealer or a private seller; we can ask:

If a private seller is selling a car, will the price be lower (or possibly higher) than if a dealer offers the car?

What kind of variables do we have? *Price* is a **quantitative** variable, and the variable *Seller* is **categorical.** This means that we are comparing a quantitative variable between groups, which is the middle top box in the *Descriptive Region* of the **Stat Land Map.** What are the appropriate analyses? For the descriptive analysis, we need a graphic and summary statistics that compare two distributions. Here they are for the BMW 3 Series.

It does appear that the cars being sold by private sellers are generally cheaper than the cars being sold by the dealers on www.cars.com. Is this a big enough difference to be beyond sampling variation? If we had random samples, we could use a hypothesis test for **comparing two independent means**. Since we have a large sample total size ($n = 755$), and since the *Price* distributions are not extremely skewed, the normality condition is not a problem although we must remember that we are not certain that we have a random sample. Here is the Fathom output; the *p*-value is very low; what we see is unlikely to be just sampling variability.

However, although we have shown that on average the cars sold by private sellers cost less than those sold by dealers, we could ask: are the cars comparable? Perhaps the private sellers are selling older cars, and for that reason the cars have lower prices.

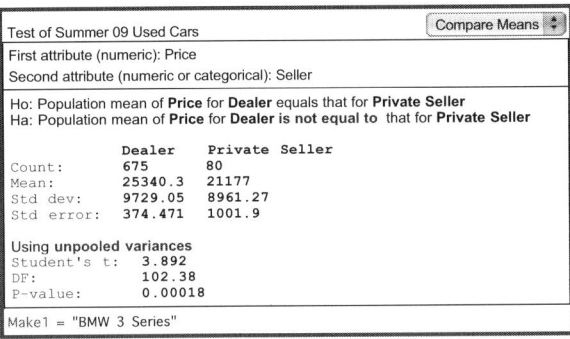

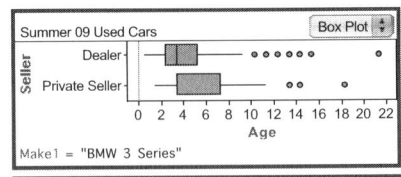

A second statistical question:

If a private seller is selling a car, will the age be older than if a dealer offers the car?

Again, we are comparing one quantitative variable (*Age*) between two groups, so we will do the same kind of analysis but with *Age* instead of *Price*. The descriptive analysis here appears to show that it is true that the cars being sold by private sellers are older on average than the cars being sold by dealers, and this may account for the lower prices of the private seller used BMW 3 Series. It may also be that there is a different mix of body styles sold.

A statistical question with two categorical variables

Do the proportions of coupes, sedans, and convertibles differ by the variable Seller?

We should immediately recognize that the variables *Body* (indicating the body style of the car) and *Seller* are both *categorical*. (On the **Stat Land Map** we look at the box nearest the "river" in the *Descriptive Region*.) For descriptive analysis, we need a table showing the number of the various body styles being offered for sale by dealers and by private sellers.

Summer 09 Used Cars

	Body					Row Summary
	Convertible	Coupe	Hatchback	Sedan	Wagon	
Seller Dealer	55	91	2	554	12	714
Private Seller	19	11	0	22	2	54
Column Summary	74	102	2	576	14	768

S1 = count( )
make1 = "BMW 3 Series"

Since we are regarding *Seller* as the explanatory variable, we will want to compare the proportion of convertibles (for example) according to whether they are being sold by a dealer or a private seller. So we will calculate $P(Conv \mid Dealer) = \frac{55}{714} \approx 0.077$ compared with $P(Conv \mid Private\ Seller) = \frac{19}{54} \approx 0.352$ as well as $P(Sedan \mid Dealer) = \frac{554}{713} \approx 0.776$ compared with $P(Sedan \mid Private\ Seller) = \frac{22}{143} \approx 0.407$, and we are surprised! Our results tell us that the private sellers are much more likely to be selling convertibles than the dealers and less likely to be selling sedans. The difference is big enough that we suspect that the inferential analysis will be statistically significant. The relevant test here will be a **chi-square test of independence** since the categorical variable *Body* has more than two categories. Here is the Fathom output. Notice that we have done an *Add Filter* to do the analysis only for the coupes, sedans, and convertibles since there were too few wagons and hatchbacks for the conditions for the inferential procedure to be met. Notice also that the chi-square test of independence does not tell us the details of the differences between the dealers and the private sellers; the test just tells us that there *is* a difference and that, whatever difference it is, it cannot be explained by sampling variability. The direction of the differences is surprising because we think that convertibles are generally *more expensive* than sedans, other things being equal. This time, we want to include the variable Age as well as the variable Price in our analysis, so we will look at the *relationship* between price and age for the convertibles and non-convertibles (coupes, sedans, wagons, and hatchbacks).

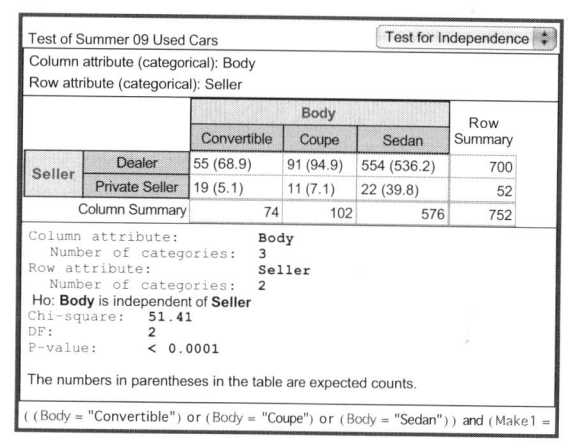

Test of Summer 09 Used Cars — Test for Independence
Column attribute (categorical): Body
Row attribute (categorical): Seller

	Body			Row Summary
	Convertible	Coupe	Sedan	
Seller Dealer	55 (68.9)	91 (94.9)	554 (536.2)	700
Private Seller	19 (5.1)	11 (7.1)	22 (39.8)	52
Column Summary	74	102	576	752

```
Column attribute:       Body
   Number of categories: 3
Row attribute:          Seller
   Number of categories: 2
Ho: Body is independent of Seller
Chi-square:   51.41
DF:           2
P-value:      < 0.0001
```
The numbers in parentheses in the table are expected counts.

((Body = "Convertible") or (Body = "Coupe") or (Body = "Sedan")) and (Make1 =

A statistical question about the relationship between two quantitative variables

Is the relationship between the variables Age and Price different for convertible and non-convertible BMWs?

Since we have *two quantitative variables*, and we are examining the relationship between them, we will look at a scatterplot of the data and see if it is appropriate to fit a linear model to the data. On the **Stat Land Map** we are still in the Descriptive Region (above the "river") and on the right-hand side. The Fathom output is shown below with two least squares regression lines, one for the convertible BMWs and the other one for all the other body styles—the coupes, sedans, etc. What we see is that the slopes of the linear models for the convertibles and non-convertibles are nearly the same; for each year a BMW ages, the value goes down about 2,900 dollars by our calculations. However, the least squares regression line for the convertibles is always about five thousand dollars above the least squares regression line for the non-convertibles. We also see some scatter in the data (not much, however) and some curviness to the plot, both of which make the linear model not fit well. Still, the *Coefficients of Determination* R^2 ($= r^2$) show that for the convertibles, 65% of the variability in *Price* is accounted for by the linear model on *Age*, and for the non-convertibles, 78% of the variability in *Price* is accounted for by the linear model on *Age*.

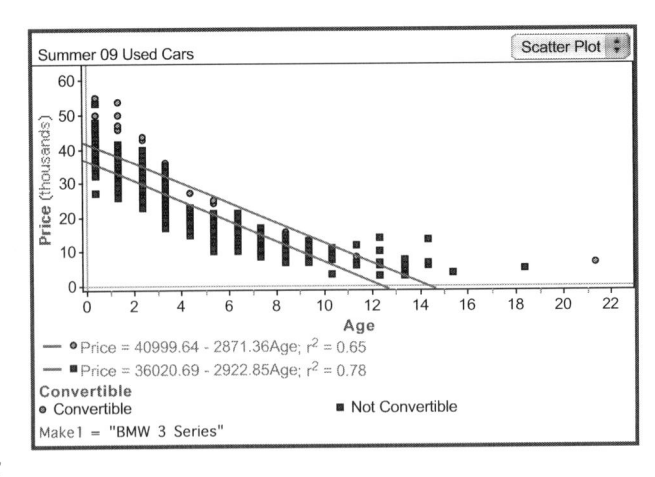

The End of This Story: What is the best analysis? We started out asking whether BMWs sold by private sellers are more (or less) expensive than BMWs sold by dealers. The answer overall, on average, is that the BMWs sold by private sellers are less expensive. We then tried to find out why this may be. It turns out that the private sellers have a tendency to sell cars that are older (and therefore cheaper) than the dealers. However, it also turns out that the private sellers are far more likely to sell convertibles, and convertibles are generally more expensive! However, the best way to answer our question would be to do an analysis of *matched pairs* of cars. We were able to match each of the private seller car with a dealer car of the same place (Boston, Chicago, Dallas, or SF), age, and whether a convertible or not, and chose the car with the nearest miles driven (there were, of course, far more dealer-sold cars than private seller cars). Here is the dot plot showing the difference of price. The variable *DiffPrice* = *Price2* − *Price1*, where *Price 2* is the price for the dealer car, and *Price1* is the price for the private seller car. If there were no difference, we would expect a distribution to have a mean of zero. (In the **Stat Land Map** we are on the Descriptive Region on the top left-hand box.) Here are the summary statistics. From a descriptive analysis, it does appear that the dealer cars sell for more. The relevant inferential procedure is a paired-comparison test. Shown here is the Fathom output for this test, giving us evidence that a car offered by a dealer is more expensive than a private seller car.

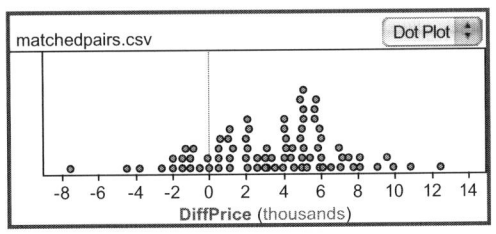

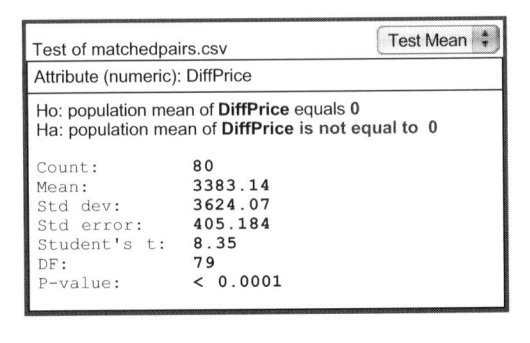

One Collection or Paired Comparison on One Collection

Type	Confidence Interval	Hypothesis Tests	Comments (Mostly Technical Conditions)
One Sample Problem	$CI = \bar{x} \pm t^* \dfrac{s}{\sqrt{n}}$ Get the t^* using the t distribution Table for the row corresponding to the df and the column for the level of significance. $df = n - 1$	1. Hypotheses: $H_0 : \mu = \mu_0$ $H_a : \mu \begin{cases} < \mu_0 \\ > \mu_0 \\ \neq \mu_0 \end{cases}$ 3. Test Statistic $t = \dfrac{\bar{x} - \mu_0}{\dfrac{s}{\sqrt{n}}}$ 3. Get an estimate of the p value using the t distribution Table with $df = n - 1$	The sample must be an SRS (= Sample random sample) from the population. In the population from which the sample is drawn, the variable x must be normal. But see the comments on robustness below. Our judgment of the normality of the population can often be made from the sample distribution.
Paired Comparisons	$CI = \bar{x}_{Diff} \pm t^* \dfrac{s_{Diff}}{\sqrt{n}}$ where $\bar{x}_{Diff} = \bar{x}_1 - \bar{x}_2$, but s_d must be calculated directly from the differences. $s_d \neq s_1 - s_2$ $df = n - 1$	$H_0 : \mu_d = 0$ $H_a : \mu_d \begin{cases} < 0 \\ > 0 \\ \neq 0 \end{cases}$ $t = \dfrac{\bar{x}_{Diff} - \mu_0}{\dfrac{s_{Diff}}{\sqrt{n}}}$ Get the p value using $df = n - 1$	Distribution of differences is normal in the population; but see comments on robustness below. Our judgment of the normality of the population of differences can often be made from the sample distribution of differences.

Robustness:

- By "robust" we mean that the procedures will give reliable results even if the one of the technical conditions fails to be met. If the population distribution is not Normal, as it should be under the technical conditions, the t procedures are "robust" if the sample size is sufficiently large. Use the 15/40 Rule to determine whether the sample size is large enough after you look at the shape of the distribution.

  n under 15: Sample Distribution should be nearly Normal; Otherwise, abandon the procedure.

  n between 15 and 40: Some skewness allowed, but not extreme skewness or extreme outliers. Consider a transformation for right skewness.

  n over 40: The Robustness of the t procedures can handle skewness for sample sizes greater than 40.

- The t procedures are not robust if we do not have a random sample. Robustness only applies to the Normality Condition.

Two Independent Collections

Type	Confidence Interval	Hypothesis Tests	Comments (Mostly Conditions)
Independent Collections:	$CI = (\bar{x}_1 - \bar{x}_2) \pm t^* \sqrt{\dfrac{s_1^2}{n_1} + \dfrac{s_2^2}{n_2}}$ $df = \dfrac{\left(\dfrac{s_1^2}{n_1} + \dfrac{s_2^2}{n_2}\right)^2}{\dfrac{1}{n_1 - 1}\left(\dfrac{s_1^2}{n_1}\right)^2 + \dfrac{1}{n_2 - 1}\left(\dfrac{s_2^2}{n_2}\right)^2}$	$H_0 : \mu_1 - \mu_2 = 0$ $H_a : \mu_1 - \mu_2 \begin{cases} \neq 0 \\ < 0 \\ > 0 \end{cases}$ $t = \dfrac{(\bar{x}_1 - \bar{x}_2) - 0}{\sqrt{\dfrac{s_1^2}{n_1} + \dfrac{s_2^2}{n_2}}}$ df as in CI	The samples must be independently drawn and they must be SRS samples from the populations. The population distribution of populations 1 and 2 must each be normally distributed; but see comments on robustness above. Use the guideline for the sum of the sample sizes: $n_1 + n_2$

Two or More Collections:

ANOVA:

Hypotheses:

$H_0: \mu_1 = \mu_2 = \cdots = \mu_k$

$H_a:$ At least one of the k means *differs from the others*

where k indicates the number of different groups in the test.

Test Statistic:

$$F = \frac{MSG}{MSE} = \frac{\text{Variation between the means of the groups}}{\text{Variation within the groups}}$$

where:

$$MSG = \frac{n_1(\bar{x}_1 - \bar{\bar{x}})^2 + n_2(\bar{x}_2 - \bar{\bar{x}})^2 + \cdots + n_k(\bar{x}_k - \bar{\bar{x}})^2}{k-1}$$

and

$$MSE = \frac{(n_1 - 1)s_1^2 + (n_2 - 1)s_2^2 + \cdots + (n_k - 1)s_k^2}{N-k}$$

and where:

$\bar{\bar{x}}$ is the mean for all of the data taken together;

$\bar{x}_1, \bar{x}_2, \cdots, \bar{x}_k$ are the means of the 1, 2, . . . , k groups;

$n_1, n_2, \cdots, n_k$ are the sample sizes in each of the groups, and

$N = n_1 + n_2 + \cdots + n_k$ is the total sample size for all the groups.

Conditions:

- The samples must be k *independent simple random samples*.
- The populations from which each of the k samples is drawn must be *normally distributed*.
- The populations from which each of the k samples is drawn must have the *same standard deviation*.

Rule-of-Thumb for equal standard deviations:

- ANOVA has been found to give reliable results if the largest of the k standard deviations is no greater than twice the smallest of the k standard deviations.

440

§5.6 Exercises on What To Do

1. Used Mercedes Benzes (Convertible or Not)

The parts of this exercise will present you with statistical questions and you must determine what statistical technique to use. You will not be told what to do; to decide what to do is part of the exercise. Then you will either use Fathom to carry out that technique or, if it is simpler, do it by hand.

- Open the Fathom file **Summer 09 Mercedes.ftm** and get a **case table.** These are part of the used car data collected in the summer of 2009 for cars being sold in www.cars.com. The important variables are:

Place1	indicates the place the car was advertised: Boston, Chicago, Dallas, and SF Bay Area.
Price	the price listed on the website, in dollars
Miles	the reported miles that the car had been driven
Age	the age of the used car, in years
Convertible	whether the car being sold was a convertible or not
Body	indicates the body style of the car: convertible, coupe, sedan, wagon

  Summer 09 Mercedes C-class

	Make1	Place1	Price	Miles	Age	Convert...	Body	Seller	Distance	NoPrice
1	Mercede...	Boston	54987	16507	1.33333	Not Conv...	Coupe	Dealer	6	Price Given
2	Mercede...	Boston	52897	3860	0.416667	Not Conv...	Sedan	Dealer	13	Price Given
3	Mercede...	Boston	50377	31415	3.33333	Not Conv...	Sedan	Dealer	3	Price Given
4	Mercede...	Boston	47000	18500	2.33333	Convertible	Convertible	Private S...	108	Price Given
5	Mercede...	Boston	44899	29029	3.33333	Not Conv...	Sedan	Dealer	105	Price Given

  a. List the quantitative variables shown here and the categorical variables shown here.
  b. We want to know the differences in the proportions of convertibles being sold in the four places. Describe the kind of graphic, or table, or measure you will get to do a *descriptive* (not *inferential*) analysis of this question about the proportions of Mercedes convertibles being sold in Boston, etc. What is it that you want to get?

- Use Fathom to do the analysis you chose to do in part b.
- Either by hand or using Fathom, get conditional probabilities to help answer your question.
  c. Express the proportion of convertibles using proper conditional probability notation. Note the place with the highest proportion and the place with the lowest proportion.
  d. Our statistical question is: *"Is there a difference in the places (as populations) in the proportions of convertibles being sold?"* What *inferential* technique will you use to answer this question about the four places? (Comparing proportions, chi-square, ANOVA, compare means?) Give a reason for your choice.

- Get Fathom to do the *inferential* analysis you decided on in part d and make a copy.
  e. Are the conditions for the inferential procedure you have chosen met?
  f. What are the null and alternate hypotheses for the inferential technique you have chosen?
  g. Interpret the hypothesis test in the context of the question about whether the proportion of convertibles being sold in the four places differs.

h. Our statistical question is: *"Is there a difference in the ages of the convertibles being sold compared with the age of the non-convertibles?"* Describe what *descriptive* (not *inferential*) techniques you will use to answer this question. Think: do the convertibles tend to be younger, older, more diverse in their ages than the non-convertibles? Describe the kinds of descriptive measures or graphics you will get.

- Get Fathom to get the graphics (dot plots or box plots are a good idea) or the summary statistics you need. Make certain that you get relevant numbers (statistics) as well as graphics.

  i. Write up a coherent explanation of what the graphics and the numbers tell you in the context of the data.

  j. Our question is: *"Is there a statistically significant difference in average age between convertibles and non-convertibles?"* What *inferential* technique will you use? (Comparing proportions, chi-square, ANOVA, compare means?) Give a reason for your answer.

  k. Before embarking on the analysis, are the two conditions for the inferential technique that you have chosen met? Explain how they are or, if you think not, why you have reservations.

  l. Set up, using the correct notation, the null and alternate hypotheses for your inferential procedure.

  m. Confused Conrad has $H_0 : p_{NC} = p_C$ and Forgetful Fiona has $H_0 : \bar{x}_{NC} = \bar{x}_C$ for their answers to part m. Is either correct? What errors are they making?

- Get Fathom to do the calculations of the *inferential* procedure you chose; make a copy for your answer.

  n. Interpret the results of your *inferential* procedure in the context of the question.

  o. Our question is: *"Estimate how much older the non-convertibles are compared with the convertibles."* What inferential procedure should be used to get an estimate of how much older the non-convertibles that are being sold are compared to the convertibles?

- Get Fathom to do the calculations and make a copy for your answer.

  p. Give a good interpretation of the results of your *inferential* procedure.

- With the Collection icon selected, go to Object>Add Filter and use the palette to Not and then Body= "Wagon". There are only nine wagons, and we want to exclude them. You should see what is pictured here.

  q. Our question is: *"Is there a difference in mean age by body style (convertible, coupe, and sedan) in the population that these used cars are from?"* Choose both *descriptive* and *inferential* techniques to answer this question and describe them.

- Get Fathom to do the calculations for the procedures you chose. .

  r. Give a good interpretation of the results of your analysis in the context of the statistical question.

2. **Year 2006 Mercedes Benzes** This question uses the same data as question 1 but only looks at the cars for model year 2006. Our question is: *"Do drivers of convertibles drive less than drivers of other body styles of Mercedes Benzes being sold?"* Here are some Fathom analyses.

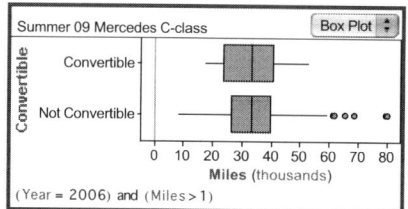

    a. Is this a one-sided or two-sided test? Explain your answer.

    b. Are the conditions met for the hypothesis test? Explain your answer.

    c. Show how the test statistic was calculated using the numbers given. (Show where the numbers go in the formula.)

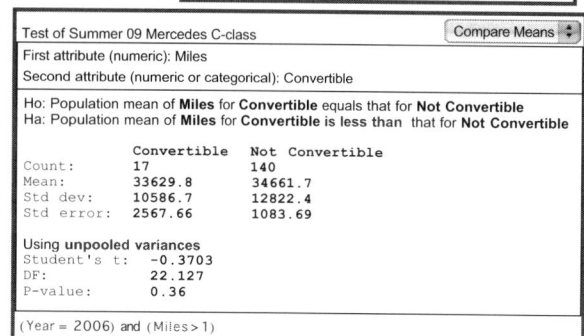

    d. Explain how the hypothesis test gives an answer to the statistical question posed above.

    Summer 09 Mercedes C-class — Box Plot
    Convertible / Not Convertible
    Miles (thousands): 0 10 20 30 40 50 60 70 80
    (Year = 2006) and (Miles > 1)

    Test of Summer 09 Mercedes C-class — Compare Means
    First attribute (numeric): Miles
    Second attribute (numeric or categorical): Convertible
    Ho: Population mean of Miles for Convertible equals that for Not Convertible
    Ha: Population mean of Miles for Convertible is less than that for Not Convertible

	Convertible	Not Convertible
Count:	17	140
Mean:	33629.8	34661.7
Std dev:	10586.7	12822.4
Std error:	2567.66	1083.69

    Using unpooled variances
    Student's t: -0.3703
    DF: 22.127
    P-value: 0.36
    (Year = 2006) and (Miles > 1)

3. **Which Airport?** This exercise looks at the differences between Oakland and SFO airports for flights to LAX.

    - Open the file **OntimeCombLAXSample.ftm**. This is a simple random sample of the flight data from the SF Bay Area to LAX.

    a. What are the cases for these data?

    Sample of OntimeCombLAX

	Carrier_	Airport	Sched...	Actua...	Departure_De...	Taxiout...	DelayOv...	DiffActS...
10	WN	OAK	75	71	8	8	Departur...	-4
11	WN	OAK	75	61	16	4	Departur...	-14
12	UA	SFO	82	77	-3	9	Departur...	-5
13	UA	SFO	84	87	17	19	Departur...	3
14	UA	SFO	85	85	19	19	Departur...	0
15	UA	SFO	82	72	-5	14	Departur...	-10
16	WN	SFO	85	77	-3	19	Departur...	-8
17	WN	SFO	80	65	143	9	Departur...	-15

    Our first statistical question is: *"Is the duration of the flight to LAX (the ActualDuration in minutes) from the SF Bay Area different from SFO compared with Oakland International (OAK)?"*

    b. Here is a Fathom Summary Table showing means and standard deviations for the variable *ActualDuration* for flights from SFO and Oakland. What other *descriptive* analyses will be useful?

    Sample of OntimeCombLAX

		ActualDuration
Airport	OAK	71.6444
		6.79008
		225
	SFO	79.295
		7.81752
		200
	Column Summary	75.2447
		8.22539
		425

    S1 = mean ( )
    S2 = s ( )
    S3 = count ( )

    - Use Fathom to do this additional descriptive analysis (graphs or whatever you chose).

    c. Interpret the descriptive analysis in the context of the question. Interpret both the graphics and the numbers in the summary table.

    d. What kind of *inferential* analyses are implied by the statistical question? Give a reason for your answer.

    e. Use your descriptive analysis and the sample sizes to assess whether the conditions are met for the *inferential* analysis you chose.

    - Get Fathom to do the *inferential* analysis you chose.

    f. You can actually answer this question with either a hypothesis test or a confidence interval. If you chose a hypothesis test, show how Fathom calculated the test statistic. If you chose a confidence interval, show (approximately) how Fathom calculated that. (Fathom used a $df = 396.84$.)

g. Interpret the results of your analysis in the context of our statistical question. Use good stat language but also put the answer into a form an "ordinary" airline traveler would understand.

Our statistical second question is:

"Does departure delay differ between the two airports?"

Here is another summary table for the variable that asks whether the departure delay for a flight is over fifteen minutes, or fifteen minutes or less.

Sample of OntimeCombLAX		Airport		Row Summary
		OAK	SFO	
DelayOver15	Departure Delay 15 min or Less	187	145	332
	Departure Delay Over 15 min	38	55	93
	Column Summary	225	200	425

S1 = count ( )

h. Calculate probabilities, using good notation and in the correct direction, to answer our question in a descriptive sense. Write your conclusion.

i. Which *inferential* analysis will be appropriate for our statistical question? (Your answer should be something like, "A hypothesis test/confidence interval for..." that is, you need to specify the type of procedure and which of the procedures that we have done. There is more than one correct answer, and there is more than one incorrect answer.) Give a reason for your answer.

j. Are the conditions met for your procedure?

- Get Fathom to do the *inferential* analysis you chose.

k. Interpret the results of the inferential procedure in the context of the statistical question posed.

4. **Which Airport, continued.** This exercise looks at the differences between Oakland and SFO airports for flights to LAX.

- Open the Fathom file **OntimeCombLAXSample.ftm**, which is a simple random sample of the flight data from the SF Bay Area to LAX, and get a case table.

Sample of OntimeCombLAX							
	Carrier_...	Airport	Sched...	Actua...	Departure_De...	Taxiout...	DelayOv...	DiffActS...
10	WN	OAK	75	71	8	8	Departur...	-4
11	WN	OAK	75	61	16	4	Departur...	-14
12	UA	SFO	82	77	-3	9	Departur...	-5
13	UA	SFO	84	87	17	19	Departur...	3
14	UA	SFO	85	85	19	19	Departur...	0
15	UA	SFO	82	72	-5	14	Departur...	-10
16	WN	SFO	85	77	-3	19	Departur...	-8
17	WN	SFO	80	65	143	9	Departur...	-15

One of the variables that is measured for every flight is the *Taxiout_TimeMinutes*, the time it takes before the plane actually leaves the ground. Our statistical question is:

"Do Taxiout_TimeMinutes differ between the two airports?"

a. Besides the graphic shown, what other *descriptive* analyses will be useful to answer our question?

- Use Fathom to get the additional descriptive analysis.

b. Interpret the *descriptive* analysis in the context of the question. You should have gotten some numerical summaries to add to the graphic. Comment on what the numbers and the graphic tell you.

c. What kind of *inferential* analyses are implied by the statistical question? Give a reason for your answer.

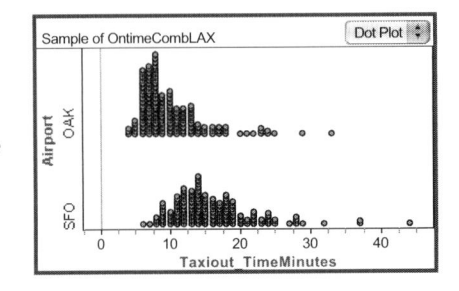

d. Use your descriptive analysis and the sample sizes to assess whether the conditions are met for the *inferential* analysis you chose.

- Get Fathom to do the *inferential* analysis you chose.

444

e. You can actually answer this question with either a hypothesis test or a confidence interval. If you chose a hypothesis test, show how Fathom calculated the test statistic. If you chose a confidence interval, show (approximately) how Fathom calculated that. (The word is "approximately" because the t^* Fathom used may come from a degrees of freedom calculated by the complicated formula shown in §5.4.)

f. Interpret the results of your analysis in the context of our statistical question. Use good stat language but also put the answer into a form an "ordinary" airline traveler would understand.

Another statistical question is how the *Taxiout_TimeMinutes* is related to the *ActualDuraton* of the flight. You would think that the longer the taxi out time, the longer the flight.

g. What kind of variables are *Taxiout_TimeMinutes* and *ActualDuraton:* categorical or quantitative?

h. What kind of analysis is appropriate to the question asked above about the relationship between *Taxiout_TimeMinutes* and *ActualDuraton*? Give a reason for your answer.

- Use Fathom to do the analysis you chose. You can compare OAK to SFO by dragging the variable *Airport* to the "body" of your graphic.

i. You should have a linear model to answer the question. Interpret the slopes (for the two airports) in the context of the question.

j. Interpret the coefficient of determination for the two airports in the context of the question.

[k. Optional but good question: You should have found (in these two exercises) that SFO has both a greater likelihood of a departure delay and a longer average taxi-out time for flights to LAX. To a resident of the Bay Area who sometimes flies, does this make sense? Can you think of some reasons for these findings?]

Standard Normal Probabilities

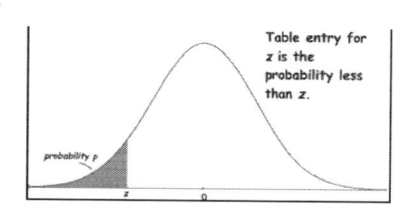

Table entry for z is the probability less than z.

z	.00	.01	.02	.03	.04	.05	.06	.07	.08	.09
-3.8	.0001	.0001	.0001	.0001	.0001	.0001	.0001	.0001	.0001	.0001
-3.7	.0001	.0001	.0001	.0001	.0001	.0001	.0001	.0001	.0001	.0001
-3.6	.0002	.0002	.0001	.0001	.0001	.0001	.0001	.0001	.0001	.0001
-3.5	.0002	.0002	.0002	.0002	.0002	.0002	.0002	.0002	.0002	.0002
-3.4	.0003	.0003	.0003	.0003	.0003	.0003	.0003	.0003	.0003	.0002
-3.3	.0005	.0005	.0005	.0004	.0004	.0004	.0004	.0004	.0004	.0003
-3.2	.0007	.0007	.0006	.0006	.0006	.0006	.0006	.0005	.0005	.0005
-3.1	.0010	.0009	.0009	.0009	.0008	.0008	.0008	.0008	.0007	.0007
-3.0	.0013	.0013	.0013	.0012	.0012	.0011	.0011	.0011	.0010	.0010
-2.9	.0019	.0018	.0018	.0017	.0016	.0016	.0015	.0015	.0014	.0014
-2.8	.0026	.0025	.0024	.0023	.0023	.0022	.0021	.0021	.0020	.0019
-2.7	.0035	.0034	.0033	.0032	.0031	.0030	.0029	.0028	.0027	.0026
-2.6	.0047	.0045	.0044	.0043	.0041	.0040	.0039	.0038	.0037	.0036
-2.5	.0062	.0060	.0059	.0057	.0055	.0054	.0052	.0051	.0049	.0048
-2.4	.0082	.0080	.0078	.0075	.0073	.0071	.0069	.0068	.0066	.0064
-2.3	.0107	.0104	.0102	.0099	.0096	.0094	.0091	.0089	.0087	.0084
-2.2	.0139	.0136	.0132	.0129	.0125	.0122	.0119	.0116	.0113	.0110
-2.1	.0179	.0174	.0170	.0166	.0162	.0158	.0154	.0150	.0146	.0143
-2.0	.0228	.0222	.0217	.0212	.0207	.0202	.0197	.0192	.0188	.0183
-1.9	.0287	.0281	.0274	.0268	.0262	.0256	.0250	.0244	.0239	.0233
-1.8	.0359	.0351	.0344	.0336	.0329	.0322	.0314	.0307	.0301	.0294
-1.7	.0446	.0436	.0427	.0418	.0409	.0401	.0392	.0384	.0375	.0367
-1.6	.0548	.0537	.0526	.0516	.0505	.0495	.0485	.0475	.0465	.0455
-1.5	.0668	.0655	.0643	.0630	.0618	.0606	.0594	.0582	.0571	.0559
-1.4	.0808	.0793	.0778	.0764	.0749	.0735	.0721	.0708	.0694	.0681
-1.3	.0968	.0951	.0934	.0918	.0901	.0885	.0869	.0853	.0838	.0823
-1.2	.1151	.1131	.1112	.1093	.1075	.1056	.1038	.1020	.1003	.0985
-1.1	.1357	.1335	.1314	.1292	.1271	.1251	.1230	.1210	.1190	.1170
-1.0	.1587	.1562	.1539	.1515	.1492	.1469	.1446	.1423	.1401	.1379
-0.9	.1841	.1814	.1788	.1762	.1736	.1711	.1685	.1660	.1635	.1611
-0.8	.2119	.2090	.2061	.2033	.2005	.1977	.1949	.1922	.1894	.1867
-0.7	.2420	.2389	.2358	.2327	.2296	.2266	.2236	.2206	.2177	.2148
-0.6	.2743	.2709	.2676	.2643	.2611	.2578	.2546	.2514	.2483	.2451
-0.5	.3085	.3050	.3015	.2981	.2946	.2912	.2877	.2843	.2810	.2776
-0.4	.3446	.3409	.3372	.3336	.3300	.3264	.3228	.3192	.3156	.3121
-0.3	.3821	.3783	.3745	.3707	.3669	.3632	.3594	.3557	.3520	.3483
-0.2	.4207	.4168	.4129	.4090	.4052	.4013	.3974	.3936	.3897	.3859
-0.1	.4602	.4562	.4522	.4483	.4443	.4404	.4364	.4325	.4286	.4247
0.0	.5000	.4960	.4920	.4880	.4840	.4801	.4761	.4721	.4681	.4641

Standard Normal Probabilities

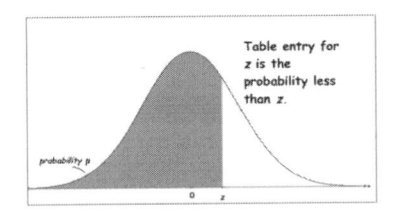

Table entry for z is the probability less than z.

z	.00	.01	.02	.03	.04	.05	.06	.07	.08	.09
0.0	.5000	.5040	.5080	.5120	.5160	.5199	.5239	.5279	.5319	.5359
0.1	.5398	.5438	.5478	.5517	.5557	.5596	.5636	.5675	.5714	.5753
0.2	.5793	.5832	.5871	.5910	.5948	.5987	.6026	.6064	.6103	.6141
0.3	.6179	.6217	.6255	.6293	.6331	.6368	.6406	.6443	.6480	.6517
0.4	.6554	.6591	.6628	.6664	.6700	.6736	.6772	.6808	.6844	.6879
0.5	.6915	.6950	.6985	.7019	.7054	.7088	.7123	.7157	.7190	.7224
0.6	.7257	.7291	.7324	.7357	.7389	.7422	.7454	.7486	.7517	.7549
0.7	.7580	.7611	.7642	.7673	.7704	.7734	.7764	.7794	.7823	.7852
0.8	.7881	.7910	.7939	.7967	.7995	.8023	.8051	.8078	.8106	.8133
0.9	.8159	.8186	.8212	.8238	.8264	.8289	.8315	.8340	.8365	.8389
1.0	.8413	.8438	.8461	.8485	.8508	.8531	.8554	.8577	.8599	.8621
1.1	.8643	.8665	.8686	.8708	.8729	.8749	.8770	.8790	.8810	.8830
1.2	.8849	.8869	.8888	.8907	.8925	.8944	.8962	.8980	.8997	.9015
1.3	.9032	.9049	.9066	.9082	.9099	.9115	.9131	.9147	.9162	.9177
1.4	.9192	.9207	.9222	.9236	.9251	.9265	.9279	.9292	.9306	.9319
1.5	.9332	.9345	.9357	.9370	.9382	.9394	.9406	.9418	.9429	.9441
1.6	.9452	.9463	.9474	.9484	.9495	.9505	.9515	.9525	.9535	.9545
1.7	.9554	.9564	.9573	.9582	.9591	.9599	.9608	.9616	.9625	.9633
1.8	.9641	.9649	.9656	.9664	.9671	.9678	.9686	.9693	.9699	.9706
1.9	.9713	.9719	.9726	.9732	.9738	.9744	.9750	.9756	.9761	.9767
2.0	.9772	.9778	.9783	.9788	.9793	.9798	.9803	.9808	.9812	.9817
2.1	.9821	.9826	.9830	.9834	.9838	.9842	.9846	.9850	.9854	.9857
2.2	.9861	.9864	.9868	.9871	.9875	.9878	.9881	.9884	.9887	.9890
2.3	.9893	.9896	.9898	.9901	.9904	.9906	.9909	.9911	.9913	.9916
2.4	.9918	.9920	.9922	.9925	.9927	.9929	.9931	.9932	.9934	.9936
2.5	.9938	.9940	.9941	.9943	.9945	.9946	.9948	.9949	.9951	.9952
2.6	.9953	.9955	.9956	.9957	.9959	.9960	.9961	.9962	.9963	.9964
2.7	.9965	.9966	.9967	.9968	.9969	.9970	.9971	.9972	.9973	.9974
2.8	.9974	.9975	.9976	.9977	.9977	.9978	.9979	.9979	.9980	.9981
2.9	.9981	.9982	.9982	.9983	.9984	.9984	.9985	.9985	.9986	.9986
3.0	.9987	.9987	.9987	.9988	.9988	.9989	.9989	.9989	.9990	.9990
3.1	.9990	.9991	.9991	.9991	.9992	.9992	.9992	.9992	.9993	.9993
3.2	.9993	.9993	.9994	.9994	.9994	.9994	.9994	.9995	.9995	.9995
3.3	.9995	.9995	.9995	.9996	.9996	.9996	.9996	.9996	.9996	.9997
3.4	.9997	.9997	.9997	.9997	.9997	.9997	.9997	.9997	.9997	.9998
3.5	.9998	.9998	.9998	.9998	.9998	.9998	.9998	.9998	.9998	.9998
3.6	.9998	.9998	.9999	.9999	.9999	.9999	.9999	.9999	.9999	.9999
3.7	.9999	.9999	.9999	.9999	.9999	.9999	.9999	.9999	.9999	.9999
3.8	.9999	.9999	.9999	.9999	.9999	.9999	.9999	.9999	.9999	.9999

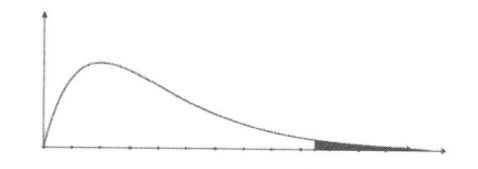

Chi Square Distributions

Probability α in the tail of the distribution

df	0.25	0.20	0.15	0.10	0.05	0.025	0.02	0.01	0.005	0.0025	0.001	0.0005
1	1.32	1.64	2.07	2.71	3.84	5.02	5.41	6.63	7.88	9.14	10.83	12.12
2	2.77	3.22	3.79	4.61	5.99	7.38	7.82	9.21	10.60	11.98	13.82	15.20
3	4.11	4.64	5.32	6.25	7.81	9.35	9.84	11.34	12.84	14.32	16.27	17.73
4	5.39	5.99	6.74	7.78	9.49	11.14	11.67	13.28	14.86	16.42	18.47	20.00
5	6.63	7.29	8.12	9.24	11.07	12.83	13.39	15.09	16.75	18.39	20.52	22.11
6	7.84	8.56	9.45	10.64	12.59	14.45	15.03	16.81	18.55	20.25	22.46	24.10
7	9.04	9.80	10.75	12.02	14.07	16.01	16.62	18.48	20.28	22.04	24.32	26.02
8	10.22	11.03	12.03	13.36	15.51	17.53	18.17	20.09	21.96	23.77	26.12	27.87
9	11.39	12.24	13.29	14.68	16.92	19.02	19.68	21.67	23.59	25.46	27.88	29.67
10	12.55	13.44	14.53	15.99	18.31	20.48	21.16	23.21	25.19	27.11	29.59	31.42
11	13.70	14.63	15.77	17.28	19.68	21.92	22.62	24.73	26.76	28.73	31.26	33.14
12	14.85	15.81	16.99	18.55	21.03	23.34	24.05	26.22	28.30	30.32	32.91	34.82
13	15.98	16.98	18.20	19.81	22.36	24.74	25.47	27.69	29.82	31.88	34.53	36.48
14	17.12	18.15	19.41	21.06	23.68	26.12	26.87	29.14	31.32	33.43	36.12	38.11
15	18.25	19.31	20.60	22.31	25.00	27.49	28.26	30.58	32.80	34.95	37.70	39.72
16	19.37	20.47	21.79	23.54	26.30	28.85	29.63	32.00	34.27	36.46	39.25	41.31
17	20.49	21.61	22.98	24.77	27.59	30.19	31.00	33.41	35.72	37.95	40.79	42.88
18	21.60	22.76	24.16	25.99	28.87	31.53	32.35	34.81	37.16	39.42	42.31	44.43
19	22.72	23.90	25.33	27.20	30.14	32.85	33.69	36.19	38.58	40.89	43.82	45.97
20	23.83	25.04	26.50	28.41	31.41	34.17	35.02	37.57	40.00	42.34	45.31	47.50
21	24.93	26.17	27.66	29.62	32.67	35.48	36.34	38.93	41.40	43.78	46.80	49.01
22	26.04	27.30	28.82	30.81	33.92	36.78	37.66	40.29	42.80	45.20	48.27	50.51
23	27.14	28.43	29.98	32.01	35.17	38.08	38.97	41.64	44.18	46.62	49.73	52.00
24	28.24	29.55	31.13	33.20	36.42	39.36	40.27	42.98	45.56	48.03	51.18	53.48
25	29.34	30.68	32.28	34.38	37.65	40.65	41.57	44.31	46.93	49.44	52.62	54.95
26	30.43	31.79	33.43	35.56	38.89	41.92	42.86	45.64	48.29	50.83	54.05	56.41
27	31.53	32.91	34.57	36.74	40.11	43.19	44.14	46.96	49.64	52.22	55.48	57.86
28	32.62	34.03	35.72	37.92	41.34	44.46	45.42	48.28	50.99	53.59	56.89	59.30
29	33.71	35.14	36.85	39.09	42.56	45.72	46.69	49.59	52.34	54.97	58.30	60.73
30	34.80	36.25	37.99	40.26	43.77	46.98	47.96	50.89	53.67	56.33	59.70	62.16
40	45.62	47.27	49.24	51.81	55.76	59.34	60.44	63.69	66.77	69.70	73.40	76.09
50	56.33	58.16	60.35	63.17	67.50	71.42	72.61	76.15	79.49	82.66	86.66	89.56
60	66.98	68.97	71.34	74.40	79.08	83.30	84.58	88.38	91.95	95.34	99.61	102.70
80	88.13	90.41	93.11	96.58	101.88	106.63	108.07	112.33	116.32	120.10	124.84	128.26
100	109.14	111.67	114.66	118.50	124.34	129.56	131.14	135.81	140.17	144.29	149.45	153.17

t-Distribution Critical Values

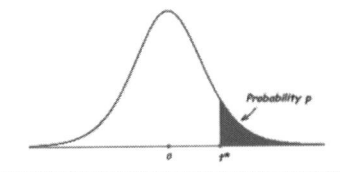

	Probabilities in the right tail of the distribution									
df	0.25	0.15	0.10	0.05	0.025	0.02	0.01	0.005	0.001	0.0005
1	1.000	1.963	3.078	6.314	12.706	15.895	31.821	63.657	318.309	636.619
2	0.816	1.386	1.886	2.920	4.303	4.849	6.965	9.925	22.327	31.599
3	0.765	1.250	1.638	2.353	3.182	3.482	4.541	5.841	10.215	12.924
4	0.741	1.190	1.533	2.132	2.776	2.999	3.747	4.604	7.173	8.610
5	0.727	1.156	1.476	2.015	2.571	2.757	3.365	4.032	5.893	6.869
6	0.718	1.134	1.440	1.943	2.447	2.612	3.143	3.707	5.208	5.959
7	0.711	1.119	1.415	1.895	2.365	2.517	2.998	3.499	4.785	5.408
8	0.706	1.108	1.397	1.860	2.306	2.449	2.896	3.355	4.501	5.041
9	0.703	1.100	1.383	1.833	2.262	2.398	2.821	3.250	4.297	4.781
10	0.700	1.093	1.372	1.812	2.228	2.359	2.764	3.169	4.144	4.587
11	0.697	1.088	1.363	1.796	2.201	2.328	2.718	3.106	4.025	4.437
12	0.695	1.083	1.356	1.782	2.179	2.303	2.681	3.055	3.930	4.318
13	0.694	1.079	1.350	1.771	2.160	2.282	2.650	3.012	3.852	4.221
14	0.692	1.076	1.345	1.761	2.145	2.264	2.624	2.977	3.787	4.140
15	0.691	1.074	1.341	1.753	2.131	2.249	2.602	2.947	3.733	4.073
16	0.690	1.071	1.337	1.746	2.120	2.235	2.583	2.921	3.686	4.015
17	0.689	1.069	1.333	1.740	2.110	2.224	2.567	2.898	3.646	3.965
18	0.688	1.067	1.330	1.734	2.101	2.214	2.552	2.878	3.610	3.922
19	0.688	1.066	1.328	1.729	2.093	2.205	2.539	2.861	3.579	3.883
20	0.687	1.064	1.325	1.725	2.086	2.197	2.528	2.845	3.552	3.850
21	0.686	1.063	1.323	1.721	2.080	2.189	2.518	2.831	3.527	3.819
22	0.686	1.061	1.321	1.717	2.074	2.183	2.508	2.819	3.505	3.792
23	0.685	1.060	1.319	1.714	2.069	2.177	2.500	2.807	3.485	3.768
24	0.685	1.059	1.318	1.711	2.064	2.172	2.492	2.797	3.467	3.745
25	0.684	1.058	1.316	1.708	2.060	2.167	2.485	2.787	3.450	3.725
26	0.684	1.058	1.315	1.706	2.056	2.162	2.479	2.779	3.435	3.707
27	0.684	1.057	1.314	1.703	2.052	2.158	2.473	2.771	3.421	3.690
28	0.683	1.056	1.313	1.701	2.048	2.154	2.467	2.763	3.408	3.674
29	0.683	1.055	1.311	1.699	2.045	2.150	2.462	2.756	3.396	3.659
30	0.683	1.055	1.310	1.697	2.042	2.147	2.457	2.750	3.385	3.646
35	0.682	1.052	1.306	1.690	2.030	2.133	2.438	2.724	3.340	3.591
40	0.681	1.050	1.303	1.684	2.021	2.123	2.423	2.704	3.307	3.551
45	0.680	1.049	1.301	1.679	2.014	2.115	2.412	2.690	3.281	3.520
50	0.679	1.047	1.299	1.676	2.009	2.109	2.403	2.678	3.261	3.496
60	0.679	1.045	1.296	1.671	2.000	2.099	2.390	2.660	3.232	3.460
70	0.678	1.044	1.294	1.667	1.994	2.093	2.381	2.648	3.211	3.435
80	0.678	1.043	1.292	1.664	1.990	2.088	2.374	2.639	3.195	3.416
90	0.677	1.042	1.291	1.662	1.987	2.084	2.368	2.632	3.183	3.402
100	0.677	1.042	1.290	1.660	1.984	2.081	2.364	2.626	3.174	3.390
125	0.676	1.041	1.288	1.657	1.979	2.075	2.357	2.616	3.157	3.370
150	0.676	1.040	1.287	1.655	1.976	2.072	2.351	2.609	3.145	3.357
175	0.676	1.040	1.286	1.654	1.974	2.069	2.348	2.604	3.137	3.347
200	0.676	1.039	1.286	1.653	1.972	2.067	2.345	2.601	3.131	3.340
250	0.675	1.039	1.285	1.651	1.969	2.065	2.341	2.596	3.123	3.330
400	0.675	1.038	1.284	1.649	1.966	2.060	2.336	2.588	3.111	3.315
1000	0.675	1.037	1.282	1.646	1.962	2.056	2.330	2.581	3.098	3.300
∞	0.674	1.036	1.282	1.645	1.960	2.054	2.326	2.576	3.091	3.291
	50%	70%	80%	90%	95%	96%	98%	99%	99.5%	99.9%

Confidence Level C

Made in the USA
San Bernardino, CA
16 November 2013